Louis R. Caplan (Ed.)

Brain Ischemia

Basic Concepts and Clinical Relevance

With 79 Figures

Springer-Verlag
London Berlin Heidelberg New York
Paris Tokyo Hong Kong
Barcelona Budapest

Louis R. Caplan, MD
Professor and Chairman
Department of Neurology
New England Medical Center Hospitals
750 Washington Street
NEMCH #314
Boston
Massachusetts 02111
USA

ISBN-13: 978-1-4471-2075-9 **e-ISBN-13: 978-1-4471-2073-5**
DOI: 10.1007/978-1-4471-2073-5

British Library Cataloguing in Publication Data
A catalogue record for this book is available from the British Library

Library of Congress Cataloging-in-Publication Data
A catalog record for this book is available from the Library of Congress

Softcover reprint of the hardcover 1st edition 1995

Typeset by Concept Typesetting Ltd, Salisbury

12/3830-54210 Printed on acid-free paper

Preface

Before a reader begins, or at least dives deeply into a book, especially a technical one, he or she may be helped to understand the book better if given some insight into the author's motivations and goals. Why was the book written? What type of subjects and materials will be included? Who are the contributors? How will the book be written? Is there anything special or unique that the author thinks or hopes will distinguish this book from others on the same or similar topics?

We are now at a very exciting and pivotal time in the history of the care of stroke patients and stroke research. During the 19th and the first half of the 20th centuries, interest in stroke patients centered around brain and vascular pathology and clinicopathological correlation. Physicians were mostly interested in how the brain worked and in what symptoms and signs were present in patients with softenings and hemorrhages at various locations in the brain. The nature of the brain and blood vessel pathology was also important in understanding disease and in prognosticating recovery. Until the last three decades there was little effective therapy, and little technology able to aid physicians in making very precise diagnoses of the vascular lesions during life.

This was before the advent of CT scanning. The introduction and dissemination of CT during the mid-1970s led to a dramatic change. Now, during life, and often very soon after stroke symptoms began, physicians could tell what was happening within the brain and where, and ischemia could be reliably seperated from hemorrhage. However, CT was only the beginning of a technological revolution; since then, MRI has provided even better definition of the brain abnormalities in stroke patients, and the introduction of various ultrasound techniques and their constant improvements has made it possible to safely image and analyze the flow properties of the major extracranial and intracranial arteries that supply the brain. Magnetic resonance angiography (MRA), developed within the past 6 years, is another addition to the non-invasive methods capable of imaging cerebrovascular structures. Brain functions and metabolism can also be studied using radioisotopes and electrophysiologic techniques. Also, during the past two decades there have been great advances in the technology available to study the heart and the blood. We now know much more about the cardiac disorders that lead to stroke, the concurrence of stroke and cardiac disease, and the coagulation and serological and cellular elements of the blood.

This new technology has moved stroke care very rapidly into the modern era. Physicians now have the ability to rapidly and safely acquire all the data necessary to make logical and rational treatment decisions, but this field is now moving so fast that physicians not directly involved have not been able to keep pace. The basic science aspects of brain, blood, vascular and cardiac functions have suddenly become very important to understand, since they have the potential to be applied to the daily care of stroke patients both at the bedside and in the clinic. The basic science now relates directly to practical treatment decisions.

This book seeks to bring the clinicians who treat stroke patients up to date on these basic science aspects, and to merge the information as much as possible with the important clinical issues. Even a cursory glance at the chapters will quickly make readers realize that basic sciences still do not have many of the answers to key questions. What causes brain tissue to die? What precipitates an occlusive thrombus? Why do atherosclerotic plaques develop and why do they have predilections for certain locations? Why does vascular disease vary with race and sex? Knowledge of the present state of the art may help to stimulate research, both clinical and basic, that might provide some of the answers.

Although much has been written about the basic science aspects of cerebrovascular diseases, the material is not easy for clinicians to access, digest and interpret. Much is written to be read by other basic investigators and scientists, not by clinicians. Hematologic, vascular, cardiologic, physiologic, blood flow and metabolic studies are often published in different journals. As a clinician long exposed to basic science information and research, I feel qualified to try to collate and mold the basic information into a volume that will be useful for clinicians. I think that I know well what clinicians will want to know and how to translate the material into a digestible form.

I have chosen the topics and the contributors very carefully, to cover all the important and diverse topics that relate to stroke. These include brain morphology and function: metabolism, cell death, edema, metabolic functions, growth factors and potential toxins; eye ischemia; blood coagulation, fibrinolysis and immunology; the blood vessels: innervation, atherosclerotic disease, microvascular pathology and function, endothelial functions and vasoconstriction; blood flow: physics, rheology and control; the heart and heart–brain interactions; and epidemiological and registry data. In order to contain the size of the volume, the authors and I have tried to be eclectic and selective, covering only the most important material as concisely as possible.

The great majority of contributors have or have had major clinical experience and responsibility in caring for patients. The others have had long-term very close relationships with clinical departments and clinicians. I know all of them personally: they have been selected because of the depth and breadth of their knowledge and experience, and their ability to communicate well. They are all acknowledged experts in their fields.

I plan to edit this book quite differently than most. I take my cue and method from the British writer, Henry Fielding. In his classic book *Tom Jones,* which was written in the middle of the eighteenth century, Fielding frequently inserted himself into the novel. He compared his role as author to a host setting a feast for his guests. The host would describe the feast to come—'It hath been usual with the honest and well-meaning host to provide a bill of fare which all persons may peruse at their first entrance into the house'.[1] Then, as the meal (story) began and continued, the host (author), at various times, would interject, comment, and explain to ensure that the guests (readers) derived the maximum value from the offerings. Similarly, I plan to play a more active role than is customary for editors. Before each major section, I will introduce the subject of that section, and at the end of each will add my comments, touching on areas that may have been omitted, emphasizing the key points and those areas in which knowledge is scanty or controversial. I will also try to focus on the clinical impact of the basic information contained in the section.

I would like to thank all who made this book possible. The contributors, who put up with my cajoling, carping and criticism, deserve medals. The people at Springer-Verlag were very helpful at every stage of the process. My loyal secretary, Pauline Dawley, helped greatly with the transcription and management of the material. If in some small measure it helps with the care of stroke patients, I will consider the effort worthwhile.

Louis R. Caplan
Boston, December 1992

Reference

1. Fielding Henry (1950) The History of Tom Jones, a Foundling. New York, Random House, p. 1

Contents

Section III. The Blood

Section IV: The Blood Vessels

Section V: Blood Flow, Blood Pressure and Intracranial Pressure

Section VI: The Heart

Section VII: Clinical Research: Epidemiology, Statistics, Databases and Trials

Contributors

A. S. Ali
Center for Stroke Research
Department of Neurology (K–11)
Henry Ford Hospital and Health Sciences Center
2799 West Grand Blvd
Detroit
Michigan 48202–2689
USA

Jean-Claude Baron
Institut National de la Santé et de la Recherche Médicale
Boulevard H. Becquerel
BP 5027
14021 Caen Cedex
France

Louis R. Caplan
Department of Neurology
New England Medical Center Hospitals
750 Washington Street
NEMCH #314
Boston
Massachusetts 02111
USA

Dennis W. Choi
Department of Neurology
Box 8111
Washington University School of Medicine
660 S. Euclid Avenue
St Louis
Missouri 63110
USA

Gregory J. del Zoppo
Department of Molecular and Experimental Medicine
The Scripps Research Institute
10666 North Torrey Pines Road
La Jolla
California 92037
USA

William M. Feinberg
Department of Neurology
Health Sciences Center
The University of Arizona
Tucson
Arizona 85724
USA

Seth P. Finklestein
CNS Growth Factor Research Laboratory
Department of Neurology
Massachusetts General Hospital
Boston
Massachusetts 02114
USA

Marc Fisher
Department of Neurology
The Medical Center of Central Massachusetts and the University of Massachusetts Medical School
Worcester
Massachusetts 01605
USA

Mark Fisher
Department of Neurology: Stroke Research
University of Southern California School of Medicine
1333 San Pablo Street
MCH 246
Los Angeles
California 90033
USA

Anthony J. Furlan
Cerebrovascular Disease Center
Department of Neurology
Cleveland Clinic Foundation
One Clinic Center
9500 Euclid Avenue
Cleveland
Ohio 44195–5001
USA

Julio H. Garcia
Henry Ford Hospital (K–6)
2799 West Grand Boulevard
Detroit
Michigan 48202–2689
USA

Eugene V. Golanov
Division of Neurobiology
Cornell University Medical College
411 East 69th Street
New York 10021
USA

Philip B. Gorelick
Rush Medical College
1725 West Harrison Suite 755
Chicago
Illinois 60612–3864
USA

Françoise Gray
Department of Pathology (Neuropathology)
Hôpital Henri Mondor
94010 Creteil Cedex
France

James C. Grotta
Department of Neurology
University of Texas Medical School
6431 Fannin #7044
Houston
Texas 77030
USA

Joseph P. Hanna
Cerebrovascular Disease Center
Department of Neurology
Cleveland Clinic Foundation
One Clinic Center
9500 Euclid Avenue
Cleveland
Ohio 44195–5001
USA

Thomas R. Hedges III
Department of Neuroophthalmology
New England Medical Center
750 Washington Street
Boston
Massachusetts 02111
USA

Donald D. Heistad
Departments of Internal Medicine and Pharmacology
College of Medicine
Department of Internal Medicine
The University of Iowa
Iowa City
Iowa 52242
USA

Daniel B. Hier
Department of Neurology (M/C 796)
Neuropsychiatric Institute
912 South Wood Street Room 855N
Chicago
Illinois 60612
USA

Marc Hommel
Stroke Unit
Department of Clinical and Biological Neurosciences
University Hospital
BP 217
38043 Grenoble Cedex 9
France

Steven R. Levine
Center for Stroke Research
Department of Neurology (K–11)
Henry Ford Hospital and Health Sciences Center
2799 West Grand Bòulevard
Detroit
Michigan 48202–2689
USA

Robert Macfarlane
Department of Neurosurgery
Royal London Hospital
Whitechapel
London E1 1BB
UK

Michael A. Moskowitz
Stroke Research Laboratory
Harvard Medical School
Massachusetts General Hospital
32 Fruit Street
Boston
Massachusetts 02114
USA

M. D. O'Brien
Department of Neurology
Guy's and St Thomas' Hospital
St Thomas Street
London SE1 9RT
UK

K. Perktold
Institut für Mathematik
Technische Universität Graz
Steyreergasse 30
A–8010 Graz
Austria

Donald J. Reis
Division of Neurobiology
Cornell University Medical College
411 East 69th Street
New York 10021
USA

Allan H. Ropper
Division of Neurology
St Elizabeth's Hospital
736 Cambridge Street
Boston
Massachusetts 02135
USA

Una S. Ryan
T Cell Sciences, Inc
38 Sidney Street
Cambridge
Massachusetts 02139
USA

H. Schmid-Schönbein
Institut für Physiologie
Klinikum der RWTH
Pauwelsstrasse 30
D–5100 Aachen
Germany

Michael A. Sloan
Departments of Neurology, Radiology, Epidemiology and Preventive Medicine
University of Maryland School of Medicine
22 S. Greene Street
Baltimore
Maryland 21201
USA

Jeffrey D. Trachtenberg
Department of Surgery
Washington University School of Medicine
Washington University Medical Center
Box 8109
4960 Audubon Avenue
St Louis
Missouri 63110
USA

SECTION I

The Brain

Prefatory Comments

New Strategies to Treat Brain Ischemia

During the past decade a dramatic change has occurred in the strategies for treating patients with brain ischemia. The development of modern technology that has the capability to image brain lesions and characterize the causative cardiovascular process safely and quickly now allows treating physicians to tailor treatment to the individual patient's problem. At the same time, research in the laboratory has begun to clarify the anatomical and biochemical changes that occur within ischemic brain.

For the greater part of the 20th century, treatment strategies were simple and general. Goals were entirely prevention- and rehabilitation-orientated. Doctors sought to prevent new strokes by modifying known risk factors such as hypertension. They also attempted to prevent ordinary medical complications of stroke such as pneumonia, bed sores and urosepsis by careful attention to the skin, bladder and respiratory systems of patients who had had strokes. Rehabilitation provided various mechanical devices such as braces and wheelchairs, and sought to maximize function despite neurologic deficits. In the Museum of Science in Chicago, as late as 1980, the exhibit on stroke featured only devices used in stroke rehabilitation. There is little solace for stroke survivors and their families in knowing they can be fitted for a wheelchair if they become paralyzed. No wonder therapeutic nihilism prevailed.

During the second half of the 20th century emphasis began to be placed on thrombosis and its prevention. Since obstruction of arteries was nearly always related to occlusive thrombosis, often with embolization of thrombi or clot fragments, attention was appropriately directed to the coagulation process. Heparin and warfarin were given prophylacticly to prevent thrombosis when risk factors for thromboembolism were present. These risks included mitral valve disease with atrial fibrillation and recent myocardial infarction. When thrombi were suspected, or known to be present, these drugs were used to prevent the formation of fresh thrombus on the tail of clots and to prevent the propagation and embolization of thrombi. Later, when it became clear that platelet adhesion and aggregation played a prominent role in the thrombotic process, drugs that affected platelet functions (secretion, adhesion, aggregation, etc.) were introduced to treat patients with threatened or actual myocardial and brain ischemia. Because information about the patient's vascular lesions was difficult to obtain and often required invasive techniques, physicians often treated patients characterized only by the temporal features of ischemia, e.g. transient ischemic attack, progressive stroke, 'completed stroke', etc. Trials of various antithrombotic agents were and are still being performed in which patients are enrolled according to the time course of ischemia. I have commented elsewhere on the futility of this strategy (Caplan 1983, 1988, 1992, 1993).

Temporal patterns of ischemia do not predict prognosis, vascular etiology, or even the presence or absence of infarction. Anticoagulant treatments are also, in a sense, general since they were often used without requiring definition of the vascular lesion.

During the last decade, attention has begun to move in other directions while maintaining a major focus on anticoagulant treatments. The two newer areas of therapeutic emphasis are reperfusion of the ischemic brain and brain resuscitation, that is, increasing the brain's resistance to ischemia and reversing the ischemic changes within the brain. Reperfusion can be accomplished by surgery (direct endarterectomy or surgical bypass), by mechanical means (such as angioplasty) using various devices placed into arteries through endovacular techniques, and using thrombolytic drugs such as recombinant tissue plasminogen activator (rt-PA), streptokinase and urokinase. These reperfusion techniques usually require specific knowledge of the nature, location and severity of the occlusive vascular process. Ultrasound (cardiac, extracranial and transcranial) and magnetic resonance angiography now allow definition of the vascular lesion in most patients. In others, standard catheter angiography can show lesions more safely than in the past. Reperfusion has been referred to as the 'plumbing' approach to the treatment of ischemic stroke: find the pipes that are blocked and open them.

Brain protection or resuscitation has been stimulated by research in the laboratory using experimental animal models of ischemia, tissue cultures and in vitro measurements. As is the trend in all areas of modern medicine, this research has become microscopic, that is, directed to biochemical and molecular changes at a cellular level rather than to macroscopic changes visible grossly to the eye. The key questions are: What causes irreversible ischemic damage and cell death? Are there natural body defences that act to minimize or reverse damage? Are some cells and regions more vulnerable or more resistant than others? Is there selective vulnerability of cells, and what explains it?

Reperfusion

Quite early in its development, doctors recognized that reperfusion could be futile, or even worse, potentially harmful. Using the analogy of watering a lawn, bringing irrigation and water to regions where the grass is dead and gone will be a useless waste of water and energy. Too much water could flood the lawn and so cause damage. Similarly, reperfusion of dead brain tissue is not helpful: bringing blood into injured tissues could lead to hemorrhage and edema.

Since reperfusion is not always wise, a number of important questions come readily to mind. Some of these concern timing. When does brain tissue become irreversibly injured by the interruption of its circulatory supply? When, after ischemia, are complications of reperfusion most likely to occur? These questions refer to the time 'window of opportunity' that is available to minimize ischemia by reperfusion strategies. The vascular stresses causing ischemia in individual patients are quite heterogeneous and depend on its nature, severity, location, rapidity of development, lesions in other arteries, blood flow and coagulability and general circulatory factors, to mention only some of the important variables. Thus it is unlikely that simple answers such as '60 minutes' will apply to all instances of ischemia, or in all parts of evolving ischemic lesions. The answers will be average estimates.

Since vulnerability and survival of the brain tissue vary greatly in individual patients, doctors will want to know the state of the ischemic brain at the time of planned reperfusion or other treatment. Is the brain normal, having escaped or recovered completely from the ischemic insult? Or is the tissue dead, that is, irreversibly injured? In many, perhaps the majority, of instances tissue viability lies somewhere on a continuum between normal and dead. Injured tissue can be non-functional ('stunned') but resuscitable, that is, reversibly injured but with the potential to return to normal function and structure. How can doctors tell if the brain tissue in their patient is normal, stunned or irreversibly injured?

The questions clinicians ask then are: When? – the timing of brain viability and ischemia reversibility – the window of opportunity; What? – the causes at cellular and tissue levels that lead to irreversible damage; and Where? – the location of regions at risk and whether parts of the lesion are differentially involved and selectively vulnerable. This section of the book seeks to provide answers to these questions.

Animal Models and Laboratory Research in Stroke

As I have mentioned, much of our knowledge about the biochemical and molecular aspects of ischemic injury come from basic animal and in vitro research, not from clinical in vivo observations in human stroke patients. I feel impelled to offer some comments on basic stroke research, since otherwise it will be difficult for readers to place the results in perspective. Very recently, three editorials were published in the journal *Stroke* concerning the relevance of animal models to the treatment of stroke patients in the clinic (Wiebers et al. 1990; Zivin and Grotta 1990; Millikan 1992).

Animal models do not replicate the situation in stroke patients. As a result, treatments that work in animal models have not proven effective in humans (Wiebers et al. 1990), for the following reasons:

1. Animal brain and vascular anatomy is different from that in humans. The craniocerebral vascular system in rats, rabbits, gerbils and even cats is quite different from humans, and we know from comparative anatomy that the brains of these animals are different from each other and from our brains.

2. The vascular lesion models are usually quite different from lesions occurring in natural circumstances. Surgically placed lesions often involve operative vascular trauma and require anesthesia. Many lesions, such as the tying-off of all large arteries and bleeding the animals to cause hypotension, are completely unlike human thromboembolic strokes and even unlike the situation in systemic hypoperfusion secondary to cardiac arrest. Most induced experimental lesions are made instantly, whereas clinical insults are often more gradual.

3. The animals chosen often do not replicate human stroke victims. My Finnish colleague, Marcu Kaste, is fond of saying that 'they use healthy young rats for experiments while his patients are old and in lousy shape'. Many stroke patients have multiple vascular lesions and past and present illnesses and risks, such as hypertension, smoking, diabetes and hypercholesterolemia, which have affected their vascular systems before their stroke and which greatly modify the outcome and response to treatment.

4. Many important deficits are not testable in animals. Vision, sensation and cognitive abnormalities such as speech and memory are not testable. Results are usually defined in morphologic and biochemical terms.

5. The timing of various treatments used in animal models often does not reflect that possible in humans. Some treatments are given to animals before, concurrent with, or shortly after vascular insults. Frustratingly, many stroke patients do not seek medical care quickly. When they do so there are often long delays before they are evaluated by specialists who are experienced and knowledgeable in stroke treatment.

Leaving the nature of the models aside, there are other problems with laboratory research:

1. When considering the role of a putative substance in the ischemic process, other likely culprits are ignored or kept constant. One major asset of research in the laboratory is the ability to study one variable while keeping all other potential variables or confounding influences constant. In the clinic, multiple substances and variables may in fact interact, making it difficult to define the nature of the actual process. For example, in his contribution in this section, Choi comments on the interactions between *N*-methyl-D-aspartate (NMDA) channels, oxygen free radicals, C^{2+} and local acidosis. Complex interactions are not easy to quantify and define, either in models or in the clinic.

2. New, potential, putative substances are discovered all the time. A new potentially active toxic substance (e.g. nitric oxide, leukotrienes, tumor necrosis factor) or helpful substances (various growth factors) are discovered almost every one or two years. I like to call this problem the Agatha Christie effect: when reading a mystery novel, readers are often introduced to many possible suspects during the initial chapters. Readers ponder 'who dunnit' but at times it turns out that the true culprit is introduced only in the last chapters and did not appear early in the novel. How was the reader to know it was Sam's lover who killed Harry if he had not been told of the exsistence of that individual? Similarly, research done 10 years ago would not have implicated excitotoxins, since no-one was particularly aware of their involvement. Nothing can or should be done to retard progress, but caution should make us aware we do not yet know everything. We cannot study or measure substances that we are unaware of.

3. In a similar vein, the study of various biochemical agents in treament is, by definition, pharmacology. In my opinion, pharmacology is a very inexact science. The practical effects of drugs on stucture and function can be studied and often measured, but why and how the drugs work is often speculative. We become aware of some functions of drugs but others are unknown until discovered later. Aspirin was used solely as an antipyretic and anti-inflammatory drug until

its effect on platelet function was discovered by chance. Amantadine was used for its antiviral effect, and it was discovered by accident that it had some action against Parkinsonian symptoms. I am cautious in interpreting the explanation for the effectiveness of a drug in a given situation, since not all its effects may be known and tested.

These remarks are not meant to criticize or discourage laboratory or animal research. They are meant to caution readers to place the animal and in vitro results in perspective, and to explain why things that work experimentally often are ineffective in real patients. Laboratory testing is very valuable in generating ideas and putative variables and substances, and in devising ways to measure their effects. Laboratory testing is also necessary for safety. Some toxicities are unsuspected before animal trials and some toxic effects are not readily foreseeable from the drawing board.

Contributions in this Section

I have tried to select topics and authors that will give a broad and diverse overview of changes in the brain during ischemia. Garcia, a very experienced experimental and clinical neuropathologist, will describe the morphological and biochemical changes in the ischemic brain. He will discuss the issues of selective vulnerability and reversible ischemia, and the factors that relate to cell death. He will also outline events in the microvasculature that are important in determining reperfusion. Choi, a neurologist and basic researcher, will review one very important aspect of neuronal injury, that of excitotoxicity. He discusses various putative excitotoxins and their receptors and cites evidence for their role in causing ischemic injury. Choi also reviews the effects of other chemical changes known to occur in ischemia, such as local changes in Ca^{2+}, lactic acid, oxyen free radicals and the interactions of these substances with the excitotoxic cascade. We often forget that there are potent natural factors and substances which limit injury. Finklestein, a stroke neurologist, long active in the basic research laboratory, will discuss recent knowledge about various growth factors and their potential role in preventing, limiting and reversing ischemic damage. We need to identify and promote the positive influences (such as growth factors) and neutralize the negative factors (such as toxins) and alter the equilibrium to 'accentuate the positive and eliminate the negative'.

The next contribution relates to one aspect of brain injury, cerebal edema. O'Brien, a neurologist clinician, long active in research on brain edema, will review the various fluid compartments in the brain and the development of edema within cells (cytotoxic edema) and fluid accumulation in the extracellular space (vasogenic edema). The pathophysiology of fluid shifts and the pathogenesis of brain edema related to ischemia will be discussed in detail. O'Brien will consider the role of many factors mentioned by Garcia, Choi and Finklestein. These four contributions all concern microscopic and cellular changes. Finally, Baron, a neurologist and experienced research scientist in the field of brain metabolism , will discuss brain ischemia from a more macroscopic view, that provided by positron emission tomography (PET). PET has given physicians a window on biochemical, metabolic and blood flow changes in areas of ischemia and into events within various parts of ischemic territories.

References

Caplan LR (1983) Are terms such as completed stroke or RIND of continued usefulness? Stroke 14:431–433
Caplan LR (1988) TIAs – We need to return to the question, "What is wrong with Mr. Jones?" Neurology 38:791–793
Caplan LR (1992) What's wrong with Mr. Jones, a 58-year-old man with cerebrovascular disease? Heart Dis Stroke 1:252–254
Caplan LR (1993) Terms describing brain ischemia by tempo are no longer useful: a polemic (with apologies to Shakespeare). Surg Neurol 40:91–95
Millikan C (1992) Animal stroke models. Stroke 23:795–797
Wiebers DO, Adams HP, Whisnant J (1990) Animal models of stroke: are they relevant to human disease? Stroke 21:1–3
Zivan JA, Grotta JC (1990) Animal stroke models. They are relevant to human disease. Stroke 21:981–983

1. Mechanisms of Cell Death in Ischemia

J. H. Garcia

Intracranial ischemia secondary to arterial occlusion is encountered daily in patients undergoing endarterectomy for the treatment of symptomatic carotid atherosclerosis. This intervention involves the unilateral interruption of circulation (for up to 30 minutes) through three major vessels: the common, internal and external carotid arteries (Sundt et al. 1981; Morawetz et al. 1984). The resulting circulatory alterations are reflected in ipsilateral changes in the electroencephalographic (EEG) tracings and significant drops in regional cerebral blood flow (rCBF) values (Sundt et al. 1981; Zampella et al. 1991).

Some patients undergoing endarterectomy develop focal neurologic deficits in the contralateral limbs, attributable to the iatrogenic hemispheric ischemia, but almost invariably the EEG abnormalities and the neurologic deficits disappear within hours of starting reperfusion (Sundt et al. 1981). Comparable observations have been recorded in awake non-human primates, undergoing experimental occlusion of the middle cerebal artery (MCA). In these animals, the appearance of neurologic deficits with abrupt drop in rCBF values coincided with the arterial occlusion. Reopening the artery 1–2 hours later resulted in significant improvement of the neurologic deficits and progressive recovery of the rCBF values over a period of 24–48 hours (Garcia et al. 1983).

Collectively, these observations suggest that the functional effects of focal ischemia develop synchronously with the arterial occlusion, but there is a period of time generally known as 'the window of opportunity' during which the deficits so created might be partly or completely reversed, either by reperfusing the ischemic tissue or by administering yet-to-be-indentified therapeutic agent(s). This reversal of the effects of focal cerebal ischemia may be signified by recovery of the lost function; however, the duration of this period of grace remains undefined.

The development of a focal neurologic deficit, such as unilateral hemiparesis, which spontaneously disappears within 24 hours is a relatively common event clinically indentified as a transient ischemic attack (TIA) (Whisnant et al. 1973). This designation implies that when an artery becomes occluded the resulting focal ischemia produces a neurologic deficit; a presumed spontaneous reperfusion of the ischemic territory reverses the symptoms. This interpretation extrapolates observations made in patients with transient monocular blindness (amaurosis fugax). Funduscopic observations during the symptomatic period have shown, in a few of these patients, that the episodic blindness is acompanied by occlusion (presumably embolic) of either the central retinal artey or one of its main branches. Disintegration of the occlusive material and reopening of the vessel lumen are accompanied by recovery of vision (Appen et al. 1975). Whether all patients who have episodic neurologic deficits have also transient arterial occlusions is not known. A small percentage of patients initially diagnosed as having TIAs do not have arterial occlusions or areas of brain ischemia; instead, they may have selected metabolic disturbances (such as profound hypoglycemia), focal demyelinating disease (such as multiple sclerosis) or brain tumors (Harrison 1982). Moreover, in a significant number of patients who have TIAs, computed tomography (CT) demonstrates focal brain hypodensities (infarcts?) at sites appropriately located to account for the transient neurologic deficit (Koudstaal et al. 1991).

Mechanisms of Cell Death

This review of mechanisms of cell death after focal ischemia aims to answer the questions: What heralds the end of the period of reversibility?, and What factor(s) created by an arterial occlusion make the process irreversible?

Among the various hypotheses that may explain the transition from reversible ischemia to infarct, the following should be considered:

1. Occluding a large artery sets in motion 'microcirculatory disturbances' (i.e. the formation of thromboemboli) in the distal vessels: these presumed microvascular occlusions progressively increase in number and continue to impair blood perfusion.
2. Local tissue concentration of noxious metabolites (e.g. lactic acid and glutamate, among others) originating from a few necrotic neurons increases as a function of time and spreads to adjacent areas where these metabolites kill more neurons.
3. Endothelial cells (in response to changes in perfusion pressure) interact with circulating cells (e.g. polymorphonuclear (PMN) leukocytes and platelets) and begin to generate neurotoxic products, such as free radicals.
4. PMN leukocytes participate in the process of brain necrosis.
5. Several of the above factors acting either sequentially or together may possibly induce brain necrosis in areas supplied by an occluded artery (Garcia and Anderson 1989).

Cell Death

After reviewing fundamental definitions of cellular death, this chapter summarizes selected studies dealing with possible ways in which neurons die in foci of ischemia created by an arterial occlusion.

Cellular death can be broadly defined as the loss of functions essential for preservation. Ehrlich cells maintained in incubation chambers are considered alive so long as mitochondria respire and the capacity of the cells to maintain volumetric control remains intact when challenged with either hypertonic or hypotonic solutions. In tisue cultures, the passive entry of a dye into the cytosol is considered a sign of cell death. Both of these are tests of the plasma membrane functions, which include maintaining ionic homeostasis and preventing the entry of extraneous substances. The cell's ability to synthesize proteins and to divide is further evidence of life (Trump and Arstila 1975). In vivo testing of these functions in cerebral neurons, for proof of viability, would be extremely difficult. Instead, neuronal viability is estimated by measuring the amplitude and frequency of either spontaneous or evoked action potentials (Zampella et al. 1991). Two caveats are in order: (1) the EEG tracings recorded at a given brain site reflect the action potentials generated by a large number of neurons; therefore, EEG amplitude measurements lack sensitivity. This obviously does not apply to recordings derived from individual neurons (Heiss 1992); and (2) the absence of EEG activity does not denote irreversibility. An EEG tracing may remain isoelectric for as long 30 minutes after a carotid artery is occluded, yet normal EEG amplitude can be re-established by reperfusion (Zampella et al. 1991). Recovery of action potentials as late as 30 minutes after reopening the artery was demonstrated in experiments involving transient occlusion of the middle cerebral artery (MCA) in cats (Rosner et al. 1986).

The morphologic criteria of cellular death are based on assessing the integrity of the cellular plasma and mitochondrial membranes; cardiac myocytes made ischemic by occluding a coronary artery fail to recover their contractility and disintegrate once the plasma membrane loses its biochemical/structural integrity; almost simultaneously, electron-dense deposits of calcium salts appear in the mitochondria (Jennings and Reimer 1991). These important markers of individual cell death are very useful and necessary to evaluate acute ischemic lesions; this is because after an arterial occlusion the necrotic cells may be widely scattered (Jennings and Reimer 1991).

Focal ischemia injures cells in a manner dependent on both degree and duration of the ischemia as demonstrated by studies in which the level of ischemia and its duration were controlled and measured (Heiss 1992). For example, after exposing the brain to the effects of regional ischemia (via occlusion of an artery), the CBF values at various regions of the ischemic territory can be measured with oxygen-sensitive electrodes; thereafter, examination of multiple histologic samples from the various areas at risk can establish the extent to which the proportion of necrotic cells correlates with the regional CBF values obtained at variable periods of time (Garcia et al. 1983; de Girolami et al. 1984).

Studies conducted in hepatic, renal, myocardial and other tissues in which circulatory conditions can be easily controlled determined that individual cells dying among living neighbors develop microscopic alterations designated as coagulation necrosis. After this type of injury, the dead cells remain viable for variable periods of time without eliciting an obvious inflammatory response (Farber 1982). In contrast, when cells are fatally injured by non-ischemic mechanisms (e.g. bacterial infections) the necrotic cells are transformed into pus through a process called liquefaction necrosis (Cotran et al. 1989).

In studies of ischemic injury, the fatally injured cells are easily recognized under the microscope because they undergo pyknosis (marked nuclear chromatin condensation) and loss of hematoxylinophilia. This type of cell injury (originally called eosinophilic degeneration) was first described by Spielmeyer (1922) in human brains injured by transient episodes of global ischemia, i.e. hypotensive crises or reversible cardiac arrest for unstated time intervals preceeding death. Spielmeyer (1922) correctly surmised that the appearance of red (eosinophilic) neurons was a somewhat belated result of ischemic injury. Profound systemic hypotension leading to death within a few minutes does not cause neuronal changes; however, survival of several days will induce pyknosis/eosinophilia in isolated neuronal groups. Evidence connecting the pyknotic red neurons with previous ischemic events had to wait until the 1950s, when Kety applied the Fick principle governing the exchange of gases and thus measured human CBF in vivo for the first time (Kety 1948).

Spielmeyer (1922) suggested the term 'ischemic cell necrosis' (ICN) to designate the eosinophilic or red neurons; this designation is not entirely appropriate, and is indeed inaccurate because it implies that ICN is the only reaction elicited in neurons by ischemia. As previously mentioned, ischemia is not an all-or-nothing phenomenon: rather, it is a dynamic phenomenon strongly influenced by degree and duration; therefore, the neuronal responses to ischemia could vary significantly as a function of the time when the observation is made. Furthermore, the term ICN mistakenly implies that red neurons develop solely in response to ischemic injuries, but they are known to occur also in herpes simplex encephalitis.

The red neuron, as seen in preparations stained with hematoxylin and eosin, has been traditionally accepted as a reliable indicator of cell necrosis, but experimental evidence demonstrating that pyknotic/eosinophilic cells are indeed irreversibly injured did not appear until the early 1980s. In experiments of hepatic ischemia, Farber et al. (1981) noted that the appearance of necrotic cells could be accelerated by reperfusion, and that cytologic alterations similar to those of coagulation necrosis could be mediated through either viruses or chemicals (Farber et al. 1981).

A more advanced stage (than pyknosis/eosinophilia) of cell injury, in which the nuclear chromatin is either extruded from the nucleus or loses its affinity for hematoxylin, is known as karyolysis. This type of cell injury is reflected in the histologic appearance of ghost cells (i.e. cells in which neither nuclear nor cytoplasmic nucleic acids are stainable by hematoxylin). Both the red/pyknotic and the pale/karyoltic neurons have leaky plasma membranes that allow the entry of ambient Ca^{2+} into both the cytosol and the mitochondria (Farber 1982). At the ultrastructural level, necrosis has been defined by discontinuities in the nuclear/plasma membranes and the presence of calcium precipitates in the mitochondria (Trump and Arstila 1975; Farber 1982; Jennings and Reimer 1991).

The previously described responses apply to cells derived from any organ, including those belonging to the CNS. In addition, two somewhat idiosyncratic responses to ischemia have been noted in neurons: selective vulnerability and the maturation phenomenon.

Selective Vulnerability

The broad definition of this term applies to the variability of responses that the same injury (e.g. systemic ischemia) elicits among the constituent cells of organs such as the liver, kidney, heart and brain. In response to a systemic injury such as transient cardiac arrest, only selected groups of cells at predictable locations become necrotic. Thus, after a period of severe transient hypotension (followed by adequate survival time), necrosis affects only the central part of the liver lobules; a comparable injury to the heart results in extensive necrosis primarily involving subendocardial myocytes (Cotran et al. 1989). The distribution of these selective types of necrosis is attributed to local conditions created by topographic or angioarchitectonic factors; hepatocytes in the center of the lobule become necrotic first, presumably because they are located furthest from the portal triad, the source of arterial blood. The same explanation accounts for the necrosis limited to the subendocardial sector of the heart wall; myocytes in this location are furthest away from the epicardial branches of the coronary arteries, and during hypotensive crises the decrease in blood flow is probably most severe in the subendocardial territory.

Angioarchitectonic factors also influence the topographic distribution of neuronal necrosis that develops after transient episodes of global ischemia. Among patients who survive either severe hypotensive crises or episodic cardiac arrest, brain necrosis is confined mainly to the arterial border, or boundary zones (ABZ), of the brain and spinal cord; these are sometimes also called watershed areas (Howard et al. 1987). Such ABZ in the cerebral hemispheres include arterial territories located between the terminal branches of the middle and anterior cerebral arteries, the globus pallidus and the periventricular white matter. Brierley et al. (1980) have reproduced ischemic necrosis in the ABZ of primates subjected to controlled hypotensive crises followed by several days' survival.

After global ischemic injuries to the brain, necrosis initially involves neuronal groups at locations where angioarchitectonic factors cannot explain the helter-skelter distribution of the necrotic cells (Spielmeyer 1922; Vogt and Vogt 1925; Scholz 1953). These neuronal groups, known to be extremely sensitive to ischemia, are traditionally identified as being located in the pyramidal-cell layer of the hippocampus and the Purkinje-cell layer of the cerebellar cortex, among others. Neurons at these sites are said to be selectively vulnerable to ischemia, and this attribute is thought to be explained by intrinsic neuronal features which are not reflected in their morphology; some authors suggest that after transient ischemic events the most susceptible neurons are those especially rich in a given excitatory neurotransmitter, such as glutamate (Benveniste et al. 1989).

One of the clearest examples of selective neuronal vulnerability is found in the unique necrosis involving the Sommer's sector of the hippocampus. This lesion, called hippocampal sclerosis when it reaches the state of glial proliferation, is common among patients with a history of systemic hemodynamic crises or repeated epileptic seizures (Sommer 1880; Schmidt-Kastner and Freund 1991). Necrosis selectively involving small groups of hippocampal neurons (CA-1 sector) also develops in non-human primates after systemic hypotensive episodes (Brierley et al. 1969) and in rodents (rats and gerbils) subjected to transient forebrain ischemia by means of bilateral carotid clamping (Pulsinelli and Brierley 1979).

Hippocampal neurons selectively vulnerable to short ischemic periods may be the same cells that are unusually susceptible to the effects of an amino acid such as glutamate. The interstitial concentration of glutamate increases in the hippocampus, following forebrain ischemia (Benveniste et al. 1984); excessive interstitial glutamate is known to be neurotoxic because of its effect on the NMDA-sensitive dendritic receptors. The ultimate effect of this excessive availability of glutamate is to facilitate Ca^{2+} entry into the cells injured by ischemia; in this way glutamate can induce neuronal necrosis. Excessive Ca^{2+} entry into the cell is the ultimate step in the set of events that most researchers believe is necessary to cause ischemic necrosis (Uematsu et al. 1991).

Maturation Phenomenon

This designation has been applied to the belated necrosis that has been observed in rodent brains exposed to short-term episodes of forebrain ischemia followed by several days' reperfusion; in these rodents the delayed neuronal death appears to be primarily confined to the hippocampus (Ito et al. 1975; Kirino et al. 1984).

Hepatocytes injured during a few minutes of global ischemia adopt the features of coagulation necrosis (i.e. pyknosis/eosinophilia) only or mostly after the cells are reperfused (Farber et al. 1981). Interestingly, this necrosis, secondary to the effects of ischemia/reperfusion, can be completely forestalled if Ca^{2+} availability is either blocked or drastically reduced (Farber 1982). Thus, the role of excessive Ca^{2+} entry in ischemic injuries has been considered crucial, and the possible beneficial effects of blocking Ca^{2+} entry into cells injured by short periods of ischemia has been explored in numerous experimental models (Uematsu et al. 1991).

The feasibility of transplanting cadaveric human organs completely ischemic for several hours suggests, paradoxically, that the effects of no flow are better tolerated than those of residual flow (Garcia 1992). The circulatory conditions created by occluding a brain artery include alternating areas of increased, decreased or normal CBF values (depending on the time when these measurements are made) (Garcia et al. 1983; Nagasawa and Kogure 1989); therefore, the type of ischemia induced by an arterial occlusion is incomplete. After an artery has been occluded the residual circulation is maintained in the ischemic area by the end-to-end arterial anastomoses that exist both on the brain surface and in the superficial brain parenchyma; these arterial connections have been well described by van der Eecken (1959).

In rodents, short periods (5 min) of forebrain ischemia (secondary to bilateral carotid artery ligation) followed by unmodified reperfusion produce a progressive increase in the number of necrotic (red/eosinophilic) neurons in the CA-1 segment of the hippocampus. Brains from animals killed one day after the ischemic/reperfusion injury may have only 10% necrotic neurons in the hippocampus, whereas those allowed to survive for 4 days have as much as a threefold increase in the number of necrotic neurons (Kirino et al. 1984). This curious phenomenon, whereby neurons which appear normal on day 1 become necrotic on day 4, despite an arterial circulation no longer impeded, has been designated delayed neuronal death, or 'maturation phenomenon' (Ito et al. 1975). Based on the similarity with the observations made by Farber (1982) in ischemic/reperfused liver, there might be a connection between these two phenomena: 'reperfusion' injury and 'maturation' phenomenon.

Comparing the results of sucessful functional recovery in an organ, such as the kidney or heart, deprived of circulation for several hours, with the progressive cellular necrosis observed in the gerbil hippocampus after 5 minutes of incomplete ischemia followed by 4 days' reperfusion, has led some to conclude that reperfusion alone may not be sufficient to protect ischemic neurons.

Nevertheless, at least two conditions are significantly different in the case of transplanted organs, and both may influence their functional recovery after complete ischemia: (1) the transplanted organ is 'protected' during the period of ischemia by hypothermia; and (2) there may be circulating factor(s) or cytokines in the residual blood flow of an incompletely ischemic organ than progressively contribute(s) to the effects of ischemia. Among the prime candidates are interleukin 1 (IL-1) and tissue necrosis factor α (TNF-α).

Residual Blood Flow as a Potential Contributor to Ischemic Injury

The moment an artery is occluded, the altered blood flow may elicit various responses from at least two cell types: the endothelium and circulating elements. Depending on the time elapsed after the occlusion, the interactions between endothelial and circulating cells may either ameliorate or aggravate the impact of ischemia on the parenchymal cells.

Ames and colleagues (1968) described the 'no-reflow' phenomenon in a rabbit model of transient global cerebral ischemia. The injury, induced by the simultaneous occlusion of arterial and venous blood vessels, resulted in the incomplete patency of cerebral microvessels once the cervical vessels were reopened. The incomplete reperfusion of the cerebral microvasculature became apparent when patches of pallor ('no-reflow') were revealed in areas where microvessels failed to be filled with a circulating marker (carbon black) (Ames et al. 1968). Hossmann (1990) referred to this phenomenon as 'postischemic hypoperfusion' and suggested that the inability to perfuse previously ischemic organs in a homogeneous manner may be caused by one or more factors, such as the squeezing effects on microvessels of perivascular astrocytic swelling. Astroglial swelling is one of the earliest cell changes induced by single-artery occlusion (Garcia 1983). Other factors include parenchymal bleeding; intraluminal formation of microthrombi; detachment of endothelial cell fragments; platelet aggregation; endothelial cell swelling; PMN leukocyte adherence to endothelial cells; and vasospasm.

After occluding an artery such as the MCA, rCBF values in isolated foci fall to levels below the thresholds necessary to maintain plasma membrane integrity (Garcia et al. 1983; Heiss 1992). There is general acceptance that local CBF values are significantly heterogeneous in various regions of the area at risk (Nagasawa and Kogure 1989), but there is lack of agreement on whether the individual local rCBF values continue to deteriorate as a function of time elapsed after the arterial occlusion (Hakim et al. 1992).

Could postischemic hypoperfusion develop after occlusion of an artery, and could this hypoperfusion be overcome by reopening the artery? The beneficial effects of administering thrombolytic agents to patients with coronary-artery occlusion (Braunwald 1988) sparked considerable interest in applying similar strategies to the treatment of intracranial artery occlusion (del Zoppo et al. 1992). The effects of experimental ischemia on myocardium, kidney, skeletal muscle, liver and intestine can be ameliorated by inducing neutropenia (Jaëschke et al. 1990; Dereski et al. 1992). Leukocytes (in particular PMN neutrophils) may interact with the endothelial lining and cause mechanical microvascular occlusion as well as endothelial injury. This effect could develop through the activation of free radicals, not only during the ischemic episode but perhaps even more during reperfusion. This prompted Braunwald (1988) to suggest that myocardial reperfusion is a double-edged weapon: applied at the wrong time, reperfusion could be more harmful than ischemia alone.

The chronology of PMN neutrophils arriving in the ischemic brain, and their distribution within an area of focal brain ischemia, has begun to be defined. In two separate models of permanent MCA occlusion in rats, Dereski et al. (1992) and Garcia et al. (1993) documented the arrival of small numbers of intravascular PMN neutrophils in the ischemic territory as early as 15–30 minutes after the arterial occlusion. The neutrophils begin transmigrating into the area of ischemia after 6 hours, and their numbers peak at about 12–24 hours. Interestingly, red (necrotic) neurons do not appear in large numbers until 12 hours after the arterial occlusion (Garcia et al. 1993). This poses the question of whether massive leukocytic infiltration of the ischemic brain parenchyma is a cause or a consequence of neuronal necrosis?

The relative contribution of PMN leukocytes to the ("no-reflow") phenomenon in focal ischemia and the importance of the microvasculature as the target of these events has been demonstrated recently. Working with a model of transient (3 hour) MCA occlusion followed by 1 hour of reperfusion in baboons, del Zoppo et al. (1991) have shown microvascular plugging by PMN leukocytes and erythrocytes. This effect is more severe in the territory of the lenticulostriate arteries (LSA) than in other regions within the area at risk. At the striatum PMN leukocytes completely obstructed 20% of the capillaries, and most leukocytes were directly apposed to venular endothelium.

Moreover, microvascular patency in the territory supplied by the lenticulostriate arteries is much better preserved in animals treated with a murine antihuman CD-18 monoclonal antibody than in those receiving a placebo (Mori et al. 1992). The CD-11B/CD-18 (Mac-1) complex is the principal endothelial adherence receptor on PMN leukocytes. The adhesion of this very specific complex to the endothelial surface receptors (ICAMs, or intercellular adhesion molecules) is essential for the PMN leukocytes to first stop in the ischemic territory and then migrate across the vessel wall into the ischemic tissue (McEver 1991).

Precise information concerning the chemotactic mechanisms that may become operative upon the occlusion of an intracranial artery is not currently available. By analogy to events surrounding ischemia/ reperfusion in other organs, it is posible that cytokines such as TNF or IL-1 may activate endothelial cells and set in motion leukocytic attraction and adhesion to the endothelial surface. Endothelial activation (as opposed to stimulation) involves the synthesis of proteins and therefore it is hours, or even days, before it becomes apparent (Pober and Cotran 1990). Both platelet-activating factor (PAF) and IL-1 can be synthesized by endothelial cells, and the stimulus to release these proteins could be either a decrease in shear force or another circulatory event generated by the arterial occlusion.

A second potential contributor to the defects in microvascular perfusion, or 'no-reflow' phenomenon, could be the formation of fibrin. Tissue factor (TF), the most potent endogenous procoagulant, is a transmembrane glycoprotein found primarily in the gray matter of the brain and around blood vessels larger than 10 μm in diameter (del Zoppo et al. 1992). Blockade of TF function by a murine antihuman TF monoclonal antibody before MCA occlusion significantly increases microvascular patency (Thomas et al. 1993). These observations and the finding of platelets, fibrin and PMN leukocytes in microvascular thrombi (del Zoppo et al. 1991) suggest that the causes of 'no-reflow' are complex, but involve various mechanisms influenced by these circulating cells. The results of these studies have suggested new therapeutic approaches to preserve microvascular patency in focal ischemia.

Dynamic Nature of the Brain Lesion Created by an Arterial Occlusion

The availability of an experimental model of brain infarct in which the MCA can be consistently occluded without opening the skull (Koizumi et al. 1986; Zea-Longa et al. 1989) has provided the means to study the chronology of the brain lesion at frequent intervals, beginning at 30 minutes and ending at 7 days after occlusion of the artery (Garcia et al. 1993). The lesion, initially involving less than 5% of the hemisphere's volume, incorporates 39% and almost 75% of the ipsilateral hemisphere after 6 hours and 72 hours respectively, (Fig. 1.1). The normal microscopic features of the rat brain (Fig. 1.2) are replaced by acute neuronal changes. Moreover, the predominant acute neuronal changes (shrinkage, scalloping, swelling) observed during the first 6 hours (Fig. 1.3) are abruptly replaced some time between 6 and 12 hours by delayed changes (Fig. 1.4). These include pyknosis/eosinophilia or pallor/karyolysis, i.e. coagulation necrosis, as defined by Farber (1982). The possibility that some of the acute neuronal changes might be reversible offers the hope of therapeutic interventions during the period between 30 minutes and 6 hours after occlusion. Such therapeutic intervention would aim to either slow or halt the progression to necrosis.

The very dynamic time-dependent aspects of the brain lesion initiated by an arterial occlusion in rats is expressed in at least two different ways: (1) there is progressive extension of the lesion into adjacent regions of the area at risk; the pattern is highly predictable: it involves first the preoptic area, next the striatum, and finally the cerebral cortex; (2) a maturation or ripening effect is clearly demonstrable in an increasing number of neurons. Despite the fact that the circulatory conditions created by the arterial occlusion are not iatrogenically modified by systemic factors (such as changes in systemic blood pressure, alterations in body temperature or modifications in serum glucose), there is a relentless progression that affects an increasing number of neurons. Starting with an essentially normal appearance at 30 minutes, neurons become slightly modified (acute changes) during the next 6 hours and then abruptly adopt the features of coagulation necrosis (pyknosis/eosinophilia); these delayed changes become manifest in large numbers of neurons some time during the period between 6 and 12 hours after the arterial occlusion (Garcia et al. 1993).

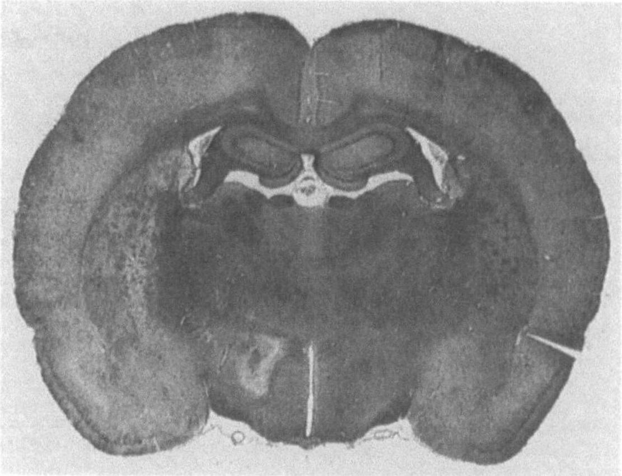

a

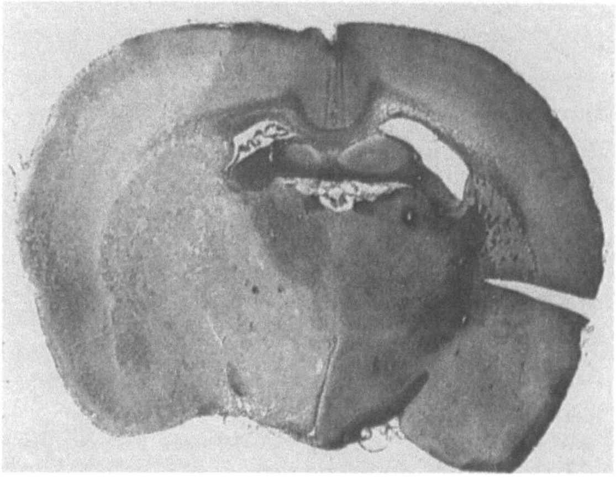

b

Fig. 1.1. Rat brain. **a** Lesion in right hemisphere (arrow) secondary to occlusion of the MCA, for 2 hours. **b** Right hemispheric lesion secondary to MCA occlusion (72 hours). (H&E)

The possibility that small focal alterations develop in the territory of the occluded artery, and that these microvascular changes may influence the progression to neuronal necrosis, has been explored by quantifying two biological phenomena: the number of plugged or occluded microvessels (diameter ‹15 μm), and the degree of astrocytic swelling; both of these are frequently mentioned as likely contributors to the continued deterioration of the circulatory conditions in an area of focal ischemia (Hossmann 1990).

The microvasculature of the territory distal to the site of an arterial occlusion is subjected to abnormal conditions, such as decreased shear stress. These abnormalities could trigger endothelial cell responses which activate specific surface receptors on some of the circulating cells. Specifically, the endothelial chemotaxis of both

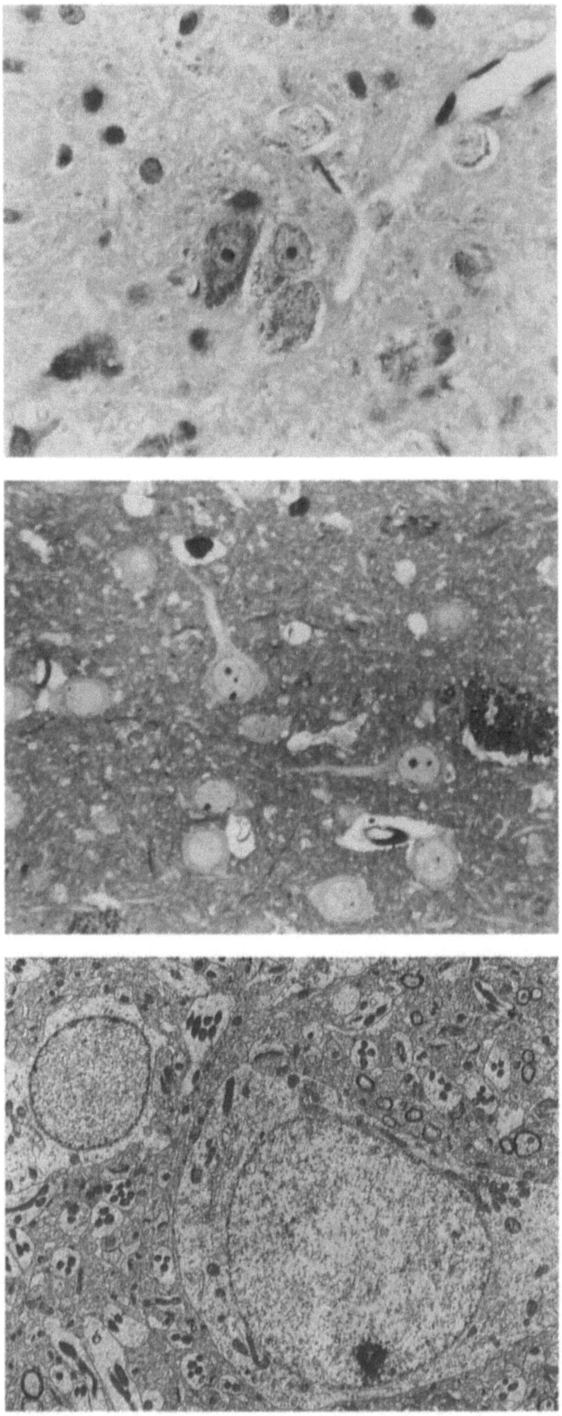

Fig. 1.2. Normal rat brain (control animals). **a** Cerebral cortex, paraffin-embedded, 6 μm-thick section. (H&E; original magnification x 130) **b** Putamen, araldite-embedded, toluidine blue stained 1μm-thick section. (Original magnification x 130) **c** Cortex; astrocyte left upper corner close to a large neuronal perikaryon in the right lower corner. (x 3700)

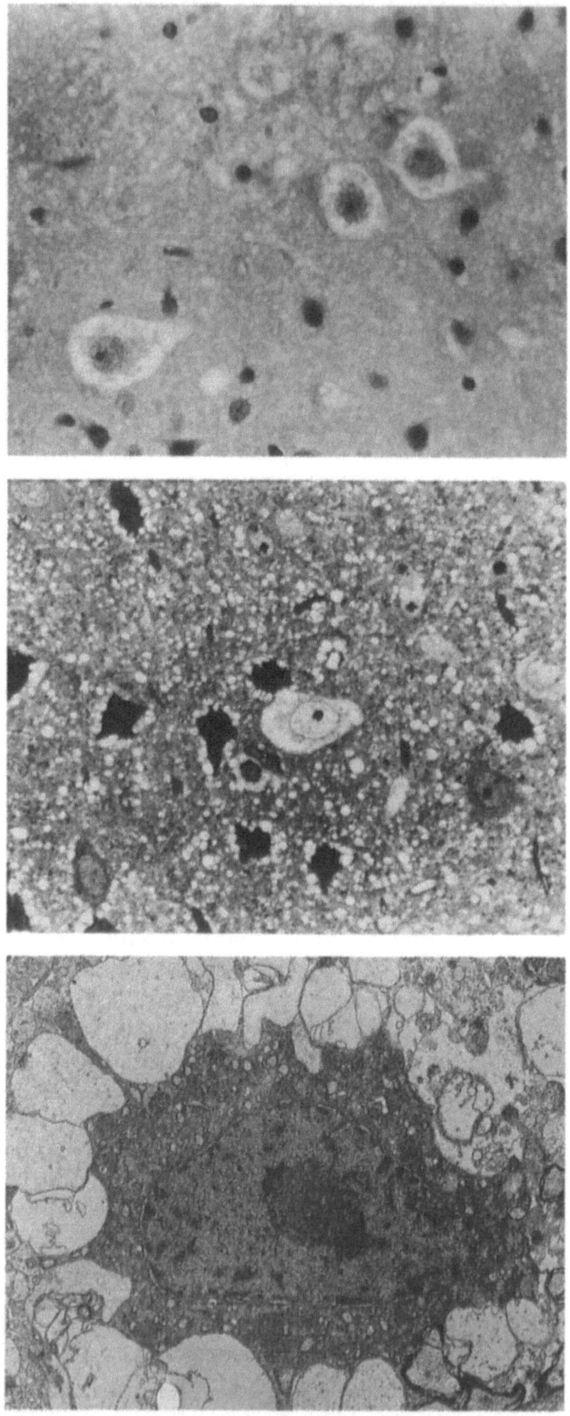

Fig. 1.3. Acute neuronal changes in the territory of an occluded artery (Wistar rat). **a** Neuronal swelling (cerebral cortex) 90 minutes after MCA occlusion. (H&E; original magnification x 130) **b** Neuronal shrinkage, swelling and scalloping in the cerebral cortex, 4 hours after MCA occlusion. (Toluidine blue; original magnification x 130) **c** Shrunken (scalloped) neuron, cerebral cortex, same area as in b. (x 5550)

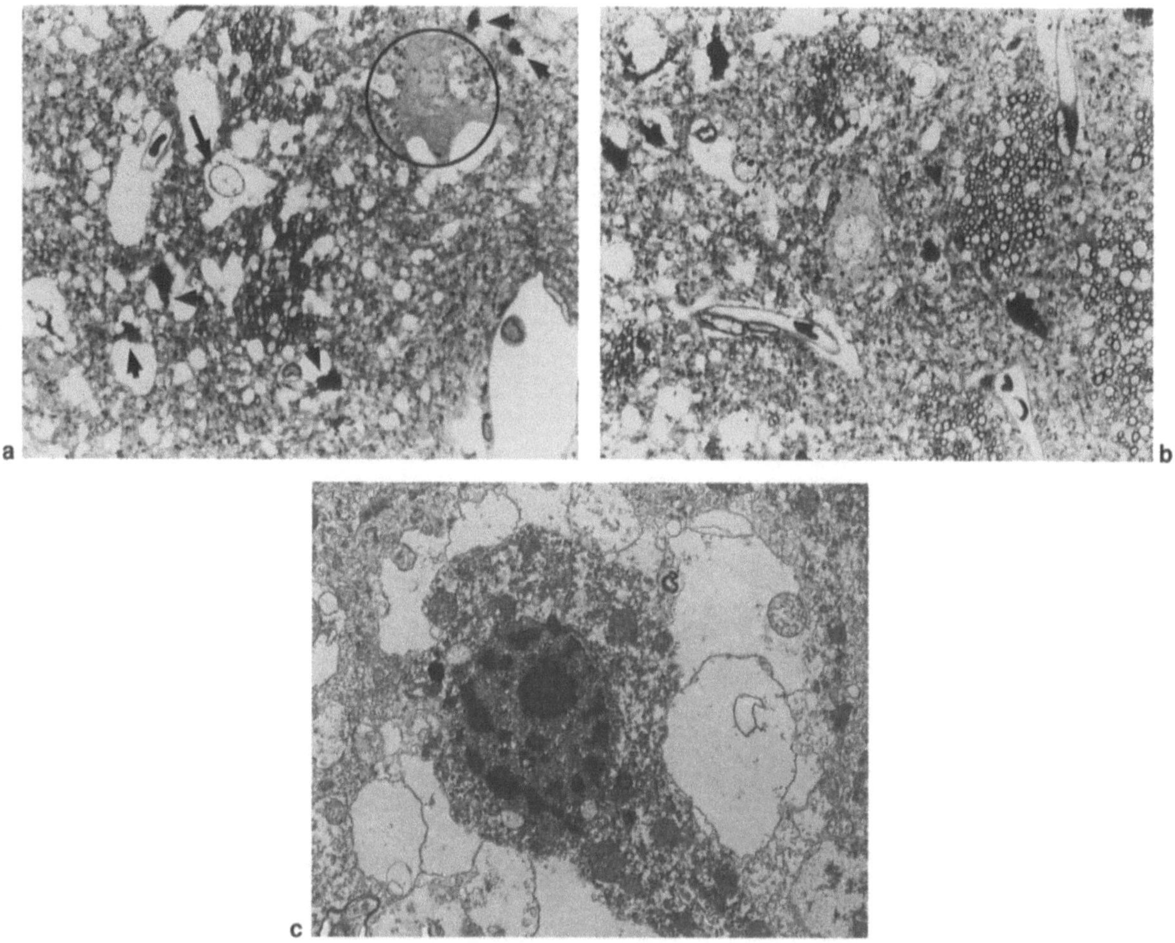

Fig. 1.4. Delayed neuronal changes in Wistar rat with MCA occlusion. **a** Striatum; 24 hours after arterial occlusion most small neurons (arrowheads) are necrotic, a large neuron (circle) shows acute changes, an astrocyte (arrow) is markedly swollen and an inflammatory cell is visible inside a venule (right lower corner). (Toluidine blue; original magnification x 130) **b** Striatum, same sample as in a. Myelin sheaths are still intact. A large neuron in the center of the field is minimally injured. (Toluidine blue; original magnification x 130) **c** Striatum; small necrotic neuron showing breaks in nuclear and plasma membranes as well as dense mitochondrial densities. The perineuronal 'halo' is made of swollen processes, mostly astrocytic. (x 6370)

leukocytes and platelets could contribute to the continuing microcirculatory deterioration. These effects could be a consequence of the mechanical interference to the circulation offered by adhered leukocytes, platelet aggregates or even fibrin thrombi (Fig 1.5). Deleterious effects on the brain parenchyma can also be the result of the activation of both PMN leukocytes and platelets.

In the rat model of MCA occlusion, the number of plugged microvessels (<15 μm) within the territory of the occluded artery progressively increases as a function of time; the percentage of plugged microvessels varies according to region, but after 4 hours close to 100% of the microvessels in the supraoptic area are plugged. (A plug was defined in this study as any circulating cell (or fragment) visible within the lumen of microvessels.) The controls, in these studies, included microvessels in non-occluded arterial territories. As late as 7 days after the arterial occlusion less than 5% of the microvessels in the territory of the anterior cerebral artery showed microvascular plugs (Garcia et al. 1993).

Three cell types plugged the lumen of blood vessels less than 15 μm in caliber. Erythrocytes, PMN leukocytes and platelets were most commonly found, but there was a correlation between circulating cells and blood vessel type. Leukocytes, mostly PMN neutrophils, were found in contact with capillary venular endothelium; in contrast, platelets were primarily apposed against the surface of arterial/arteriolar

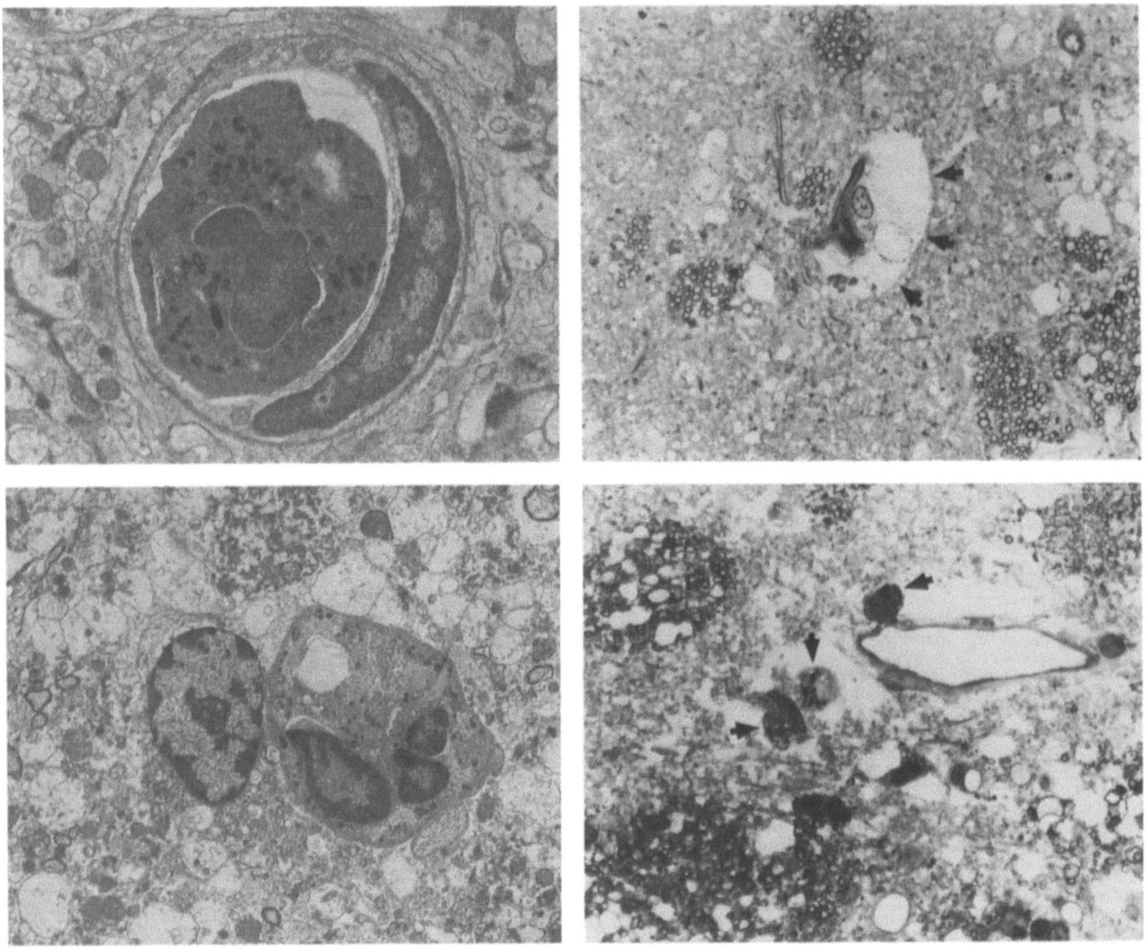
a c b d

Fig. 1.5. Astrocytic and leukocytic responses to MCA occlusion (Wistar rat). **a** Striatum; PMN leukocyte inside venule, 1 hour after the arterial occlusion. (x 5560) **b** Cortex; PMN leukocyte in area of brain necrosis, 48 hours after arterial occlusion. (x 4100) **c** Striatum; massively swollen astrocyte (arrowhead) next to a collapsed microvessel, 24 hours after arterial occlusion. (Toluidine blue; x 130) **d** Striatum; several macrophages (arrowheads) are visible in the putamen 72 hours after the arterial occlusion. (Toluidine blue; x 130)

endothelium. Additional evidence for the dynamic nature of this biological phenomenon was noted as follows: until about 6 hours all leukocytes were intravascular; thereafter, these cells were seen in the brain parenchyma. The numbers of PMN leukocytes, counted in serial sections of the ischemic hemisphere, peaked at about 48–72 hours after the arterial occlusion, but as late as 7 days after MCA occlusion they were still identifiable in the now fully mature (or healing) brain infarct.

Whether PMN leukocytes have a deleterious effect on focal brain ischemia, or are a part of the reaction to neuronal necrosis, is still unclear; however, most are found in the ischemic cerebral hemisphere when the number of necrotic (red) neurons peaks (48–72 hours).

Astrocytic responses to the conditions created by a single-artery occlusion vary with the location. In the core of the ischemic area, soon after arterial occlusion (3 hours), astrocytes disintegrate even before signs of neuronal necrosis are visible; in the marginal areas of the ischemic region, nuclear astrocytic diameter increases by as much as 16% (within 2 hours of MCA occlusion) compared to the non-ischemic areas of the brain. This astrocytic swelling, together with a hypothetical arteriolar vasospasm, could contribute to the decrease in the microvascular luminal surface and impair vascular patency in a progressive manner. At the interphase between ischemic and non-ischemic parenchyma, astrocytes become 'reactive': their cytoplasm becomes eosinophilic and glial fibrillary protein is expressed. This biologic change appears to be irreversible, but its significance in terms of the eventual outcome of the lesion is unclear. Reactive astrocytes become visible in this model of focal brain ischemia about 12–18 hours after the arterial occlusion (Garcia et al. 1993a).

In summary, the nature and the sequence of cellular events that attend the development of a brain infarct (following the occlusion of a single intracranial artery) are beginning to be defined. Attempts to modify some of these events by, for example, the suppression of leukocytic responses and alterations in hemostasis, may provide additional clues that could help answer the questions: What brings about irreversibility to a lesion initiated by the occlusion of an intracranial artery?; What explains the fact that it takes between 6 and 12 hours after an aterial occlusion before pan-necrosis appears?; and finally What are the differences between the microvascular and parenchymal events that occur following sustained occlusion of a single artery as opposed to a situation of occlusion and reperfusion?

Understanding the biological phenomena of focal ischemia is especially difficult in the brain because of the heterogeneous responses that each cell type may exhibit, and also because of the influence of tissue edema on an organ encased in a non-expandable container. Edema is probably not as important in the progression of the ischemic lesion in organs such as the heart and the kidney.

Acknowledgement: I am grateful to Drs S. M. Aronson (Providence, RI), G. J. del Zoppo (La Jolla, CA), and J. C. Geer (Birmingham, AL) for their helpful comments on preliminary versions of this manuscript.

References

Ames A. III, Wright LW, Kowada M, Thurston JM, Majors G (1968) Cerebral ischemia: II. The no-reflow phenomenon. Am J Pathol 52: 437–453

Appen RE, Wray SH, Cogan DG (1975) Central retinal artery occlusion. Am J Ophthalmol 79:374–381

Benveniste H, Dreir J, Schousboe A, Diemer NH (1984) Elevation of the extracellular concentration of glutamate aspartate in rat hippocampus during transient cerebral ischemia monitored by intracellular microdialysis. J Neurochem 43:1369–1374

Benveniste H, Horgenson MB, Sandberg M, Christensen T, Hagberg H, Diemer NH (1989) Ischemic damage in hippocampal CA1 is dependent on glutamate release and intact innervation from CA3. J Cerebr Blood Flow Metab 9:629–639

Braunwald E (1988) Thrombolytic reperfusion of acute myocardial infarction: resolved and unresolved issues. J Am Coll Cardiol 12 (Suppl): 85A

Brierley JB, Brown AW, Excell BJ, Meldrum BS (1969) Brain damage in the Rhesus monkey resulting from profound arterial hypotension. I. Its nature, distribution and general physiological correlates. Brain Res 13:68

Brierley JB, Prior PF, Calverly J, Jacksin SJ, Brown AW (1980) The pathogenesis of ischemic neuronal damage along the cerebral arterial boundary zones in *Papio anubis*. Brain 103:929

Cotran RS, Kumar V, Robbins SL (1989) Cellular injury and adaptation In: Cotran RS et al. (eds) Robbin's pathologic basis of disease. WB Saunders, Philadelphia, pp 1–38

de Girolami U, Crowell RM, Marcoux FN (1984) Selective necrosis and total necrosis in focal cerebral ischemia. Neuropathologic observations on experimental cerebral artery occlusion in the macaque monkey. J Neuropathol Exp Neurol 43:57–71

del Zoppo GJ, Schmid-Schönbein GW, Mori E et al. (1991) Polymorphonuclear leukocytes occlude capillaries following middle cerebral artery occlusion and reperfusion. Stroke 22:1276–1283

del Zoppo GJ, Poeck K, Pessin MS et al. (1992) Recombinant tissue plasminogen activator in acute thrombotic and embolic stroke. Ann Neurol 32:78–86

Dereski MO, Chopp M, Knight RA, Chen H, Garcia JH (1992) Focal cerebral ischemia in the rat: temporal profile of neutrophil responses. Neurosci Res Commun 11:179–186

Farber JL (1982) Biology of disease. Membrane injury and calcium homeostasis in the pathogenesis of coagulative necrosis. Lab Invest 47:114–123

Farber JL, Chien KR, Mittnacht S Jr (1981) The pathogenesis of irreversible cell injury in ischemia. Am J Pathol 102:271–281

Garcia JH (1983) Ischemic injuries of the brain: morphologic evolution. Arch Pathol Lab Med 107:157–162

Garcia JH (1992) The evolution of brain infarcts. A review. Neuropathol Exp Neurol 51:387–393

Garcia JH, Anderson ML (1989) Physiopathology of cerebral ischemia: a review. CRC Crit Rev Neurobiol 4:303–324

Garcia JH, Mitchem HL, Briggs L et al. (1983) Transient focal ischemia in subhuman primates: neuronal injury as a function of local cerebral blood flow. J Neuropathol Exp Neurol 42:44–60

Garcia JH, Yoshida Y, Chen H, Li Y, Zhang ZG et al. (1993a) Progression from ischemic injury to infarct following middle-cerebral-artery occlusion in the rat. Am J Pathol 142:623–635

Garcia JH, Yoshida Y, Lian J et al. (1993b) Experimental occlusion of a middle cerebral artery is accompanied by early polymorphonuclear leukocyte infiltration. Stroke 24:21 [Abstract]

Hakim AM, Hogan MJ, Carpenter S (1992) Time course of cerebral blood flow and histological outcome after focal cerebral ischemia in rats. Stroke 23:1138–1144

Harrison MJG (1982) Pathogenesis. In: Warlow C, Morris PJ (eds) Transient ischemic attacks. Marcel Dekker, New York, pp.21–46

Heiss WD (1992) Experimental evidence of ischemic thresholds and functional recovery. Stroke 23:1668–1672

Hossmann KA (1990) Hemodynamics of post ischemic reperfusion of the brain. In: Weinstein PR, Faden AI (eds) Protection of the brain from ischemia. Williams and Wilkins, Baltimore, pp 21–36

Howard R, Trend P, Russell RW (1987) Clinical features of ischemia in cerebral arterial border zones after periods of reduced cerebral blood flow. Arch Neurol 44:934–940
Ito U, Spatz M, Walker JT, Klatzo I (1975) Experimental cerebral ischemia in Mongolian gerbils. I. Light microscopic observations. Acta Neuropathol 32:209–215
Jaëschke H, Farhood A, Smith W (1990) Neutrophils contribute to ischemia reperfusion in rat liver in vivo. FASEB J 4:3355–3359
Jennings RB, Reimer KA (1991) The cell biology of acute myocardial ischemia. Ann Rev Med 42:225–246
Jiang Q, Zhang ZG, Chopp M, Helpern JA, Ordidge RJ, Garcia JH et al. (1993) The application of diffusion-weighted NMR imaging as a predictor of brain injury following MCA occlusion in the rat. J Neurol Sci 120:123–130
Kety SS (1948) Quantitative measurement of cerebral blood flow in man. Meth Med Res 1:204–217
Kirino T, Tamura A, Sano K (1984) Delayed neuronal death in rat hippocampus following transient forebrain ischemia. Acta Neuropathol 64:139–146
Koizumi J, Yoshida Y, Nakazawa T, Ooneda G (1986) Experimental studies of ischemic brain edema. 1. A new experimental model of cerebral embolism in rats in which recirculation can be introduced in the ischemic area. Jpn J Stroke 8:1–8
Koudstaal PJ, Van Gijn J, Lodder J, Frenken WGW, Vermeulen M et al. (1991) Transient ischemic attacks with and without a relevant infarct on computed tomographic scans cannot be distinguished clinically. Arch Neurol 48:916–920
McEver RP (1991) Selectins. Novel receptors that mediate leukocyte adhesion during inflammation. Thromb Haemost 65:223–228
Morawetz RB, Zeiger HE, McDowell HA, McKay RD et al. (1984) Correlation of cerebral blood flow and EEG during carotid occlusion for endarterectomy (without shunting) and neurologic outcome. Surgery 96:184–189
Mori E, del Zoppo GJ, Chambers JD, Copeland BR, Arfors KE (1992) Inhibition of polymorphonuclear leukocyte adherence suppresses no-reflow after focal cerebral ischemia in baboons. Stroke 23:712–718
Nagasawa H, Kogure K (1989) Correlation between cerebral blood flow and histologic changes in a new rat model of middle cerebral artery occlusion. Stroke 20:1037–1043
Pober JS, Cotran RS (1990) Overview: the role of endothelial cells in inflammation. Transplantation 50:537–544
Pulsinelli WA, Brierley JB (1979) A new model of bilateral hemispheric ischemia in the unanesthetized rat. Stroke 10:267–272
Rosner G, Graf R, Kataoka K, Heiss WD (1986) Selective functional vulnerability of cortical neurons following transient MCA occlusion in the cat. Stroke 17:76–82
Schmidt-Kastner R, Freund TF (1991) Selective vulnerability of the hippocampus in brain ischemia. Neuroscience 40:599–636
Scholz W (1953) Selective neuronal necrosis and its topistic patterns in hypoxemia and oligemia. J Neuropathol Exp Neurol 12:249–261
Sommer W (1880) Erkrankung des Ammonshorns als aetiologisches Moment der Epilepsie. Arch Psychiat Nervenkr 10:631–675
Spielmeyer W (1922) Histopathologie des Nervensystems. 19. Allgem. Teil. Julius Springer, Berlin
Sundt TM, Sharbrough FW, Piepgras DG, Kearns TP et al. (1981) Correlation of cerebral blood flow and electroencephalographic changes during carotid endarterectomy with results of surgery and hemodynamics of cerebral ischemia. Mayo Clin Proc 56:533–543
Thomas WS, Mori E, Copeland BR, Yu J-Q, Morrisey JH, del Zoppo GJ (1993) Tissue factor contributes to microvascular defects following cerebral ischemia. Stroke (in press)
Trump BF, Arstila AV (1975) Cellular reaction to injury. In: LaVia MF, Hill RB (eds) Principles of pathobiology, 2nd edn. New York, Oxford University Press, pp 9–96
Uematsu D, Araki N, Greenberg JH, Sladky J, Reivich M (1991) Combined therapy with MK-801 and nimodipine for protection of ischemic brain damage. Neurology 41:88–94
van der Eecken HM (1959) The anastomoses between the leptomeningeal arteries of the brain. C.C. Thomas, Springfield
Vogt C, Vogt O (1922) Erkrankungen der Grosshirnrinde im Lichte der Topistik, Pathoklise und Pathoarchitektonik. J Psychol Neurol 28:1–189
Vogt O, Vogt C (1925) Der Begriff der Pathoklise. J Psychol Neurol 31:245–255
Whisnant JP, Matsumoto N, Elveback LR (1973) Transient cerebral ischemic attacks in a community. Rochester, Minnesota, 1955 through 1969. Mayo Clin Proc 48:194–198
Zampella E, Morawetz RB, McDowell HA, Zeiger E et al. (1991) The importance of cerebral ischemia during carotid endarterectomy. Neurosurgery 29:727–731
Zea Longa E, Weinstein PR, Carlson S et al. (1989) Reversible middle cerebral artery occlusion without craniectomy in rats. Stroke 20:84–91

2. Positron Emission Tomography (PET) in Acute Ischemic Stroke: Pathophysiological and Clinical Implications

J. C. Baron

Experimental studies in animal stroke models have shown that, at variance with global ischemia from cardiac arrest, the abrupt occlusion of a cerebral artery (e.g. the trunk of the middle cerebral artery, MCA) triggers complex regulatory mechanisms that tend to compensate for the decrease in perfusion pressure, such as distal bed vasodilatation and subsequent 'opening' of pial anastomoses. This results in a gradient of reduction in cerebral blood flow (CBF) and tissue oxygenation from the ischemic core, where perfusion is most reduced or may even be arrested, towards the borders of the affected vascular territory, where CBF may be only mildly reduced (oligemia) or even normal (autoregulated) (Pulsinelli 1992). The concept of the 'ischemic penumbra' dictates that, in each block of tissue, the level of residual perfusion will determine to what degree the cascade of ischemic-dependent biochemical events is triggered. These events in turn determine both the extent of neuronal dysfunction acutely and, eventually (depending on the duration of the arterial occlusion), the outcome of the tissue (Lassen et al. 1991; Heiss et al. 1992). Thus, knowledge of the local cerebral perfusion and residual cell function is important in acute stroke to assess the status of brain tissue, whether still ischemic or already reperfused, and whether still viable or already irreversibly compromised. Such information could provide the basis for individually tailored, best medical management.

Positron emission tomography (PET) is well suited to provide such information as it is the only method presently available that simultaneously measures the CBF, cerebral blood volume (CBV), cerebral oxygen extraction fraction (OEF) and cerebral metabolic rate of oxygen ($CMRO_2$) in both the human and the non-human primate (Baron et al. 1989). Indeed, this method allows assessment in each block of tissue of both the oxygen supply (determined by the CBF) and the oxygen demand (determined by the residual mitochondrial activity, which mainly reflects synaptic activity at neuron terminals). It uses tracer amounts of CO_2, O_2 and CO gases labeled with the short-lived ($T_{1/2} = 2$ min) positron-emitting isotope ^{15}O (Baron et al. 1989). For each of these variables quantitative parametric maps are obtained for several axial cuts of the brain, with a spatial resolution of about 5–10 mm and a temporal resolution of about 1 hour.

Terminology and Basic Science

The Normal Brain

Physiologically, there exists a matching of local values of CBF, $CMRO_2$ and CBV, according to linearly proportional relationships. The matching is such that local CBF and CBV values are highest in areas of highest $CMRO_2$ and vice versa, and the OEF image is uniform; this describes the metabolic regulation of the cerebral circulation. Thus, in the normal resting human brain, the local CBF reliably reflects the prevailing $CMRO_2$.

Profiles of CBF and $CMRO_2$ Interrelationships in Stroke

Three main profiles have been determined by PET in acute stroke, in two of which the CBF may become uncoupled from the $CMRO_2$ and the OEF focally altered. These are hemodynamic failure, luxury perfusion and remote metabolic depression.

Hemodynamic failure. Four stages of hemodynamic failure of increasing severity have been defined; a) the stage of CBF *autoregulation*, where CBF is maintained via an increase in CBV; b) the stage of *oligemia*, where CBF is reduced, OEF is increased and the $CMRO_2$ is unaltered; c) the stage of *ischemia*, where CBF is reduced, OEF is still increased and $CMRO_2$ is reduced; and d) the stage of presumably *irreversible damage*, where both CBF and $CMRO_2$ are profoundly reduced to near-zero levels. 'Misery perfusion' describes, across stages b, c and d, the situation of reduced CBF with increased OEF.

Luxury perfusion. This profile is defined by a focally reduced OEF, which indicates a perfusion in excess of local oxygen demand. There are two different types of luxury perfusion: a) *early luxury perfusion* (postischemic reperfusion), seen within hours of stroke onset, which is characterized by hyperperfused tissue regardless of $CMRO_2$ (which may range from normal to profoundly reduced); and b) *subacute luxury perfusion*, seen days to weeks after stroke onset and always affecting necrotic tissue; it is characterized by markedly reduced $CMRO_2$ in the face of excess but variable (increased, normal or reduced) perfusion.

Remote metabolic depression. Brain areas with moderately but proportionately reduced CBF, CBV and $CMRO_2$ and located remote from the ischemic zone are frequently revealed by PET during the very acute stage of stroke. These remote effects are widely explained as depressed synaptic activity at sites distant from, but neurally connected (either directly or across several relays) to, the damaged area. Although they are often referred to collectively as 'diaschisis' (Feeney and Baron 1986), this word conceals a host of distinct cellular derangements (from reversible hypofunction to evolving degeneration) which have the same PET expression (Baron 1991). Yet some of these effects may represent, in part or in whole, truly functional (i.e. potentially recoverable) transsynaptic derangement and, as such, may participate in both the acute clinical expression of, and the subsequent recovery from, stroke.

Three main types of remote metabolic depression have been reported: a) *intrahemispheric*, such as thalamic hypometabolism after cortical damage, which reflects in part thalamocortical neuron retrograde death, and neocortical hypometabolism after deep-seated (e.g. thalamic) infarcts, which is well correlated to, and recovers in parallel with, cognitive impairment (e.g. aphasia) (Baron et al. 1986; Karbe et al. 1989; Metter et al. 1989); b) *contralateral hemispheric*, still a matter of controversy (Wise et al. 1986; Dobkin et al. 1989), which could explain instances of global brain dysfunction after unilateral stroke, and whose subsequent alleviation could contribute to functional recovery; and c) *contralateral cerebellum* (Baron et al. 1980): this effect, which is present in over 50% of MCA stroke patients acutely and only inconsistently recovers, probably reflects a functional transneuronal effect along the glutamatergic corticopontocerebellar mossy-fiber pathway; up until now, its clinical significance has remained elusive.

Practically, PET is ideally suited to differentiate between the three main profiles of flow–metabolism interrelationships just described. For instance, as reduced CBF is a condition that belongs in all three profiles, tools such as SPECT, which measures only CBF, are less adequate than PET in the evaluation of the pathophysiology of acute stroke.

PET Findings in Acute MCA Territory Stroke

Ischemic Tissue

The main finding that has emerged from PET studies carried out during the acute stage of MCA territory stroke, and one that has radically expanded our view of the 'window' of therapeutic opportunity in stroke, has been the demonstration, hours into the episode, of still *criti-*

cally ischemic tissue (defined according to the criteria stated above). Thus, wide zones of cortex with reduced CBF (often below the 'penumbral' threshold of 20 ml/100 g/min) and massively increased OEF (often above 0.80), yet with relatively preserved $CMRO_2$ (above the generally accepted threshold for irreversibility of 1.4 ml/100 g/min) (Fig. 2.1) have been reported in over 50% of the patients studied within 9 hours of onset, in up to 25% of cases at 24 hours and, at least in one instance, up to 30 hours post-onset (Ackerman et al. 1981, 1989; Wise et al. 1983; Baron et al. 1983; Powers et al. 1984, 1985b; Hakim et al. 1989; Sette et al. 1989). These alterations would be consistent with at-risk but still potentially recuperable (i.e. penumbral) tissue, although proof that it is indeed recuperable is not yet available. The temporal profile of this finding among patients suggests that the incidence of such at-risk tissue declines steadily from onset, but may still occasionally be present until 24–36 hours after clinical onset.

Oligemic Tissue

Oligemic areas often surround the critically ischemic tissue. Such areas (which presumably represent tissue with efficiently compensated perfusion pressure drop via pial collaterals) can already be observed in the hyperacute stage and throughout the entire acute stage until the 4th day (Baron et al. 1983). Although simply oligemic, these areas still are inadequately oxygenated and may sustain damage either from any further reduction in perfusion pressure (systemic or local) or perhaps from sustained oligemia.

Implications of OEF in Managing the Acute Stroke Patient

Demonstration of high OEF in the setting of acute stroke implies that the autoregulation of CBF has been overriden in the affected territory. This finding is especially important in view of the frequent occurrence of elevated arterial pressure acutely after stroke. Thus, any lowering of the systemic blood pressure is likely to further reduce the cerebral perfusion pressure and, in turn, the CBF, in the affected tissue, and such reductions of CBF can have potentially damaging effects if the tissue was already in the penumbra. Conversely, if low OEF with hyperperfusion is found, management of arterial hypertension may be warranted, particularly if early edema is demonstrated by CT or MRI, since experimental studies show that hyperperfusion in necrotic tissue may promote the development of malignant brain swelling.

Irreversibly Damaged Tissue

Presumably irreversible damage, characterized by profoundly reduced $CMRO_2$ is often seen to very early affect the deep MCA territory, in association with cortical misery perfusion (Fig. 2.1). Presumably, because of its poor collaterals, this area, in contrast to the cerebral cortex, constitutes the core of ischemia and rapidly suffers irreversible damage; in baboons subjected to MCA occlusion, PET studies have shown a reduced $CMRO_2$ in this area as early as 1–2 hours postocclusion (Pappata et al. 1993).

In other patients this apparently irreversible damage is already extensive, involving wide cortical areas, only hours after stroke onset (Marchal et al. 1993a) (Fig 2.2).

Luxury Perfusion in Acute Stroke

Absolute luxury perfusion with hyperperfusion, indicating recanalization of the occluded artery, is observed in about one-third of patients seen after the 6th hour and half of patients seen after 16 hours (Ackerman et al. 1989; Marchal et al. 1993a). Hyperperfusion is associated with variable $CMRO_2$, either normal or reduced, presumably reflecting the extent of cell damage incurred during the period of ischemia (Baron et al. 1983) (Fig. 2.2). Hyperperfusion may be sustained and, after 48 hours, may affect both the already irreversibly damaged area (as defined by late CT scanning) and its viable surroundings, which are differentiated by profoundly reduced $CMRO_2$ in the former and essentially preserved $CMRO_2$ in the latter zone.

Relative luxury perfusion, with normal or reduced CBF, is rare during the acute stage but becomes progressively more prevalent with elapsing time, affecting the necrotic area, which expresses a very low $CMRO_2$ (Ackerman et al. 1981; Baron et al. 1983; Wise et al. 1983; Hakim et al. 1989).

Evolution of PET Changes With Time

The evolution of initially ischemic areas towards frank infarction entails a rapid decline in $CMRO_2$

despite stable CBF, suggesting the transition from penumbral ischemia to infarction (Wise et al. 1983); this process is strikingly illustrated by the associated dramatic fall in the OEF, from very high to very low values, which indicates exhaustion of the tissue's oxygen needs. However, in a few cases, a similar situation of marked initial ischemia did not lead to actual infarction (Powers et al. 1985b; Baron et al. 1987); this variable final outcome of ischemia as viewed by PET would be consistent with a potentially reversible 'ischemic penumbra'in humans.

In areas already profoundly hypoperfused/hypometabolic at initial PET study, follow-up imaging shows either a lack of evolution or a variable increase in CBF, but without parallel increases in $CMRO_2$ (and thus also with a marked drop in OEF), indicating variable reperfusion in an already necrotic tissue.

Quantitative Thresholds of Infarction

The relationships between quantitative values of CBF and $CMRO_2$ and final tissue outcome have shown that, in areas with initial misery perfusion, well-defined thresholds separated viable from necrotic areas, demonstrating values of 11 and 1.7 ml/100 g/min for CBF and $CMRO_2$, respectively (55% and 70% of contralateral values, respectively) (Baron et al. 1983, 1987). Powers et al. (1985a) reported that the $CMRO_2$ values in the infarcted areas were below 1.3 ml/100 g/min in 80% of the patients; corresponding analysis of CBF, CBV and OEF did not reveal definable thresholds in this study. Ackerman et al. (1989) found that areas with $CMRO_2$ below 1.5 ml/100 g/min in the hyperacute stage always went on to infarction on late CT scan, while those with $CMRO_2$ above 2.5 ml/100 g/min were always intact; values with intermediate $CMRO_2$ displayed either infarcted or intact tissue. Overall, the above studies indicate a threshold for $CMRO_2$ at about 1.5 ml/100 g/min below which, when measured later than 2–6 hours after onset, necrosis is constantly observed; above this value there may well exist a 'gray-zone' for $CMRO_2$ in which tissue outcome is uncertain, corresponding to the concept of 'penumbra'. Whether prompt therapeutic intervention could prevent infarction in these areas remains to be proven, but could be tested by experimental models adapted to PET research (Pappata et al. 1993; Touzani et al. 1993).

Mapping Already Irreversible Damage in Acute Stroke

In a clinical application of the above findings, Marchal et al. (1993b) showed that the volume of profoundly hypometabolic tissue (i.e. with $CMRO_2$ ‹1.4 ml/100 g/min), as assessed with PET 5–18 hours post-onset of stroke, was highly linearly correlated to infarct volume, as measured by CT scan 1 month later; the former, however, generally underestimated the latter. Thus, the mapping of profoundly hypometabolic tissue in the acute stage of stroke provides an early assessment of already established damage and predicts a minimum volume of final infarction. As the results were almost similar when the volume of profoundly hypoperfused (CBF ‹12 ml/100 g/min) rather than profoundly hypometabolic tissue was assessed, quantitative mapping of CBF by techniques more widely available than PET, e.g. SPECT, may be useful.

Patterns of CBF and $CMRO_2$ in Acute Stroke: Relation to Clinical Outcome

In an attempt to determine whether acute-stage PET could be useful to predict outcome, Marchal et al. (1993a) studied 18 patients with first-ever MCA territory stroke and examined the relationships between the changes in CBF and $CMRO_2$,

→

Fig. 2.1 Parametric PET images representing cerebral blood flow (CBF, upper left), metabolic rate of oxygen ($CMRO_2$, upper right), blood volume (CBV, bottom right) and oxygen extraction fraction (OEF, bottom left) in a patient studied 12 hours after onset of stroke in the right middle cerebral artery (MCA) territory who also had left hemiparesis. The images are all from a single cut through the basal ganglia. They are displayed according to a pseudocolor scale (shown on the right) which ranges from black (zero value) to white (maximum value, which is set to 50 ml/100 g/min, 4 ml/100 g/min, 1 and 8 ml/100 g for CBF, $CMRO_2$, OEF and CBV, respectively). The images show a striking decrease in CBF in the entire right MCA territory, more severe in the deep than in the cortical territory; the OEF is markedly increased in the same areas, and the $CMRO_2$, is unaltered in the frontal cortex (oligemia), moderately reduced in the temporal cortex (ischemia) and profoundly reduced in the caudate nucleus (presumably irreversible damage).

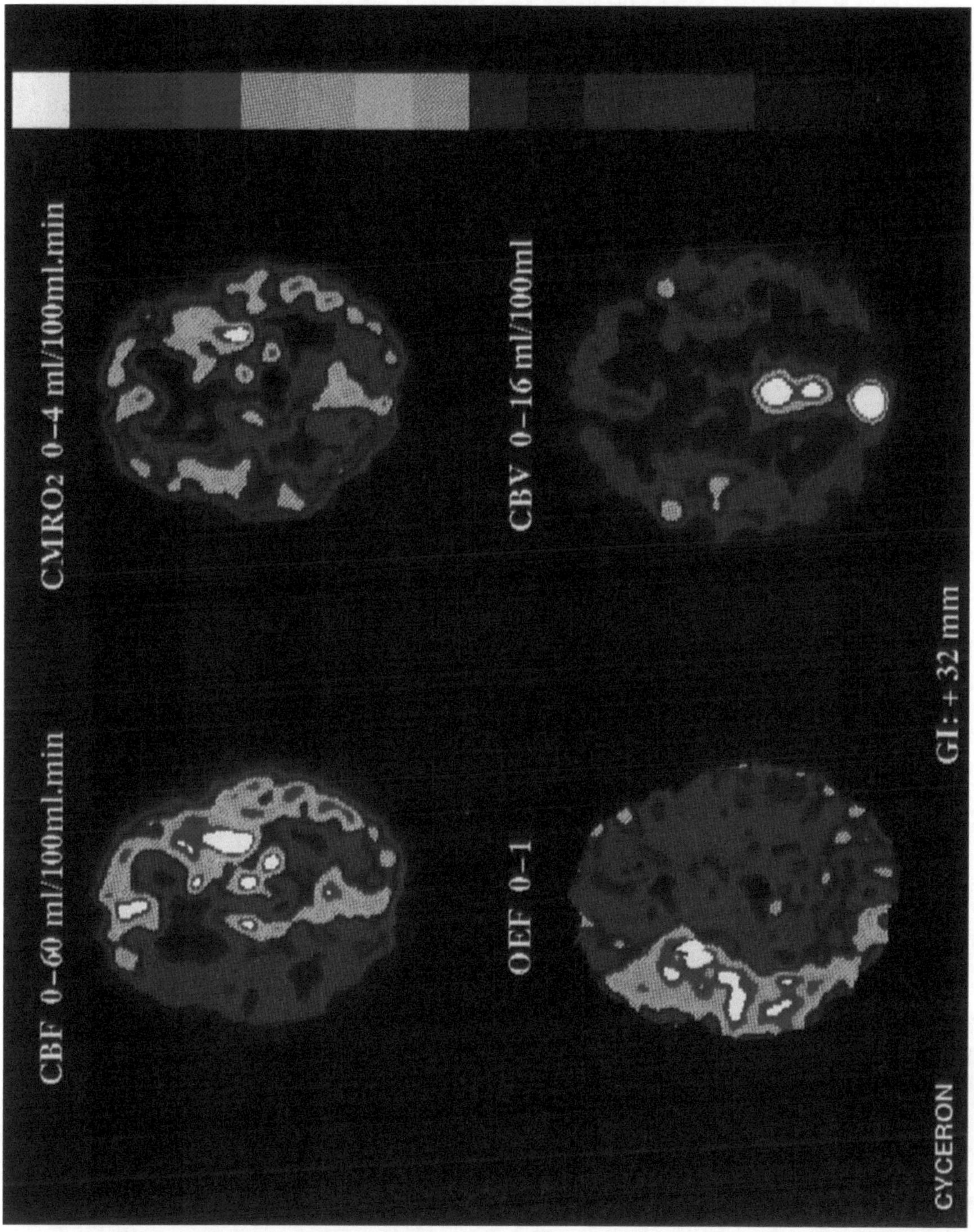
CBF 0–60 ml/100ml.min
CMRO2 0–4 ml/100ml.min
OEF 0–1
CBV 0–16 ml/100ml
GI: + 32 mm
CYCERON

Fig. 2.1.

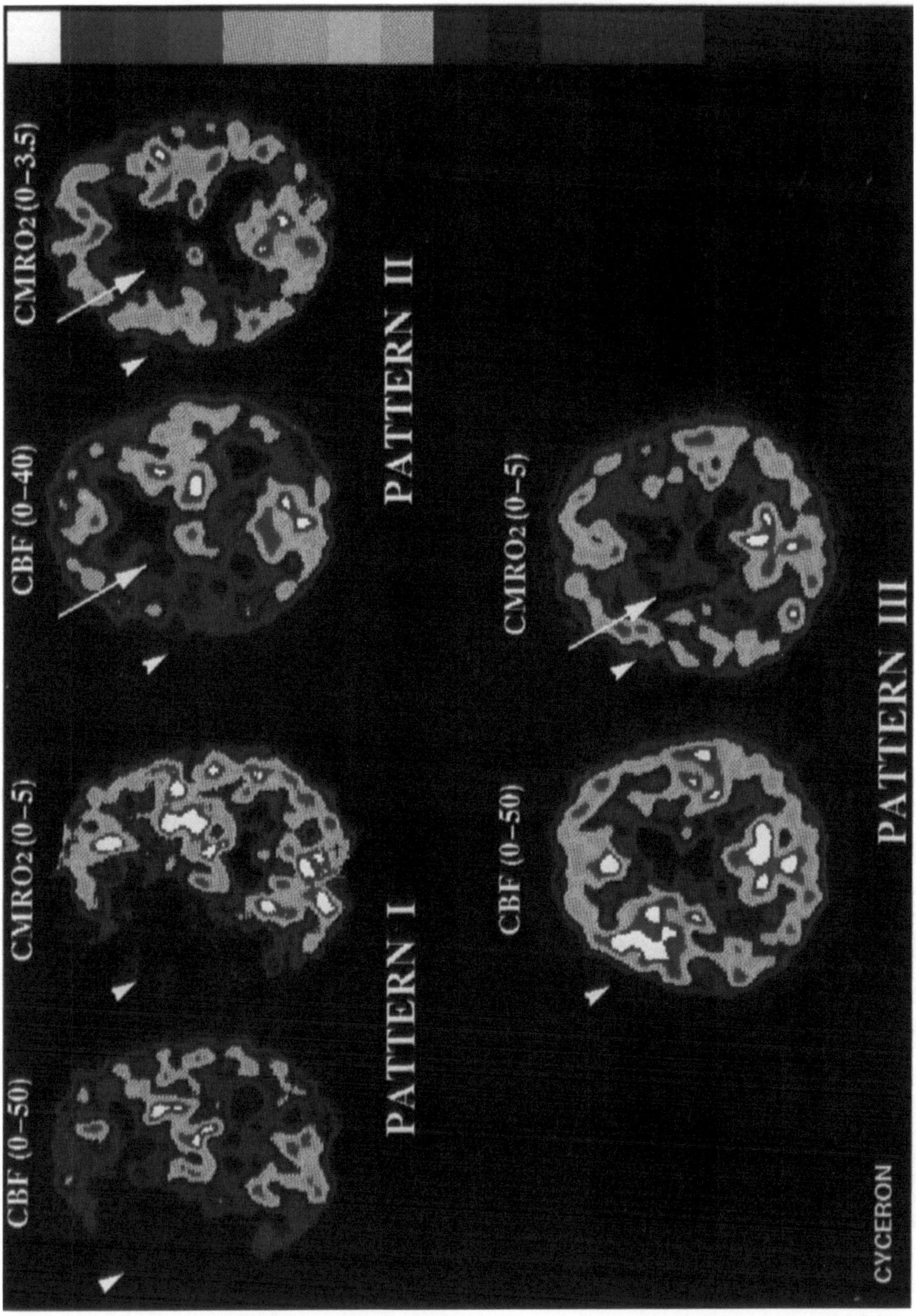
CBF (0–50)
CMRO2 (0–5)
CBF (0–40)
CMRO2 (0–3.5)
PATTERN I
PATTERN II
CBF (0–50)
CMRO2 (0–5)
PATTERN III
CYCERON

Fig. 2.2.

observed 5–18 hours post-onset of stroke, and the subsequent spontaneous neurological course over 2 months, quantified with validated stroke scales. They were able to identify three different patterns of PET changes (Fig. 2.2) into which each patient could be classified: pattern 1 was characterized by a large area of profoundly depressed CBF and $CMRO_2$, suggesting already extensive necrosis (presumably as a result of inadequate pial collaterals); pattern 2 was characterized by a moderately to markedly reduced CBF but a normal or relatively preserved $CMRO_2$ (and thus with an increased OEF), except possibly in a small and often deeply seated area; this pattern reflects ongoing ischemia with still limited necrosis, possibly as a result of good cortical collaterals; pattern 3 was characterized by an increased CBF, with either essentially normal $CMRO_2$, or only a very limited area of profound hypometabolism; this pattern reflects early spontaneous reperfusion with only limited damage incurred during arterial occlusion (again presumably thanks to adequate collaterals). There was a highly statistically significant relationship between PET patterns and neurological outcome. Thus, all patients classified as pattern 1 did poorly (massive infarcts with either death or recovery index $<25\%$), whereas all patients classified as pattern 3 did well (small infarcts and recovery index $>75\%$); patients classified as pattern 2 had a very variable course, ranging from death to full recovery.

These results of Marchal et al. (1993a) suggest the following. First, early spontaneous reperfusion appears to be beneficial, which is at variance with the prevalent belief that sudden tissue reoxygenation might exacerbate damage (Nelson et al. 1992) but consistent with a large body of experimental literature showing reduction of infarction size by early reopening of the occluded MCA (Lassen et al. 1991; Heiss 1992). Furthermore, recent data from other sources also suggest that, in humans, early recanalization is associated with smaller infarcts and rapid recovery, whereas larger infarcts and massive brain swelling are associated with persistence of MCA trunk occlusion (Minematsu et al. 1992; Ringelstein et al. 1992; Wardlaw et al. 1993). Secondly, PET imaging performed 5–18 hours post-onset of stroke seems able to accurately predict outcome for patterns 1 and 3 patients, more accurately than the initial neurological status; as neither pattern 1 nor pattern 3 patients would be likely to benefit from therapy to any meaningful extent, their inclusion in drug trials not only would seem unjustified but may also blur any beneficial effect on other patient categories. For instance, entering such patients in a trial of thrombolysis would appear to be not only useless but also hazardous, especially in pattern 1. Thirdly, the unpredictable outcome in pattern 2 suggests that the ischemic (penumbral?) tissue progressed to necrosis in some cases but not in others, presumably depending on the occurrence of spontaneous reperfusion early after the PET study. Reasoning along these lines also provides a pathophysiological rationale for retrospectively determined clinical classifications, for example in 'deteriorating', 'non-recovering' or 'rapidly recovering' stroke. Finally, pattern 2 patients may represent the subgroup most likely to benefit from therapeutic trials. Based on the incidence of high OEF in acute stroke detailed above, most patients in the 0–6 hour interval might be amenable to therapy, but this might still be true for at least a fraction of the patients in the 6–18 hours interval.

Progressive Metabolic Deterioration of Ischemic Tissue in Relation to Final Infarction: Implications for the Concepts of Penumbra and the Window of Therapeutic Opportunity

Although Wise et al. (1983) and Hakim et al. (1989) described the occurrence of metabolic deterioration in acutely ischemic tissue at PET

←

Fig. 2.2. Typical examples of the three patterns of alteration in CBF and $CMRO_2$ observed with PET 5–18 hours after onset of MCA territory stroke (Marchal et al. 1993a). This figure displays three sets of CBF and $CMRO_2$ images from a single most illustrative cut in each patient. Each parametric image is displayed according to the pseudocolor scale shown on the right, with zero corresponding to black and maximal pixel value (shown above each image) to white. Arrowheads point to the typical changes that define the three patterns. Upper left, profound decrease in both CBF and $CMRO_2$ over most of the right MCA territory (pattern 1); upper right, marked decrease in CBF over the whole right MCA territory with essentially preserved $CMRO_2$ (pattern 2); note the profound decrease in both CBF and $CMRO_2$ in the deep MCA territory (long arrow), indicating already irreversible damage in a small area; bottom left, marked and extensive hyperperfusion in the anterior and deep portions of the right MCA territory, with preserved $CMRO_2$ (pattern 3), except in the deep MCA territory (long arrow), suggesting irreversible damage in a small area.

follow-up a few days later, the relationships between the metabolically deteriorating tissue and final tissue outcome, whether infarcted or structurally intact, was left unaddressed until recently. Also, the time-course of metabolic deterioration could not be determined in these studies.

PET studies of experimental MCA occlusion in baboons are well suited to answer the latter question, as they allow both very early and sequential investigations which are not possible ethically in man. Pappata et al. (1993) reported that, despite stable perfusion in the 'penumbral' range, the affected tissue $CMRO_2$ remained only slightly reduced or even sometimes unaltered during the first 2 hours post-occlusion, to decline only by the 4th hour in the entire (and especially in the deep) MCA territory. These experimental data indicated that the ultimately hypometabolic areas exhibit only little $CMRO_2$ decline initially, suggesting that 'penumbral' regions may acutely be only mildly impaired metabolically, if at all. In an extension of this work, Touzani et al. (1993) found this metabolic deterioration to progress both in degree and in volume, and to take place not just over the first few hours post-occlusion but up until at least 24 hours. These studies therefore suggest a prolonged therapeutic window for penumbral tissue.

Consistent with these baboon studies, Heiss et al. (1992) found in humans that, surrounding the already profoundly hypometabolic area, there existed acutely a thin rim of tissue with still relatively preserved $CMRO_2$ and increased OEF, and in which the $CMRO_2$ significantly fell at follow-up PET; however, in this study most initial PET scans were done later than 18 hours, and the relationships of the metabolic deterioration with CT-defined infarct topography was not assessed. This was done recently by Marchal et al. (1993c) in a sample of 8 patients in whom the initial PET was performed before 18 hours and who survived to be restudied 2–4 weeks later. In 6 of these 8 patients, substantial parts of the ultimately infarcted tissue initially exhibited $CMRO_2$ values well above the threshold for irreversible damage, together with 'penumbral' ranges of reduced perfusion and increased OEF; at follow-up PET study, all these areas showed $CMRO_2$ <1.4 ml/100 g/min. This study therefore provided further evidence for the persistence, up to 18 hours post-onset of stroke, of still viable but deteriorating (penumbral?) tissue, a finding with potentially important therapeutic implications.

References

Ackerman RH, Correia JA, Alpert NM, Baron JC, Gouliamos A, Grotta JC, Brownell GL, Taveras JM (1981) Positron imaging in ischemic stroke disease using compounds labelled with oxygen-15. Arch Neurol 38:537–543

Ackerman RH, Lev MH, MacKay BC, Katz PM, Babikian VL, Alpert NM, Correia JA, Panagos PD, Senda M (1989) PET studies in acute stroke: findings and relevance to therapy. J Cerebr Blood Flow Metab 9 (Suppl 1):S359

Baron JC (1991) Testing cerebral function: will it help the understanding or diagnosis of central nervous system disease? Exploring brain functional anatomy with positron tomography. Wiley, Chichester (Ciba Foundation symposium 163): 250–264

Baron JC, Bousser MG, Comar D, Castaigne P (1980) Crossed cerebellar diaschisis in human supratentorial brain infarction. Trans Am Neurol Assoc 105:459–461

Baron JC, Rougemont D, Bousser MG, Lebrun-Grandie P, Iba-Zizen MT, Chiras J (1983) Local CBF, oxygen extraction fraction (OEF) and $CMRO_2$: prognostic value in recent supratentorial infarction in humans. J Cerebr Blood Flow Metab 3 (Suppl 1):S1–S2

Baron JC, D'Antona R, Pantano P, Serdaru M, Samson Y, Bousser MG (1986) Effects of thalamic stroke on energy metabolism of the cerebral cortex. Brain 109:1243–1259

Baron JC, Samson Y, Pantano P, Chiras J, Derouesne C, Bousser MG (1987) Interrelationships of local CBF, OEF and $CMRO_2$ in ischemic areas with variable outcome: further PET studies in humans. J Cerebr Blood Flow Metab 7 (Suppl 1): 41

Baron JC, Frackowiak RSJ, Herholz K, Jones T, Lammertsma AA, Mazoyer B, Wienhard K (1989) Use of PET methods for measurement of cerebral energy metabolism and hemodynamics in cerebrovascular disease. J Cerebr Blood Flow Metab 9:723

Dobkin JA, Levine RL, Lagoze HL, Dulli DA, Nickles RJ, Rowe BR (1989) Evidence for transhemispheric diaschisis in unilateral stroke. Arch Neurol 46:1333–1336

Feeney D, Baron JC (1986) Diaschisis. Stroke 17:817–830

Hakim AM, Evans AC, Berger L, Kuwabara H, Worsley K, Marchal G, Biel C, Pokrupa R, Diksic M, Meyer E, Gjedde A, Marret S (1989) The effect of nimodipine on the evolution of human cerebral infarction studied by PET. J Cerebr Blood Flow Metab 9:523–534

Heiss WD (1992) Experimental evidence of ischemic thresholds and functional recovery. Stroke 23:1668–1672

Heiss WD, Huber M, Fink GR, Herholz K, Pietryk U, Wagner R, Wienhard K (1992) Progressive derangement of periinfarct viable tissue in ischemic stroke. J Cerebr Blood Flow Metab 12:193–203

Karbe H, Herholz K, Szelies B, Pawlik G, Wienhard K, Heiss WD (1989) Regional metabolic correlates of token test results in cortical and subcortical left hemispheric infarction. Neurology 39:1087–1088

Lassen NA, Fieschi C, Lenzi GL (1991) Ischemic penumbra and neuronal death: comments on the therapeutic window in acute stroke with particular reference to thrombolytic therapy. Cerebrovasc Dis 1 (Suppl 1):32–35

Marchal G, Serrati C, Rioux P, Petit-Taboue MC, Viader F, De La Sayette V, Le Doze F, Lochon P, Derlon JM, Orgogozo JM, Barron JC (1993a) PET imaging of cerebral perfusion and oxygen consumption in acute ischaemic stroke: relation to outcome. Lancet 341:925–926

Marchal G, Beaudouin V, Furlan M, Serrati C, Viader F, Le Doze F, De La Sayette V, Rioux P, Lochon P, Derlon JM, Baron JC (1993b) Correlations of low flow and matabolic impairment indices with neurological outcome and infarct size: a PET study in the acute stage of stroke. J Cerebr Blood Flow Metab 13 (Suppl 1):S347

Marchal G, Beaudouin V, Furlan M, Lochon P, Onfroy MC, Petit-Taboue MC, Derlon JM, Baron JC (1993c) Demonstration of metabolically deteriorating ischemic tissue within the eventually infarcted volume by pixel integration-threshold analysis: an acute and follow-up PET study. J Cerebr Blood Flow Metab 13 (Suppl 1): S274

Metter EJ, Kempler D, Jackson C et al. (1989) Cerebral glucose metabolism in Wernicke's, Broca's and conduction aphasia. Arch Neurol 46:27–34

Minematsu K, Yamaguchi T, Omae T (1992) 'Spectacular shrinking deficit': rapid recovery from a major hemispheric syndrome by migration of an embolus. Neurology 42:157–162

Nelson CW, Wei Ep, Povlishock JT, Kontos HA, Moskowitz MA (1992) Oxygen radicals in cerebral ischemia. Am J Physiol (Heart Circ Physiol) 263:H1356–1362

Pappata S, Fiorelli M, Rommel T, Hartmann A, Dettmers C, Yamaguchi T, Chabriat H, Poline JB, Crouzel C, Di Giambernardino L, Baron JC (1993) PET study of changes in local brain hemodynamics and oxygen metabolism after unilateral middle cerebral artery occlusion in baboons. J Cerebr Blood Flow Metab 13:416–424

Powers WJ, Grubb RL, Raichle ME (1984) Physiological response to focal cerebral ischemia in humans. Ann Neurol 16:546–557

Powers WJ, Grubb RL, Darriet D, Raichle ME (1985a) Cerebral blood flow and cerebral metabolic rate of oxygen requirements for cerebral function and viability in humans. J Cerebr Blood Flow Metab 5:600–608

Powers WJ, Grubb RL, Baker RP, Mintun MA, Raichle ME (1985b) Regional cerebral blood flow and metabolism in reversible ischemia due to vasospasm. J Neurosurg 62:539–546

Pulsinelli W (1992) Pathophysiology of acute ischaemic stroke. Lancet 339:16–19

Ringelstein EB, Biniek R, Weiller C, Ammeling B, Nolte PN, Thron A (1992) Type and extent of hemispheric brain infarction and clinical outcome in early and delayed middle cerebral artery recanalization. Neurology 42:289–298

Sette G, Baron JC, Mazoyer B, Levasseur M, Pappata S, Crouzel C (1989) Local brain hemodynamics and oxygen metabolism in cerebrovascular disease: positron emission tomography. Brain 112:931–951

Touzani O, Young AR, Marchal G, Beaudouin V, Ravenel N, Mezenge F, Rioux P, Derlon JM, Baron JC, Mackenzie ET (1993) Progression of severely hypometabolic tissue after permanent occlusion of the middle cerebral artery (MCAO): PET studies in anaesthetized baboons. J Cerebr Blood Flow Metab 13 (Suppl 1): S797

Wardlaw JM, Dennis MS, Lindley RI, Warlow CP, Sandercock PAG, Sellar R (1993) Does early reperfusion of a cerebral infarct influence cerebral infarct swelling in the acute stage or the final clinical outcome? Cerebrovasc Dis 3:86–93

Wise RJS, Bernardi S, Frackowiak RSJ, Legg NJ, Jones T (1983) Serial observations on the pathophysiology of acute stroke. The transition from ischaemia to infarction as reflected in regional oxygen extraction. Brain 106:197–222

Wise RJS, Gibbs J, Frackowiak RSJ, Marshall J, Jones T (1986) No evidence for transhemispheric diaschisis after human cerebral infarction. Stroke 17:853–860

3. Excitotoxicity and Stroke

D. W. Choi

The development of effective interventional treatments for stroke has been an important research goal for many years. Major effort to date has appropriately focused on insult limitation, for example, utilizing thrombolytic therapy in the minutes to hours after the onset of symptoms to restore blood flow in a clogged cerebral artery. However, a complementary approach is to reduce the intrinsic vulnerability of brain parenchyma to hypoxic–ischemic damage. This latter parenchymal approach has gained both momentum and specific form from recent data suggesting that the toxic overactivation of neuronal glutamate receptors – 'excitotoxicity' – may contribute to the pathogenesis of hypoxic–ischemic neuronal death (Meldrum 1985; Rothman and Olney 1987; Choi 1988). Pharmacological agents directed at antagonizing excitotoxicity may provide a new class of neuroprotective drugs useful in reducing the brain damage associated with acute stroke.

Hypoxia–Ischemia as a Trigger for Excitotoxicity

In keeping with its widespread role in central excitatory neurotransmission, glutamate is present at millimolar levels throughout CNS gray matter. Under normal conditions, much of this glutamate is stored inside synaptic terminals, and energy-dependent cellular uptake rapidly clears synaptically released glutamate from the extracellular space. However, when tissue energy stores are compromised by hypoxia–ischemia, the combination of lost cellular glutamate uptake, together with increased glutamate efflux, leads to a buildup of extracellular glutamate sufficient to destroy neurons.

Microdialysis measurements have indicated that extracellular glutamate during hypoxia–ischemia (Benveniste et al. 1984; Globus et al. 1988) can attain the hundred-micromolar levels demonstrated to be rapidly neurotoxic to cortical neurons in vitro (Choi 1992). Much lower concentrations of glutamate may be neurotoxic if glial uptake is disabled due to energy depletion (Rosenberg and Aizenman 1989). Other animal experiments have established that excitatory pathway deafferentation can reduce hypoxic neuronal injury, that the pathology of hypoxic injury and excitotoxic injury is similar, and that neuronal calcium overload occurs prominently in both conditions (see below).

Although this chapter will discuss the possible role of excitotoxicity in stroke, the idea that the toxic overstimulation of postsynaptic receptors by endogenous glutamate may contribute significantly to pathological neuronal death is not unique to stroke, or even to hypoxia–ischemia in general. Rather, growing evidence suggests that excitotoxicity may also contribute to brain or spinal cord neuronal cell loss after several other acute insults, including hypoglycemia (Wieloch 1985), prolonged seizures (Collins and Olney 1982), and trauma (Faden and Simon 1988; Faden et al. 1989). The latter is usually accompanied by hypoxia–ischemia in vivo, but study in vitro has suggested that excitotoxicity can exacerbate neuronal death following a purely mechanical insult (Tecoma et al. 1989).

Glutamate Receptors and Excitotoxicity

Glutamate activates three major families of ionophore-linked receptors classified by their preferred agonists: *N*-methyl-D-aspartate (NMDA), kainate and α-amino-3-hydroxy-5-methyl-4-isoxazolepropionic acid (AMPA) (Watkins and Olverman 1987; Watkins et al. 1990). Information derived from the recent cloning of multiple functional receptor subunits from each family (Hollman et al. 1989; Moriyoshi et al. 1991; Sommer and Seeburg 1992) will surely lead to refinements in this basic scheme.

The channels gated by all three receptor subtypes are permeable to both Na^+ and K^+. Channels gated by NMDA receptors, but only a small subset of channels gated by AMPA or kainate receptors (see below), additionally possess high permeability to Ca^{2+} (MacDermott et al. 1986). Glutamate also activates a family of metabotropic receptors that activate second-messenger systems rather than directly gating ion channels (Masu et al. 1991; Tanabe et al. 1992). The best-characterized member of this group is a quisqualate-preferring receptor that is linked via G proteins to phospholipase C. Activation of this receptor induces the hydrolysis of phosphatidylinositol-4,5-bisphosphate (PIP_2) to liberate the dual messengers inositol 1,4,5-*tris* phosphate (IP_3) and diacylglycerol (Sladeczek et al. 1985; Nicoletti et al. 1986).

These different glutamate receptor subtypes do not participate equally in excitotoxicity (Choi 1992). Antagonist experiments suggest that most of the neuronal death associated with brief intense glutamate exposure requires NMDA receptor activation. Death can be almost completely blocked by selective blockade of NMDA receptors, although predictably (since AMPA/kainate receptors can still be activated) glutamate-induced excitation and acute neuronal swelling are only partially attenuated by the same blockade (Choi et al. 1988; Manev et al. 1989; Michaels and Rothman 1990). In contrast, selective blockade of AMPA/kainate receptors has only a small effect on late neuronal death (Koh and Choi 1991). Only when both NMDA and AMPA/kainate receptors are blocked is acute glutamate-induced neuronal swelling eliminated.

However, AMPA/kainate receptors do mediate excitotoxic death. Whereas brief exposure to high concentrations of kainate or AMPA produces little death (Koh et al. 1990), kainate and AMPA are both highly neurotoxic if exposure time is extended for hours (Choi et al. 1988; Frandsen et al. 1989). With 24-hour exposure, 10 micromolar concentrations of either kainate or AMPA can induce widespread death of cultured cortical neurons (Koh et al. 1990).

Thus glutamate receptor-mediated neuronal injury, at least to cortical neurons in vitro, occurs in two main patterns: rapidly triggered excitotoxicity induced by the brief intense stimulation of large numbers of NMDA receptors; and slowly triggered excitotoxicity induced by the prolonged stimulation of AMPA/kainate receptors (or the low-level stimulation of NMDA receptors). The former is critically dependent on the presence of extracellular Ca^{2+}, consistent with the idea that rapidly triggered toxicity is initiated by excessive Ca^{2+} influx through the Ca^{2+}-permeable NMDA receptor-gated channel.

Slowly triggered AMPA/kainate receptor-mediated excitotoxicity may also be initiated by excessive Ca^{2+} influx. Most channels gated by AMPA or kainate receptors have limited Ca^{2+} permeability, but Ca^{2+} influx can occur indirectly, for example through voltage-gated Ca^{2+} channels activated by the Na^+ influx and membrane depolarization associated with AMPA/kainate receptor stimulation. Other indirect pathways for Ca^{2+} entry may include reverse operation of the Na^+–Ca^{2+} exchanger, membrane stretch-activated conductances, or leak conductances activated by cell swelling (Choi 1988). In addition, recent studies have indicated that some AMPA receptors may gate channels permeable to Ca^{2+} (Iino et al. 1990), possibly reflecting a molecular composition lacking the Glu-R2/Glu-RB subunit (Hume et al. 1991; Verdoorn et al. 1991). Recent study suggests that a minority of cortical neurons may possess large numbers of AMPA receptor-gated Ca^{2+} channels and, perhaps as a direct result, are especially vulnerable to damage mediated by AMPA receptor overactivation (Turetsky et al. 1992).

The effect of metabotropic receptor stimulation upon excitotoxic injury has not been well defined, due to the limitations of current pharmacology: good antagonists have not yet been identified. Unexpectedly, the selective metabotropic receptor agonist *trans*-1-aminocyclopentyl-1,3-dicarboxylate (*trans*-ACPD) has been reported to reduce rapidly triggered excitotoxicity (Koh et al. 1991). This is surprising, since the ability of *trans*-ACPD to induce the liberation of intracellular Ca^{2+} stores and contribute to increases in intracellular free Ca^{2+} might be expected to increase excitotoxic injury. Further study will be needed in this important area.

Chain of Events Triggered by Overexposure to Glutamate

Glutamate neurotoxicity can be modeled as a three-stage process: induction, amplification and expression (Choi 1990a). First, the prolonged availability of extracellular glutamate is transduced by neuronal membrane receptors into potentially lethal intracellular derangements, in particular intracellular Ca^{2+} overload. Secondly, modulatory events amplify these derangements, increasing their intensity and promoting the involvement of other neurons. Finally, these derangements trigger the cytotoxic cascades responsible for subsequent delayed neuronal disintegration, the expression of cell death.

It is the author's hypothesis that abnormal movements of Ca^{2+} are the primary mediators of injury in both rapidly triggered excitotoxicity and slowly triggered excitotoxicity. Relevant Ca^{2+} movements include, in particular, the excess entry of extracellular Ca^{2+}, but may also include the release of Ca^{2+} from intracellular stores. As a result of these events, intracellular free Ca^{2+}, $[Ca^{2+}]_i$, and total cellular Ca^{2+} achieve a lasting elevation, eventually causing neuronal degeneration.

Induction

Elevated extracellular glutamate activates its array of neuronal membrane receptors, leading directly to the development of an initial set of intracellular changes – increases in cytoplasmic Ca^{2+}, Na^+, Cl^-, water, IP_3 and diacylglycerol – that serve as triggers for subsequent events. Although these changes are potentially lethal, they precede truly irreversible steps.

Amplification

Following the induction of glutamate toxicity, a multiplicity of events may act to further amplify resultant internal derangements, especially local elevations in $[Ca^{2+}]_i$ (Choi 1990a). These amplification events may include:

1. Additional influx of extracellular Ca^{2+} through voltage-gated Ca^{2+} channels, reverse operation of the Na^+–Ca^{2+} exchanger, Ca^{2+}-activated Ca^{2+} channels, or stretch-activated cation channels.
2. The release of Ca^{2+} from intracellular stores, for example mediated by elevations in IP_3.
3. The activation of key enzyme families – including C kinases and calmodulin-regulated enzymes – and genes, which together may lead to a lasting enhancement of certain processes relevant to excitotoxicity, including Ca^{2+}-dependent glutamate release from nerve terminals, postsynaptic receptor responsiveness to glutamate, and Ca^{2+} influx through voltage-gated channels. The activation of phospholipase A_2 and nitric oxide synthase increases the formation of the intercellular messengers arachidonic acid and nitric oxide, which can induce proexcitotoxic changes in the glutamate system. Once formed in postsynaptic neurons, these messengers can diffuse freely across lipid membranes to affect neighboring neurons and glia.
4. Glutamate efflux from injured neurons, which may induce further excitotoxic injury, a positive-feedback loop that might both intensify existing injury and help propagate excitotoxic injury on to neurons spared by the initial insult.

The net contribution of these amplification steps to neuronal death may be a function of the strength of initial induction events. Following intense activation of NMDA receptors, enough Ca^{2+} may enter neurons directly through NMDA receptor-gated channels to produce lethal effects without subsequent amplification. Amplification may therefore be more important to slowly triggered excitotoxicity than to rapidly triggered excitotoxicity.

Expression

Sustained elevations in $[Ca^{2+}]_i$ can be expected to trigger destructive cascades capable of bearing the ultimate responsibility for neuronal degeneration. Many specific Ca^{2+}-triggered destructive events may be common to rapidly triggered excitotoxicity and slowly triggered excitotoxicity.

One important class of expression cascade may be the unleashing of Ca^{2+}-activated catabolic enzymes. The Ca^{2+}-activated neutral protease calpain has been specifically linked to glutamate receptors in the rat hippocampus (Siman et al. 1989), and can degrade major neuronal structural proteins. Calpain inhibitors have been recently found to be neuroprotective in gerbils subjected to global ischemia (Lee et al. 1991). Elevated cytosolic Ca^{2+} also activates phospholipases capable of breaking down cell membranes and liberating arachidonic acid, and endonucleases capable of breaking down genomic DNA.

Another important class of expression cascade may involve oxygen free radicals. These reactive molecules can initiate many destructive processes, including lipid peroxidation (Braughler and Hall 1989; Siesjö 1989); once formed, they may promote further excitotoxic injury by promoting glutamate release (Pellegrini-Giampietro et al. 1988). 21-aminosteroid inhibitors of lipid peroxidation can reduce the death associated with both rapidly triggered (Monyer et al. 1990) and slowly triggered (Hartley and Choi, unpublished observations) excitotoxicity on cortical neurons. Free radical production might be linked to loss of cellular Ca^{2+} homeostasis in several ways, including Ca^{2+} activation of phospholipase A_2, leading to the liberation of arachidonic acid which, upon further metabolism, leads to free radical production (Chan et al. 1985); Ca^{2+} triggering the conversion of xanthine dehydrogenase to xanthine oxidase, a rich enzymatic source of free radicals (Dykens et al. 1987); and stimulation of NMDA receptors, leading to the activation of nitric oxide synthase and the release of nitric oxide (Garthwaite et al. 1988; Dawson et al. 1991), which in turn can react with superoxide to form peroxynitrite and promote the production of hydroxyl radicals (Beckman et al. 1990). Indeed, pharmacological inhibition of nitric oxide synthase, or removal of the nitric oxide precursor arginine from the culture medium, was observed to reduce rapidly triggered excitotoxicity in rat cortical cultures (Dawson et al. 1991), leading those authors to suggest that this third mechanism is of primary importance. In our culture system, arachidonic acid metabolism may be a major source of free radical generation. Lipoxygenase inhibitors such as nordihydroguaiaretic acid or baicalein can reduce NMDA- or AMPA-induced slowly triggered excitotoxicity, and exposure to arachidonic acid itself is cytotoxic (Rose et al. 1990; also unpublished results).

Neuroprotective Effects of NMDA Antagonists

The neuroprotective action of NMDA antagonists against hypoxia has been straightforward to demonstrate in vitro, where such insults can be delivered to brain cells under highly controlled conditions. These in-vitro experiments have established that NMDA antagonists can exert neuroprotective effects directly on brain cells, without requiring changes in systemic metabolism or in blood flow. Several classes of NMDA antagonist all appear to be neuroprotective, including competitive antagonists, which block the glutamate-binding site itself (for example CGS 19755); PCP site ligands, which bind to a site in the NMDA receptor-activated channel and interfere with current through the channel (for example, MK-801, ketamine or dextrorphan); and glycine site antagonists (for example 7-chlorokynurenate). Other putative modulatory sites on the NMDA receptor complex that might be targets for pharmocological manipulation include polyamine site(s), zinc sites, pH site(s) (see below), or oxidation–reduction site(s).

The situation is, as would be expected, more complicated in vivo, but most studies of NMDA antagonists in animal models of focal hypoxia–ischemia have demonstrated reductions in infarct (or ischemic injury) volume ranging from 18% to 88% over untreated controls, with a majority showing over 50% reduction (Albers et al. 1989; Buchan 1990). In contrast, results in animal models of global ischemia – presumably a better reflection than stroke of events associated with cardiac arrest – have been variable. NMDA antagonists probably have their most important protective actions in the ischemic penumbra, present only in focal ischemia. The penumbra is an area where insult severity is submaximal, and yet NMDA receptors may be briskly activated. In global ischemia, several factors, including extracellular acidosis, may specifically limit the contribution of NMDA receptors to resultant injury (Choi 1990b).

The extracellular acidosis which accompanies brain ischemia in vivo has been widely postulated to contribute to ischemic brain damage. In experimental animals, administration of sufficient glucose to worsen ischemic lactic acidosis is associated with potentiation of brain damage. Glia in particular may be vulnerable to injury by extracellular acidity (Norenberg et al. 1987; Goldman et al. 1989; Giffard et al. 1990a); Plum and colleagues (1983) have suggested that acidosis may play a key role in the pathogenesis of ischemic infarction. However, a possible basis for a beneficial effect of acidosis on ischemic neuronal survival was provided by Morad and colleagues (1988), who found that moderate acidity (pH 6.6) could markedly reduce NMDA receptor-activated currents in cultured hippocampal neurons. Moderate extracellular acidity (pH 6.5) also markedly reduced the inward whole-cell current induced by NMDA on cultured cortical neurons, and reduced the cortical neuronal injury caused by combined oxygen and glucose deprivation (Giffard et al. 1990b). Similar neuroprotective effects of extracellular acidity were observed independently by Tombaugh and Sapolsky (1990) in hippocampal cultures. Acidosis may thus be a two-edged sword: reducing neuronal loss due to NMDA receptor-mediated excitotoxicity, but also directly damaging cells, especially glia. Under conditions of global ischemia in vivo, extracellular acidity might reduce NMDA receptor activation, thereby unmasking a slowly triggered, AMPA/kainate receptor-mediated component of excitotoxic injury. It is interesting that the selective AMPA/kainate receptor antagonist NBQX has been found to reduce neuronal loss in a rodent model of global ischemia (Sheardown et al. 1990). Whereas an AMPA/kainate antagonist alone does not protect cultured cortical neurons from hypoxic damage, substantial additional protection is observed when an AMPA/kainate antagonist is added to NMDA receptor blockade (Kaku et al. 1991).

Towards an Antiexcitotoxic Therapy for Stroke?

The observations and ideas outlined above justify continued consideration of antiexcitotoxic strategies as a possible interventional therapy for stroke patients (Choi 1990a,b; Albers et al. 1992). NMDA antagonist drugs are under current development by several pharmaceutical companies, and may be among the first strategies to reach phase II clinical trial, although reservations about efficacy have been expressed (Buchan 1992).

It must be noted that NMDA antagonists have potential safety risks, including acute psychotomimetic effects, disruption of synaptic plasticity and the induction of neuronal vacuolization in a subset of neurons localized to the retrosplenial and cingulate cortices (Olney et al. 1989). However, the first two risks are likely to be small, given acute limited-duration therapy and an adult patient population. The last risk, of direct brain toxicity, presents reason for pause, but the changes observed to date have been restricted anatomically and, in large part, transient and reversible. Furthermore, coadministration of certain other drugs, such as muscarinic cholinergic antagonists or benzodiazepines, can be used to block the vacuolization induced by NMDA antagonists (Olney et al. 1991). Balanced against the potential substantial benefit in stroke, the known risks of NMDA antagonist therapy may be acceptable.

In addition to NMDA antagonists, several other logical antiexcitotoxic approaches are also under active development (Table 3.1):

1. Presynaptic approaches directed at reducing pathological glutamate release, such as hypo-

Table 3.1. Antiexcitotoxic strategies presently under development

Excitotoxicity stage	Event targeted	Strategy
Induction	Glutamate release from nerve terminals	Hypothermia N or T type Ca^{2+} channel antagonists Adenosine agonists Mg^{2+}
Induction	Glutamate receptor activation	NMDA antagonists including Mg^{2+} AMPA/kainate antagonists
Amplification	Additional Ca^{2+} influx	L type Ca^{2+} channel antagonists
Amplification	? improve Ca^{2+} homeostasis	Growth factors
Amplification	??	Gangliosides
Expression	Lipid peroxidation	21-aminosteroids or other radical scavengers
Expression	Activation of neutral proteases	Calpain inhibitors

thermia, blockers of non-L type (possible N or P type) Ca^{2+} channels on presynaptic nerve terminals, or adenosine agonists. Administration of high doses of Mg^{2+} may attenuate glutamate release, as well as NMDA receptor activation.

2. AMPA/kainate antagonists, as discussed above. Approaches directed towards the reduction of presynaptic glutamate release will effectively attenuate both NMDA and AMPA/kainate receptor activation.

3. Approaches directed at reducing the amplification of excitotoxic Ca^{2+} overload. The most mature strategy in this area may be the use of antagonists of L-type Ca^{2+} channels to attenuate Ca^{2+} influx additional to that occurring directly through glutamate receptor-gated channels. In addition, gangliosides have been found to attenuate excitotoxic injury on cerebellar granule cells and forebrain neurons; their mechanism of action is unknown, but blockade of protein kinase C activity has been postulated (Manev et al. 1989). Considerable interest has focused on growth factors, subsequent to the work of Mattson et al. (1989), who found that fibroblast growth factor can reduce excitotoxic injury on hippocampal neurons, perhaps by improving recovery from Ca^{2+} overload.

4. Approaches directed at blocking the expression cascades bearing direct responsibility for neuronal degeneration, such as calpain inhibitors or 21-aminosteroid inhibitors of lipid peroxidation.

Whichever of the above is chosen, it is likely that time will be of the essence. For example, available evidence suggests that NMDA antagonists may no longer reduce infarct size if given more than 2–3 hours after the onset of ischemia. The present leisurely manner in which stroke patients are handled by the health care system, from transport to the hospital to initial evaluation, will hopefully be replaced by rapid management protocols similar to those used for myocardial infarction. Many antiexcitotoxic therapies may be suitable for either hemorrhagic or ischemic stroke, and hence might be given by paramedics before full neurological evaluation.

Combinations of certain antiexcitotoxic approaches may have special merit; for example, additive benefits against ischemic insults have been reported with the combination of an NMDA antagonist and a dihydropyridine calcium-channel antagonist (Uematsu et al. 1991). The further addition of other therapeutic modalities such as thrombolytic therapy, or therapies directed at improving astrocyte survival, is clearly an attractive area for future exploration.

References

Albers GW, Goldberg MP, Choi DW (1989) *N*-methyl-D-aspartate antagonists: ready for clinical trial in brain ischemia? Ann Neurol 25:398–403

Albers GW, Goldberg MP, Choi DW (1992) Do NMDA antagonists prevent neuronal injury? Yes. Arch Neurol 49:418–420

Beckman JS, Beckman TW, Chen J, Marshall PA, Freeman BA (1990) Apparent hydroxyl radical production by peroxynitrite: implications for endothelial injury from nitric oxide and superoxide. Proc Natl Acad Sci USA 87:1620–1624

Benveniste H, Drejer J, Schousboe A, Diemer NH (1984) Elevation of the extracellular concentrations of glutamate and aspartate in rat hippocampus during transient cerebral ischemia monitored by intracerebral microdialysis. J Neurochem 43:1369–1374

Braughler JM, Hall ED (1989) Central nervous system trauma and stroke. I. Biochemical considerations for oxygen radical formation and lipid peroxidation. J Free Radic Biol Med 6:289–301

Buchan AM (1990) Do NMDA antagonists protect against cerebral ischemia: are clinical trials warranted? Cerebrovasc Brain Metab Rev 2:1–26

Buchan AM (1992) Do NMDA antagonists prevent neuronal injury? No. Arch Neurol 49:420–421

Chan PH, Fishman RA, Longar S, Chen S, Yu A (1985) Cellular and molecular effects of polyunsaturated fatty acids in brain ischemia and injury. Prog Brain Res 63:227–235

Choi DW (1988) Glutamate neurotoxicity and diseases of the nervous system. Neuron 1:623–634

Choi DW (1990a) Methods for antagonizing glutamate neurotoxicity. Cerebrovasc Brain Metab Rev 2:105–147

Choi DW (1990b) Cerebral hypoxia – some new approaches and unanswered questions. J Neurosci 10:2493–2501

Choi DW (1992) Excitotoxic cell death. J Neurobiol 23:1261–1276

Choi DW, Koh J, Peters S (1988) Pharmacology of glutamate neurotoxicity in cortical cell culture: attenuation by NMDA antagonists. J Neurosci 8:185–196

Collins RC, Olney JW (1982) Focal cortical seizures cause distant thalamic lesions. Science 218:177–179

Dawson VL, Dawson TM, London ED, Bredt DS, Snyder SH (1991) Nitric oxide mediates glutamate neurotoxicity in primary cortical cultures. Proc Natl Acad Sci USA 88:6368–6371

Dykens JA, Stern A, Trenkner E (1987) Mechanism of kainate toxicity to cerebellar neurons in vitro is analogous to reperfusion tissue injury. J Neurochem 49:1222–1228

Faden AI, Simon RP (1988) A potential role for excitotoxins in the pathophysiology of spinal cord injury. Ann Neurol 23:623–626

Faden AI, Demediuk P, Panter SS, Vink R (1989) The role of excitatory amino acids and NMDA receptors in traumatic brain injury. Science 244:798–800

Frandsen A, Drejer J, Schousboe A (1989) Direct evidence that excitoxicity in cultured neurons is mediated via *N*-methyl-D-aspartate (NMDA) as well as non-NMDA receptors. J Neurochem 53:297–299

Garthwaite J, Charles SL, Chess-Willians R (1988) Endothe-

lium-derived relaxing factor release on activation of NMDA receptors suggests role as intercellular messenger in the brain. Nature 336:385–388
Giffard RG, Monyer H, Choi DW (1990a) Selective vulnerability of cultured cortical glia to injury by extracellular acidosis. Brain Res 530:138–141
Giffard RG, Monyer H, Christine CW, Choi DW (1990b) Acidosis reduces NMDA receptor activation glutamate neurotoxicity and oxygen-glucose deprivation neuronal injury in cortical cultures. Brain Res 506:339–342
Globus MY, Busto R, Dietrich WD, Martinez E, Valdes I, Ginsberg MD (1988) Effect of ischemia on the in vivo release of striatal dopamine, glutamate and gamma-aminobutyric acid studied by intracerebral microdialysis. J Neurochem 51:1455–1464
Goldman SA, Pulsinelli WA, Clarke WY, Kraig RP, Plum F (1989) The effects of extracellular acidosis on neurons and glia in vitro. J Cerebr Blood Flow Metab 9:471–477
Hollman M, O'Shea-Greenfield A, Rogers SW, Heinemann S (1989) Cloning by functional expression of a member of the glutamate receptor family. Nature 342:643–648
Hume RI, Dingledine R, Heinemann SF (1991) Identification of a site in glutamate receptor subunits that controls calcium permeability. Science 253:1028–1031
Iino M, Ozawa S, Tsuzuki (1990) Permeation of calcium through excitatory amino acid receptor channels in cultured rat hippocampal neurons. J Physiol (Lond) 424:151–165
Kaku DA, Goldberg MP, Choi DW (1991) Antagonism of non-NMDA receptors augments the neuroprotective effect of NMDA receptor blockade in cortical cultures subjected to prolonged deprivation of oxygen and glucose. Brain Res 554:344–347
Koh J, Choi DW (1991) Selective blockade of non-NMDA receptors does not block rapidly triggered glutamate-induced neuronal death. Brain Res 548:318–321
Koh J, Goldberg MP, Hartley DM, Choi DW (1990) Non-NMDA receptor-mediated neurotoxicity in cortical culture. J Neurosci 10:693–705
Koh JY, Palmer E, Cotman CW (1991) Activation of the metabotropic glutamate receptor attenuates *N*-methyl-D-aspartate neurotoxicity in cortical cultures. Proc Natl Acad Sci USA 88:9431–9435
Lee KS, Frank S, Vanderklish P, Arai A, Lynch G (1991) Inhibition of proteolysis protects hippocampal neurons from ischemia. Proc Natl Acad Sci USA 88:7233–7237
MacDermott AB, Mayer ML, Westbrook GL, Sith SJ, Barker JL (1986) NMDA-receptor activation increases cytoplasmic calcium concentration in cultured spinal cord neurons. Nature 321:519–522
Manev H, Favaron M, Guidotti A, Costa E (1989) Delayed increase of Ca^{2+} influx elicited by glutamate: role in neuronal death. Mol Pharmacol 36:106–112
Masu M, Tanabe Y, Tsuchida K, Shigemoto R, Nakanishi S (1991) Sequence and expression of a metabotropic glutamate receptor. Nature 349:760–765
Mattson MP, Murrain M, Guthrie PB, Kater SB (1989) Fibroblast growth factor and glutamate: opposing roles in the generation and degeneration of hippocampal neuroarchitecture. J Neurosci 9:3728–3740
Meldrum B (1985) Possible therapeutic applications of antagonists of excitatory amino acid neurotransmitters. Clin Sci 68:113–122
Michaels RL, Rothman SM (1990) Glutamate neurotoxicity in vitro: antagonist pharmacology and intracellular calcium concentrations. J Neurosci 10:283–292
Monyer H, Hartley DM, Choi DW (1990) 21-Aminosteroids attenuate excitotoxic neuronal injury in cortical cell cultures. Neuron 5:121–126
Morad M, Dichter M, Tang CM (1988) The NMDA activated current in hippocampal neurons is highly sensitive to $[H^+]_0$. Soc Neurosci Abs 14:791
Moriyoshi K, Masu M, Ishii T, Shigemoto R, Mizuno N, Nakanishi S (1991) Molecular cloning and characterization of the rat NMDA receptor. Nature 354:31–37
Nicoletti F, Wroblewski JT, Novelli A, Alho H, Guidotti A, Costa E (1986) The activation of inositol phospholipid metabolism as a signal-transducing system for excitatory amino acids in primary cultures of cerebellar granule cells. J Neurosci 6:1905–1911
Norenberg MD, Mozes LW, Gregorios JB, Norenberg LB (1987) Effects of lactic acid on astrocytes in primary culture. J Neuropathol Exp Neurol 46:154–166
Olney JW, Labruyere J, Price MT (1989) Pathological changes induced in cerebrocortical neurons by phencyclidine and related drugs. Science 244:1360–1362
Olney JW, Labruyere J, Wang G, Wozniak DF, Price MT, Sesma MA (1991) NMDA antagonist neurotoxicity: mechanism and prevention. Science 254:1515–1518
Pellegrini-Giampietro DE, Cherici G, Alesiani M, Carla V, Moroni F (1988) Excitatory amino acid release from rat hippocampal slices as a consequence of free-radical formation. J Neurochem 51:1960–1963
Plum F (1983) What causes infarction in ischemic brain? The Robert Wartenberg Lecture. Neurology 33:222–233
Rose K, Bruno VMG, Oliker R, Choi DW (1990) Nordihydroguaiaretic acid (NDGA) attenuates slow excitatory amino acid-induced neuronal degeneration in cortical cultures. Soc Neurosci Abs 16:288
Rosenberg PA, Aizenman E (1989) Hundred-fold increase in neuronal vulnerability to glutamate toxicity in astrocyte-poor cultures of rat cerebral cortex. Neurosci Lett 103:162–168
Rothman SM, Olney JW (1987) Excitotoxicity and the NMDA receptor. Trends Neurosci 10: 299–302
Sheardown MJ, Nielsen EO, Hansen AJ, Jacobsen P, Honore T (1990) 2,3,-Dihydroxy-6-nitro-7-sulfamoyl-benzo (F) quinoxaline: a neuroprotectant for cerebral ischemia. Science 247:571–574
Siesjö BK (1989) Free radicals and brain damage. Cerebrovasc Brain Metab Rev 1:165–211
Siman R, Noszek JC, Kegerise C (1989) Calpain I activation is specifically related to excitatory amino acid induction of hippocampal damage. J Neurosci 9:1579–1590
Sladeczek F, Pin JP, Recasens M, Bockaert J, Weiss S (1985) Glutamate stimulates inositol phosphate formation in striatal neurons. Nature 317:717–719
Sommer B, Seeburg PH (1992) Glutamate receptor channels: novel properties and new clones. Trends Neurosci 13:291–296
Tanabe Y, Masu M, Ishii T, Shigemoto R, Nakanishi S (1992) A family of metabotropic glutamate receptors. Neuron 8:169–179
Tecoma ES, Monyer H, Goldberg MP, Choi DW (1989) Traumatic neuronal injury in vitro is attenuated by NMDA antagonists. Neuron 2:1541–1545
Tombaugh GC, Sapolsky RM (1990) Mild acidosis protects hippocampal neurons from injury induced by oxygen and glucose deprivation. Brain Res 506:343–345
Turetsky DM, Goldberg MP, Choi DW (1992) Kainate-activated cobalt uptake identifies a subpopulation of cultured cortical neurons that are preferentially vulnerable to kainate-induced damage. Soc Neurosci Abs 18:81
Uematsu D, Araki N, Greenberg JH, Sladky J, Reivich M (1991) Combined therapy with MK-801 and nimodipine for protection of ischemic brain damage. Neurology 41:88–94

Verdoorn TA, Burnashev N, Monyer H, Seeburg PH, Sakmann B (1991) Structural determinants of ion flow through recombinant glutamate receptor channels. Science 252:1715–1718

Watkins JC, Olverman JH (1987) Agonists and antagonists for excitatory amino acid receptors. Trends Neurosci 10:265–272

Watkins JC, Krogsgaard-Larsen P, Honore T (1990) Structure–activity relationships in the development of excitatory amino acid receptor agonists and competitive antagonists. Trends Pharmacol Sci 11:25–33

Wieloch T (1985) Hypoglycemia-induced neuronal damage prevented by an *N*-methyl-D-aspartate antagonist. Science 230:681–683

4. Growth Factors in Stroke

S. P. Finklestein

Polypeptide growth factors are proteins that, at very low concentrations act as 'signaling' molecules to initiate and sustain complex cellular processes, including processes of cell survival, growth and division. An increasing number of such factors have been identified during the last decade, most falling into several known gene 'superfamilies,' including the neurotrophin, fibroblast growth factor (FGF), transforming growth factor α and β (TGF-α and TGF-β), platelet-derived growth factor (PDGF), and insulin-like growth factor (IGF) families. Growth factors exert their effects through interaction with specific cell-surface (and intracellular) receptors; the ensuing 'signal transduction' cascade results in the activation of second- (and higher-order) messenger systems, with subsequent initiation of new gene expression and protein synthesis leading to new cell function.

A number of identified growth factors are localized in the mammalian brain, and have been shown to exert potent survival-promoting ('trophic') effects on brain neurons and other cell types. For the most part, the effects of growth factors on brain cells have been characterized in tissue culture systems; a major challenge is to understand the role of these factors in the mammalian brain in vivo. This review will focus on two families about which most is known in regard to function in the brain, namely the neurotrophin and fibroblast growth factor (FGF) families. The neurotrophins include the homologous polypeptide nerve growth factor (NGF), brain-derived growth factor (BDNF) and neurotrophins 3, 4, and 5 (NT-3, 4 and 5; for reviews see Barde 1989; Gospodarowicz 1989; Logan 1990; Snider and Johnson 1989; Thoenen 1991; Sabel et al. 1992). At least three high-affinity and one low-affinity neurotrophin receptor molecules have now been identified (Bothwell 1991). The prototype neurotrophin, NGF, was discovered more than 40 years ago by virtue of its potent trophic properties on peripheral sympathetic and sensory neurons. More recently, NGF has been localized in the mammalian brain, and has been shown to support the survival and outgrowth of central cholinergic neurons. NGF is synthesized in brain regions that are targets of ascending cholinergic pathways (including cortex and hippocampus), and is conveyed via retrograde axonal transport to cholinergic cell bodies in the basal forebrain. NGF expression increases in target tissues following axotomy of cholinergic inputs, and exogenously administered NGF prevents the degeneration of cholinergic cell bodies following axotomy (Collins and Crutcher 1985; Varon et al. 1991). Although less is known about BDNF, NT-3 and other members of the neurotrophin family, the biological spectra of activity of these factors appear to be similar (although not identical) to NGF (Barde 1989; Thoenen 1991).

The FGF family includes acidic and basic FGF (aFGF and bFGF, respectively) as well as several homologous oncogene products (for reviews see Klagsbrun 1989; Baird and Klagsbrun 1991; Sabel et al. 1992). Acidic and basic FGF and their receptors are widely distributed in brain, where these factors have potent multipotential trophic effects. Both aFGF and bFGF promote the survival and outgrowth of a wide variety of brain neurons in vitro (Walicke 1988). They also promote the proliferation of brain glial cells (including astroglia and oligodendroglia) as well as brain capillary endothelial and smooth muscle cells in culture (Pettman et al. 1985; Gospodarowicz et al. 1986). Basic FGF has also been shown to promote brain glial and capillary proliferation in vivo (Barotte et al. 1989; Puumala et al. 1990). Like NGF, FGF levels are increased in brain following experimental injury (Finklestein et al. 1990b), and bFGF has been shown to prevent secondary neuronal degeneration following axotomy (Anderson et al. 1988).

Because of their localization in brain, their potent trophic effects on brain cells and their apparent role following mechanical brain injury, it is likely that polypeptide growth factors also play an important role following brain ischemia. The following review will focus on two themes: (1) that endogenous growth factor expression plays a role in wound healing and brain reorganization after ischemia, and (2) that exogenously administered polypeptide growth factors may be useful as 'neuroprotective' agents to limit the extent of initial neural damage after stroke.

Endogenous Growth Factor Expression in Wound Healing and Functional Recovery After Ischemia

Following ischemic injury to the mammalian brain, there ensues a cascade of cellular reactions that contributes to wound healing and to functional recovery. After focal cerebral infarction, there is breakdown of the blood–brain barrier, followed by infiltration of inflammatory cells (Giulian and Robertson 1990). Glial and vascular proliferation – neovascularization – begins at the borders of focal infarcts during the first week, and lasts for 2–3 weeks after stroke (duBois et al. 1985). Neural sprouting, both from the ends of damaged axons (regenerative sprouting) and from intact axons into denervated synapses (collateral sprouting), begins within the first few weeks and may last for months after infarction. Such sprouting appears to be associated with functional recovery in some model systems (Cotman and Nieto-Sampedro 1984). On the other hand, some varieties of sprouting, e.g. 'sympathetic sprouting', may be deleterious to functional recovery (Goldstein and Davis 1990). Apart from inflammatory cell infiltration, the cellular changes seen after global cerebral ischemia (including glial proliferation and axonal sprouting) are similar to those seen after focal ischemia.

Polypeptide growth factors, cytokines and other cell 'signaling' molecules probably play important roles in orchestrating the complex sequence of cellular reactions occurring after brain ischemia. Indeed, several recent studies have addressed changes in growth factor expression after stroke. Lindvall et al. (1992) found increased expression of mRNAs for the neurotrophins NGF and BDNF in hippocampus (a 'selectively vulnerable' brain region) following global cerebral ischemia in rats. Neurotrophin mRNA expression reached a peak at 2–4 hours after ischemia, then returned to baseline levels before frank histological evidence of cell death occurred in this model. Moreover, NGF and BDNF mRNA levels were highest in the dentate gyrus of the hippocampus, a region relatively resistant to ischemia. In another study, Lorez et al. (1989) found increased NGF levels in selectively vulnerable brain regions (including cortex, striatum and hippocampus) 5 days after global cerebral ischemia in rats, although Shigeno et al. (1991) reported decreased hippocampal NGF levels following global ischemia in gerbils. The precise role of altered neurotrophin expression in vulnerable brain regions (e.g. hippocampus) after global ischemia is currently unclear. As noted above, NGF appears to exert its major trophic effects on cholinergic neurons in the brain, whose cell bodies are located in the basal forebrain and striatum (Barde 1989; Snider and Johnson 1989; Thoenen 1991). Although intrinsic hippocampal neurons are not normally responsive to NGF, they might 'sprout' NGF receptors after injury or ischemia. Moreover, NGF produced in ischemic hippocampus might be transported retrogradely by cholinergic afferents to distal cell bodies of origin, preventing secondary neuronal degeneration. The available evidence suggests that neurotrophin expression may occur in two phases after ischemia: an early phase, occurring before cell death, perhaps contributing to the protection of some neuronal populations, and a late phase, occurring after cell death, perhaps contributing to axonal sprouting and synaptic reorganization.

Recent studies in our own and other laboratories have focused on changes in the expression of fibroblast growth factors (FGFs) after stroke. As noted above, these factors are widely distributed in brain, and have potent trophic effects on a wide variety of brain neurons, as well as mitogenic effects on brain glia and vascular endothelial cells (Klagsbrun 1989; Baird and Klagsbrun 1991). In a study of focal cerebral ischemia in rats, we found a steady increase in FGF levels in tissue surrounding focal cortical infarcts during the first 3 weeks after stroke (Finkestein et al. 1990a). Heparin-affinity high-performance liquid chromatography (HPLC) showed that, at 3 weeks after infarction, increased FGF levels were due largely to increased levels of bFGF. In a study of global cerebral ischemia in rats, Kiyota et al. (1991) found increased bFGF immunoreactivity in selectively vulnerable brain regions (including hippocampus) at 10 days after ischemia. Immunoreactive cells had the appearance of 'reactive' astroglia, in agreement with other studies showing increased bFGF immunoreactivity in reactive glia following focal mechanical brain injury (Finklestein et al. 1988). Taken together, these findings of increased bFGF expression during the first month after stroke are consistent with the presumed role of this growth factor in one or more of the cellular changes occurring after ischemia, including glial and vascular proliferation and neural sprouting. Indeed, exogenously administered bFGF has been reported to promote capillary proliferation in the ischemic brain (Lyons et al 1991). Clearly, further studies are required to fully define changes in the levels and cellular distribution of FGFs and their receptors after stroke, and to directly examine the effects of FGFs in the ischemic brain.

Exogenous Growth Factors as Neuroprotective Agents After Stroke

As a corollary to their neuronotrophic (survival-promoting) effects, it has become increasingly recognized that polypeptide growth factors can also protect neurons against the effects of various neurotoxins, including excitatory amino acids (EAAs), elevated intracellular calcium and hypoglycemia. Mattson and colleagues showed that, at a very low concentration (5 ng/ml), bFGF protects cultured rat hippocampal neurons against glutamate neurotoxicity (Mattson et al. 1989). Similar findings have been obtained for cultured striatal, cerebellar and cortical neurons (Caday et al. 1991; Fernandez et al. 1992; Freese et al. 1992). Both bFGF and NGF protect cultured neurons against hypoglycemia in vitro (Cheng and Mattson 1991). The neuroprotective effects of bFGF against EAA toxicity in culture require at least 5–12 hours of preincubation with bFGF to become maximal, are associated with decreased intraneuronal Ca^{2+} concentrations, and depend on new neuronal gene expression and protein synthesis (Mattson et al. 1989; Freese et al. 1992).

EAA release, with consequent activation of NMDA receptors and massive Ca^{2+} entry into cells, appears to be an important mechanism of cell death after cerebral ischemia (Choi 1990). The observations of growth factor neuroprotection against EAA toxicity in vitro have thus led to the hypothesis that these factors may be useful as pharmacological agents to limit the extent of neural damage after cerebral ischemia. Indeed, several recent studies have shown that exogenously administered neuronotrophic growth factors reduce neural death in models of both global and focal cerebral ischemia. Buchan et al. (1990) found that NGF, administered by continuous intraventricular infusion (1.2 μg/day) for 10 days before global cerebral ischemia in rats significantly reduced cell loss in CA-1 hippocampus. More recently, Shigeno et al. (1991) found that a single intraventricular injection of NGF (10 μg) just before or 15 minutes after global cerebral ischemia in gerbils markedly reduced CA-1 neuronal death. Ortiz et al. (1990) reported that repeated intraventricular injections of NGF after focal cerebral ischemia in rats reduced behavioural dysfunction in these animals at 2 weeks after infarction, but no histopathological findings were given. Overall, these findings, although intriguing, are also somewhat puzzling, since NGF is not normally considered a trophic factor for intrinsic hippocampal or cortical neurons. On the other hand, as noted above, ischemic neurons may 'sprout' NGF receptors; NGF may also prevent secondary neuronal degeneration after ischemia.

Recent studies in our own and other laboratories have examined the potential neuroprotective effects of FGFs in in vivo models of EAA toxicity and cerebral ischemia. As noted above, FGFs support survival and block EAA toxicity for a number of neuronal types in vitro. Moreover, previous studies showed that endogenous

FGF expression does not reach a peak for 3 weeks after stroke (Finklestein et al. 1990a), suggesting that earlier exogenous administration of these factors might be useful in limiting neural damage after ischemia. Nozaki et al. (1993) found that intraperitoneally administered bFGF (100 μg/kg) reduced striatal lesions produced by direct NMDA injection or systemic hypoxia/ischemia in neonatal rats. Neonatal animals were chosen for these studies to allow easier entry of systemically administered growth factor across the immature blood–brain barrier. In recent preliminary studies in mature rats, Koketsu et al. (1994) found that continuous intraventricular administration of bFGF (1.2 μg/day) for 3 days before focal cerebral infarction reduced infarct size by 25%–30%. Infusion of bFGF was not associated with changes in several physiological parameters (including blood pressure, temperature, blood gases, glucose or hematocrit), but was associated with a small increase in post-operative weight loss in some animals. In another study, Yamada et al. (1991) administered bFGF by repeated intracisternal injection (1 μg/week for 4 weeks), starting 24 hours after focal cerebral infarction in mature rats. Although these investigators found no effect on infarct size, they did find a marked reduction in retrograde neuronal degeneration in the ipsilateral thalamus. The neuron-sparing effects of bFGF in the above studies may have been due to direct protective effects on vulnerable neurons. On the other hand, the neuroprotective effects of bFGF may have been indirect, through activation of brain glial cells (in turn secreting other neuronotrophic factors), through stimulation of vascular proliferation ('angiogenesis'), or through direct effects on cerebrovascular tone. Given at higher doses systemically, bFGF lowers blood pressure through vasodilation (Cuevas et al. 1991), and may produce changes in cerebral blood flow or its distribution. Two other recent reports suggest that intraventricularly administered aFGF and bFGF reduce CA-1 neuronal death after global cerebral ischemia (Oomura et al. 1990; Berlove et al. 1991), although another study using lower doses of bFGF injected directly into the hippocampus failed to show such an effect (Hara et al. 1991).

In summary, evidence from several recent studies is consistent with the principle that neuronotrophic growth factors can limit the extent of neural damage after stroke. However, it must be emphasized that in all of the published studies to date growth factors have been administered at times preceding (or just after) the onset of cerebral ischemia and, in the case of mature animals, have been given directly into the brain. Clearly, any potential clinical application of these findings must await the demonstration that growth factors can be effective when given after ischemia by a clinically practicable route. Although systemically administered polypeptides are not expected to cross the intact blood–brain barrier, they may cross the damaged blood–brain barrier after stroke. Moreover, it may be possible to 'engineer' shorter active peptides, or to complex growth factors to other molecules that might more readily cross the blood–brain barrier. Regardless of the direct clinical usefulness of growth factor in the treatment of stroke, further studies of this phenomenon will undoubtedly shed light on molecular mechanisms governing neuronal survival versus death following cerebral ischemia.

References

Anderson KJ, Dam D, Lee S, Cotman CW (1988) Basic fibroblast growth factor prevents death of lesioned cholinergic neurons in vivo. Nature 332:360–361

Baird A, Klagsbrun M (eds) (1991) The fibroblast growth factor family. Annals of the New York Academy of Sciences, Vol. 638

Barde Y-A (1989) Trophic factors and neuronal survival. Neuron 2:1525–1534

Barotte C, Enclancher F, Ebel A, Labourdette G, Sensenbrenner M, Will B (1989) Effects of basic fibroblast growth factor (bFGF) on choline acetyltransferase activity and astroglial reaction in adult rats after partial fimbria transection. Neurosci Lett 101:197–202

Berlove DJ, Caday CG, Moskowitz MA, Finklestein SP (1991) Basic fibroblast growth factor (bFGF) protects against ischemic neuronal death in vivo. Soc Neurosci Abstr 17:1267

Bothwell M (1991) Keeping track of neurotrophin receptors. Cell 65: 915–918

Buchan AM, Williams L, Bruederlin B (1990) Nerve growth factor: pretreatment ameliorates ischemic hippocampal neuronal injury. Stroke 21:77

Caday CG, Kemmou A, Finklestein SP (1991) Basic fibroblast growth factor (bFGF) protects against glutamate toxicity in cultured cerebrocortical neurons. J Cell Biol 15:419a

Cheng B, Mattson MP (1991) NGF and bFGF protect rat hippocampal and human cortical neurons against hypoglycemic damage by stabilizing calcium homeostasis. Neuron 7:1–20

Choi DW (1990) Cerebral hypoxia: some new approaches and unanswered questions. J Neurosci 10:2493–2501

Collins F, Crutcher KA (1985) Neurotrophic activity in the adult rat hippocampal formation: regional distribution and increase after septal lesion. J Neurosci 5:2809–2814

Cotman CW, Nieto-Sampedro M (1984) Cell biology of synaptic plasticity. Science 225:1267–1294

Cuevas P, Carceller F, Ortega S, Zazo M, Nieto I, Gimenez-Gallego G (1991) Hypotensive activity of fibroblast growth factor. Science 254:1208–1210
duBois M, Bowman PD, Goldstein GW (1985) Cell proliferation after ischemic injury in gerbil brain. Cell Tissue Res 242:17–23
Fernandez MT, Torreblanca A, Gascon S, Novelli A (1992) Differentiation independent protection of basic fibroblast growth factor from NMDA and non-NMDA receptor mediated neurotoxicity in cerebellar neurons. J Cell Biol 115:419a
Finklestein SP, Apostolides PJ, Caday CG, Prosser J, Philips MF, Klagsbrun M (1988) Increased basic fibroblast growth factor (bFGF) immunoreactivity at the site of focal brain wounds. Brain Res 460:253–259
Finklestein SP, Caday CG, Kano M, Berlove DJ, Hsu CY, Moskowitz M et al. (1990a) Growth factor expression after stroke. Stroke 21:III-122–III-124
Finklestein, SP, Fanning PJ, Caday CG, Powell PP, Foster J, Clifford EM, Klagsbrun M (1990b) Increased levels of basic fibroblast growth factor (bFGF) following focal brain injury. Rest Neurol Neurosci 1:387–394
Freese A, Finklestein SP, DiFiglia M (1992) Basic fibroblast growth factor protects striatal neurons in vitro from NMDA-receptor mediated excitotoxicity. Brain Res 575:351–355
Giulian D, Robertson C (1990) Inhibition of mononuclear phagocytes reduces ischemic injury in the spinal cord. Ann Neurol 27:33–42
Goldstein LG, Davis JN (1990) Restorative neurology: drugs and recovery following stroke. Current Concepts Cerebrovasc Dis 25:19–24
Gospodarowicz D (1989) Neurotrophic factors. In: Neuroimmune networks: physiology and diseases. Alan R. Liss, New York, pp 163–172
Gospodarowicz D, Massoglia S, Cheng J, Fujii DK (1986) Effect of fibroblast growth factor and lipoproteins on the proliferation of endothelial cells derived from bovine adrenal cortex, brain cortex, and corpus luteum capillaries. J Cell Physiol 127:121–136
Hara H, Onodera H, Kawagoe J, Kogure K (1991) Failure of basic fibroblast growth factor to prevent postischemic neuronal damage in the rat. Eur J Pharmacol 209:195–198
Kiyota Y, Takami K, Iwane M, Shino A, Miyamoto M, Tsukuda R et al. (1991) Increase in basic fibroblast growth factor-like immunoreactivity in rat brain after forebrain ischemia. Brain Res 545:322–328
Klagsbrun M (1989) The fibroblast growth factor family: structural and biological properties. Prog Growth Factor Res 1:207–235
Koketsu N, Berlove DJ, Moskowitz MA, Kowa NW, Caday CG, Finklestein SP (1994) Pretreatment with intraventricular basic fibroblast growth factor (bFGF) decreases infarct size following focal cerebral ischemia in rats. Ann Neurol (in press)
Lindvall O, Enfors P, Bengzon J, Kokaia Z, Smith M-L, Siesjo BK et al. (1992) Differential regulation of mRNAs for nerve growth factor, brain-derived neurotrophic factor, and neurotrophin 3 in the adult rat brain following cerebral ischemia and hypoglycemic coma. Proc Natl Acad Sci (USA) 89:648–652
Logan A (1990) CNS growth factors. Br J Hosp Med 43:429–435
Lorez H, Keller F, Ruess G, Otten U (1989) Nerve growth factor increases in adult rat brain after hypoxic injury. Neurosci Lett 98:339–344
Lyons MK, Anderson RE, Meyer FB (1991) Basic fibroblast growth factor promotes in vivo cerebral angiogenesis in chronic forebrain ischemia. Brain Res 558:315–320
Mattson MP, Murrain M, Guthrie PB, Kater SB (1989) Fibroblast growth factor and glutamate: opposing roles in the generation and degeneration of hippocampal neuroarchitecture. J Neurosci 9:3728–3740
Nozaki K, Finklestein SP, Beal MF (1993) Basic fibroblast growth factor protects against hypoxia–ischemia and NMDA neurotoxicity in neonatal rats. J Cerebr Blood Flow Metab 13:221–228
Oomura Y, Sasaki K, Muto T, Suzuki K, Hanai K, Tooyama I et al. (1990) Physiological actions of fibroblast growth factor (FGF) in central nervous system. Soc Neurosci Abstr 16:516
Ortiz A, MacDonall JS, Mahakik SP, Karpiak SE (1990) Nerve growth factor treatment of cortical focal ischemia. Soc Neurosci Abstr 16:942
Pettman B, Weibel M, Sensenbrenner M, Labourdette G (1985) Purification of two astroglial growth factors from bovine brain. FEBS Lett 189:102–108
Puumala M, Anderson RE, Meyer FB (1990) Intraventricular infusion of HBGF-2 promotes cerebral angiogenesis in the Wistar rat. Brain Res 534:283–286
Sabel BA, Vantini G, Finklestein SP (1992) The role of neurotrophic factors in the treatment of neurological disorders. In: Vecsei L, Freese A, Swarz KJ, Beal MF (eds) Neurological disorders: novel experimental and therapeutic strategies. Ellis Horwood, New York, pp 113–180
Shigeno T, Mima T, Takakura K, Graham DI, Kato G, Hashimoto Y et al. (1991) Amelioration of delayed neuronal death in the hippocampus by nerve growth factor. J Neurosci 11:2914–2919
Snider WD, Johnson EM (1989) Neurotrophic molecules. Ann Neurol 6:489–506
Thoenen H (1991) The changing scene of neurotrophic factors. TINs 14:165–170
Varon S, Hagg T, Manthorpe M (1991) Nerve growth factor in CNS repair and regeneration. In Timiras PS (ed) Plasticity and regeneration of the nervous system. Plenum Press, New York, pp 267–276
Walicke PA (1988) Basic and acidic fibroblast growth factors have trophic effects on neurons from multiple CNS regions. J Neurosci 8:2618–2627
Yamada K, Kinoshita A, Kohmura E, Sakaguchi T, Taguchi J, Kataoka et al. (1991) Basic fibroblast growth factor prevents thalamic degeneration after cortical infarction. J Cerebr Blood Flow Metab 11:472–478

5. Ischemic Cerebral Edema

M. D. O'Brien

Cerebral edema is defined as a relative increase in the water content of the brain, and this excess water may be in either the intracellular or the extracellular compartments, or both. There are two other causes of brain swelling, vascular congestion and hydrocephalus; these are not due to edema, but may be both associated and confused with it. The early literature distinguished between brain swelling or dry edema (Hirnschwelling), which corresponds to intracellular edema, and brain or wet edema (Hirondem), which corresponds to extracellular edema. The differentiation between these two forms depended on the appearance of the cut surface of the brain (Reichardt 1904/5).

Intracranial Fluid Compartments

The intracranial fluid compartments are intracellular, extracellular including the cerebrospinal fluid, and intravascular. These compartments are separated by membrane barriers, which maintain their interior milieu. The extracellular space and the cerebrospinal fluid are in continuity, but the considerable discrepancy in size between these spaces makes it useful to consider them as separate compartments. The blood-brain barrier has many functions, but as far as edema formation is concerned this discussion will be limited to the transendothelial resistance to water diffusion, the passage of micro- and macromolecules, and the state of the endothelial tight junctions. Other properties of the blood–brain barrier, such as active and facilitated transport, will not be discussed although they play an important part in the passage of osmotically active molecules, such as the sugars.

Water is freely diffusible between all these compartments and its movement is entirely passive following osmotic and hydrostatic pressure gradients. The rate of equilibration depends on the hydraulic conductivity of the various membranes, and this is much less for the blood–brain barrier than for other vascular beds. The cerebrovascular endothelial tight junctions considerably reduce the permeability to water. The amount of edema formation depends on tissue compliance, which is a composite of the compliance of all the compartments. White-matter tissue compliance is much greater than that of gray matter, and this difference is one of the reasons why white matter usually becomes more edematous than the overlying cortex. Furthermore, there are regional differences in compliance, with greater compliance in the deep white matter than in the subcortical white matter (Kuroiwa et al. 1990).

The blood hydrostatic pressure is the source of all intracranial hydrostatic pressures and is the principal driving force in the formation of extracellular edema. The intra- and extracellular hydrostatic pressures can be considered together as the tissue hydrostatic pressure, although these two compartments may alter volume separately under differing osmotic pressures. If the hydrostatic pressure is zero, as occurs in complete ischemia, there may be a shift of water from the extracellular space to the intracellular space due to membrane failure, without an increase in total brain tissue water, and this is of course not cerebral edema, since there is no overall increase in brain water content, even though there is intracellular edema. This theoretical situation does not occur in practice, since in all but the smallest infarcts there is always some movement of water from the vascular compartment to the extracellular space, and therefore a rise in parenchymal water content. Furthermore, in most of the territory of an infarct there is some residual blood flow, which is the source of additional fluid.

These fluid compartments are all interdependent, but it is convenient to consider the factors which alter the volume of each compartment separately.

The Intracellular Compartment

An increase in the intracellular water content has been called 'metabolic' or 'cytotoxic' edema. It is caused by anything that damages cell metabolism, particularly the sodium pump mechanism which is highly energy dependent and produces a change in the osmotic balance across the cell membrane. Metabolic injury can be produced experimentally with a wide range of poisons, such as cyanide and hexachlorophene, and by specific sodium pump inhibitors such as ouabain (Tanaka et al. 1977). Direct damage to the cell due to trauma, including cold injury, also results in the destruction of cell membranes. Ischemia produces cytotoxic edema in parallel with failure of cellular metabolism and the cell's inability to maintain its membrane function.

The Extracellular Compartment

Brain swelling due to extracellular edema has been called 'vasogenic edema' (Klatzo 1967). This can be conveniently divided into edema with an intact blood–brain barrier and edema with a damaged blood–brain barrier.

Edema with an intact blood–brain barrier is an ultrafiltration of plasma, and may occur with a moderate or slow rise of blood pressure into a dilated or disautoregulated vascular bed. Edema can be produced quite easily in experimental animals by a rise in blood pressure with hypercapnia (Meinig et al. 1972) and following application of pressure to the cortex to produce loss of autoregulation. These mechanisms explain edema formation in the brain surrounding cerebral tumors. Vascular congestion and extracellular edema are the principal causes of the rapid swelling of the brain which may follow surgical removal of a large intracranial mass, such as meningioma. If the blood–brain barrier is damaged – the next stage in severity – there is protein extravasation and osmotic extracellular edema in addition to hydrostatic edema. Hypertensive encephalopathy is a good clinical example. It may be produced experimentally with a rapid rise in blood pressure alone, or at lower levels of blood pressure, with loss of autoregulation (Johansson 1976). Vasogenic edema with blood–brain barrier damage can also be induced by trauma, either by freezing or by pressure.

Extracellular edema spreads extensively through the white matter by bulk flow along hydrostatic pressure gradients, and not by diffusion, since it has been shown that substances with different diffusion coefficients travel at the same speed and further than they would by diffusion alone (Reulen 1976). Edema fluid is cleared by a number of mechanisms, including drainage into the CSF, with which it is in continuity, so that CSF pressure has some effect on edema resolution. Edema fluid is also cleared by absorption into the blood vessels (Naruse et al. 1990). Klatzo et al. (1980) have shown that the resolution of extracellular edema is paralleled by the intraglial uptake of extravasated serum proteins, and suggested that the removal of this osmotic effect allowed the clearance of the excess fluid. However, this mechanism of clearance is not a particularly important factor in ischemic cerebral edema, except in large infarcts with surrounding extracellular edema.

It is possible to have the blood–brain barrier open to macromolecules without the formation of edema, and this has been demonstrated following infusions of hypertonic solutions. It was postulated that the hypertonic solution caused shrinkage of endothelial cells and the opening of tight junctions (Rapoport et al. 1971); the finding of horseradish peroxide in the tight junctions was said to support this view (Klatzo 1972). However, the tight junctions have never been observed to open and it now seems clear that endothelial cells merely become more permeable to macromolecules, which traverse the cells by pinocytosis (Rapoport 1976a; Nag et al. 1977; Petito 1979). It is possible that edema does not develop because the osmotic gradient is not sufficiently altered by the extravasation of protein in the presence of a hyperosmolar vascular compartment, but the relationship between extracellular protein and osmolarity is not linear and there may well be a threshold before significant edema can develop (Rapoport 1976b). This dissociation between macromolecule extravasation and edema formation occurs in epileptic seizures, where the seizure-induced hypertension causes most of the damage to the blood–brain barrier and the enhanced micropinocytosis (Petito et al. 1976). This dissociation also occurs in brain infarcts (O'Brien et al. 1974a) but for a different reason: in stroke there is cellular disruption before the blood–brain barrier is disturbed.

An ultrafiltrate edema with an intact blood–brain barrier, which requires an increase in hydrostatic pressure to capillaries through disautoregulated and maximally dilated arteries and arterioles, might occur in stroke if the collateral circulation was exceptionally good, or particularly if an embolus impacts, fragments and moves off leaving a patent vessel to feed a disautoregulated vascular bed, which is then exposed to systemic pressures. Such reperfusion may result in damage to the blood–brain barrier and massive edema or hemorrhage. Reperfusion edema is a predictable complication following thrombus dissolution with tissue plasminogen activator, urokinase or streptokinase, particularly if clot dissolution occurs more than 6–8 hours after the stroke.

The Cerebrospinal Fluid Compartment

The CSF hydrostatic pressure is one determinant of the overall cerebral perfusion pressure and is a measure of the average tissue and venous pressures. The CSF hydrostatic pressure is seldom an important parameter in the mechanisms of ischemic cerebral edema.

The Intravascular Compartment

Brain swelling due to vascular congestion occurs when cerebral blood vessels are fully dilated by anoxia or hypercapnia, provided that there is an adequate input under sufficient pressure to distend small blood vessels, and this may require some elevation in blood pressure. These changes occur during the first stage of high-altitude 'cerebral edema'. Initially there is brain swelling due to vascular congestion because of the anoxic stimulus aggravated by an elevation of blood pressure with exertion. In experimental animals considerable congestion can be achieved with anoxia or hypercapnia alone. Endothelial cells are relatively resistant to hypoxia and tight junctions are preserved. An increase in the size of the vascular pool usually occurs in and around an infarct, since some flow usually exists via collateral circulation, and ischemic vessels are maximally dilated.

Obstruction of Venous Outflow

Venous sinus thrombosis may cause a marked rise in intracranial pressure with gross distension of cortical veins and produce venous infarction characterized by small hemorrhages and ischemic edema.

Effect of Edema on Cerebral Function

Obviously, in metabolic edema function is impaired at the outset since the edema is due to cellular malfunction, but extracellular or vasogenic edema appears to have remarkably little effect on cell function, as shown by electrical activity and evoked responses (Sutton et al. 1980), until raised intracranial pressure causes blood flow reduction to critical levels (Marshall et al. 1976). This is shown clinically in patients with benign intracranial hypertension, who may remain quite alert despite considerable cerebral edema. Focal cerebral edema may be more serious, with the production of a rapidly expanding mass effect and intracranial herniation.

Ischemic Cerebral Edema

The degree of ischemia is not homogeneous throughout an infarct. Furthermore, the development of an infarct is a rapidly evolving process, with different parts of the lesion developing at different rates. All the various mechanisms of edema formation may occur in stroke, and will vary not only throughout the lesion but also with time.

The important factors which determine the formation of ischemic cerebral edema are the rate of development; the duration and severity of ischemia, which determine the changes in cell metabolism; the residual blood flow; and the extent to which the systemic blood pressure is transmitted to the infarct. In addition, there are the secondary effects of the ischemic process on the hydrostatic and osmotic pressures of each fluid compartment.

Absolute Ischemia

If ischemia is absolute and the blood flow zero, the electrical activity of the neurons stops within a few seconds, followed by a failure of the sodium pump and glucose depletion. Intracellular edema, mostly derived from the extracellular space, is evident after a few minutes and there is a massive rise in tissue lactate. The amount of the edema is proportional to the lactate formation, and there is a pH threshold of 5.6 (Staub et al. 1990). Up to this stage the process is potentially reversible, but thereafter a stepwise sequence of organelle failure occurs as thresholds for survival are reached and passed.

If there is no blood flow there is no source of additional fluid and true cerebral edema cannot occur, apart from the very small amount of fluid that can be absorbed from the vascular compartment, and this may be associated with some swelling of astrocytic foot processes (Little 1976). There is, however, a shift of water from the extracellular to the intracellular space, causing intracellular edema without brain edema. Hossman (1976) showed that, after 1 hour of total ischemia in cats, the extracellular space diminished from 18.9 vol% to 8.5 vol%, whereas the total water content and intracranial pressure remained unchanged.

The Preservation of Some Residual Flow

Ischemia is, however, less than complete, and flows of around 25% of normal have been measured in the vascular bed distal to middle cerebral artery occlusion in monkeys (Symon 1967; Symon et al. 1974). This residual flow slows down the development of the infarct and prolongs the ischemic cascade, and of course the process can be arrested at any stage. There are flow thresholds for both function and survival, and these are determined by a combination of the depth of ischemia and its duration (Heiss and Rosner 1983). Acute experiments in animals have shown that there is a failure of electrical function, as measured by evoked responses at blood flow around 18 ml/100 g/min, whereas potassium flux, which is a marker of cell disruption, does not occur until the blood flow falls to around 5–8 ml/100 g/min (Astrup et al. 1977). However, other cell functions, such as protein synthesis, are affected at flow rates not associated with acute cell disruption and this leads to delayed cell death (Mies et al. 1991). Flow thresholds for the development of ischemic edema obtained from animal experiments vary according to the species and the experimental protocol, but the earliest changes reflect the osmotic effect of electrolyte shifts (Hossman 1989).

Animal experiments have also shown that the presence or absence of venous obstruction is an important factor (Hossman and Olesen 1970), since recovery is poorest and the damage greatest if both arteries and veins are occluded at the same time. The situation is better if the veins are not obstructed, and better still if the brain is perfused with saline during arterial occlusion. These experiments imply that both stasis, with the involvement of platelets and leukocytes, and the accumulation of metabolites, are important factors in ischemic damage, and therefore in the development of edema.

Whether or not the blood–brain barrier opens to macromolecules depends on the blood hydrostatic pressure at capillary level as well as the duration and depth of ischemia, and numerous animal experiments have clearly shown threshold phenomena for all these parameters. However, protein extravasation is not usually a significant factor in ischemic cerebral edema. O'Brien et al. (1974b) studied the distribution of water in the brains of cats from 4 hours to 20 days after occlusion of the middle cerebral artery. In this model the water content reached a maximum at 2 days. At the same time the blood–brain barrier function was studied with ^{99}Tc pertechnetate and ^{131}I-labeled albumin (O'Brien et al. 1974a). The greatest extravasation of these tracers did not occur until 3 or 4 days after occlusion, remained high for the duration of the study, and showed continued high levels long after the edema resolved, showing that blood–brain barrier damage is not associated with an additional shift of water. Since endothelial cells are among the structures most resistant to ischemia, and there is necrosis of brain parenchyma before the endothelial cells are affected, it may be that the leak of proteins does not produce a sufficient osmotic effect in the presence of necrotic brain cells. This is in accord with the clinical experience that edema is maximal between 1 and 3 days after a stroke, whereas technetium pertechnetate brain scans or enhanced CT scans do not show a maximum lesion to background ratio until 7–10 days later.

Reperfusion

The reperfusion of an ischemic area may result in the restoration of normal function if it occurs sufficiently soon after the event, and again, clear time thresholds have been demonstrated in a wide range of animal models. Hossman (1976) showed that reperfusion after 1 hour of total cerebral ischemia in cats caused a rapid restitution of the extracellular fluid, with a corresponding increase in total brain water content and an inevitable rise in intracranial pressure, the resulting state being a combination of extracellular and intracellular edema. The degree of edema depends critically on the perfusion pressure, and is therefore patchy where parts of an infarct are affected by the no-reflow phenomenon (Branston et al. 1980).

Effect of Alteration of Physiological Parameters

Biological and Therapeutic

Hydrostatic Pressure

The effect of changes in systemic pressure depends almost entirely on the extent to which this is transmitted to the capillaries. If the major arterial supply to an infarcted region is occluded, the blood supply to the infarct is from the collateral circulation and, at least in the acute stage, the flow will be pressure-dependent because the arterioles will be maximally dilated by the anoxic ischemic stimulus. This loss of autoregulation is proportional to the degree of ischemia, particularly for falls in blood pressure (Symon et al. 1976). If the collateral circulation is particularly tenuous, as may occur in bilateral carotid occlusion, there may be some autoregulation against a rise in systemic pressure in responsive vessels sufficiently proximal to the ischemic area to be unaffected by it, and this may reduce the perfusion pressure distally (O'Brien et al. 1983). An increase in blood pressure may increase the flow at the periphery of an infarct, but may also be associated with an increase in cerebral edema, so there may be no net benefit (Hatashita et al. 1990).

If the main arterial supply to an infarct is patent, either because an embolus has fragmented or as a result of therapeutic intervention, or if the collateral circulation is exceptionally good, particularly with small lesions, the infarct is exposed to systemic pressures and any increase in blood pressure will immediately drive the formation of extracellular edema. Numerous animal experiments have demonstrated this point using cold injury, where the lesion is exposed to systemic pressures from the start, and following the removal of temporary arterial clips after varying intervals of ischemia. Once again, a threshold for the formation of edema can be demonstrated. A rise in blood pressure may increase blood flow, but false autoregulation may also occur, that is, a rise in pressure that does not appear to be accompanied by a rise in flow in a disautoregulated vascular bed. When this phenomenon was first observed it was thought to be due to an increase in edema, and therefore a rise in the local tissue pressure which limited vasodilation. A similar explanation was postulated for the finding that reactive hyperemia appeared to be less evident in edematous brain. However, these observations are an artefact of the method of measurement and not a consequence of a squeeze on cerebral vessels, increasing their vascular resistance. It is due to the increase in volume of edematous tissue, and blood flow is calculated as a perfusion rate per unit mass of tissue. If the blood flow measurements are based on dry weight, there is no significant change in flow (Hossman and Bloink 1981).

A reduction in blood pressure can be equally harmful. It might retard the formation of cerebral edema if the ischemic area was sufficiently exposed to systemic pressures, but in the majority of patients with stroke, whose ischemic core is protected from systemic pressures, a reduction in pressure would cause a reduction of flow, since the vessels in the periphery of an infarct are disautoregulated, and this could well have a critical effect on cellular metabolism. In most clinical circumstances hypotension is more critical than hypertension.

Effect of Alteration in Vessel Caliber

Cerebral blood vessels autoregulate to changes in blood pressure, becoming maximally dilated when perfusion pressure falls below a mean arterial blood pressure of about 60 mmHg. Anoxia without ischemia also causes cerebral vasodilation, and this occurs at high altitude. However, the arterial Pco_2 has the most powerful effect on vessel caliber. Hypercapnia produces cerebral vasocongestion in normal brain by dilation of responsive vessels, and although some Pco_2 responsiveness is preserved in most infarcts, it would have little effect on blood vessels already dilated by an anoxic ischemic stimulus. However, hypocapnia is probably the quickest method of producing a rapid reduction in intracranial pressure, and might prevent intracranial herniation with a large infarct, but it is necessarily a short-term procedure.

Alteration of Blood Osmolarity

Osmotic agents can only reduce the tissue water content of perfused brain; they only work if the endothelial and cell membranes are intact, and can therefore reduce the water content of normal brain. If a large infarct produces a mass effect and there is significant extracellular edema in the normal surrounding brain due to disautoregulation, or in areas of ischemia without infarction, osmotic agents may have some effect but they have little or no effect in the core of ischemic infarcts, partly because of a failure to deliver the osmotic agent to the infarct and partly because of disruption of cell membranes. Perfusion might be increased in some areas by a reversal of the swelling of perivascular astrocytic foot processes (Garcia and Kamijyo 1974; Little 1976).

Possible Metabolic and Biochemical Effects on Ischemic Cerebral Edema

A considerable body of information is accumulating about the many complex metabolic and biochemical changes which occur as a result of ischemia, and this is mostly derived from animal experiments. Caution should be exercised in extrapolating some of these phenomena to stroke in humans. It is often difficult to determine whether biochemical changes which are found to be proportional to the degree of ischemia and edema are the direct consequence of the ischemia, or independently contribute to the formation of edema.

Leukocytes and Platelets

The evidence on the role of leukocytes is conflicting, and varies according to the experimental model. Increased edema has been observed with neutropenia using the cold injury model (Schurer et al. 1990). This suggests that leukocytes, which accumulate around infarcts, occlude the vessels, thereby protecting the infarct from the effect of systemic pressure. However, leukopenia induced by whole-body irradiation reduced edema surrounding a simulated intracerebral hemorrhage (Kane et al. 1990). This may be due to an improvement of flow in small vessels (Dutka et al. 1989) and to the release of vasoactive substances from activated leukocytes.

Sodium, Potassium and Idiogenic Osmoles

Intracellular accumulation of sodium can be measured from the onset of ischemia, and rises progressively. This progress starts at flow thresholds above that associated with membrane failure, and is not immediately associated with loss of intracellular potassium. This may be because sodium channels are opened by the action of excitatory neurotransmitters. The resulting increase in intracellular osmotic pressure draws in water and chloride from the extracellular space, which then shrinks. However, the amount of water cannot be entirely explained by the rise in intracellular sodium, suggesting that ischemia can produce some rise in intracellular osmotic pressure that is not due to a shift of electrolytes, and this has given rise to the concept of 'idiogenic osmoles', which are probably cations. The massive influx of calcium and loss of potassium from the intracellular space marks the death of the cell (Hossman 1989).

Calcium and Nitric Oxide

Considerable calcium influx into the cell occurs at the same time as the massive rise in extracellular potassium which is usually considered to be a marker of cell membrane disruption.

However, some accumulation of calcium occurs before the loss of cell membrane function which contributes to the rise in intracellular osmotic pressure. Calcium entry blockers can limit some of this movement and reduce the resultant intracellular metabolic edema (Abe et al. 1988; Yamamoto et al. 1990). Ischemia stimulates the formation of nitric oxide in endothelial cells, and this may have an effect on the permeability of the endothelial cells to the diffusion of water, but the role of nitric oxide in the development of extracellular edema has yet to be determined. Ischemia also considerably increases the formation of intracellular nitric oxide, which appears to be associated with the rise in intracellular calcium.

Free Radicals

Oxygen-derived free radicals are produced in ischemic brain and have been implicated both in the formation of vasogenic edema due to changes in vascular permeability (Chan et al. 1984), and also in the production of intracellular edema by neuronal cell membrane damage (Siesjo et al. 1985; Chan et al. 1990). These effects can be ameliorated by the use of free radical scavengers (Ikeda and Long 1990; Imaizumi et al. 1990; Oh and Betz 1991).

Arachidonic Acid

Arachidonic acid is released from damaged brain tissue during ischemia and the ensuing cascade of its metabolites is highly vasoactive, increasing vascular permeability and contributing to the formation of vasogenic brain edema (Katayama et al. 1990).

Excitotoxins

Cerebral ischemia is associated with an increased release of excitatory neurotransmitters, such as glutamate. It has been postulated that excessive release of excitatory neurotransmitters may be associated with sodium and calcium accumulation and neuronal damage, as well as disrupting blood–brain barrier function, all of which contribute to the development of both vasogenic and cytotoxic edema (Saito et al. 1990). Some of these effects can be blocked by the use of specific excitatory amino acid antagonists such as MK801 (Shinohara et al. 1990).

Conclusion

Cerebral ischemia involves a very complex cascade of events in which both the vascular system and cellular function are affected, both of which contribute to the formation of ischemic cerebral edema. However, the movement of water is passive and follows osmotic and hydrostatic pressure gradients, and is therefore more a marker than a cause of brain dysfunction. The sequence of events in the formation of edema due to cerebral ischemia is a follows: first, there is a shift of water from the extracellular to the intracellular space. Some extracellular edema may occur, probably associated with increased pinocytosis if the local perfusion pressure is adequate. Only much later does the blood–brain barrier open to macromolecules if the ischemia is sufficient to cause cellular disruption and there is sufficient blood hydrostatic pressure at capillary level, but this is not associated with additional cerebral edema.

Ischemic cerebral edema is not an important factor in determining the outcome following a stroke, except where a large infarct with edema produces a mass effect.

References

Abe K, Kogure K, Watanabe T (1988) Prevention of ischaemic and post-ischaemic brain oedema by a novel calcium antagonist. J Cerebr Blood Flow Metab 8: 436–439

Astrup J, Symon L, Branston NM, Lassen NA (1977) Cortical evoked potential and extracellular K^+ and H^+ at critical levels of brain ischaemia. Stroke 8:51–57

Branston NM, Bell BA, Hunstock A, Symon L (1980) Time and flow factors in the function of postischaemic oedema in the primate cortex. Adv Neurol 28:291–298

Chan PH, Schmidley JW, Fishman RA, Longar SM (1984) Brain injury, edema and vascular permeability induced by oxygen derived free radicals. Neurology 34: 315–320

Chan PH, Fishman RA, Wesley MA, Longar SM (1990) Pathogenesis of vasogenic oedema in focal cerebral ischaemia, the role of superoxide radicals. Adv Neurol 52:177–183

Dutka AJ, Cochanek PM, Hallenbeck JM (1989) The influence of granulocytopenia on canine cerebral ischaemia induced by air embolism. Stroke 20:390–395

Garcia JH, Kamijyo Y (1974) Cerebral infarction: evolution of histopathological changes after occlusion of a middle cerebral artery in primates. J Neuropathol Exp Neurol 33:408–421

Hatashita T, Ito M, Miyaoka M, Ishii S (1990) Chronological alterations of regional cerebral blood flow, glucose utilization and edema formation after focal ischemia in hypertensive and normotensive rats. Adv Neurol 52:29–37

Heiss WD, Rosner G (1983) Functional recovery of cortical neurones as related to degree and duration of ischaemia. Ann Neurol 14:294–301

Hossman KA (1976) Development and resolution of ischemic brain swelling. In: Pappius HM, Feindel W (eds) Dynamics of brain edema. Springer-Verlag, New York, pp 219–227
Hossman KA (1989) The pathophysiology of experimental brain edema. Neurosurg Rev 12:263–280
Hossman KA, Bloink M (1981) Blood flow and regulation of blood flow in experimental peritumoral edema. Stroke 12:211–217
Hossman KA, Olesen Y (1970) Suppression and recovery of humoral function in transient cerebral ischaemia. Brain Res 22:313–325
Ikeda Y, Long DM (1990) Comparative effects of direct and indirect hydroxyl radical scavengers on traumatic brain oedema. Acta Neurochir Suppl. 51:74–76
Imaizumi S, Woolworth V, Kinouchi H, Chen SF, Fishman RA, Chan PH (1990) Liposome-entrapped superoxide dismutase ameliorates infarct volume in focal cerebral ischaemia. Acta Neurochir Suppl. 51: 236–238
Johansson BB (1976) Water content of rat brain in acute arterial hypertension. In: Pappius HM, Feindel W (eds) Dynamics of brain edema. Springer-Verlag, New York, pp 28–31
Kane PJ, Modha P, Strachan RD, Mendelow AD, Cook S, Chambers IR (1990) The effect of immunosuppression with whole body and regional irradiation on the development of cerebral oedema in a rat model of intracerebral haemorrhage. Acta Neurochir Suppl. 51:52–54
Katayama Y, Shimizu J, Suzuki S, Memezawa H, Kashiwagi F, Kamiya T et al. (1990) The role of arachidonic acid metabolism on ischemic brain edema and metabolism. Adv Neurol 52:105–108
Klatzo I (1967) Neuropathological aspects of brain edema. J Neuropathol Exp Neurol 26:1–14
Klatzo I (1972) Pathophysiological aspects of brain edema. In: Reulen HJ, Schurmann K (eds) Steroids in brain edema. Springer-Verlag, Heidelberg, pp 1–8
Klatzo I, Chui E, Fujiwara K, Spatz M (1980) Resolution of vasogenic brain oedema. In: Cervos-Navarro J, Ferszt R (eds) Advances in neurology 28: Brain edema. Raven Press, New York, pp 359–373
Kuroiwa T, Yokfujita J, Kaneko H, Okeda R (1990) Accumulation of oedema fluid in deep white matter after cerebral cold injury. Acta Neurochir Suppl. 51:84–86
Little JR (1976) Microvascular alterations and edema in focal cerebral ischemia. In: Pappius HM, Feindel W (eds) Dynamics of brain edema. Springer-Verlag, New York, pp 236–243
Marshall LF, Bruce DA, Graham DI, Langfitt JW (1976) Alterations in behaviour, brain electrical activity, cerebral blood flow and intracranial pressure produced by triethyl tin sulphate induced cerebral edema. Stroke 7:21–25
Meinig G, Reulen HJ, Hadjidimos A, Siemon C, Bartko D, Schurmann K (1972) Induction of filtration oedema by extreme reduction of cerebrovascular resistance associated with hypertension. Eur Neurol 8:97–103
Mies G, Ishimaru S, Xie Y, Seo K, Hossman KA (1991) Ischaemic thresholds of cerebral protein synthesis and energy states following middle cerebral artery occlusion in rat. J Cerebr Blood Flow Metab 11:753–761
Nag S, Robertson DM, Dinsdale HB (1977) Cerebral cortical changes in acute experimental hypertension. Lab Invest 36:150–161
Naruse H, Tanaka K, Nishimura S, Fugimoto K (1990) A microstructural study of oedema resolution. Acta Neurochir Suppl. 51:87–89
O'Brien MD, Jordan MM, Waltz AG (1974a) Ischemic cerebral edema and the blood brain barrier. Arch Neurol 30:461–465
O'Brien MD, Waltz AG, Jordan MM (1974b) Ischemic cerebral edema. Arch Neurol 30:456–460
O'Brien MD, Halsey JH, Strong ER (1983) The effect of hypertension on ischaemic cerebral oedema in spontaneously hypertensive rats. Neurol Res 5:83–93
Oh SM, Betz AL (1991) Interaction between free radicals and excitatory amino acids in the formation of ischaemic brain oedema in rats. Stroke 22:915–921
Petito CK (1979) Early and late mechanisms of increased vascular permeability following experimental cerebral infarction. J Neuropathol Exp Neurol 38:222–234
Petito CK, Schafer JA, Plum F (1976) The blood brain barrier in experimental seizures. In: Pappius HM, Feindel W (eds) Dynamics of brain edema. Springer-Verlag, New York, pp 38–42
Rapoport SI (1976a) In: Pappius HM, Feindel W (eds) Dynamics of brain edema. Springer-Verlag, New York, p 382 (discussion)
Rapoport SI (1976b) Blood brain barrier in physiology and medicine. Raven Press, New York
Rapoport SI, Hori M, Klatzo I (1971) Reversible osmotic opening of the blood brain barrier. Science 173: 1026–1028
Reichardt M (1904/5) Zur Entsenhung des Hirndrucks. Dtsch Zeitschr Nervenheilk 28:306
Reulen HK (1976) Vasogenic brain oedema. Br J Anaes 48:721–752
Saito N, Chang C, Kawai K, Joo F, Nowak TS, Mies G et al. (1990) The role of neuroexcitation on development of blood brain barrier and oedematous changes following cerebral ischaemia and traumatic brain injury. Acta Neurochir Suppl. 51:186–188
Schurer L, Prugner U, Kempski O, Arfors K-E, Baethmann A (1990) Effects of antineutrophil serum on post-traumatic brain oedema in rats. Acta Neurochir Suppl. 51:49–51
Shinohara Y, Yamamoto M, Haida M, Yazaki K, Kurita D (1990) Effect of glutamate and its antagonist on shift of water from extra to intracellular space after cerebral ischaemia. Acta Neurochir Suppl. 51:198–200
Siesjo BK, Bendek G, Koide T, Westerberg E, Wieloch T (1985) Influence of acidosis on lipid peroxidation in brain tissue in vitro. J Cerebr Blood Flow Metab 5:253–258
Staub F, Baethmann A, Peters J, Kempski O (1990) Effects of lactacidosis on volume and viability of glial cells. Acta Neurochir Suppl. 51:3–6
Sutton LN, Bruce DA, Welsh FA, Jaggi JL (1980) Metabolic and electrophysiologic consequences of vasogenic edema. Adv Neurol 28:241–254
Symon L (1967) A comparative study of middle cerebral artery pressure in dogs and macaques. J Physiol 191: 449–465
Symon L, Pasztor E, Branston NM (1974) The distribution and density of reduced cerebral blood flow following acute middle cerebral artery occlusion. An experimental study by the technique of hydrogen clearance in baboons. Stroke 5:355–364
Symon L, Branston NM, Strong AJ (1976) Autoregulation in acute focal ischemia. Stroke 7:547–554
Tanaka R, Tanimura K, Veki K (1977) Ultrastructural and biochemical studies on ouabain induced oedematous brain. Acta Neuropathol (Berlin) 37:95–100
Yamamoto M, Haida M, Taniguchi R, Yazaki K, Kurita D, Fukuzaki M et al. (1990) Suppression of water shift into intracellular space by TA3090 measured with NMR. Acta Neurochir Suppl. 51:204–206

Concluding Comments

As in succeeding sections of this book, I will not attempt to review all aspects covered in the chapters, but instead will select isolated topics.

Reperfusion Injury

Garcia, Choi and O'Brien all raise the daunting possibility that reperfusion of the ischemic brain might adversely affect recovery. The very possibility that correcting a lack of flow might be bad flies in the face of common sense. Neurologists invariably take a neurocentric view of the body. After all, the heart and circulatory system serve to supply blood which carries fuel (mostly oxygen and sugar) to the brain, which cannot function without a nearly constant energy supply. When circulation to a brain region is blocked, even temporarily, that brain area is deprived of fuel and stops working; if the fuel deprivation lasts long enough, the starving brain region wilts, softens and eventually dies. Garcia has described in detail the morphological and biochemical changes that occur within the ischemic zone. If the problem is lack of fuel delivery, then the most logical and sensible response would be to bring in more fuel. This could be done by unblocking the obstruction, somehow augmenting flow around the blockage, or increasing the energy content of the blood that is delivered.

Rationality dictates that correcting an important deficiency must be good, and in fact, the everyday observations cited by Garcia in humans and experimental animals show that, in ordinary circumstances, reopening blocked arteries saves tissue. Garcia cites the circumstances of opening the carotid artery after temporary clipping in patients undergoing endarterectomy. In over 95% of individuals, the temporary clipping leads to no damage. In Garcia's animals, the longer the MCA is clipped, the more likely and more severe the ensuing brain damage. Only then under certain circumstances (e.g. after prolonged clipping) is there potential harm to reperfusion, but in that case, how do we know that reperfusion itself was harmful? Could it not have been a simple case of too little too late – that is, the situation was so far gone that new blood did not help much.

Reperfusion injury in the brain is a hypothesis, not a fact or an observation. Garcia comments that the original idea of reperfusion injury came from the study of tissues other than the brain. Babbs (1988), reviewing his theory of reperfusion injury, cited a study of reperfusion in rat hearts: the release of creatinine kinase occurred only after oxygen was restored, leading the authors to conclude that myocardial injury occurred during the period of reperfusion, but other interpretations are also possible. In reference to the nervous system, the notion of delayed damage began with the observation of Plum et al. (1962) that delayed brain damage occurred in 5 patients who had been rendered temporarily anoxic by a variety of insults. All were initially rendered comatose, awakened, and then later deteriorated, often severely and fatally. Dooling and Richardson (1976) made a similar observation of delayed worsening after strangulation. Delayed worsening after carbon monoxide poisoning is a well known and accepted phenomenon; in these examples, the injury was mostly biochemical and did not involve a blocked vessel.

There are data favoring the idea that a little blood flow is worse than no flow (Hossman and Kleihues 1973; Rehncrona et al. 1981; Plum 1983). The experimental situations showing that animals subjected to partial ischemia did worse than those with no circulation were those of global ischemia with hypoxia, not locally blocked arteries. Brain hypoxia without circulatory reduction does not reproduce the cellular changes produced by ischemia (Rehncrona et al. 1981) and the brain pathology caused by cardiac and circulatory arrest is quite different from that found in patients with thrombotic or embolic brain infarcts.

The discerning reader has probably discovered by now that I am not a devotee of the reperfusion injury idea, since I know of no observations that support its occurrence in human brain, and much common sense and many observations argue against the idea. Baron, in his chapter, cites PET data showing that in acute stroke patients reperfusion usually reverses blood flow and oxygen metabolism abnormalities in the ischemic region, and does not worsen the situation. Let us, however, for argument's sake, accept that reperfusion is at times bad. How might this be so? The possibilities are listed in Table 1. Among these only the first, second and fourth seem plausible to me. In the others, unwanted substances are brought into the ischemic zone, but what comes in also probably goes out. Recirculation and reperfusion also help remove and wash out 'bad humors'. In hypothetical reason 4, it is likely that more sugar causes more lactate to form, but reperfusion might help to distribute and drain away that metabolite.

Table 1. Putative effects of reperfusion (modified from Caplan 1991a)

Adverse Effects
Causes rupture of small vessels damaged by ischemia
Causes edema formation
Brings particles and cells into the area, including platelets, cholesterol crystals, clots etc.
Brings substrates into the ischemic zone, such as oxygen and sugar which will be metabolized to lactate and free radicals
Brings in and circulates chemicals formed within the ischemic zone, such as glutamate, oxygen free radicals, etc.
Brings Ca^{2+} and K^{+} into the ischemic zone, causing influx of these substances into cells through damaged membranes
Putative benefits
Brings in needed fuel and energy
Brings in growth factors and medicines given by physicians
Washes away potentially toxic substances

It seems to me that the only real risks of reperfusion are edema formation and hemorrhage. Fisher and Adams (1951) have shown in careful morphological studies at necropsy that the opening of arteries previously blocked by emboli causes reperfusion hemorrhage. We know that embolic infarcts are more often hemorrhagic than those due to in-situ thrombosis. The other situation in which infarcts become hemorrhagic is circulatory arrest with later reinstitution of effective blood flow. The initial arterial occlusion causes ischemia to capillaries, arterioles and vascular walls, in addition to the effect on neurons. Then, if blood flow is reinstituted, the fragile ischemic vessels could and do rupture, causing hemorrhage. Reperfusion occurs through passage of the embolus downstream or by effective collateralization. A dramatic increase in flow ('flooding'), as occurs when opening a very tight carotid stenosis, or removing an AVM that was receiving a high flow, increases the risk of bleeding. Increased blood pressure and decreased coagulability probably increase the risk of bleeding. Iatrogenic reperfusion using thrombolytic agents or mechanical means (surgery or angioplasty) carries a real risk of hemorrhage when brain (and presumably vascular) injury has already occurred.

Edema formation is another tangible risk of reperfusion. O'Brien, in his chapter, reviews the mechanisms by which sudden increases in flow could lead to vasogenic edema. In fact, my colleagues and I have recently seen several patients with severe unilateral brain edema after ipsilateral carotid endarterectomy (Breen et al. 1993). Edema can also occur after removal of an AVM, by exposing to high flow vessels that have been unaccustomed to such exposure in the past. Leaky vessels are also seen in other circumstances of rapid changes in blood flow or blood pressure, such as severe hypertension with encephalopathy, renal failure, eclampsia, etc.

I believe that ischemia, if severe enough, can be self-perpetuating. A cascade of events, once set in motion, can perpetuate itself. Perpetuation is likely to be biochemical and relate to the changes in excitotoxins, Ca^{2+}, pH and free radicals, as described by Garcia, Choi and O'Brien in their chapters. Blockage of the microcirculation also occurs, as described by Garcia. Once the events are set in sufficient motion, reperfusion is inadequate to stem the tide. I see no evidence that reperfusion is bad for the chemical events or the microcirculation: it just arrives too late. The fact that damage is noted during reperfusion does not necessarily mean that reperfusion is the cause. I do believe that the reperfusion of injured brain does carry a real risk of hemorrhage and edema. Fortunately, significant bleeding and edema should be easily detectable with modern imaging technology (such as CT and MRI). The subtle biochemical events that are suggested to occur because of reperfusion are much harder to study and measure. I believe that many have been seduced by the illogical and unsupported theory of reperfusion injury. I continue to be a supporter of opening the 'pipes' as soon as possible before it is too late, but I do fear and watch for the two known complications, brain hemorrhage and brain edema.

Microvascular Changes in the Ischemic Zone

The original ischemic insult, usually precipitated by the blockage of a large extracranial or intracranial artery, can cause progressive neuronal injury, ultimately with tissue necrosis. Garcia, in his contribution, uses the terms 'maturation phenomenon' and 'delayed neuronal death' to describe the progression of damage, despite reperfusion either by opening ('recanalization') of the originally blocked artery or by opening of effective collateral circulation to supply the ischemic zone. Most researchers attribute this delayed damage to biochemical events within the ischemic zone. A cascade of chemical changes and reactions, once set in motion, develops an impetus of its own which proceeds despite reperfusion. Choi elaborates on the most popular current theory of the biochemical events: the excitotoxin hypothesis of cell injury. This hypothesis considers known local changes in chemistry characterized by increased concentrations of local neurotransmitters, especially glutamate, local production of free radicals and passage of Ca^{2+}, Na^{+} and water into cells and passage of K^{+} out of cells into the intracellular fluid. Cytotoxic edema and loss of vital cell energy functions ensue, leading ultimately to neuronal death.

Often lost sight of in this biochemical scenario are the local changes that develop in the microvasculature within and adjacent to the ischemic zone. Most of the blood flow to cells within the brain resides in the small arterioles and capillary network. These vessels are beyond the resolution of the human eye and are not visible on angiograms. Garcia, in his chapter, emphasizes the occurrence of progressive changes in the microvasculature that prevent effective reperfusion of brain despite reopening of the major arterial channels. Blockage of these small vessels is due to occlusive microthrombi which form in situ. The ischemic insult releases tissue thromboplastins which activate the coagulation cascade. Endothelial damage and tissue substances also activate platelets, which begin to adhere to each other and to the vascular endothelium. Platelets become admixed with fibrin to form small fibrin–platelet thrombi ('white clots') which stick to vessels. Hypoperfusion with poor flow potentiates the formation of occlusive platelet-rich and red blood cell and fibrin-rich ('red clots') thrombi. These changes in platelet and coagulation functions will be discussed in more detail in the section devoted to blood.

Leukocytes may also have an important role in contributing to both microvascular blockage and cell injury. Leukocytes begin to migrate into ischemic zones soon after ischemic injury, and may peak at 48–72 hours after ischemia onset. Using tagged polymorphonuclear leukocytes and single photon emission computed tomography (SPECT) in experimental models, radiolabeled white blood cells can be seen to congregate in and around infarcts during the acute phase of ischemia (Wang et al. 1993). At necropsy, neutrophils can be found surrounding blood vessels and abutting on pial surfaces and within infarcts in patients dying of acute ischemic strokes. There may, at times, be a release of polymorphonuclear leukocytes into the CSF, usually causing a modest, but sometimes acute, pleocytosis (>100 neutrophils/ml). Neutrophils may have procoagulant effects: the release of proteolytic enzymes can also contribute to cell damage. Leukocytes, being larger than erythrocytes, can also add to mechanical changes in flow in microvessels.

Blocked microvascular flow probably explains why some techniques for imaging cerebral blood flow may continue to show regions of decreased blood flow despite wide patency of the arteries supplying these zones. Positron emission tomography (PET), SPECT, xenon-enhanced CT scans and diffusion-weighted and perfusion MRI scans can show large regions of decreased regional cerebral blood flow, despite normal arteries to these zones on angiography. This situation is most often found in patients with brain embolism in whom the embolus has blocked a feeding artery long enough to cause persistent ischemia and then has moved downstream. While the embolus resided within the feeding artery, the microvasculature became obstructed and remained so even after the major supply artery recanalized.

Blockage of the microvasculature explains the 'no-reflow' phenomenon observed by Ames et al. (1968) and described by Garcia in his chapter. The lack of rCBF in the ischemic zone not only accentuates ischemia but creates a therapeutic problem. How will pharmacological agents, growth factors and vital nutrients be delivered to the 'stunned' but reversibly injured cells in order to resuscitate them?

Clinical Determination of Perfusion and Metabolic Functions

In order for doctors to intervene to reperfuse brain in the clinic, they will need to know the location, nature and severity of obstruction of arteries in the macrocirculation, which includes the major extracranial and intracranial arteries, and the state of the brain – normal, stunned (that is, reversibly injured) or irreversibly damaged. In order to evaluate the likely success of reperfusion, knowledge of the state of the microcirculation is also very important. Large vessels may open spontaneously or after treatment, but effective reperfusion will not occur unless rCBF to the ischemic zone is augmented. Thus there are three important variables: patency of the large arteries; rCBF, which is determined by both macrocirculatory and microcirculatory factors; and the viability and state of brain function. These variables should be assessed at the beginning of treatment and can be monitored periodically thereafter.

Clinical Examination and History

The first step in evaluation is clinical. In many patients the treating clinician should be able to localize the symptomatic region to the right cerebral hemisphere – anterior circulation; the left cerebral hemisphere – anterior circulation; or the brain stem, cerebellum and posterior cerebral artery territories. In others, although localization to one of these zones is not possible on clinical grounds, one location can be eliminated. For example, a patient has had transient weakness of the left limbs but is normal by the time of the examination. The causative lesion could not be in the left hemisphere but must be localized to either the right hemisphere or the posterior circulation. In patients with unclear localization, CT, MRI or EEG may clarify which of the two territories is symptomatic.

Patency of Major Arteries

Once the symptomatic territory(ies) are identified, the next step is determining the patency of the large extracranial and intracranial arteries that supply the region. Ultrasound of the neck using Duplex scans (combined B-mode and pulsed Doppler) can usually, with a high degree of sensitivity and specificity, show important, severe, occlusive disease of the ICA and VA origins in the neck. These are the most frequent sites for severe atherosclerotic disease in white men. Continuous-wave Doppler exploration of the pharyngeal portion of the ICA and the second and third portions of the VA can determine direction of blood flow and suggest dissections or other diseases of this portion of the arteries (Hennerici et al. 1989; Touboul et al. 1987). Magnetic resonance angiography (MRA) is also useful in screening for significant extracranial disease, but MRA can overestimate degrees of stenosis and the VA origins are sometimes not well shown.

The intracranial arteries can be assessed well with transcranial Doppler (TCD) and MRA. TCD can detect occlusive lesions of the large basal intracranial arteries as well as giving information about the hemodynamic effects of extracranial occlusive lesions (Caplan et al. 1990; Hennerici et al. 1987). Both the anterior circulation arteries (ICA siphon, ophthalmic arteries, MCA, ACA) and the posterior circulation arteries (VAs, basilar artery) can be assessed by TCD and MRA, which are complementary. In some patients, standard dye injection catheter angiography will be needed to clarify and quantify vascular lesions. Echocardiography (preferably transesophageal) and cardiac rhythm monitoring will also be required in many patients, especially those with cardiac symptoms or abnormal heart findings or abnormal ECGs, and those in whom no vascular lesion is found. The aorta should also be considered as a possible embolic source and studied with transesophageal echocardiography (TEE) (Amarenco et al. 1992).

If an occlusive lesion is found, it can be monitored by repeated ultrasonography and MRA examinations. Ultrasound is more mobile, less expensive, and simpler to perform repeatedly. A common example would be embolism to the MCA arising from the heart or ipsilateral ICA in the neck. Monitoring by TCD should be capable of detecting passage of the embolus by comparison of sequential flow velocities.

Microvascular Flow

Blood flow in the microvasculature is more difficult to assess. As discussed by Baron in his review, PET can determine rCBF quite accurately. Unfortunately, PET is very expensive, requires extensive input from physicists and chemists, and is difficult to mobilize acutely. For these reasons, PET is not likely to gain widespread usage. Single photon emission computed tomography (SPECT) is less expensive and easier to maintain, and is much more likely to be widely disseminated. Using several available radionuclides (iodine-123-isopropyl iodoamphetamine (IMP), ^{99}Tc ethyl cysteinate dimer (ECD)), rCBF can be imaged and followed sequentially (Caplan 1991b; Holman and Devous 1992). Xenon-enhanced CT scans (Gur et al. 1982) and newer magnetic resonance techniques such as diffusion-weighted MR and perfusion MR (Fisher et al. 1992; Belliveau et al. 1991) are also capable of imaging rCBF, and also can be performed sequentially.

Status of Brain Function and Viability

The clinical examination provides an important guide to brain function. The presence of a hemiparesis or aphasia, for example, tells the clinician the brain is not working normally and suggests the likely involved area. The concordance of clinical and neuroimaging results is especially helpful in assessing the status of the brain. The absence of symptoms, normal neurological examination and normal imaging suggest full brain recovery. Neurological signs and symptoms more severe than would be predicted by the imaging findings suggest that some brain tissue is stunned but is not infarcted. Severe neurological findings and a proportionately sized ischemic lesion on CT or MRI suggests infarction, i.e. irreversible damage. Electrophysiological tests, such as EEG, brain-mapping EEG techniques and evoked response testing, can also document abnormal function (Nuwer et al. 1987; Duffy 1989).

PET scanning, as discussed by Baron, is an excellent but impractical method of directly assessing brain metabolism. Magnetic resonance spectroscopy can also provide a method of detecting various metabolites, such as lactic acid and *N*-acetylaspartate within the core and borders of ischemic lesions (Chopp et al. 1987; Graham et al. 1992). Of course, a thorough knowledge of the metabolic changes occurring during ischemia is needed to interpret the changes on MR spectroscopy.

The advent of a host of new technologies has now made it possible to test in vivo safely and sequentially the phenomena that basic scientists or pathologists and experimental physiologists have been studying in vitro and in experimental animals for many years. We should now be able to learn more about the natural history of ischemia and be able to study and document the effects of various treatments as compared to each other and to the natural history. A new era is dawning in the treatment of stroke patients.

References

Amarenco P, Duyckaerto C, Tzourio C et al. (1992) The prevalence of ulcerated plaques in the aortic arch in patients with stroke. New Engl J Med 326:221–225

Ames A III, Wright RL, Kowada M et al. (1968) Cerebral ischemia. II The no-reflow phenomenon. Am J Pathol 52:437–453

Babbs C (1988) Reperfusion injury of post ischemic tissues. Ann Emerg Med 17:1148–1157

Belliveau JW, Cohen MS, Weisskoff R et al. (1991) Functional studies of the human brain using high-speed magnetic resonance imaging. J Neuroimag 1:36–41

Breen JC, Caplan LR, Belkin M et al. (1993) Severe white matter edema as part of the hyperperfusion syndrome after carotid operation. Ann Neurol 34:289

Caplan LR (1991a) Reperfusion of ischemic brain: why and why not? In: Hacke W, del Zoppo G, Hirschberg M (eds) Thrombolytic therapy in acute stroke. Springer-Verlag, Berlin, pp 36–45

Caplan LR (1991b) Question driven technology assessment: SPECT as an example. Neurology 41:187–191

Caplan LR, Brass LM, DeWitt LD et al. (1990) Transcranial Doppler ultrasound: present status. Neurology 40:676–700

Chopp M, Frinak S, Walton DR et al. (1987) Intracellular acidosis during and after cerebral ischemia: in vivo nuclear magnetic resonance study of hyperglycemia in cats. Stroke 18:919–923

Dooling EC, Richardson EP (1976) Delayed encephalopathy after strangling. Arch Neurol 33:196–199
Duffy FH (1989) Clinical value of topographic mapping and quantified neurophysiology. Arch Neurol 46:1133–1134
Fisher CM, Adams RD (1951) Observations on brain embolism with special reference to hemorrhagic infarction. J Neuropathol Exp Neurol 10:92–94 Also published in 1987, in: Furlan AJ (ed) The heart and stroke. Springer-Verlag, New York, pp 17–36
Fisher M, Sotak CH, Minematsu K, Li L (1992) New magnetic resonance techniques for evaluating cerebrovascular disease. Ann Neurol 32:115–122
Graham G, Blamire AM, Howseman AM et al. (1992) Proton magnetic resonance spectroscopy of cerebral lactate and other metabolites in stroke patients. Stroke 23:333–340
Gur D, Wolfson S, Yonas H et al. (1982) Progress in cerebrovascular disease: local cerebral blood flow by xenon-enhanced CT. Stroke 13:752–758
Hennerici M, Rautenberg W, Schwartz A (1987) Transcranial Doppler ultrasound for the assessment of intracranial arterial flow velocity: II. Evaluation of intracranial arterial disease. Surg Neurol 27:523–532
Hennerici M, Steinke W, Rautenberg W (1989) High-resistance Doppler flow patterns in extracranial ICA dissection. Arch Neurol 46:670–672
Holman BL, Devous MD (1992) Functional brain SPECT: the emergence of a powerful clinical method. J Nucl Med 33:1888–1904
Hossman K-A, Kleihues P (1973) Reversibility of ischemic brain damage. Arch Neurol 29:375–384
Nuwer MR, Jordan SE, Ahn SS (1987) Evaluation of stroke using EEG frequency analysis and topographic mapping. Neurology 37:1153–1159
Plum F (1983) What causes infarction in ischemic brain. The Robert Wartenberg Lecture. Neurology 33:222–233
Plum F, Posner JB, Hain RF (1962) Delayed neurological deterioration after anoxia. Arch Neurol 110:18–25
Rehncrona S, Rosen I, Siesjo BK (1981) Brain lactic acidosis and ischemic cell damage. I. Biochemistry and neurophysiology. J Cerebr Blood Flow Metab 1:297–311
Touboul PJ, Mass JL, Bousser M-G, Laplane D (1987) Duplex scanning in extracranial vertebral artery dissection. Stroke 18:116–121
Wang P-Y, Kao C-H, Mui M-Y, Wang S-J (1993) Leukocyte infiltration in acute hemispheric ischemic stroke. Stroke 24:236–240

SECTION II

The Eye

Prefatory Comments

The eye is a common target of ischemic damage. Transient or persistent loss of vision in one eye is a common and distinctive manifestation of occlusive vascular disease. Occasionally, both eyes are involved together or sequentially, causing temporary or even permanent blindness. I include a chapter on eye ischemia in this text because the anatomy and physiology of the retina and optic nerve are quite distinctive from those of the brain; the anatomy of the arterial supply of the orbital contents is important clinically and is essential for an understanding of clinical eye ischemia; the distribution of vascular diseases that affect the eye is different from those that involve the brain; some vascular diseases affect the eye in specific and diagnostic patterns and the strategy and technology used to evaluate patients with eye ischemia differ from those used for brain ischemia.

Eye versus Brain Origin of Visual Symptoms

Separating visual loss of ocular origin from ischemia to the visual cortex and visual pathways within the brain is not always simple or obvious. Most patients are unaware of the distinction, and do not try to separate monocular visual symptoms from visual field defects by closing each eye or otherwise localizing the visual loss accurately. 'Man sees what he knows', said Goethe. Individuals unaware of how vision works will usually not be able to make the needed observations. Since most eye ischemia is transient, the physician is usually completely dependent on the patient's description of the symptoms. Colored patterns, especially grids in one part of vision, an accompanying sensation of discomfort or pain in the eye, altitudinal patterns of visual loss involving an upper or lower field, and abnormal funduscopy favor an ocular origin. A dark or neutral void to one side of vision, seeing half of objects, individuals or writings, visual neglect to one side accompanying motor, sensory or reflex abnormalities on the same side as the visual loss, and concurrent cognitive or behavioral symptoms or signs favor a central lesion of the visual pathways or striate cortex within the brain. Changes in the size and shape of objects (metamorphopsia), angulation of objects, visual perseverations and after-images (palinopsia) also favor brain lesions (Critchley 1951; Jacobs 1980).

Eye Ischemia versus Other Ocular Disorders

Seeing is all that the eye does. Almost any serious disease of the retina or anterior media can affect vision. Those ocular disorders that begin abruptly or cause transient or fluctuating visual loss can be easily confused with eye ischemia. In fact, because of the containment of the arterial supply within the optic nerve sheath, any swelling or compression of the nerve can impair the blood supply and lead to secondary optic nerve ischemia. This is the presumed mechanism of the transient visual obscurations reported in patients with papilledema, especially that due to pseudotumor cerebri. Congenital anomalies of the optic disc (e.g. staphylomas near the papilla) can be associated with visual loss lasting usually seconds to minutes (Seybold and Rosen 1977; Seybold 1988). Intraorbital masses, such as orbital cavernous angiomas or osteomas, can cause gaze-related transient loss of vision in the involved eye (Brown and Shields 1981). Retinal detachments and hemorrhages within the retina or fluid media can cause sudden-onset visual loss.

The most common and most important ocular disorder causing visual loss is glaucoma. In patients with glaucoma, the pressure and flow in the arteries feeding the retina and choroid must be sufficient to overcome the increased vascular resistance caused by the raised intraocular pressure. Glaucoma itself often causes eye ischemia. Knowledge of the intraocular pressure is always vital in interpreting measurements of pressure in the ophthalmic and central retinal arteries. Low perfusion pressures may be adequate when the intraocular pressure is low or average, but insufficient when glaucoma with raised pressures is present. Visual loss in patients with glaucoma is usually accompanied by some pain and redness of the eye. Visual loss in the dark when the pupil dilates is also characteristic of glaucoma. Careful funduscopic examination of the retina, optic nerve, iris and anterior media and measurement of intraocular pressure are essential in all patients with transient or persistent visual loss.

Eye Ischemia

Ischemia of the eye can be categorized according to laterality (monocular vs. binocular), time course (acute, progressive or chronic) and severity (transient and reversible vs. persistent) and by anatomy within the eye (retina, optic nerve or anterior media). Key to understanding the clinical presentations of eye ischemia is an awareness of the anatomy and physiology of the circulation of the eye. Hedges, in his chapter, reviews and illustrates the arterial supply of the retina, which is predominantly from the central retinal artery (CRA), and of the optic nerve, which is mostly from the ciliary arteries. This subject is also reviewed in more detail elsewhere (Hayreh 1963, 1988). In the clinic, retinal ischemia and optic nerve ischemia are different and are even given different names. Retinal infarcts are usually described as central retinal artery occlusions (CRAO) or branch retinal artery occlusions (BRAO). In contrast, optic nerve infarcts are usually referred to as anterior ischemic optic neuropathy (AION), or posterior ischemic optic neuropathy when the arterial lesion is retrobulbar. In CRAO and BRAO infarcts, the patient cannot see but the doctor can usually seen an abnormality with the ophthalmoscope. In ischemic optic neuropathy the patient cannot see, and the doctor may not be able to see the arterial lesion but can infer the cause from the appearance of the optic disc and the circumstances and abruptness of the visual loss and accompanying medical illnesses and risk factors. Hedges describes in detail what types of material can be seen in the retinal circulation. He also describes present laboratory investigations that can document and differentiate various sites and causes of eye ischemia. Of particular importance is color Doppler imaging, with its ability to show and quantify flow within the blood supply to the eye.

References

Brown GC, Shields JA (1981) Amaurosis fugax secondary to presumed cavernous hemangioma of the orbit. Ann Ophthalmol 13:1205–1209
Critchley M (1951) Types of visual perseveration, palinopsia, and illusory visual spread. Brain 74:267–299
Hayreh SS (1963) Arteries of the orbit in the human being. Br J Surg 50:938–953
Hayreh SS (1988) Arterial blood supply of the eye. In: Bernstein E (ed) Amaurosis fugax. Springer-Verlag, New York, pp 1–23
Jacobs L (1980) Visual allesthesia. Neurology 30:1059–1063
Seybold M (1988) Nonembolic sources of amaurosis fugax. In: Bernstein E (ed) Amaurosis fugax. Springer-Verlag, New York, pp 168–173
Seybold M, Rosen PN (1977) Peripapillary staphyloma and amaurosis fugax. Ann Ophthalmol 9:1139–1141

6. Ocular Ischemia

T. R. Hedges

The two relatively separate circulations to the eye, one supplied by the central retinal artery, the other derived from posterior ciliary arteries, tend to be affected in different ways by different diseases. More importantly, different visual signs and symptoms occur with the involvement of either vascular territory. In general, the retinal vessels are more often affected by embolic disease, which may lead to transient monocular blindness or retinal infarction. When the posterior ciliary vessels, which supply the optic nerve head, are affected by small-vessel disease, acute altitudinal loss of vision results from anterior ischemic optic neuropathy. Arteritis affects both circulations, but giant-cell arteritis more often affects the posterior ciliary arteries. In this chapter several pathophysiologic mechanisms of ocular vascular disease will be considered, following comments regarding normal ocular vascular anatomy and physiology.

Normal Ocular Vascular Anatomy and Physiology

The ophthalmic artery supplies the two major arterial systems of the eye, the central retinal artery and the ciliary arteries. The ophthalmic artery arises from the internal carotid artery, occasionally within the cavernous sinus (10%), but usually it leaves the internal carotid artery inside the dura just above the cavernous sinus and below the optic nerve (Hayreh 1963). The intracranial segment of the ophthalmic artery is 2–7 mm in length and sends small branches to the optic nerve and chiasm before it penetrates the dura as it enters the optic canal. It runs below the optic nerve within the canal or, on rare occasions, through its own canal. After the artery has entered the orbit, about 15–20 mm behind the globe, it may cross over the nerve laterally (80%) (Hayreh 1963) and travel medially, or it may cross under and loop up the medial aspect of the optic nerve before coursing medially. Just after the ophthalmic artery has crossed the optic nerve, it sends a large lacrimal artery laterally and a small trochlear branch medially (Fig. 6.1). The external carotid artery supplies collateral branches to the ophthalmic artery medially through the anterior and posterior ethmoidal and supratrochlear arteries. External carotid anastomoses with the lateral orbit occur by way of the lacrimal and meningeal arteries through the superior orbital fissure.

Just before the ophthalmic artery crosses the optic nerve it gives off the central retinal artery, which enters into the inferior aspect of the optic nerve 10–15 mm posterior to the globe. The central retinal artery briefly crosses the subarachnoid space, penetrates the optic nerve and travels into the eye. (Fig. 6.2). It sends a few twigs to the pial vascular network surrounding the nerve and a few radial branches to the retrobulbar optic nerve before entering the eye, where it divides into superior and inferior branches which radiate out to supply the inner two-thirds of the retina. The central retinal vein travels with the central retinal artery through the optic nerve into the orbit, where it eventually joins the superior ophthalmic vein.

The posterior ciliary arterial system is more complex than the retinal system. However, it is just as important because it supplies the optic nerve head, the choroid and the anterior segment of the eye. Two or three short posterior ciliary arteries leave the ophthalmic artery as it crosses the optic nerve 15–20 mm posterior to the globe. The short posterior ciliary arteries penetrate the sclera, where the dural sheath of the optic nerve joins the globe (Fig. 6.2). Here, major branches send blood into the choroid, which has a large volume of blood flow.

Choroidal blood flow has been calculated to be 2000 ml/min/100 g, compared to 100 ml/min/100 g in the retinal arteries (Alm and Bill 1973). This large amount of blood flow provides a cooling effect as well as oxygen and nutrition to the outer one-third of the retina, including the rods and cones. Other branches of the posterior ciliary arteries reach around the laminar portion of the optic nerve, where they incompletely anastomose with each other. Just behind the globe the posterior ciliary arteries change their medial and lateral orientations so that one short posterior ciliary artery tends to supply the superior optic nerve and the other supplies the inferior optic nerve (Olver et al. 1990). At the level of the optic nerve they also anastomose with the pial plexus of the optic nerve sheath and the choroid.

The superficial optic nerve head receives blood from the recurrent arterioles of the central retinal artery, whereas the main portions of the anterior optic nerve head receive blood almost entirely from the short posterior ciliary arteries (Fig. 6.2). In about 50% of individuals the short posterior ciliary arteries provide a cilioretinal branch, which usually penetrates the sclera temporal to the optic nerve to supply the retina in the papillomacular region. The long posterior ciliary arteries leave the ophthalmic artery just anterior to the point where they cross the optic nerve. They penetrate the globe laterally and medially to supply the ciliary body. The anterior ciliary arteries travel with the rectus muscles to supply the anterior surface of the eye.

The retinal arterioles are endarteries, and lack anastomoses once they enter the eye. However, retinal ischemia is tolerated somewhat better than brain ischemia. Irreversible retinal damage does not occur for at least 1 hour after ligation of the central retinal artery (Hayreh and Weingast 1980) whereas irreversible damage to the brain occurs after 6–8 minutes of vascular occlusion. There is a blood–retinal barrier, similar to the blood–brain barrier, which is partly due to tight junctions between endothelial cells of the retinal capillaries (Cunha-Vaz et al. 1966), although retinal capillaries tend to be smaller than cerebral capillaries (Cogan and Kuwabara 1984).

One of the unique aspects of the ocular circulation concerns the variable pressure gradients through which the various blood vessels supplying and draining the eye pass. Once the ophthalmic artery leaves the intracranial portion of the internal carotid artery, it travels through the orbit under very little pressure. The central retinal artery travels through the optic nerve sheath, where it is subjected to spinal fluid pressure, and when it enters the eye it is surrounded by intraocular pressure, which is usually in the range of 10–20 mmHg. The pressure within the arteries in the eye has been estimated at 65–70 mmHg in the upright position (Bill 1968). Retinal perfusion pressure is maintained primarily by autoregulation (Riva and Loebl 1977), whereas choroidal blood flow is primarily regulated by a rich supply of autonomic vasoactive nerves (Ruskell 1971). The central retinal artery receives some sympathetic input within the optic nerve up to the lamina cribrosa, but not within the eye (Laties 1967). Central retinal arterial autoregulation may be influenced by aging (Tachibana et al. 1982).

A variety of physical conditions change uveal blood flow, particularly within the ciliary body and the iris. However, the regulatory mechanisms involved in the retinal circulation are organized to maintain a very constant blood flow (Alm and Bill 1970), particularly with respect to changes in blood pressure and intraocular pressure. Increased concentrations of blood oxygen will produce vasoconstriction within the retina. Increasing concentrations of carbon dioxide in the blood cause a considerable amount of dilation of low-resistance retinal vessels, which can have a significant impact on retinal blood flow (Alm and Bill 1970). Carbon dioxide in the blood also increases choroidal blood flow (Dollery et al. 1966). It is presumed that the effects of carbon dioxide can be used in the management of patients with central retinal artery occlusion.

Clinical Methods of Investigating Ocular Blood Flow

Clinical methods of measuring blood supply to the eye include fundus fluorescein angiography, laser Doppler blood flow measurements of the retinal blood vessels, ocular plethysmography (OPG) and, more recently, color Doppler imaging with spectral analysis. Ophthalmodynamometry (ODN) is based on the appearance of pulsation and cessation of blood flow in the central retinal artery as pressure is applied to the eye with the instrument. This results in reduced blood flow through the ophthalmic artery, and the readings relate more to pressure in the internal carotid rather than in the retinal arteries (Alm 1992). This test has been used in the past to attempt to predict carotid artery stenosis. However, significant internal carotid artery disease may not result in a significant drop in ophthalmic artery blood pressure because of collateral supply from the external carotid artery (Hedges and Weinstein 1964).

Oculoplethysmography (OPG) and ocular pneumoplethysmography (OP-Gee) measure the ocular pulse, which can be used to determine the ophthalmic artery systolic pressure. Air-filled cups are applied laterally to the sclera of both eyes. A partial vacuum applied to the cups raises intraocular pressure. The vacuum is then slowly released until a pulse amplitude appears which can be measured. Pulse pressure differences between the two eyes may reflect carotid artery stenosis. Ophthalmic artery pressures are assumed to be at least 66% of normal systolic blood pressure (Gee 1976).

Intra-arterial angiography does not show the orbital circulation well. Although it can be visualized in many patients, inability to demonstrate the ophthalmic artery does not necessarily indicate an abnormality within that system. The central retinal and ciliary vessels are too small to be seen radiographically.

Color Doppler imaging now allows for visualization of the ophthalmic, central retinal and posterior ciliary arteries (Fig. 6.3). The central retinal, vortex and superior ophthalmic veins can also be seen. To date, the size of these vessels cannot be accurately measured, but velocities of blood flow can be quantified using spectral analysis (Erickson et al. 1989; Guthof et al. 1991; Lieb et al. 1991a). A variety of conditions that affect orbital blood vessels have been studied with Doppler imaging, including carotid artery occlusions with reversal of ophthalmic artery flow (Lieb et al. 1991b) (Fig. 6.3), carotid cavernous sinus fistulas (Flaherty et al. 1991) and central retinal artery occlusion (Hedges et al. 1993) (Fig. 6.4). It is expected that this technique will provide more understanding of the effects of a variety of disorders of ocular blood flow.

Mechanisms and Causes of Eye Ischemia

Embolism

Many different types of embolic material can enter the ocular circulation (Table 6.1). If they travel into the retinal blood vessels their characteristics can be determined using an ophthalmoscope.

Usually, cholesterol emboli do not cause occlusion of the retinal vessels, although they may be accompanied by platelet–fibrin material which can cause branch or central retinal artery occlusion (Hollenhorst 1961). Following impaction of a cholesterol embolus within a retinal arteriole, progressive sclerosis of a branch retinal artery with subsequent retinal infarction may occur (Dark and Rizk 1967; Muci-Mendoza et al. 1980). Platelet–fibrin material can occlude multiple branch retinal arteries and may be seen to move through the retina during a period of up to 1 hour (Fisher 1959). Calcific emboli usually lead to occlusion of a large branch retinal artery at one of the bifurcations or in the prelaminar area, with subsequent retinal infarction (Arruga and Sanders 1982). Prelaminar calcific embolization can be diagnosed using B-scan ultrasonography and color Doppler imaging (Sadun et al 1991; Sergott et al. 1992; Hedges et al. 1993) (Fig. 6.4).

Emboli to the posterior ciliary arteries usually cause no visible ischemic changes in the optic nerve head. In one pathologically documented case, in which the central artery was occluded by a calcific embolus with subsequent retinal infarction, a calcific embolus was seen to lodge within one posterior ciliary artery without causing optic nerve damage (Baghdassarian et al. 1970). Only a few examples of patients with presumed anterior ischemic optic neuropathy due to emboli have been reported (Lieberman et al. 1978; Tomsak 1985).

Symptoms associated with embolization to retinal arteries include acute and permanent loss of vision, and monocular visual field defects

Table 6.1. Retinal emboli

Material	Source	Ophthalmoscopic appearance
Cholesterol	Atheroma from carotid artery and aortic arch	Glistening, bright yellow Hollenhorst plaques, usually without retinal injury
Platelet–fibrin	Stenosis of carotid, ophthalmic artery Abnormal heart valve Dyskinetic myocardium (artrial fibrillation/infarct) Defects in heart wall	Gray-white, conform to lumen, usually with retinal damage
Calcium	Calcified aortic/mitral valves	White, discrete, usually with retinal infarct
Antigen–antibody complex	Endocarditis, fungemia, AIDS	Cotton-wool spots, dot hemorrhages, Roth's spot
Fat	Fractured long bones	Multiple small retinal infarcts
Other: Tumor (cardiac myxoma), air, mercury, metal, periocular depot steroid, talc, amniotic fluid, angiogram catheter tips, silicone, glass beads, cloth particles		

corresponding to the area of infarcted retina. Emboli are seen in 11%–40% of patients with central retinal artery occlusion and 60%–70% of patients with branch retinal artery occlusion (Savino et al. 1977; Wilson et al. 1979). When the emboli have the appearance of cholesterol plaques carotid artery disease is considered most likely, although aortic atheromas contain similar material. Carotid artery disease is associated with retinal stroke in a majority of patients (45%–85%) (Hedges et al. 1985; Merchut et al. 1988). Usually these are patients with branch retinal artery occlusion. Central retinal artery occlusion is associated with carotid artery stenosis less frequently than branch retinal artery occlusion (11%–46%) (Lorentzen 1969; Kollarits et al. 1972; Appen et al. 1975; Tomsak et al. 1979; Savino et al. 1977). In other patients, the possibility of arteriosclerosis directly involving the central retinal artery or the ophthalmic artery remains a possibility.

We have identified four non-arteritic types of central retinal artery occlusion, using color Doppler imaging (Hedges et al. 1993). The first is emboli from the heart (Fig. 6.4); the second type is due to emboli associated with carotid stenosis; the third is associated with reversal of blood flow through the ophthalmic artery in patients with complete carotid artery occlusion (Fig. 6.5); the fourth type is due to focal intrinsic disease within the central retinal and ophthalmic arteries without carotid or cardiac disease. The role of the ophthalmic artery in ocular occlusive disease remains elusive, but hopefully color Doppler imaging will provide more insight into the role of the ophthalmic artery in a variety of conditions. Central retinal artery occlusion may occur by at least two other known mechanisms: inflammation, especially in patients with giant-cell arteritis, and vasospasm.

Transient monocular blindness (TMB), lasting from seconds to minutes, most often occurs from emboli to the retinal blood vessels (Amaurosis Fugax Study Group 1990). This is based on observation of platelet–fibrin emboli traveling through the eye in patients with TMB clinically (Fisher 1959) and pathologically (McBrien et al. 1963), as well as cholesterol plaques in retinal arteriolar bifurcations clinically (Hollenhorst 1961) and pathologically (David et al. 1963). Transient embolic occlusion of the retinal blood vessels is certainly likely in patients with repeated spells of varying duration and affecting different parts of their visual field. It is still not clear whether emboli can cause repeated spells of visual loss of exactly the same duration affecting the exact same portion of the visual field. Other mechanisms must be considered.

Hemodynamic–Hypoperfusion Syndromes

Transient hemostasis remains a possible pathophysiological mechanism for some spells of transient visual loss, especially those which are repetitive and stereotyped. This idea is based on similar reasoning to that of Fisher, who listed 37 reasons why cerebral TIAs of the same duration occurring in the same arterial distribution could

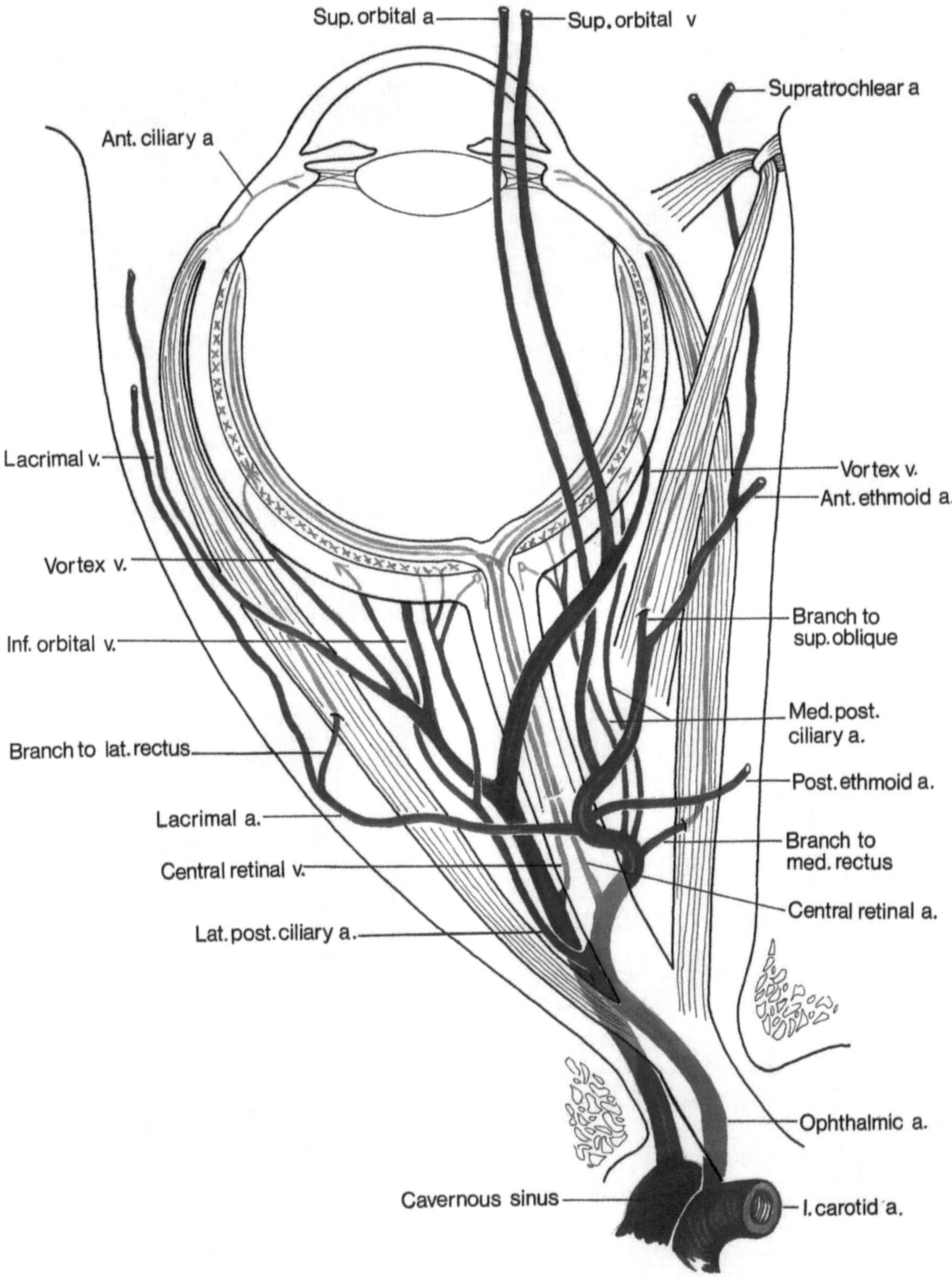

Fig. 6.1. Orbital blood vessels.

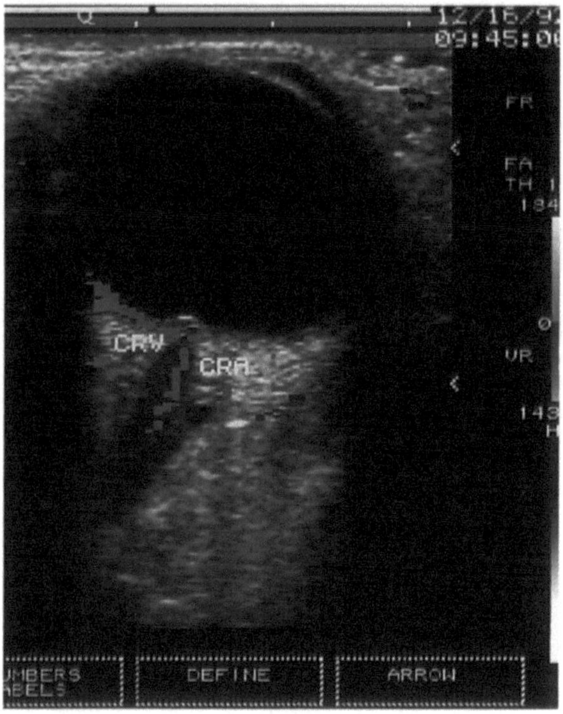

Fig. 6.3.

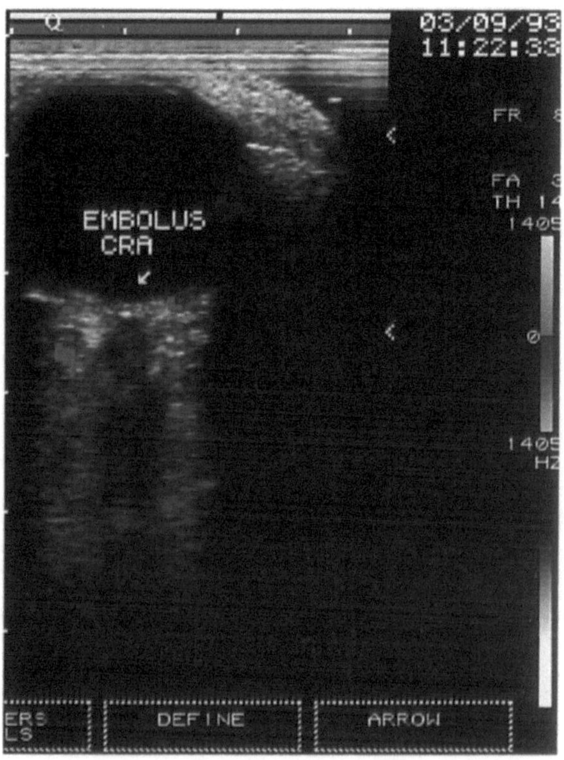

Fig. 6.4.

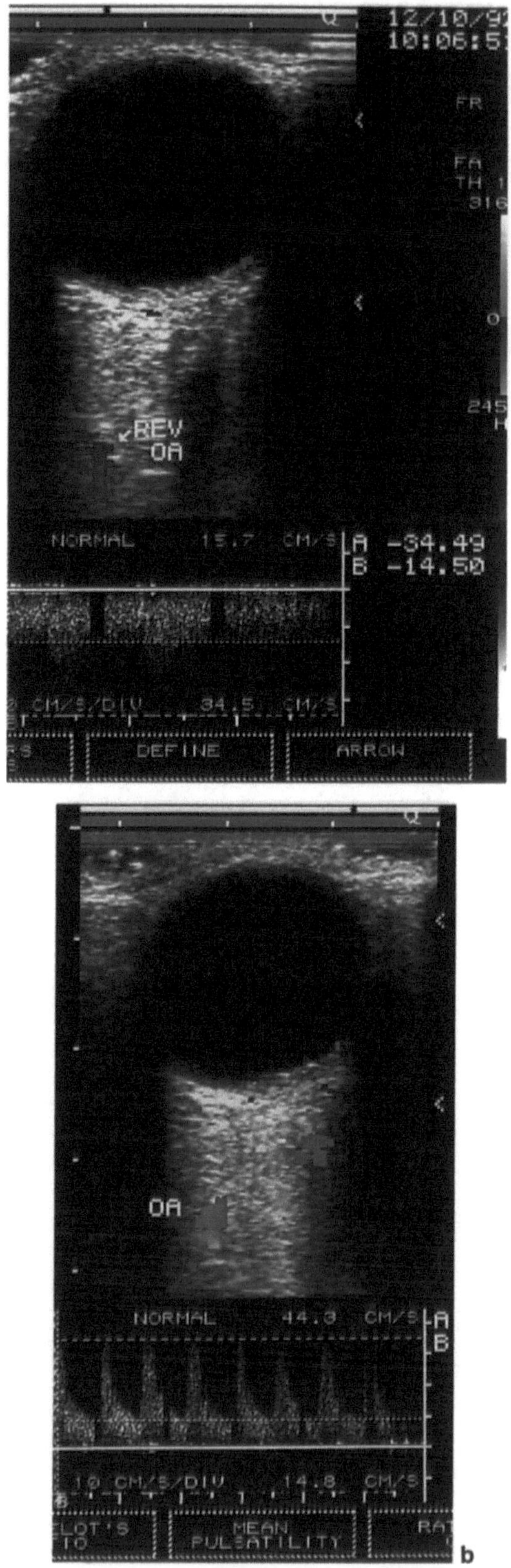

Fig. 6.5.

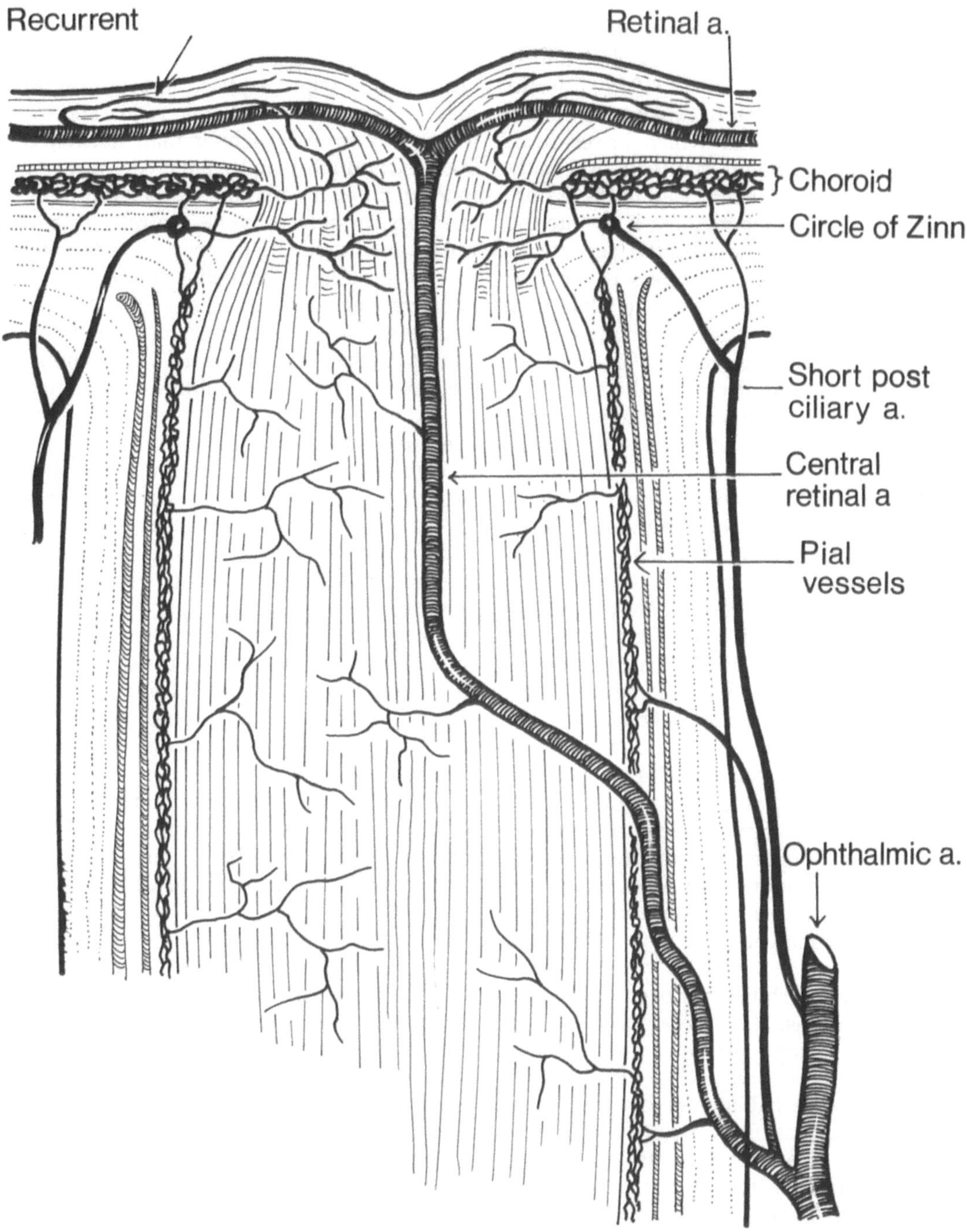

Fig. 6.2. Blood supply to the anterior portion of the optic nerve.

Fig. 6.3. Color Doppler image of a normal orbit. CRA, central retinal artery; CRV, central retinal vein.

Fig. 6.4. Color Doppler image showing embolus in the central retinal artery.

Fig. 6.5. Color Doppler image **a** with spectral waveform indicating reversal of flow in the ophthalmic artery in a patient with preocclusive carotid artery stenosis who had transient monocular blindness. **b** Restoration of normal flow and spectral waveform in the ophthalmic artery following carotid endarterectomy.

not be embolic (Fisher 1976). The absence of embolic material in retinal vessels during episodes of TMB has been reported by Hollenhorst (1965) as well as Dyll and David (1966). Carotid artery stenosis, in addition to causing embolization of thrombus material which arises from irregularities within the arterial wall, especially in areas of ulceration, can also lead to the propagation of thrombus which may reach up into the siphon and send embolic material repeatedly into the ophthalmic and retinal vessels. However, this mechanism does not readily explain why patients have identical episodes in one eye on many repeated occasions. Some variability of symptoms would be predicted if emboli were the cause of transient visual or neurologic symptoms, and indeed many patients with carotid artery disease do have variable symptoms (Marshall and Meadows 1968; Pessin et al. 1977); Goodwin et al. 1984). Embolic sources are often not found in patients with TMB, many of whom have no carotid artery lesions. In one study of 33 patients, only 19 (57%) had hemodynamically significant carotid stenosis and 8 (25%) had normal carotids (Pessin et al. 1977). What remains to be determined is whether TMB can be caused by local disease within the ophthalmic artery itself, either at its origin from the carotid or along its course in the orbit, especially within the central retinal artery. Recently Sergott et al. (1992) have shown several mechanisms of transient visual loss using color Doppler imaging, including an embolus within the central retinal artery where it enters the optic nerve, an arteriovenous malformation in the retrobulbar space, and ophthalmic artery stenosis.

Another issue which has not been resolved is whether transient visual spells can be caused by insufficiency of the optic nerve or uveal circulation, in addition to reduced retinal blood flow. It is well known that patients with very severe carotid stenosis or occlusion may have a variety of ocular findings and symptoms which may reflect hypoperfusion. In such individuals with chronic ocular ischemic syndromes, the microvascular circulation of the retina appears to break down. This leads to small hemorrhages in the midperiphery of the retina and, in some cases, neovascularization of the optic disc, retina and iris (Hedges 1963; Kearns and Hollenhorst 1963). In some ways these changes resemble the global ischemic process seen in diabetes. When iris neovascularization occurs, glaucoma may soon follow due to obstruction of aqueous outflow by new blood vessel growth from the iris on to the cornea over the trabecular meshwork. Symptoms associated with chronic ocular ischemia include TMB similar to that seen with carotid stenosis. Other patients describe dimming of the vision when exposed to bright light. This may be because the ischemic retina takes longer to recover from bleaching than normal retina. Such individuals usually have very low ophthalmodynamometry readings, and even light digital pressure may provoke central retinal artery pulsations. Such patients may also be susceptible to any stimulus which diverts blood from the collateral supply. Heating of the face in patients with complete carotid artery occlusion has been shown to cause loss of normal vision and normal electroretinographic amplitudes, presumably by diverting blood away from an underperfused eye (Ross-Russell and Page 1983).

Slight alterations in systemic blood pressure probably do not account for significant transient visual symptoms. However, some people, especially young individuals, can have binocular or monocular dimming of vision lasting less than 30 seconds upon standing quickly after bending or stooping for long period of time. These so-called transient visual obscurations resemble those seen in patients with optic disc swelling, especially papilledema. Some healthy young people who have brief 'dim-outs' appear to have congenitally crowded optic nerve heads (Sadun et al. 1984). It is of passing interest to note that patients who develop non-arteritic anterior ischemic optic neuropathy also tend to have more crowded optic nerve heads (Doro and Lessell 1985).

Severe hypotension can affect the ocular circulation. This may happen with massive blood loss due to trauma, gastrointestinal hemorrhage or abdominal aortic aneurysm repair. However, more recently, coronary artery bypass grafting has increased our awareness of this phenomenon. In some patients, central retinal artery occlusion or anterior ischemic optic neuropathy occurs, but the retrobulbar optic nerves appear to be most susceptible (Johnson et al. 1987; Rizzo and Lessell 1987). In such individuals visual loss is noted immediately or soon after recovering from surgery. The ocular fundi initially appear normal, or there may be minimal optic nerve head swelling. However, soon afterwards the optic nerves become pale. Some recovery of vision usually occurs. Whether the effect on the optic nerves is simply due to systemic hypotension or the additional effects of the pump-oxygenator, associated anemia, hypothermia or a combination of factors, remains unknown (Chisholm 1969; Sweeney et al. 1982).

Hematologic Disorders

Coagulopathy and hyperviscosity can lead to retinal and choroidal arterial occlusions (Levine et al. 1988), but retinal venous occlusions are more commonly due to disorders of the blood. However, localized abnormalities of the veins are also thought to be of primary importance. Branch retinal vein occlusion commonly occurs at arteriovenous crossing points, and is associated with arteriovenous nicking in other areas of the retina (Frangieh et al. 1982). Central retinal vein occlusion appears to be due to thrombosis in the prelaminar area in the optic nerve, where the central retinal vein is next to the central retinal artery. Contributing factors include turbulence from indentation by a nearby sclerotic central retinal artery, increased intraocular pressure and/or hematologic factors, such as increased immunoglobulin M. Pathologically, there may be inflammation within the walls of the vein (Green et al. 1981). Experimental studies of retinal vein occlusion have demonstrated the pathophysiological role of disease within the nearby central retinal artery. Occlusion of the central retinal vein alone produced venous congestion alone, whereas occlusion of both the central retinal vein and the central retinal artery was required to demonstrate the typical 'blood and thunder' appearance of retinal vein occlusion (Hayreh et al. 1978). Although the association of hematological disease and retinal vein occlusion is well established (Dodson et al. 1982; Peduzzi et al. 1962), the association of central retinal vein occlusion and carotid artery disease remains controversial (Brown et al. 1984).

Chronic hypertension is a well known predisposing factor for retinal vein occlusion, primarily with regard to the effects on retinal arterioles and nearby veins, as noted above. Acute, severe hypertension causes fibrous endarteritis, with fibrinoid necrosis of precapillary retinal arterioles (Ashton 1972). This results in multifocal retinal and optic nerve ischemia (Garner et al. 1975).

Small-Vessel Disease

Apart from diabetic retinopathy, the major consequence of small-artery disease on the eye is anterior ischemic optic neuropathy in which a segment of the optic nerve is infarcted. This is considered by many to be the ocular counterpart of brain lacunes. As mentioned earlier, the retrolaminar portion of the optic nerve may be considered a watershed zone supplied secondarily by the central retinal artery, posterior ciliary arteries and pial vessels. Risk factors for anterior ischemic optic neuropathy are the same as those for small-vessel occlusion in the brain, and include diabetes and hypertension (Beri et al. 1987). The peak incidence of anterior ischemic optic neuropathy is between 60 and 70 years of age (Boghen and Glaser 1975). The onset of dysfunction is often acute, or discovered by affected patients in the morning. The frequency of occurrence of anterior ischemic optic neuropathy in the morning may be related to a nocturnal drop in systolic blood pressure accompanied by a nocturnal increase in intraocular pressure, which occurs as a normal diurnal variation (Hayreh 1987; 1993, personal communication). This possibility has also been raised by others (Katz et al. 1990).

The typical appearance of the optic disc in anterior ischemic optic neuropathy is one of swelling and peripapillary hemorrhage. Visual field defects reflect sectorial involvement of the optic nerve head, and tend to be altitudinal. This type of visual loss reflects the altitudinal distribution of arteriolar branches of the posterior ciliary artery after they enter the retrolaminar portion of the optic disc (Olver et al. 1990).

An interesting predisposing factor for the development of anterior ischemic optic neuropathy also appears to be related to optic disc anatomy: individuals with more crowded optic nerves tend to be predisposed to anterior ischemic optic neuropathy (Beck et al. 1987).

In some patients, anterior ischemic optic neuropathy is progressive with the decline in vision occurring during a period of up to 6 weeks (Knox and Duke 1971; Borchert and Lessell 1988). In this group of individuals optic nerve sheath fenestration may improve the outcome, perhaps by preventing edema following infarction from further compromising the optic nerve blood supply and axons (Sergott et al. 1989). In 30%–40% of patients with anterior ischemic optic neuropathy, the opposite eyes become affected (Beck et al. 1987). However, recurrence within a previously affected eye is rare.

Posterior ischemic optic neuropathy may also be caused by small-vessel disease, but this is a clinical diagnosis which is difficult to make with certainty. Occasionally, elderly patients who have acute monocular loss of vision with clinical findings indicating a retrobulbar optic neuropathy may be presumed to have ischemia. However, in many patients it is difficult to distinguish this from inflammatory disease (Hayreh 1981).

Until better diagnostic imaging techniques become available, the diagnosis of posterior ischemic optic neuropathy will remain difficult to establish clinically.

Inflammatory Disease

The most important and most common vasculitis affecting the eye is temporal or giant-cell arteritis. Up to 50% of untreated patients with giant-cell arteritis lose vision, 90% due to anterior ischemic optic neuropathy and 10% due to central retinal artery occlusion (Miller 1991). Some patients with ischemic optic neuropathy may have prodromal transient visual loss. Branch retinal artery occlusion and occipital infarcts from vertebrobasilar arterial inflammation are less common forms of visual loss in patients with giant-cell arteritis.

Unlike the non-arteritic forms of anterior ischemic optic neuropathy, there have been more pathologic studies of patients with anterior ischemic optic neuropathy due to temporal arteritis (Crompton 1959; Spencer and Hoyt 1960; Henkind et al. 1970). These show inflammation of the ophthalmic, posterior ciliary and central retinal arteries. Apparently it is the posterior ciliary arteries that suffer the most, leading to anterior ischemic optic neuropathy. It is of some interest from a pathophysiologic point of view that, occasionally, visual loss can be recovered if steroids are administered promptly, but hemodynamic alterations, documented in the ophthalmic, posterior ciliary and central retinal arteries by color Doppler imaging, show no change following steroid administration (Sergott 1992, personal communication). When properly given, however, steroids effectively prevent progression and subsequent ischemic episodes of visual loss. In the future, color Doppler imaging may allow for the differentiation between non-arteritic and arteritic forms of ocular ischemia.

Other vasculitic conditions may affect vision, especially when the ocular circulation is compromised. Both anterior ischemic optic neuropathy and retinal artery occlusion can occur in patients with systemic lupus erythematosus (Lessell 1979; Jabs et al. 1986) and polyarteritis nodosa (Newman et al. 1974). A variety of inflammatory conditions affecting the ocular circulation are listed in Table 6.2.

Table 6.2. Ocular inflammatory vascular disorders

Systemic lupus erythematosus
Polyarteritis nodosa
Microangiopathy of the brain and retina
Churg–Strauss
Takayasu's arteritis
Behçet's disease
Scleroderma
Syphilis
Wegner's granulomatosis
Orbital mucormycosis
Giant-cell arteritis
Herpes zoster ophthalmicus

Vasospasm

As a cause of transient loss of vision, vasospasm of the retinal blood vessels has recently gained attention. There have been several clinical observations of transient segmental narrowing of the retinal blood vessels (Burger et al. 1991). Exercise-induced migrainous loss of vision can occur in one eye as well as in homonomous visual fields (Imes and Hoyt 1989). Young people have been reported to have retinal artery occlusion in association with migraine (Coppeto et al. 1986; Newman et al. 1989), and others have described anterior ischemic optic neuropathy associated with migraine (McDonald and Sanders 1971; Cowan and Knox 1982).

More recently, the possibility of vasospasm as a cause of TMB has been supported by cases of apparent reversal of visual loss by medication, including nitroglycerin (Kuritzky 1990) and nifedipine (Winterkorn and Teman 1991). There is experimental evidence that blood flow through the retina and choroid, along with electroretinographic responses to light, can be reduced by systemic administration of serotonin in atherosclerotic animals (Williams et al. 1989). Thus, vasospasm involving the ocular (retinal and ciliary) circulations may occur in migraine patients, and may cause TMB in some patients with atherosclerosis, especially if patients have both conditions.

Vascular Anomalies

Anomalous circulation to the eye may be associated with transient visual loss. Optic nerve anomalies, especially congenital crowding of the optic nerve, may be associated with transient visual obscuration, usually lasting less than 1 minute (Beck et al. 1987). Although anatomical effects on the axons alone may be the cause of visual loss, limitations of blood flow through the associated anomalous vasculature may be the cause.

Amaurosis fugax has been reported in association with anomalous origin of the ophthalmic artery from the middle meningeal artery (Weinberg et al. 1981). Sergott has also reported transient loss of vision in a patient with an apparent arteriovenous shunt within the orbit (Sergott et al. 1992). Central retinal artery occlusion has been observed in a patient with moya moya disease (Chace and Hedges 1984).

References

Alm A (1992) Ocular circulation. In: Hart WM Jr (ed) Adler's physiology of the eye. Saunders, Philadelphia, p 198

Alm A, Bill A (1970) Blood flow and oxygen extraction in the cat uvea at normal and high intraocular pressures. Acta Physiol Scand 80:19–28

Alm A, Bill A (1973) Ocular and optic nerve blood flow at normal and increased intraocular pressures in monkeys (*Macaca irus*): a study with radioactively labelled microspheres including flow determinations in brain and some other tissues. Exp Eye Res 15:15–29

Amaurosis Fugax Study Group (1990) Current management of amaurosis fugax. 21:201–208

Appen RE, Wray SH, Cogan DG (1975) Central retinal artery occlusion. Am J Ophthalmol 79:374–381

Arruga J, Sanders MD (1982) Ophthalmologic findings in 70 patients with evidence of retinal embolism. Ophthalmology 89:1336–1347

Ashton N (1972) The eye in malignant hypertension. Trans Am Acad Ophthalmol Otolaryngol 76:17–40

Baghdassarian SA, Crawford JB, Rathbun JE Jr (1970) Calcific emboli of the retinal and ciliary arteries. Am J Ophthalmol 69:372–375

Beck RW, Servais GE, Hayreh SS (1987) Anterior ischemic optic neuropathy. IX. Cup-to-disc ratio and its role in prognosis. Ophthalmology 94:1503–1508

Beri M, Klugman MR, Kohler JA, Hayreh SS (1987) Anterior ischemic optic neuropathy. VII. Incidence of bilaterality and various influencing factors. Ophthalmology 94:1020–1028

Bill A (1968) Capillary permeability to and extravascular dynamics of myoglobin, albumin and gammaglobulin in the uvea. Acta Physiol Scand 73:204–219

Boghen DR, Glaser JS (1975) Ischaemic optic neuropathy. Brain 98:689–708

Borchert M, Lessell S (1988) Progressive and recurrent nonarteritic anterior ischemic optic neuropathy. Am J Ophthalmol 106:443–449

Brown GC, Shah HG, Magargal LE et al. (1984) Central retinal vein obstruction and carotid artery disease. Ophthalmology 91:1627–1633

Burger SK, Saul RF, Selhorts JB, Thurston SE (1991) Transient monocular blindness caused by vasospasm. New Engl J Med 325:87–873

Chace R, Hedges TR III (1984) Retinal artery occlusion due to moyamoya disease. J Clin Neuro-ophthalmol 4:31–34

Chisholm IA (1969) Optic neuropathy of recurrent blood loss. Br J Ophthalmol 53:289–295

Cogan DG, Kuwabara T (1984) Comparison of retinal and cerebral vasculature in trypsin digest preparations. Br J Ophthalmol 68:10–12

Coppeto JR, Lessell S, Sciarra R, Bear L (1986) Vascular retinopathy in migraine. Neurology 36:267–270

Cowan CL Jr, Knox DL (1982) Migraine optic neuropathy. Ann Ophthalmol 14:164–166

Crompton MR (1959) The visual changes in temporal (giant-cell) arteritis. Report of a case with autopsy findings. Brain 82:377–390

Cunha-Vaz JG, Shakib M, Ashton N (1966) Studies on the permeability of the blood–retinal barrier. Br J Ophthalmol 50:441–453

Dark AJ, Rizk SN (1967) Progressive focal sclerosis of retinal arteries: a sequel to impaction of cholesterol emboli. Br Med J 1:270–273

David NJ, Klintworth GK, Friedberg SJ et al. (1963) Fatal atheromatous cerebral embolism associated with bright plaques in the retinal arterioles: report of a case. Neurology 13:708–713

Dodson PM, Spalton DJ, Hamilton AM et al. (1982) Retinal vein occlusion and the prevalence of lipoprotein abnormalities. Br J Ophthalmol 66:161–164

Dollery CT et al. (1966) Focal retinal ischaemia, I: Ophthalmologic and circulatory changes in focal retinal ischaemia. Br J Ophthalmol 50:283–324

Doro S, Lessell S (1985) Cup–disc ratio and ischemic optic neuropathy. Arch Ophthalmol 103:1143–1144

Dyll L, David NJ (1966) Amaurosis fugax: funduscopic and photographic observations during an attack. Neurology 16:135–138

Erickson SJ, Hendrix LE, Massaro BM et al. (1989) Color Doppler flow imaging of the normal and abnormal orbit. Radiology 173:511–516

Fisher CM (1959) Observations of the fundus oculi in transient monocular blindness. Neurology 9:333–347

Fisher CM (1976) Transient ischemic attacks – an update (Discussion). In: Scheinberg P (ed) Cerebrovascular diseases: Tenth Princeton Conference. Raven Press, New York, pp 50–53

Flaherty PM, Lieb WE, Sergott RC, Bosley TM, Savino PJ (1991) Color Doppler imaging – a new noninvasive technique to diagnose and monitor carotid cavernous sinus fistulas. Arch Ophthalmol 109:522–526

Frangieh GT, Green WR, Barraquer-Somers E et al. (1982) Histopathologic study of nine branch retinal vein occlusions. Arch Ophthalmol 100:1132–1140

Garner A, Ashton N, Tripathi R, Kohner EM, Bulpitt CJ, Dollery CT (1975) Pathogenesis of hypertensive retinopathy: an experimental study in the monkey. Br J Ophthalmol 59:3–44

Gee W, Oller DW, Wiley J (1976) Noninvasive diagnosis of carotid occlusion by ocular pneumoplethysmography. Stroke 7:18–21

Goodwin JA, Gorelick P, Helgason C (1984) Transient monocular visual loss: amaurosis fugax versus migraine. Neurology 34:246

Green WR, Chan CC, Hutchins GM et al. (1981) Central retinal vein occlusion: a prospective histopathologic study of 29 eyes in 28 cases. Retina 1:27 Trans Am Ophthalmol Soc 79:371

Guthoff RF, Berger RW, Winkler P, Helmke K, Chumbley LC (1991) Doppler ultrasonography of the ophthalmic and central retinal veins. Arch Ophthalmol 109:532–536

Hayreh SS (1963) Arteries of the orbit in the human being. Br

J Surg 50:938–953
Hayreh SS (1981) Posterior ischemic optic neuropathy. Ophthalmologica 182:29–41
Hayreh SS (1987) Anterior ischemic optic neuropathy. VIII. Clinical features and pathogenesis of post-hemorrhagic amaurosis. Ophthalmology 94:1488–1502
Hayreh SS, Weingast TA (1980) Experimental occlusion of the central artery of the retina. IV. Retinal tolerance time to acute ischemia. Br J Ophthalmol 64:818–825
Hayreh SS, vanHueven WAJ, Hayreh MS (1978) Experimental retinal vascular occlusion: I. Pathogenesis of central retinal vein occlusion. Arch Ophthalmol 96:311–330
Hedges TR (1963) Ophthalmoscopic findings in internal carotid occlusion. Am J Ophthalmol 55:1007–1012
Hedges TR, Weinstein JD (1964) Ophthalmic artery pressure response to carotid occlusion. Neurology 14:192–201
Hedges TR Jr, Giliberti O, Magargal LE (1985) Intravenous digital subtraction angiography and its role in ocular vascular disease. Arch Ophthalmol 103:666–669
Hedges TR, Reichel E, Duker JS, Puliafito CA, Heggerick PA (1993) Color Doppler imaging identifies different mechanisms of central retinal artery occlusion. Invest Ophthalmol Visual Sci 34:842
Henkind P, Charles NC, Pearson J (1970) Histopathology of ischemic optic neuropathy. Am J Ophthalmol 69:78–90
Hollenhorst RW (1961) Significance of bright plaques in the retinal arterioles. JAMA 178:23–29
Hollenhorst RW (1965) The neuro-ophthalmology of strokes. In: Smith JL (ed) Neuro-ophthalmology: symposium of the University of Miami and the Bascom Palmer Eye Institute, vol 2. Mosby-Year Book, St Louis, pp 109–121
Imes RK, Hoyt WF (1989) Exercise-induced transient visual events in young healthy adults. J Clin Neuro-ophthalmol 9:178–180
Jabs DA, Miller NR, Newman SA, Johnson MA, Stevens GG (1986) Severe retinal vaso-occlusive disease in systemic lupus erythematosus. Arch Ophthalmol 104:558–563
Johnson MW, Kincaid MC, Trobe JD (1987) Bilateral retrobulbar optic nerve infarctions after blood loss and hypotension. A clinicopathologic case study. Ophthalmology 94:1577–1584
Katz B, Weinreb RN, Wheeler DT, Klauber MR (1990) Anterior ischaemic optic neuropathy and intraocular pressure. Br J Ophthalmol 74:99–102
Kearns TP, Hollenhorst RW (1963) Venous-stasis retinopathy of occlusive disease of the carotid artery. Mayo Clin Proc 38:304–312
Knox DL, Duke JR (1971) Slowly progressive ischemic optic neuropathy: a clinicopathologic case report. Trans Acad Ophthalmol Otol 75:1065–1068
Kollarits CR, Lubow M, Hissong SL (1972) Retinal strokes. I. Incidence of carotid atheromata. JAMA 222:1273–1275
Kuritzky S (1990) Nitroglycerin to treat acute loss of vision. New Engl J Med 323:1428
Laties AM (1967) Central retinal artery innervation. Arch Ophthalmol 77:405–409
Lessell S (1979) The neuro-ophthalmology of systemic lupus erythematosus. Doc Ophthalmol 47:13–42
Levine SR, Crofts JW, Lesser GR, Floberg J, Welch KMA (1988) Visual symptoms associated with the presence of a lupus anticoagulant. Ophthalmology 95:686–691
Lieb WE, Flaharty PM, Sergott RC, Medlock RD, Brown GC, Bosley T, Savino PJ (1991a) Color Doppler imaging provides accurate assessment of orbital blood flow in occlusive carotid artery disease. Ophthalmology 98:548–552
Lieb WE, Cohen SM, Merton DA, Shields JA, Mitchell DG, Goldberg BB (1991b) Color Doppler imaging of the eye and orbit: technique and normal vascular anatomy. Arch Ophthalmol 109:527–531
Lieberman MF, Shahi A, Green WR (1978) Embolic ischemic optic neuropathy. Am J Ophthalmol 86:206–210
Lorentzen SE (1969) Occlusion of the central retinal artery. Acta Ophthalmol 47:690–703
McBrien DJ, Bradley RD, Aston N (1963) The nature of retinal emboli in stenosis of the internal carotid artery. Lancet 1:697–699
McDonald WI, Sanders MD (1971) Migraine complicated by ischaemic papillopathy. Lancet 2:521–523
Marshall J, Meadows S (1968) The natural history of amaurosis fugax. Brain 91:419–434
Merchut MP, Gupta SR, Naheedy MH (1988) The relation of retinal artery occlusion and carotid artery stenosis. Stroke 19:1239–1242
Miller NR (1991) Walsh and Hoyt's clinical neuro-ophthalmology, 4th edn. Williams and Wilkins, Baltimore, pp 2601–2627
Muci-Mendoza R, Arruga J, Edward WO et al. (1980) Retinal fluorescein evidence for atheromatous microembolism. Stroke 11:154–158
Newman NM, Hoyt WF, Spencer WH (1974) Macula-sparing monocular blackouts: clinical and pathologic investigations of intermittent choroidal vascular insufficiency in a case of periarteritis nodosa. Arch Ophthalmol 91:367–370
Newman NJ, Lessell S, Brandt EM (1989) Bilateral central retinal artery occlusions, disk drusen and migraine. Am J Ophthalmol 107:236–240
Olver JM, Spalton DJ, McCartney ACE (1990) Microvascular study of the retrolaminar optic nerve in man: the possible significance in anterior ischaemic optic neuropathy. Eye 4:7–24
Peduzzi M, Codeluppi L, Poggi M et al. (1962) Abnormal viscosity and erythrocyte deformability in retinal occlusion. Am J Ophthalmol 96:399–400
Pessin MS, Duncan GW, Mohr JP et al. (1977) Clinical and angiographic features of carotid transient ischemic attacks. New Engl J Med 296: 358–362
Riva CE, Loebl M (1977) Autoregulation of blood flow in the capillaries of the human macula. Invest Ophthalmol Vis Sci 16:568–571
Rizzo JF III, Lessell S (1987) Posterior ischemic optic neuropathy during general surgery. Am J Ophthalmol 103:808–811
Ross-Russell RW, Page NGR (1983) Critical perfusion of brain and retina. Brain 106:419–434
Ruskell GL (1971) Facial parasympathetic innervation of the choroidal blood vessels in monkeys. Exp Eye Res 12:166–172
Sadun AA, Currie JN, Lessell S (1984) Transient visual obscurations with elevated optic discs. Ann Neurol 16:489–494
Sadun AA, Green GL, Nobe JR, Cano MR (1991) Papillopathies associated with unusual calcifications in the retrolaminar optic nerve. J Clin Neuro-ophthalmol 11:175–180
Savino PJ, Glaser JS, Cassady J (1977) Retinal stroke: is the patient at risk? Arch Ophthalmol 95:1185–1189
Sergott RC, Cohen BS, Bosley TM, Savino PJ (1989) Optic nerve decompression may improve the progressive form of nonarteritic ischemic optic neuropathy. Arch Ophthalmol 107:1743–1754
Sergott RC, Flaharty PM, Lieb WE, Ho AC, Kay MD, Mittra RA et al. 1992 Color Doppler imaging identifies four syndromes of the retrobulbar calcification in patients with amaurosis fugax and central retinal artery occlusions. Trans Am Ophthalmol Soc (in press)
Spencer WH, Hoyt WF (1960) A fatal case of giant-cell arteritis (temporal or cranial arteritis) with ocular involve-

ment. Arch Ophthalmol 64:862–867
Sweeney PJ, Breuer AC, Selhorst JB, Waybright EA, Furlan AJ, Lederman RJ et al. (1982) Ischemic optic neuropathy. A complication of cardiopulmonary by-pass surgery. Neurology 32:560–562
Tachibana H, Gotoh F, Ishikawa Y (1982) Retinal vascular autoregulation in normal subjects. Stroke 13:149–155
Tomsak RL (1985) Ischemic optic neuropathy associated with retinal embolism. Am J Ophthalmol 99:590–592
Tomsak RL, Hanson M, Gutman FA (1979) Carotid artery disease and central retinal artery occlusion. Cleveland Clin Quart 46:7–11
Weinberg PE, Patronas NJ, Kim KS, Melen O (1981) Anomalous origin of the ophthalmic artery in a patient with amaurosis fugax. Arch Neurol 38:315–317
Williams K, Baumbach GL, Armstrong ML, Heistad DD (1989) Hypothesis: vasoconstriction contributes to amaurosis fugax. J Cerebr Blood Flow Metab 9:111–116
Wilson LA, Warlow CP, Ross-Russell RW (1979) Cardiovascular disease in patients with retinal arterial occlusion. Lancet 1:292–294
Winterkorn JMS, Teman AJ (1991) Recurrent attacks of amaurosis fugax treated with calcium channel blocker. Ann Neurol 30:423–424

Concluding Comments

Hedges has described in detail the anatomy of the orbital circulation, the mechanisms of eye ischemia and the various available methods of evaluation. I will comment on some of the clinical patterns of eye ischemia and their causes, and on some conditions in which examination of the eye can be diagnostic.

Transient Monocular Blindness (TMB)

Although transient loss of vision in one eye is often partial and is most accurately designated as transient monocular visual loss (TMVL), years of custom have led clinicians to keep the traditional term TMB. Visual loss is most often described by patients as a graying, darkening, blurring, fogging or dimming of vision. The whole field of vision may be affected, or only lateral or altitudinal sections are involved at the beginning or for the duration of the attack. Many patients describe a dark curtain that gradually descends or, less often, ascends or moves laterally across the eye. At times there are scintillations, colored or bright displays or streaks or shimmers (Fisher 1959; Goodwin et al. 1987) Fisher's patient, who was examined and found to have fibrin–platelet emboli during an attack of TMB, 'likened the failure of vision to the snowing up of a television screen . . . colorless snowflakes were bright, shining, and jumping' (Fisher 1959). The fact that positive, bright visual phenomena can occur in TMB due to occlusive disease is emphasized by Goodwin et al. (1987). Positive phenomena are taken by some to be strongly suggestive of a migrainous etiology of TMB, and Gautier, in his analysis of the presentation of TMB, noted that scintillations are typically absent (Gautier 1988). TMB due to occlusive carotid or ophthalmic artery disease can have either negative or positive visual phenomena. Attacks of TMB usually last from seconds to minutes, but can persist for hours.

Pain, in the head or eye, accompanying TMB is rare in patients with atherosclerotic occlusive disease, but is common in patients under 45 years of age, who probably have migraine (Tippin et al. 1989), and in patients with carotid dissection. Some patients with TMB have accompanying ipsilateral facial paresthesias (Ropper 1985). Tingling, warm flushing and a sensation as if the eye were twitching can affect the eye, eyelid, periorbital area or cheek, and either precede the visual loss or begin simultaneously with TMB (Ropper 1985). In some patients episodes of transient ipsilateral hemispheral dysfunction also occur, but usually not simultaneous with TMB (Pessin et al. 1977). The findings of attacks of both TMB and transient ipsilateral hemispheral dysfunction is virtually diagnostic of severe internal carotid artery occlusive disease (Pessin et al. 1977).

There are several different mechanisms of TMB. Hedges, in his review of the pathogenesis, has described the most common mechanisms. Wray (1988a) divided patients with TMB into four large groups depending on demographic features, clinical findings and laboratory results.

In her classification, type I TMB was due to embolism. Attacks are characterized by loss of all or a portion of vision in one eye and last from seconds to a few minutes, with full recovery. Many of the emboli come from the carotid artery in the form of cholesterol crystals, particles of red thrombi and platelet–fibrin clumps. Often, the emboli or focal regions of retinal ischemia are evident through the ophthalmoscope. Interestingly, these types of attack have been correlated with 'open' internal carotid arteries, that is, vessels with ulcerations and irregularities, but without critical degrees of narrowing (Pessin et al. 1977). Embolisms from carotid artery disease have been correlated with localized partial loss of vision in an altitudinal, lateral or sector distribution (Bruno et al. 1990). Longer attacks have also been correlated with 'open' carotids, compared to repeated, very brief attacks which are usually found in patients with severe carotid narrowing or occlusion (Goodwin et al. 1987; Pessin et al. 1977). Calcific emboli can also cause type I TMB, and probably arise most often from the heart (calcific valves, mitral annulus calcification) or the aorta. Occasionally, pieces of tumor, especially from cardiac myxomas, and insoluble particles such as talc and cornstarch derived from pills that are dissolved and injected intravenously by drug abusers, can be found in the eye and cause embolic TMB (Atlee 1972; Caplan et al. 1982).

Wray's type II TMB includes attacks which are either frequent or very brief, and are less rapid in onset and longer in duration. Recovery often occurs in a gradual manner (Wray 1988a). These attacks are due to low flow, usually caused by hemodynamically significant severe occlusive disease of the carotid or ophthalmic arteries. Most often the etiology is atherosclerosis, but severe disease of the aortic arch vessels, such as Takayasu's or giant-cell arteritis, can cause the same type of TMB. Attacks are often precipitated by orthostatic changes, changes in blood pressure or blood volume, or exposure to bright light. Central retinal artery pressures are often quite low and signs of chronic ocular ischemia are also often present (Carter 1985).

Type III TMB attacks are thought to be due to episodes of vasoconstriction ('vasospasm'). Headache and eye pain are frequent and most patients are young, most often female, and have no evidence of cardiac disease or atherosclerosis. These attacks are often 'white-outs' and may include flashing stars or lines, shimmering and grid-like patterns, and last 2–15 or 20 minutes or longer (Wray 1988a; Burger et al. 1991). Funduscopic examination during an attack shows constriction of the retinal arteries, often with segmentation in a thin and slowly moving blood column (Burger et al. 1991). Venous narrowing, retinal edema, dilated veins and delayed fluorescein filling of retinal arteries have also been found (Wray 1988a; Kline and Kelly 1980). Type IV TMB were patients who did not fit well into the other three groups. They were often young, had multiple attacks, and probably also had vasospasm. Vasoconstriction has also been described as a cause of TMB in a cocaine abuser (Libman et al. 1993).

Contrary to the expectations of most clinicians, ultrasound and angiography in large series of patients with TMB show a rather low frequency (<1/3 of patients) of significant carotid artery stenosis (Adams et al. 1983; Chawluck et al. 1988; Tippin et al. 1989). Patients who show cholesterol crystals in the fundus, and those with both TMB and hemispheral attacks, do have a much higher incidence of severe carotid occlusive disease.

Eye 'Strokes' and Chronic Ischemia

The preceding section considered transient temporary episodes of visual loss. Herein I will comment on some of the clinical aspects of persistent ocular ischemia. This subject is conveniently divided into retinal infarct, optic nerve infarct and chronic ocular ischemia.

Retinal Infarcts

Retinal infarction is usually characterized as being due to central retinal artery occlusion (CRAO), when there is diffuse monocular visual loss and branch retinal artery occlusion (BRAO) when infarcts are localized to a segment of the retina. In both instances abnormalities in the arteries are visible through the ophthalmoscope.

In patients with CRAO, the predominant symptom is the sudden onset of painless blindness with some persistent visual loss. At times, light or movements can still be preserved in parts of the visual field. Von Graefe (1859) is usually credited with the earliest full description of the condition. He described attenuated arteries and veins, a pale disc, clouding of the retina and a cherry red spot in the macula. Shortly after CRAO, so-called segmentation of the blood column is often seen, with slow streaming in the veins in the fundus near the disc. Within the arteries and veins, the blood appears dark and clear areas may alternate with clumped cells, giving the appearance of segmentation (Wray 1988b). The central cherry red appearance of the macula is due to accentuation of the normal fovea through which the choroid appears red. Retinal artery pressures, as measured by an ophthalmodynamometer, are invariably low. If CRAO lasts more than 1 hour, the retina becomes infarcted.

In BRAO, branches of the central retinal artery are occluded and the visual loss and retinal ischemia are focal and have a lateral, altitudinal or scotomatous character. Blockage of the retinal branches is often visible with the ophthalmoscope and diagnosis is facilitated by fluorescein angiography, which shows a lack of perfusion of fluorescein containing blood into the affected branch. Often the point of occlusion can be pinpointed accurately on the fluorescein study (Michelson and Friedlander 1988). The ischemic portion of the retina often has a ground-glass appearance, due to retinal ischemic edema.

In most series of patients with CRAO and BRAO the incidence of potential embologenic lesions in the carotid arteries and heart are higher than in series of TMB patients, but CRAO and BRAO patients are similar. Cardiac-origin embolic sources are common, and were found in 29 of 103 patients (28%) with retinal artery occlusions in one series (Wilson et al. 1979). In this series 12 of 18 patients had carotid irregularity or stenosis on angiography (Wilson et al. 1979). In another series of patients with either CRAO or BRAO, 29 of 34 (85%) had abnormal ICAs on angiography; of those, 12 had occlusion or severe stenosis and 17 had plaques, ulcers or ‹60% stenosis (Merchut et al. 1988). Chawluk et al. (1988) studied 17 patients with CRAO using ultrasound, and found that 24% had ICA occlusion and 36% had stenotic or ulcerated lesions. In this series, carotid artery lesions were more common in patients with retinal infarcts than in those with TMB.

The sudden onset of visual loss, observation of materials of different types in the retinal arteries, and the high incidence of cardiac and carotid arterial sources of embolism make it likely that most cases of CRAO and BRAO are embolic. Hedges discusses the use of color Doppler imaging of the ocular blood supply and its ability to image calcific and other emboli. Some retinal arterial occlusions are clearly due to hypercoagulability and in-situ thrombosis.

Optic Nerve Infarcts

Ischemic optic neuropathy is discussed in some detail by Hedges. Patients usually notice painless monocular visual loss on arising in the morning; the severity of the visual loss varies, but usually it is not total. The optic disc and retina may appear normal acutely, or the disc may show some edema with splinter-shaped hemorrhages along the disc margins. Optic atrophy is seen later. Usually the visual loss does not progress in the same eye, but often affects the contralateral eye. Hedges points out the high incidence of risk factors for small-artery disease in this group. Arteritis is a common cause, but severe carotid artery disease is rarely found.

Bogousslavsky and colleagues (1987) reported 3 patients who had simultaneous infarction of an optic nerve and the ipsilateral cerebral hemisphere. All had carotid artery occlusions. I have also seen several patients with this syndrome, all of whom had occlusion of the ICA in the siphon portion of the artery involving the ophthalmic artery origin.

Chronic Ocular Ischemia

Chronic ischemia causes a quite different clinical picture in the eye. Ischemic pain may be present and is described as a constant ache over the orbit, temple and upper face, and is occasionally better when the patient lies down. Vision may be normal or there may be an ill-defined subjective sensation of blurring; sometimes the visual image is less bright or sharp, and may seem washed-out (Carter 1985, 1988). Often there are superimposed brief repeated episodes of transient visual loss that may be precipitated by standing or exposure to bright light, or after drops in systemic blood pressure. Most patients have total occlusion of the ICA on the side of the ischemic eye or very severe critical stenosis.

The diagnostic funduscopic changes have been called venous stasis retinopathy after Hedges (1963) and Kearns and Hollenhorst (1963). Microaneurysms and small 'dot and blot' hemorrhages are found most prominently at the midperiphery of the retina. Later, dilation and darkening and irregular caliber of the retinal veins are found (Carter 1988). Neovascularization of the retina and optic disc may occur and the macula may become pale or gray due to retinal edema. The anterior ocular media are also often involved. Findings include episcleral vascular congestion, clouding of the cornea, cells and flare in the anterior chamber of the eye, and neovascularization of the iris. The pupil may be dilated and non-reactive due to iris ischemia. The intraocular pressure is often elevated. Invariably, the pressure measured in the central retinal artery is very low.

Chronic ocular ischemia is often bilateral and most such patients have severe occlusive disease, sometimes involving the common and external carotid arteries as well as the ICAs. All attempts should be made to augment perfusion of the eye in these patients.

Unusual Vascular Disorders Diagnosable by the Retinal Findings

Many consider the eye to be a very convenient window into the brain. The appearance of the retinal vasculature may, in some conditions, mirror what is occurring in the brain vessels, and in other conditions there is a peculiar predilection for involvement of the orbital vessels.

Sarcoidosis (Karmi 1979; Caplan et al. 1983)

The retinal veins often show a periphlebitis characterized by yellowish-white diffuse or focal sheathing. Hard exudates are sometimes visible nearby, and are referred to as *taches de bougie* because of their resemblance to the drippings of candle wax. The optic disc may contain sarcoid granulomas. Similar, but less severe, perivenous sheathing is often seen in multiple sclerosis patients. The uvea, conjunctiva and lacrimal glands are also often involved in sarcoidosis.

Microangiopathy of the Brain, Ear and Retina (Susac et al. 1979; Bogousslavsky et al. 1989)

In this very distinctive syndrome, which usually affects young women, progressive blindness, deafness and central nervous system findings develop. The retinal findings are distinct and pathognomonic. Some retinal arteries are amputated while others are severely narrowed or attenuated, and light streaking characterizes their thickened vascular walls. Similar non-inflammatory arterial obliteration affects the cochlea and the brain arterioles.

Drug Abuser Retinopathy (Atlee 1972)

This condition occurs in patients who mash and inject intravenously or intra-arterially (carotid) drugs made for oral consumption. Methylphenidate (Ritalin) and pentazocine (Talwin) with pyribenzamine ('T's and blues') are the most frequently reported offending drugs. Pills are often held together by fillers, including talc, microcrystalline cellulose and cornstarch, and these materials are sometimes seen in the retinal arteries of abusers.

Cogan's Syndrome (Cogan 1945; Cheson et al. 1976)

This condition includes patients with interstitial keratitis and vestibulocochlear dysfunction probably due to a restricted or generalized autoimmune vasculitis. The earliest symptoms are usually photophobia, reduced vision and a red eye. An interstitial keratitis, sometimes with uveitis, is found on careful ophthalmological examination.

Eales' Disease (Singhal and Dastur 1976; Gordon et al. 1988)

Eales described a distinctive type of vasculitis that affects both retinal arteries and veins, usually in young men. The illness is most common in India and the Middle East. Patients report floaters, cobwebs, specks, curtains and blurred vision. Veins and arteries appear sheathed and there are flame-shaped retinal and vitreous hemorrhages. Both eyes are usually involved. Strokes, uveitis and meningitis may also occur.

Other Uncommon Conditions

In other diseases, eye findings, when combined with the general features, should suggest systemic vascular diseases. Subluxation of the lens is helpful in recognizing Marfan's syndrome. Angioid streaks in the retina combined with thick grooved skin folds in the axilla and antecubital region suggest the presence of pseudoxanthoma elasticum. Recurrent uveitis, when combined with aphthous ulcers in the mouth and genitalia, suggests Behçet's disease. In hyperviscosity syndromes such as Waldenstrom's macroglobulinemia, patients may report blurred vision and the retinal veins are dilated, tortuous and may show segmentation of the blood column within the veins. In eclamptic women, retinal artery spasm, small hemorrhages and retinal edema may occur.

References

Adams HP, Putnam S, Corbett JJ et al. (1983) Amaurosis fugax: the results of arteriography in 59 patients. Stroke 14:742–744

Atlee W (1972) Talc and cornstarch emboli in the eyes of drug abusers. JAMA 219:49–51

Bogousslavsky J, Regli F, Zografos L, Uske A (1987) Optico-cerebral syndrome: simultaneous hemodynamic infarction of optic nerve and brain. Neurology 37:263–268

Bogousslavsky J, Gaio JM, Caplan LR et al. (1989) Encephalopathy, deafness, and blindness in young women. A distinct retino-cochleo-cerebral arteriopathy. J Neurol Neurosurg Psych 52:43–46

Bruno A, Corbett JJ, Biller J, Adams HP, Qualls C (1990) Transient monocular visual loss patterns and associated vascular abnormalities. Stroke 21:34–39

Burger SK, Saul RF, Selhorst JB, Thurston SE (1991) Transient monocular blindness caused by vasospasm. New Engl J Med 325:870–873

Caplan LR, Corbett J, Goodwin J et al. (1983) Neuroophthalmological signs in the angiitic form of neurosarcoidosis. Neurology 33:1130–1135

Caplan LR, Thomas C, Banks G (1982) Central nervous system complications of 'T's and blues' addiction. Neurology 32:623–628

Carter JE (1985) Chronic ocular ischemia and carotid vascular disease. Stroke 16:721–728

Carter JE (1988) Chronic ocular ischemia and carotid vascular disease. In: Bernstein EF (ed) Amaurosis fugax. Springer-Verlag, New York, pp 118–134

Chawluck JB, Kushner MJ, Bank WJ et al. (1988) Atherosclerotic carotid artery disease in patients with retinal ischemic syndromes. Neurology 38:858–863

Cheson BD, Bluming AZ, Alroy J (1976) Cogan's syndrome: a systemic vasculitis. Am J Med 60:549–555

Cogan DG (1945) Syndrome of nonsyphilitic interstitial keratitis and vestibuloauditory symptoms. Arch Ophthalmol 33:144–149

Fisher CM (1959) Observations of the fundus oculi in transient monocular blindness. Neurology 9:333–347

Gautier JC (1988) Clinical presentation and differential diagnosis of amaurosis fugax. In: Bernstein EF (ed) Amaurosis fugax. Springer-Verlag, New York, pp 24–42

Goodwin JA, Gorelick PB, Helgason C (1987) Symptoms of amaurosis fugax in atherosclerotic carotid artery disease. Neurology 37:829–832

Gordon MF, Coyle PK, Golub B (1988) Eales' disease presenting as a stroke in the young adult. Ann Neurol 24:264–266

Hedges TR (1963) Ophthalmoscopic findings in internal carotid artery occlusion. Am J Ophthalmol 55:1007–1012

Karmi A (1979) Ophthalmic changes in sarcoidosis. Acta Ophthalmol (Suppl) 141:1–94

Kearns TP, Hollenhorst RW (1963) Venous-stasis retinopathy of occlusive disease of the carotid artery. Mayo Clin Proc 38:1130–1135

Kline LB, Kelly CL (1980) Ocular migraine in a patient with cluster headaches. Headache 20:253–257

Libman RB, Masters SR, dePaolo A, Mohr JP (1993) Transient monocular blindness associated with cocaine abuse. Neurology 43:228–229

Merchut MF, Gupta SR, Naheedy MH (1988) The relation of retinal artery occlusion and carotid artery stenosis. Stroke 19:1239–1242

Michelson JB, Friedlander MH (1988) Angiography of retinal and choroidal vascular disease. In: Bernstein EF (ed) Amaurosis fugax. Springer-Verlag, New York, pp 51–71

Pessin MS, Duncan GW, Mohr JP et al. (1977) Clinical and angiographic features of carotid transient ischemic attacks. New Engl J Med 296:358–362

Ropper AH (1985) Transient ipsilateral paresthesias (TIPs) with transient monocular blindness. Arch Neurol 42:295

Singhal BS, Dastur DK (1976) Eales' disease with neurological involvement. J Neurol Sci 27:312–321; 323–345

Susac J, Hardman J, Selhorst J (1979) Microangiopathy of the brain and retina. Neurology 29:313–316

Tippin J, Corbett JJ, Kerber RE, Schroeder E, Thompson HS (1989) Amaurosis fugax and ocular infarction in adolescents and young adults. Ann Neurol 26:69–77

Von Graefe A (1859) Ueber Embolie der arteria centralis Retinae als ursache plotzlicher Erblindung. Arch Clin Exp Ophthalmol 5:136–140

Wilson LA, Warlow CP, Ross-Russell RW (1979) Cerebrovascular disease in patients with retinal artery occlusion. Lancet I: 292–294

Wray S (1988a) Visual aspects of extracranial internal carotid artery disease. In: Bernstein EF (ed) Amaurosis fugax. Springer-Verlag, New York, pp 72–80

Wray S (1988b) Occlusion of the central retinal artery. In: Bernstein EF (ed) Amaurosis fugax. Springer-Verlag, New York, pp 81–89

Section III

The Blood

Prefatory Comments

Necropsy studies of patients dying of apoplectic strokes during the 19th century (Cheyne 1812) clearly separated brain hemorrhages from brain softening. Physicians during the ensuing years recognized that bleeding due to trauma or reduced blood coagulability could lead to brain hemorrhage. The cause of brain softening, then customarily called 'encephalomalacia' or 'ramollissements' was, however, not evident until the time of Virchow, who was probably the first to show that softenings or infarcts in any organ, including the brain, were due to the occlusion of blood vessels that supplied the infarcted tissue (Virchow 1843; Fisher 1987). Yet surprisingly, until very recently there had not been much interest in how the blood vessels became occluded. Interest during the late 19th and early 20th centuries centered mostly on the anatomy of the brain's blood supply and the clinical signs that resulted from infarcts in various regions. Charles Foix, one of the earliest and most prolific writers about brain infarcts in both the anterior and posterior circulations, described mostly the anatomy of brain lesions and their clinicopathological correlations (Caplan 1990). Not until several weeks before his untimely death did he look carefully at the blood vessels leading to the infarcts, and he then discovered that sometimes the arteries were occluded but often they were not (Caplan 1990; Foix et al. 1927). More recently, studies of patients with coronary artery disease (Fuster 1992), of carotid endarterectomy specimens and of the intracranial circulation in patients with acute stroke (Fieschi et al. 1989) show conclusively that occlusive thrombosis and thromboembolism is nearly always the mechanism of the final blow that blocks the arteries and leads to deprivation of blood flow to vital brain regions.

Whereas the study of blood vessels and their pathology has mostly involved anatomy and pathology, the study of blood coagulation involves more chemistry and physiology. Physiology is studied more during life, both in test tubes and in human and animal blood vessels. The story of how, why, when and where blood clots is still a very active ongoing saga, with much still to be learned. The major equation, fibrinogen $\xrightarrow{\text{thrombin}}$ fibrin, states simply that fibrin-rich clots form when a large circulating serum protein, fibrinogen, is converted to fibrin, a process catalyzed by thrombin. Thrombin, in turn, is activated from prothrombin. Early blood research identified the key agents in the process. These were blood platelets, which can aggregate together, adhere to the endothelial membranes and secrete substances which activate the coagulation process; and serum protein coagulation factors (Factors II–XII), which are substances present in circulating blood that can be activated by factors within the vascular system (the so-called 'intrinsic pathway') and by tissue injury, which releases various tissue factors including thromboplastin (the so-called 'extrinsic pathway').

More recently it has become clear that other proteins in the blood (antithrombin III, protein C, protein S) also play important roles as natural inhibitors of thrombin formation. A deficiency or neutralization of these factors (e.g. antithrombin

III by heparin) can lead to altered coagulability. The introduction into clinical practice of agents that lyse clots has reawakened interest in the body's natural fibrinolytic system. The process of coagulation also usually naturally activates a system designed to take apart and destroy the clots which are formed. Abnormalities of the fibrinolytic system also affect the formation and lysis of clots and vascular recanalization. Calcium is an important ion in the physiology of coagulation, and fibrinogen is a key agent in the coagulation process. Erythrocytes are to some extent passive victims entrapped along with fibrin into fibrin-rich thrombi ('red clots'). Platelets are also frequently admixed in the process, and can form platelet–fibrin nidi ('white clots'). This section of the book will consider the coagulation functions of the blood. Later sections will deal with the rheological and neurological aspects of blood flow.

Feinberg describes in detail in his contribution the physiology of the clotting system and modern means of measuring thrombin activation and also fibrinolysis. The Furies (1992) have also recently reviewed the molecular and cellular biology of coagulation. Del Zoppo details the most up-to-date understanding of the body's fibrolytic system, and also reviews early data on the effectiveness of fibrinolytic treatment of patients with strokes due to thromboembolism.

Very recently a new dimension has been added. Although it has long been known that tissue injury releases factors that activate coagulation and vessel repair, the larger issue of the control of coagulation by body functions and illnesses has been terra incognita. The discovery that various products of inflammation (cytokines) have widespread effects on body functions and metabolism, and probably coagulation, has opened new doors and insights. Various products of inflammation, such as interleukins, tissue necrosis factor (TNF) and lipopolysaccharides released by bacteria, profoundly affect clotting as well as other body processes. The body's immune system and nervous system probably have the capacity to regulate coagulability in ways that are still poorly understood. The recent discovery that some patients with stroke (and some without strokes) have circulating immunoglobulins called antiphospholipid antibodies that react in vitro against phospholipid membranes, has also awakened interest in the role of immunity and autoimmunity in coagulation. Fisher, in his chapter, reviews some of the most recent information about these relationships.

References

Caplan LR (1990) Charles Foix – the first modern stroke neurologist. Stroke 21:348–356

Cheyne J (1812) Cases of apoplexy and lethargy: with observations upon comatose diseases. J Moyes, London.

Fieschi C, Argentino C, Lenzi GL, Toni D, Bozzao L (1989) Clinical and instrumental evaluation of patients with ischemic stroke within the first six hours. J Neurol Sci 91:311–321

Fisher CM (1987) The history of cerebral embolism and hemorrhagic infarction. In: Furlan AJ (ed) The heart and stroke. Springer-Verlag, New York, pp 4–16

Fisher CM, Ojemann RG (1986) A clinico-pathologic study of carotid endarterectomy plaques. Rev Neurol 142:573–589

Foix C, Hillemand P, Ley J (1927) Rélativement au ramollissement cérébral à sa fréquence et à son siège, et à l'importance rélative des oblitérations artérielles, complètes ou incomplètes dans sa pathologénie. Rev Neurol (Paris) 43:217–218

Furie B, Furie BC (1992) Molecular and cellular biology of blood coagulation. New Engl J Med 326:800–806

Fuster V (1992) Pathogenesis of coronary artery disease and the acute coronary syndromes. New Engl J Med 326:242–250; 310–318

Virchow R (1843) Ueber die akut Entzundung der Arterien. Virchows Arch Path Anat 1:272–378

7. Coagulation

W. M. Feinberg

Clinical and experimental evidence has shown that acute stroke is an atherothrombotic or thromboembolic process in 80% of cases. (Mohr et al. 1978; Sloan 1986; del Zoppo et al. 1986) Angiography of patients within 12 hours of acute stroke has shown atherothrombotic stenoses or arterial occlusions in over 80% of cases (Solis et al. 1977; del Zoppo et al. 1992). When angiograms were performed at a later time, the percentage of thrombotic occlusions was much lower (Solis et al. 1977) and the passage of emboli or spontaneous recanalization of vessels was shown (Dalal et al. 1965). Recognition of the thrombotic nature of stroke has led to a renewal of interest in exogenous thrombolytic therapy (Brott et al. 1992; del Zoppo et al. 1992).

The coagulation process is actually a complex interaction of fibrin formation, fibrinolysis and platelet activation. Because of the close interrelationship of these processes they must be considered together to fully define 'coagulation' in stroke. This is further complicated by the heterogeneous nature of the stroke syndrome. Clinicopathologic studies during the last 30 years have identified various underlying conditions which can all present clinically as a stroke. Major subtypes of ischemic stroke identified in recent series (Mohr et al. 1978; Foulkes et al. 1988; Sherman et al. 1989) include atherosclerotic disease of the large extracranial vessels (atherothrombotic), emboli from the heart (cardioembolic) and small-vessel occlusion deep within the cerebrum and brain stem (lacunar infarction). Since the underlying pathophysiology of each type of stroke differs, platelet activation, thrombosis and fibrinolysis may be involved to a differing degree in each. Studies that include a heterogeneous mix of stroke types may not allow recognition of important findings regarding stroke pathophysiology in a particular stroke type.

Fibrin Formation and Degradation

The Coagulation Cascade

The coagulation sequence is shown in Fig. 7.1 (Vermylen et al. 1986; Wessler and Gitel 1986) Traditionally, this is divided into the 'intrinsic' pathway and the 'extrinsic' pathway. In the intrinsic pathway all necessary factors are present in circulating blood, and activation occurs through the introduction of a foreign surface or collagen. The extrinsic pathway is activated by the release of tissue thromboplastin. The two pathways converge at Factor X; activated Factor Xa (in the presence of Factor V, calcium and platelet phospholipid) converts prothrombin (Factor II) to thrombin. Factor VIIa also activates Factor IX (Fig. 7.1). It is noteworthy that platelet phospholipid, made available by aggregated platelets, is required at specific steps in both the intrinsic and extrinsic pathways. The close inter-

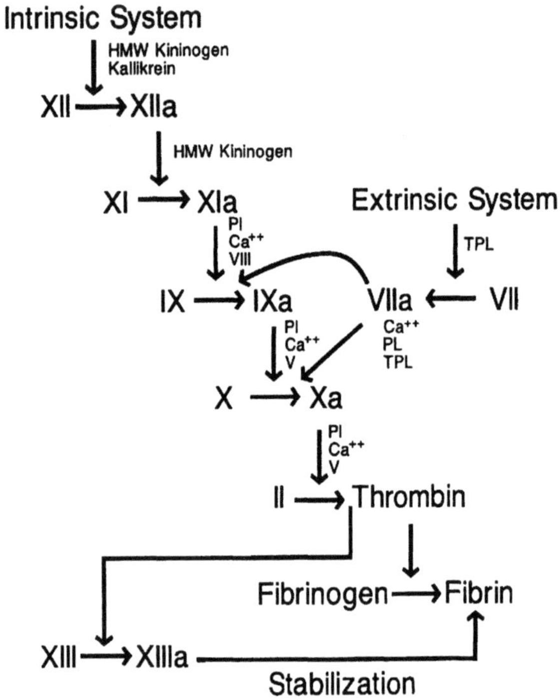

Fig. 7.1. The coagulation cascade. The coagulation sequence represents a series of linked reactions in which a zymogen is converted to a serine-protease, which in turn catalyzes the next step in the cascade. In the 'intrinsic system' clotting is initiated by factors present within the circulation, whereas the 'extrinsic system' is initiated by the release of tissue thromboplastin. See text for further details. HMW Kininogen, high molecular weight kininogen; PL, phospholipid; TPL, tissue thromboplastin. From Wessler S, Gitel SN (1984) Warfarin – from bedside to bench. Reproduced with permission of the New England Journal of Medicine 1984; 311:645–652

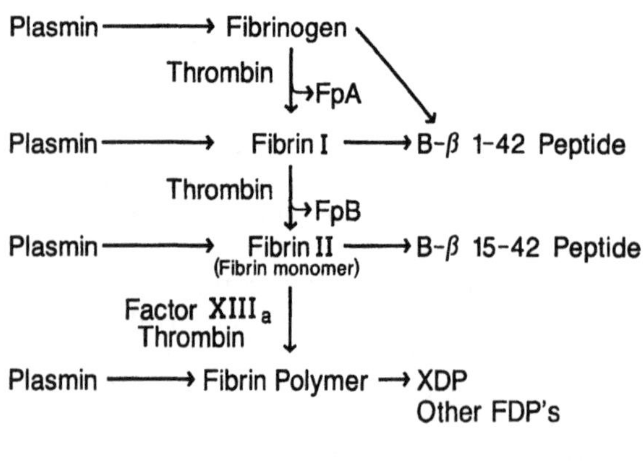

Fig. 7.2. Schematic diagram of fibrin formation and fibrin(ogen)olysis. The formation and breakdown of fibrin depends on the relative activities of thrombin and plasmin. FpA, fibrinopeptide A; FpB, fibrinopeptide B, XDP, cross-linked D-dimer.

relationship of the platelet and fibrin systems will be considered in greater detail below.

Deficiencies of coagulation proteins lead to bleeding disorders, most notably the classic hemophilias. However, there is controversy whether *increased* levels of the clotting factors may lead to a thrombotic tendency. Levels of Factors V, VII and VIII have been increased in stroke patients in small series (Estol et al. 1989; Hart and Kanter 1990; Cook and Ubben, 1990; Takano et al. 1990). Since Factors VII and VIII are acute-phase reactants, it is difficult to assess elevation of these factors after a stroke. Optimally they should be measured prospectively in populations at risk. In the Northwick Park Heart Study, increased Factor VII levels were strongly associated with subsequent ischemic heart disease events. (Meade et al. 1986) The role of increased factor levels in stroke remains uncertain, and there is as yet no evidence that measuring factor levels in an individual patient is useful.

One factor that has been demonstrated to be associated with increased stroke risk is fibrinogen (Wilhelmsen et al. 1984; Meade et al. 1986; Kannel et al. 1987; Cook and Ubben 1990; Ernst 1990). Although fibrinogen levels increase with age, blood pressure, smoking, obesity and diabetes (Kannel et al. 1987; Cook and Ubben 1990), fibrinogen is an independent risk factor after correcting for these conditions (Wilhelmsen et al. 1984; Kannel et al. 1987). In addition to serving as the substrate for fibrin formation in the thrombin pathway, fibrinogen is the primary determinant of blood viscosity and plays a crucial role in platelet aggregation. Thus, increased fibrinogen might be associated with elevated stroke risk through several interrelated processes. A number of pharmacologic agents, particular lipid-lowering agents, lower the fibrinogen level (Cook and Ubben 1990). This provides a novel approach to stroke prevention. The important role of fibrinogen in platelet aggregation is described in detail later in this chapter.

Fibrin Formation and Fibrinolysis

The conversion of fibrinogen to fibrin at the end of the coagulation cascade is actually a multistep process in itself (Nossel 1981; Alkjaersig and Fletcher 1982; Owen et al. 1983). The formation and breakdown of fibrin is shown in schematic form in Fig. 7.2. Fibrin I is formed when fibrinopeptide A (FpA), a 16-amino acid peptide, is cleaved from fibrinogen by thrombin (Nossel 1981; Owen et al. 1983). Fibrin I forms primarily thin fibers consisting of end-to-end polymers. Thrombin acts further on fibrin I to convert it to fibrin II with the removal of fibrinopeptide B (FpB), a 14-amino acid peptide. Fibrin II forms both end-to-end and side-to-side fibers, and is a thicker fiber. Plasmin is capable of degrading fibrinogen, fibrin I and fibrin II. Early degradation products are the amino terminal peptides of the B-β chain. In the case of fibrinogen and fibrin I, this is the B-β 1–42 peptide, while in the case of fibrin II it is the B-β 15–42 peptide. Once plasmin acts to release the B-β 1–42 peptide, fibrin I cannot be converted to fibrin II.

Plasmin digestion of fibrinogen and fibrin yields a number of other fragments, which are the familiar 'fibrin split products' or 'fibrin degradation products'. One fragment, called cross-linked D-dimer, is specific for the degradation of fibrin polymer (Nossel 1981; Whitaker et al. 1984; Hunt et al. 1985). The presence of D-dimer therefore indicates that fibrin polymer has formed. A monoclonal antibody specific for D-dimer is available (Whitaker et al. 1984; Hunt et al. 1985) Many laboratories are now using this in an ELISA assay for D-dimer in place of the older 'fibrin split product' assays.

As can be seen in Fig. 7.2, whether fibrin polymer is formed depends on the balance between thrombin activity and plasmin activity. By measuring FpA, D-dimer and the B-β peptides, one can theoretically assess this balance.

In addition to measuring markers of fibrin formation and degradation, one can indirectly measure the formation of thrombin itself. When Factor Xa converts prothrombin (Factor II) to thrombin, several peptide fragments are generated. Plasma levels of one of these, fragment F_{1+2}, appear to be a sensitive index of in vivo prothrombin activation (Teitel et al. 1982) Fragment F_{1+2} is elevated in acute thrombotic diatheses and can be suppressed with anticoagulants (Conway et al. 1987) Warfarin also decreases F_{1+2} levels in patients receiving anticoagulation for cerebrovascular disease (Millenson et al. 1992). Fragment F_{1+2} levels have not been reported in acute stroke patients.

Regulation of Thrombin Activity

Thrombin is under the regulatory control of several plasma proteins. These provide a physiologic feedback on thrombosis and prevent the unchecked propagation of clots. Alterations in these regulatory proteins may be important in the initiation or propagation of thrombosis.

Antithrombin III is the most important plasma serine protease inhibitor. It is an α_2-globulin of approximately 60 000 Da and is responsible for more than 85% of thrombin inhibitory activity (Wessler and Gitel 1986). Patients with deficiencies of antithrombin III are predisposed to thrombotic events. While venous thromboses are most common, arterial thrombosis including brain infarction has also been documented (Hart and Kanter 1990). Antithrombin III is essential for the anti-thrombotic action of heparin; heparin acts by accelerating the rate of antithrombin III neutralization of thrombin (Hirsch 1991). An immunoassay is available for the thrombin–antithrombin III complex, and this has been used as an additional marker for in vivo thrombin activity (Teitel et al. 1982; Takano et al. 1991, 1992).

Protein C is an important vitamin K-dependent inhibitor of coagulation (Clouse and Comp 1986; Rick 1990). Thrombin converts protein C to activated protein C (protein Ca), which inhibits the thrombotic process by inactivation of activated Factor V (Factor Va) and activated Factor VIII (Factor VIIIa) (Fig. 7.3). This is a direct feedback on thrombin activity, since thrombin converts Factors V and VIII to Factors Va and VIIIa, respectively (Wessler and Gitel 1986). Protein Ca has another important function in that it inactivates a circulating inhibitor of tissue-type plasminogen activator (t-PA), in effect increasing t-PA activity. Thus, protein Ca not only inhibits fibrin formation, but also accelerates the degradation of clot that has already formed.

Congenital and acquired deficiencies of protein C are associated with a predisposition to recurrent thrombosis (Clouse and Comp 1986; Rick 1990; Hart and Kanter 1990). Again, this is primarily venous but occasionally manifests as arterial thromboses. Stroke has been associated with protein C deficiency, but a number of these were due to cerebral venous thrombosis (Hart and Kanter 1990).

Protein C requires several cofactors for optimal activity (Fig. 7.3). Thrombomodulin is a receptor protein present on endothelial cells. In the presence of Ca^{2+} it binds thrombin; this thrombin–thrombomodulin complex increases protein Ca generation 30 000-fold (Clouse and Comp 1986). In the absence of thrombomodulin, the conversion of protein C to protein Ca by thrombin is too slow to be physiologically important. Thrombomodulin has been reported to be present in all vessels except those in the human brain (Ishii et al. 1986). This could theoretically place the cerebral circulation at a dis-

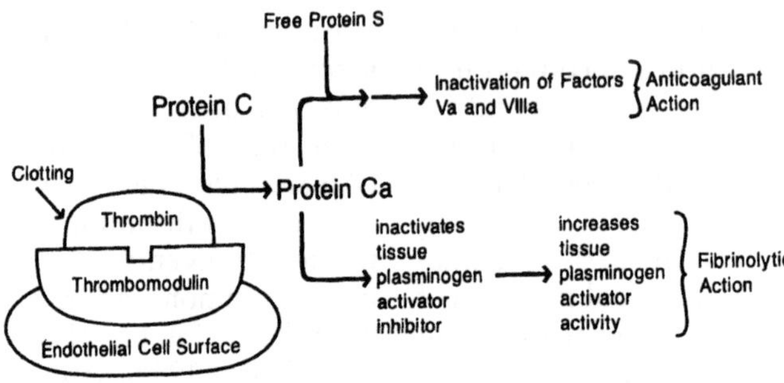

Fig. 7.3. The anticoagulant action of protein C. Thrombin binds to thrombomodulin, a receptor protein in the vascular endothelium. The thrombin–thrombomodulin complex rapidly converts protein C to activated protein Ca. Protein Ca inactivates Factors Va and VIIIa, and inactivates a plasminogen activator inhibitor. From Clouse LH, Comp PC (1986) The regulation of hemostasis: the protein C system. Reproduced with permission of the New England Journal of Medicine 1986; 314:1298–1304

advantage in terms of the natural anticoagulant systems. A recent abstract reported thrombomodulin in the brain vessels, but the authors identified marked regional variations within the brain. They postulated that this regional distribution might contribute to sites of predilection for stroke (Wong et al. 1991).

Protein S is a plasma cofactor for protein C which greatly enhances the inactivation of Factor Va by protein Ca (Clouse and Comp 1986; Rick 1990). Protein S is also a vitamin K-dependent plasma protein. Protein S forms a complex on phospholipid surfaces with protein C, which may be the mechanism of its acceleratory action. Protein S deficiency is associated with recurrent venous and, rarely, arterial thromboses (Clouse and Comp 1986; Rick 1990). There are isolated case reports of stroke associated with protein S deficiency (Sacco et al. 1989; Gómez-Aranda et al. 1992).

Fibrinolysis

Plasmin degradation of fibrin(ogen) is in itself a complex process under the control of a number of activators and inhibitors (Erickson et al. 1985; Francis and Marder 1986; Bachman 1987). Although fibrinolysis will also be considered in detail elsewhere, it will be considered briefly here since it is crucial to the balance of fibrin formation and degradation.

Fibrinolytic activity depends primarily on the conversion of plasminogen to plasmin (Erickson et al. 1985; Francis and Marder 1986; Bachman 1987). Although several activators have been identified, t-PA appears to be of primary importance, as it preferentially activates fibrin-bound rather than free plasminogen. T-PA is synthesized in endothelial cells and secreted into the circulation. It has a short half-life in the circulation of approximately 5 minutes. However, the functional half-life of t-PA activity is much longer due to fibrin-bound t-PA (Eisenberg et al 1987). T-PA is regulated by several specific inhibitors, including type 1 plasminogen activator inhibitor (PAI-1) and α_2 plasmin inhibitor. Protein C, described above, neutralizes circulating PAI-1, providing another regulatory mechanism (Clouse and Comp 1986). Finally, about 50% of circulating plasminogen is bound to an α_2 glycoprotein called histidine-rich glycoprotein (HRG) (Lijnen et al. 1981; Bachman 1987). Alterations in HRG concentration could theoretically change the amount of plasminogen available for conversion to plasmin, and ultimately affect the fibrinolytic state (Jespersen and Kluft 1982). Overall plasmin activity, and therefore the balance between fibrin formation and degradation, depends on these interrelated processes.

Platelet Activation

The crucial role of the platelet in cerebral ischemia has been recognized for a number of years. This has been underscored by clinical trials demonstrating a reduction in stroke incidence with antiplatelet agents (Antiplatelet Trialists' Collaboration 1988).

Platelet Adhesion and Aggregation

The activation of platelets in vascular disease begins with the adhesion of platelets to an exposed subendothelium. A variety of substances present in endothelium may lead to platelet adhesion, including collagen and fibronectin. Deeper collagen structures are a stronger stimulus for platelet adhesion than those that lie near the surface (Vermylen et al. 1986; Harker and Fuster 1986; Fuster et al. 1987). Platelet surface receptors are important for this process, and platelets with defective receptors may have impaired adhesion (Fuster et al. 1987).

Adhesion, which is an interaction between platelets and other surfaces, is followed by aggregation, which is a platelet–platelet interaction. Platelet membranes contain a number of receptors involved in the aggregation process (Peerschke 1985; Vermylen et al. 1986; Fuster et al. 1987; Cahill et al. 1992). Formation of an activator–receptor complex transfers a signal to the platelet interior, which activates the metabolic processes involved in aggregation and secretion.

Platelet Membrane Receptors

Fibrinogen plays a crucial role in platelet aggregation through its interaction with platelet membrane receptors. Platelet membranes contain specific receptors for fibrinogen, called glycoprotein IIb/IIIa (GPIIb/IIIa). Under normal circumstances, the majority of fibrinogen receptors are unavailable to circulating fibrinogen. Once platelet activation has occurred (by a variety of agents) the GPIIb/IIIa receptor can bind fibrinogen, fibronectin and von Willebrand's factor. Fibrinogen binding greatly accelerates platelet aggregation, and in fact is necessary for aggregation to proceed. No matter how platelet aggregation is initiated, fibrinogen binding to the GPIIb/IIIa complex is a 'final common pathway' of platelet aggregation (Cahill et al. 1992). Thus one way in which elevated fibrinogen levels may predispose to vascular disease is through an effect on platelet activation.

The importance of the platelet fibrinogen receptor has led to new therapies designed to act on these receptors. Monoclonal antibodies to the GPIIb/IIIa complex have been developed and have now reached clinical trials in humans (Cahill et al. 1992). Initial studies suggested that ticlopidine's antiplatelet activity might be through an effect on the fibrinogen receptor, but subsequent studies have shown no direct effect of ticlopidine on these receptors (Cahill et al. 1992; Hardisty et al. 1990) and its mechanism of action remains unclear. Ticlopidine inhibits adenosine diphosphate (ADP)-induced platelet aggregation, and may affect the signal transduction between the platelet membrane complex and intracellular calcium release (Hardisty et al. 1990).

Calcium-Dependent Processes

The formation of a receptor–activator complex on the platelet membrane leads to internal platelet metabolic processes which result in the discharge of Ca^{2+} from the dense tubular system. This process is primarily mediated through phospholipase C (Vermylen et al. 1986; Harker and Fuster 1986; Fuster et al 1987). Calcium mobilization leads to contraction of the platelet and the secretion of platelet granules. It also activates phospholipase A_2, which in turn activates the cyclo-oxygenase pathway, eventually resulting in the generation of thromboxane A_2. Both platelet granule release and thromboxane A_2 release greatly potentiate further platelet aggregation and amplify thrombus formation.

Another important determinant of platelet intracellular calcium release is platelet cyclic adenosine monophosphate (AMP) (Harker and Fuster 1986; Fuster et al. 1987). The amount of platelet cyclic AMP is determined by the activity of membrane adenyl cyclase. Increased levels of platelet cyclic AMP inhibit the release of intracellular calcium. Thus platelet cyclic AMP is also an important determinant of platelet secretion and aggregation. Dipyridamole increases platelet cyclic AMP levels, which may be one mechanism of its antiplatelet activity (Harker and Fuster 1986).

Platelet Granules

Platelet dense granules contain 5-hydroxytryptamine (serotonin), ADP and calcium (Fuster et al. 1987). ADP is a potent platelet aggregant in the presence of calcium ions and fibrinogen. Platelet α granules contain the proteins β-thromboglobulin (BTG), platelet factor 4 (PF4), and platelet-derived growth factor, which are not present in plasma. α granules also contain fibrinogen, fibronectin, von Willebrand's factor, albumin, Factor V and thrombospondin (Vermylen et al. 1986; Fuster et al. 1987). Thus platelet granule release provides several strong platelet aggregants as well as other chemotactic factors (such as PF4) and inducers of smooth muscle proliferation (such as platelet-derived growth factor). The secretion of α granules has a different sensitivity

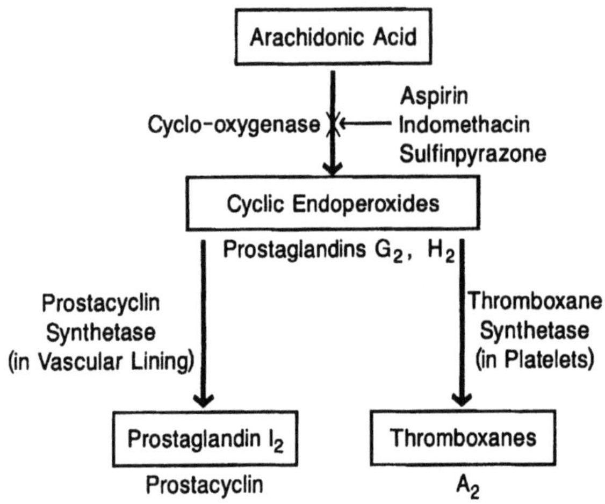

Fig. 7.4 Arachidonic acid metabolism in man. In platelets, the primary product of the arachidonic acid pathway is thromboxane A_2 whereas in the vessel wall the principal product is prostacyclin. See text for details. From Mehta J (1983) Platelets and prostaglandins in coronary artery disease. Rationale for use of platelet-suppressive drugs. Reproduced with permission of JAMA 1983; 249:2818–2823. Copyright 1983, American Medical Association.

for external stimuli than does secretion of dense granules, and they may thus serve different purposes in modulating platelet aggregation. Because BTG and PF4 are normally present in low levels in the plasma, measurement of these proteins serves as a marker of in vivo platelet activation (Dawes et al 1978; Files et al. 1981; Kaplan and Owen 1981; Owen et al. 1983).

Arachidonic Acid Metabolism

The primary product of arachidonic acid metabolism in the platelet is thromboxane A_2 (Fig. 7.4). Thromboxane A_2 is an extremely potent platelet aggregant and a vasoconstrictor (Harker and Fuster 1986; Oates et al. 1988). Platelet thromboxane release potentiates further platelet aggregation and thus amplifies clot formation. In the vessel wall, the primary product of the cyclo-oxygenase pathway is prostacyclin (PGI_2), a potent inhibitor of platelet aggregation and a vasodilator (Harker and Fuster 1986). The stable metabolites of 11-dehydrothromboxane B_2 and 2,3-di*nor*-thromboxane B_2 can be measured in the urine and serve as indices of in vivo thromboxane generation (Lawson et al. 1986; Clarke et al. 1991; Tohgi et al. 1992). Similarly urinary 2,3-di*nor*-6-keto-$PGF_{1\alpha}$ provides an index of prostacyclin production (Oates et al. 1988; Clarke et al. 1991; Tohgi et al. 1992).

Aspirin acts as an antiplatelet agent through its action on the cyclo-oxygenase pathway, by irreversibly acetylating platelet cyclo-oxygenase (Mehta 1983; Harker and Fuster 1986; Fuster et al. 1987). Because platelets are unable to synthesize new enzyme, the effect is present for the life of the platelet. Thus the effects of aspirin on platelet aggregation are detectable for 4–7 days after a single dose (Fuster et al. 1987). Although the effect on cyclo-oxygenase appears to be the primary source of aspirin's antiplatelet effect, there may be other mechanisms (Dyken et al. 1992).

Aspirin also inhibits cyclo-oxygenase in the vessel wall (Fig. 7.4). In vitro, the prostacyclin pathway in the vessel wall is much less sensitive to inhibition by aspirin than the thromboxane pathway in platelets (Oates et al. 1988). This has led to the suggestion that low doses of aspirin might inhibit thromboxane production while sparing prostacyclin generation. However, this 'biochemical selectivity' is relative and even low doses of aspirin have some effect on prostacyclin generation (Oates et al. 1988). Clarke and colleagues used a low-dose controlled-release aspirin to take advantage of the prehepatic metabolism of aspirin to salicylate, which has very weak antiplatelet activity. With this preparation they were able to demonstrate suppression of thromboxane production while leaving prostacyclin production relatively intact (Clarke et al. 1991). However, the value of 'prostacyclin sparing' remains theoretical, and the optimal dose of aspirin in patients with cerebrovascular disease remains controversial (Dyken et al. 1992; van Gijn 1992).

Platelet–Fibrin Interactions

The processes of platelet activation and fibrin formation are closely linked. A growing clot incorporates both platelets and fibrin, and positive feedback mechanisms further amplify both processes. The crucial role of fibrinogen in platelet aggregation, through the platelet membrane receptors, is described in detail above. Several other important links exist.

Thrombin is a potent activator of platelets, through a mechanism independent of ADP release or thromboxane A_2 formation (Harker and Fuster 1986). The formation of a fibrin clot may also stabilize the platelet aggregate and keep platelets in proximity with one another.

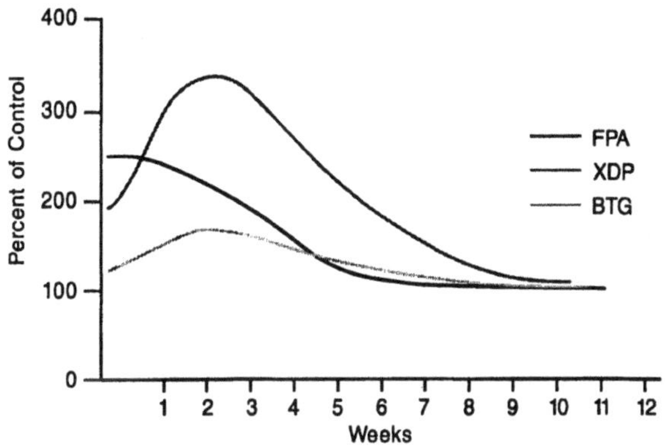

Fig. 7.5. Schematic diagram of markers of fibrin metabolism and platelet activation after acute ischemic stroke. This represents mean values of these markers in all patients. Important differences may exist among stroke subtypes; see text for details. FpA, fibrinopeptide A; XDP, cross-linked D-dimer; BTG, β-thromboglobulin.

Thrombin activates Factor XIII to Factor XIIIa (see Fig. 7.1), which stabilizes the fibrin clot and also induces platelet aggregation (Vermylen et al. 1986; Fuster et al. 1987). Thrombin generation may thus be an important element in incorporating platelets into a growing clot.

Platelet phospholipid is required for several of the steps in the clotting cascade, including the activation of Factor X by either the intrinsic or the extrinsic pathway, and the conversion of Factor II (prothrombin) to thrombin (Fig. 7.1). Thrombin generation induces more platelet activation, which provides further platelet phospholipid, allowing further clot formation.

Platelets also secrete PAI-1, but the role of platelet-derived PAI-1 is not certain (Sprengers et al. 1986; Booth et al 1988; Francis 1989). PAI-1 in platelets may exist in an inactive (latent) form (Sprengers et al. 1986), but the conditions under which this PAI-1 is activated are not clear. The release of active PAI-1 by aggregating platelets might theoretically stimulate clot formation by inhibiting plasmin generation and overall fibrinolytic activity. Platelet-derived PAI-1 may have importance locally in the area of a clot, where platelet concentration is high (Sprengers et al. 1986; Booth et al. 1988).

Thus, the platelet and fibrin pathways are interconnected by a variety of acceleratory and inhibitory processes. It seems likely that these pathways rarely operate in isolation, and anticoagulants or antiplatelet agents have effects on both systems. Although in clinical practice *either* an antiplatelet agent *or* an anticoagulant is usually used, the optimal management of cerebrovascular disease may require attention to both processes.

Measurement of Coagulation Markers Following Stoke

Markers of Fibrin Metabolism

The author's laboratory and others have utilized markers of fibrin metabolism to characterize the activation of coagulation following stroke. Levels of fibrinopeptide A (FpA) are significantly elevated following stroke (Landi et al. 1987; D'Angelo et al. 1988; Sandset and Andersson 1989; Feinberg et al. 1989; Fisher and Francis 1990; Tohgi et al. 1990; Lindahl et al. 1990). Values in acute ischemic stroke patients seen at the author's institution are shown in Table 7.1, and schematically in Fig 7.5. Mean values in these patients during the first week after stroke were 400% of control values. FpA levels decline over several weeks to months (D'Angelo et al. 1988; Feinberg et al. 1989; Fisher and Francis 1990). This is in contrast to myocardial infarction, where FpA levels decline to normal within 48 hours (Jaffe et al 1987). In one study, FpA

Table 7.1. Hemostatic markers after acute ischemic stroke

		FpA (ng/ml)	XDP (ng/ml)	BTG (IU/ml)	PF4 (IU/ml)
Controls	($n = 37$)	4.8 ± 0.9	101 ± 7	21.7 ± 2.1	3.7 ± 0.8
Week 1	($n = 70$)	15.3 ± 2.8*	243 ± 35*	25.6 ± 2.1	5.9 ± 0.8
Week 2	($n = 39$)	12.0 ± 1.9*	333 ± 60*	30.2 ± 2.6*	6.8 ± 1.1*
Week 3	($n = 26$)	10.8 ± 1.9*	330 ± 59*	25.1 ± 1.9	5.6 ± 1.1
Week 4	($n = 25$)	7.6 ± 1.4	182 ± 32*	29.1 ± 3.1	5.1 ± 0.9
3 months	($n = 28$)	7.8 ± 2.1	144 ± 18*	23.6 ± 2.0	5.2 ± 0.9

FpA, fibrinopeptide A; XDP, cross-linked D-dimer; BTG, β-thromboglobulin; PF4, platelet factor 4. Results are mean ± SEM; *Values significantly different from controls ($P < 0.05$).

levels fell more quickly when warfarin anticoagulation was used (Yasaka et al. 1990). FpA levels in the acute phase may correlate with mortality following stroke (Landi et al. 1987; D'Angelo et al. 1988; Feinberg et al. 1991b).

In contrast, mean levels of fibrin(ogen) degradation products are often normal or only slightly increased following stroke. An early study found elevated B-β 1–42 levels after stroke (Lane et al. 1983), but we (Feinberg et al. 1989) and others (Fisher and Francis 1990) have found normal or reduced values of the B-β peptides in acute stroke. Mean D-dimer levels are mildly elevated in the acute phase (150%–200% of control values) and rise in the subacute phase several weeks after stroke (Feinberg et al. 1989; Fisher and Francis 1990; Tohgi et al. 1990; Takano et al. 1992) (Fig. 7.5). The ratio of FpA to D-dimer is elevated in acute stroke (Feinberg et al 1989; Yasaka et al 1990) and declines slowly. Thus in the acute phase there is a marked elevation in FpA, normal or reduced B-β peptide levels, and a mild increase in D-dimer. Taken together, these data suggest that the acute phase following stroke is characterized by marked fibrin generation with relatively less fibrinolytic activity. Markers of endogenous fibrinolytic activity increase slowly in the subacute phase (Feinberg et al. 1989; Feinberg et al. 1991a; Takano et al. 1992).

Only a limited number of studies have measured specific activators or inhibitors of fibrinolysis in stroke. Pizzo et al. found normal or elevated levels of releasable plasminogen activator in patients with chronic strokes (Pizzo et al. 1985). Fisher and Francis (1990) found normal levels of PAI-1 antigen acutely following stroke, but elevated levels 2 months later. The author's preliminary observations have found elevated t-PA antigen acutely following stroke, but also elevated levels of both PAI-1 antigen and PAI activity (Feinberg et al. 1991a). This suggests that fibrinolytic activity may be limited by the release of PAI-1. Subgroups of patients may exist with either defective release of plasminogen activator or plasminogen activator inhibitors, as has been found in patients with recurrent deep venous thrombosis (Juhan-Vague et al. 1987). Additional studies are necessary, using assays of fibrinolytic inhibitors and activators, correlated with careful clinical stroke assessment.

Markers of Platelet Activation Following Stroke

Platelet activation has long been recognized as an important part of cerebral ischemia, but early studies of platelet activation suffered from technical limitations. These studies, which used in vitro measurements of platelet aggregation or adhesion to a variety of stimuli, demonstrated 'increased platelet aggregability' following acute stroke (Kalendovsky et al. 1975; Couch and Hassanein 1976; Dougherty et al. 1977; Mettinger et al. 1979; Hoogendijk et al. 1979; Otsuki et al. 1983; Uchiyama et al. 1983). Some investigators found this hyperaggregability returned to normal in 'chronic stroke' patients (Yamazaki et al. 1975; Dougherty et al. 1977; Mettinger et al. 1979), while others reported persistent abnormalities (Danta 1973; Kalendovsky et al. 1975; Couch and Hassonein 1976). Other direct indices of platelet activation are abnormal in patients with stroke, including shortened platelet survival time (Harker and Slichter 1972; Kalendovsky et al. 1975) and circulating platelet aggregates (Wu and Hoak 1974; Dougherty et al. 1977).

The development of assays for the platelet-specific protein β-thromboglobulin (BTG) and platelet factor 4 (PF4) advanced our ability to assess in vivo platelet activation (Dawes et al. 1978; Files et al. 1981; Kaplan and Owen 1981). As described above, these proteins are contained in α granules and released during second-phase aggregation (Fuster et al. 1987). Because of their short half-lives they serve as indices of ongoing platelet activation (Dawes et al. 1978; Files et al. 1981; Kaplan and Owen 1981). Elevated levels of platelet proteins have been found after acute stroke in most (Hoogendijk et al. 1979; Lane et al. 1981; Fisher et al. 1982; Taomoto et al. 1983; Fisher and Francis 1990; Feinberg and Bruck 1991) but not all (de Boer et al. 1982) series. The authors found that mean BTG levels reach a peak in the second week following stroke, and then decline to normal levels (Feinberg and Bruck 1991) (Fig. 7.5). Others have found both elevated (Lane et al. 1981; Taomoto et al. 1983; Fisher and Francis 1990) and normal (Cella et al. 1979) levels in patients with 'chronic' strokes. Shah et al. (1985) found that levels of BTG drawn in the first week after stroke were higher than those drawn later, but only 8 patients were evaluated after the first week (Shah et al. 1985).

Although measurement of platelet proteins represents an advance in assessing in vivo platelet activation, a number of methodological problems remain. Venepuncture technique and laboratory processing are critical, and artefactual elevations may occur (Kaplan and Owen 1981; Files et al. 1981). The anticoagulant mixture used significantly affects the measurements (Files et al.

1981). Even in carefully done studies there is a wide variation in levels of platelet proteins, and many stroke patients have normal values. Further, levels do not correlate with stroke size (Shah et al. 1985; Feinberg and Bruck 1991). Measurement of urine levels of BTG or the stable metabolites of thromboxane and prostacyclin might provide a better index of in vivo platelet activity, but these have not been carefully examined in acute stroke.

Differences Between Stroke Subtypes

Stroke is a heterogeneous entity, and generalizations regarding hemostatic markers may not hold for all stroke subtypes. Important differences have been observed among different stroke types, which may provide insights into differences in pathophysiology. These markers could eventually aid in the determination of stroke subtypes and possibly guide treatment.

Normal values of hemostatic markers are generally found in patients with small deep ('lacunar') infarction. Mean values of FpA, D-dimer, PF4 and BTG are normal in these patients (Shah et al. 1985; Feinberg et al. 1989; Fisher and Francis 1990; Feinberg and Bruck 1991; Takano et al. 1992). It is not clear whether this is simply due to the small volume of tissue infarcted or is an indication of a fundamental difference in pathophysiology. Recently, Takano et al. reported elevated 'fluorogenic prothrombin time' in patients with lacunar stroke compared to controls; values were similar to other patients with thrombotic stroke (Takano et al. 1990). The significance of this finding is still uncertain.

Elevated values of fibrin markers have been found after both cardioembolic and atherothrombotic strokes, but the time course of changes may differ (Feinberg et al. 1989; Fisher and Francis 1990; Tohgi et al. 1990; Takano et al. 1992). Takano and colleagues have suggested that D-dimer is elevated acutely only in patients with cardioembolic stroke, whereas in patients with atherothrombotic stroke D-dimer levels are normal acutely and rise after 7 days (Takano et al. 1992). We have also found that D-dimer levels during the first week are significantly greater in cardioembolic stroke than in atherothrombotic stroke, although the levels of D-dimer were not normal in our patient with atherothrombosis. High levels of D-dimer seen acutely in a stroke patient suggest a cardiogenic embolus or an intercurrent illness such as cancer, myocardial infarction or venous thrombosis.

Since D-dimer levels represent cross-linked fibrin clot which has been degraded by plasmin, in cardioembolic stroke there may have been greater clot formation (and breakdown) before a stroke occurs. It is interesting to speculate whether these higher levels represent simply a larger volume of clot, or a fundamental difference in clot formation in cardioembolic versus atherothrombotic stroke. It is tempting to conclude that in cardioembolic stroke a clot is present prior to the stroke, whereas in atherothrombotic stroke a significant volume of fibrin clot forms only after the stroke. Recently, Takano et al. reported a group of patients with cardioembolic stroke who had high levels of D-dimer in the acute stage (Takano et al. 1991). They found that patients who developed recurrent embolism or who had intracardiac thrombus had the highest levels of D-dimer, and postulated that this reflected 'local (intracardiac) derangement of the clotting process'.

Elevated levels of BTG and PF4 are seen in both atherothrombotic and cardioembolic stroke (Shah et al. 1985; Feinberg and Bruck 1991). Mean BTG levels in our patients were higher acutely in cardioembolic strokes than in atherothrombotic strokes (Feinberg and Bruck 1991). During the second week, BTG levels stayed elevated in cardioembolic stroke and rose significantly in atherothrombotic strokes. Since BTG levels did not correlate with either stroke size or severity, this again suggests differences in clot formation between cardioembolic and atherothrombotic stroke. The effect of anticoagulant or antiplatelet therapy on these markers following stroke has not been carefully examined, and might be of importance in comparing cardioembolic and atherothrombotic stroke. The elevated BTG levels in cardioembolic stroke are of interest in light of the recent Stroke Prevention in Atrial Fibrillation Study demonstrating that aspirin is effective in reducing strokes in patients with atrial fibrillation (Stroke Prevention in Atrial Fibrillation Investigators 1991). This again points out the close linkage between the platelet and fibrin pathways. Although some conditions may favor activation of one pathway over another (Cheseboro et al. 1990), many thrombotic conditions involve activation of both systems.

In both atherothrombotic and cardioembolic stroke, markers of hemostasis and platelet activation remain elevated for weeks. It is possible that the original source of thrombus remains and continues to generate clot. More likely, these elevated levels represent 'remodeling' of the clot within the cerebral circulation. Also, some of this activation of the clotting system may be secondary to the stroke, as damaged brain releases procoagulants into the blood. Whether antithrombotic therapy changes the time course of hemostatic activation has not been carefully explored. There is recent evidence of extravas cular fibrin formation after stroke, and this may also contribute to the levels of these markers (Okada et al. 1994).

Summary

Acute ischemic stroke is a thrombotic process in 80% of cases. Fibrin generation, fibrinolysis and platelet activation are linked processes which are often activated together. The balance between these processes may determine the progression and ultimate outcome of a stroke. Advances in laboratory measurement of hemostasis have allowed us to better quantify these processes. Acute stroke is characterized by brisk fibrin formation and platelet activation which continues for several weeks after the stroke. Although endogenous fibrinolysis is also activated, initially the balance is shifted towards fibrin formation. Important differences are emerging between stroke types which may help us understand the pathophysiology of stroke. These differences may eventually prove useful in diagnosis and in guiding treatment.

Whether there are chronic abnormalities in hemostasis or fibrinolysis that predispose to stroke remains less clear (Francis 1989; Gliksman and Wilson 1992). It seems logical that this must be so: coagulation mechanisms must play a role in determining why one patient with a carotid stenosis remains stable, another has an asymptomatic occlusion and a third has a large cerebral infarction. Most studies identifying chronic abnormalities of thrombosis, fibrinolysis or platelet activity have examined patients after a stroke has occurred. Studies have differed in patient selection, the definition of 'acute' or 'chronic', the types of stroke included, and even the definition of stroke. Carefully done prospective studies using modern hemostatic markers are needed to answer this most important question.

Acknowledgement: This work was supported by grants from the Arizona Affiliate of the American Heart Association, the Arizona Disease Control Research Commission, and The Robert H. Flinn Foundation.

References

Alkjaersig N, Fletcher AP (1982) Catabolism and excretion of fibrinopeptide A. Blood 60:148–156

Antiplatelet Trialists' Collaboration (1988) Secondary prevention of vascular disease by prolonged antiplatelet treatment. Br Med J 296:320–331

Bachman F (1987) Fibrinolysis. In: Verstraete M, Vermylen J, Lijnen HR, Arnout S (eds) Thrombosis and hemostasis. Leuven University Press, Leuven, Chapter 10

Booth NA, Simpson AJ, Croll A, Bennett B, MacGregor IR (1988) Plasminogen activator inhibitor in plasma and platelets. Br J Haematol 70:327–333

Brott TG, Haley EC, Levy DE et al. (1992) Urgent therapy for stroke. Part I. Pilot study of tissue plasminogen activator administered within 90 minutes. Stroke 23:632–640

Cahill M, Mistry R, Barnett DB (1992) The human platelet fibrinogen receptor: clinical and therapeutic significance. Br J Clin Pharmacol 33:3–9

Cella G, Zahavi J, de Haas HA, Kakkar VV (1979) β-thromboglobulin, platelet production time and platelet function in vascular disease. Br J Haematol 43:127–136

Chesebоro JH, Fuster V, Halperin JL (1990) Atrial fibrillation – risk marker for stroke. (Editorial) New Engl J Med 323:1556–1558

Clarke RJ, Mayo G, Price P, FitzGerald GA (1991) Suppression of thromboxane A_2 but not of systemic prostacyclin by controlled-release aspirin. New Engl J Med 325:1137–1141

Clouse LH, Comp PC (1986) The regulation of hemostasis: the protein C system. New Engl J Med 314:1298–1304

Conway EM, Bauer KA, Barzegar S, Rosenberg RD (1987) Suppression of hemostatic system activation by oral anticoagulants in the blood of patients with thrombotic diatheses. J Clin Invest 80:1535–1544

Cook NS, Ubben D (1990) Fibrinogen as a major risk factor in cardiovascular disease. Therapy Pharm Stud 11:444–451

Couch JR, Hassanein RS (1976) Platelet aggregation, stroke, and transient ischemic attack in middle-aged and elderly patients. Neurology 26:888–895

Dalal PM, Shah PM, Sheth SC, Deshpande CK (1965) Cerebral embolism: angiographic observations on spontaneous clot lysis. Lancet 1:61–64

D'Angelo AA, Landi G, D'Angelo SV et al. (1988) Protein C in acute stroke. Stroke 19:579–583

Danta G (1973) Platelet aggregation in patients with cerebral vascular disease and in control subjects. Thromb Diathes Haem 29:730–732

Dawes J, Smith RC, Pepper DS (1978) The release, distribution, and clearance of human β-thromboglobulin and platelet factor 4. Thromb Res 12:851–861

de Boer AC, Turpie ACG, Butt RW et al. (1982) Plasma beta thromboglobulin and serum fragment E in acute partial stroke. Br J Haematol 50:327–337

del Zoppo GH, Zeumer H, Harker LA (1986) Thrombolytic therapy in stroke: possibilities and hazards. Stroke 17:595–607

del Zoppo GH, Poeck K, Pessin MS et al. (1992) Recombinant tissue plasminogen activator in acute thrombotic and embolic stroke. Ann Neurol 32:78–86

Dougherty JH, Levy DE, Weksler BB (1977) Platelet activation in acute cerebral ischemia. Lancet 3:821–824

Dyken ML, Barnett HJM, Easton JD et al. (1992) Low-dose aspirin and stroke: 'It ain't necessarily so'. Stroke 23:1395–1399

Eisenberg PR, Sherman LA, Tiefenbrunn AJ et al. (1987) Sustained fibrinolysis after administration of t-PA despite its short half-life in the circulation. Thromb Haemost

57:35–40
Erickson LA, Schleef RR, Ny T, Loskutoff DJ (1985) The fibrinolytic system of the vascular wall. Clin Haematol 14:513–530
Ernst E (1990) Plasma fibrinogen – an independent cardiovascular risk factor. J Intern Med 227:365–372
Estol C, Pessin MS, DeWitt LD, Caplan LR (1989) Stroke and increased factor VIII activity. (Abstract) Neurology 39 (Suppl 1):159
Feinberg WM, Bruck DC (1991) Time course of platelet activation following acute ischemic stroke. J Stroke Cerebrovasc Dis 1:124–128
Feinberg WM, Bruck DC, Ring ME, Corrigan JJ (1989) Hemostatic markers in acute stroke. Stroke 20:582–587
Feinberg WM, Bruck DC, Jeter MA, Corrigan JJ Jr (1991a) Fibrinolysis after acute ischemic stroke. Thromb Res 64:117–127
Feinberg WM, Paullette LA, Bruck DC (1991b) Predictive value of hemostatic markers in acute stroke. (Abstract) Neurology 41 (Suppl 1):298
Files JC, Malpass TW, Yee EK et al. (1981) Studies of human platelet alpha-granule release in vivo. Blood 58:607–618
Fisher M, Francis R (1990) Altered coagulation in cerebral ischemia. Platelet, thrombin, and plasmin activity. Arch Neurol 47:1075–1079
Fisher M, Levine PH, Fullerton AL et al. (1982) Marker proteins of platelet activation in patients with cerebrovascular disease. Arch Neurol 39:692–695
Foulkes MA, Wolf PA, Price TP, Mohr JP, Hier DB (1988) The Stroke Data Bank: design, methods, and baseline characteristics. Stroke 19:547–554
Francis CW, Marder VJ (1986) Concepts of clot lysis. Annu Rev Med 37:187–204
Francis RB Jr (1989) Clinical disorders of fibrinolysis: a critical review. Blut 59:1–14
Fuster V, Badimon L, Badimon J et al. (1987) Drugs interfering with platelet functions: mechanisms and clinical relevance. In: Verstraete M, Vermylen J, Lijnen HR, Arnout J (eds) Thrombosis and haemostasis. Leuven University Press, Leuven, Chapter 15: pp 349–417
Gliksman M, Wilson A (1992) Are hemostatic factors responsible for the paradoxical risk factors for coronary heart disease and stroke? Stroke 23:607–610
Gómez-Aranda F, Dominguez JML, Fernández VR, García EM (1992) Stroke and familial protein S deficiency. (Letter) Stroke 23:299
Hardisty RM, Powling MJ, Nokes TJ (1990) The action of ticlopidine on human platelets. Studies on aggregation, secretion, calcium mobilization, and membrane glycoproteins. Thromb Haemost 64:150–155
Harker LA, Fuster V (1986) Pharmacology of platelet inhibitors. J Am Coll Cardiol 8:21B–32B
Harker LA, Slichter SJ (1972) Platelet and fibrin consumption in man. New Engl J Med 287:999–1005
Hart RG, Kanter MC (1990) Hematologic disorders and ischemic stroke. A selective review. Stroke 21:1111–1121
Hirsch J (1991) Heparin. New Engl J Med 324:1565–1574
Hoogendijk EMG, Jenkings CRP, Wijk EM, ten Cate JW (1979) Spontaneous platelet aggregation in cerebrovascular disease. II. Thromb Haemost 41:512–522
Hunt FA, Rylatt DB, Hart R, Bundesen PG (1985) Serum crosslinked fibrin (XDP) and fibrinogen/fibrin degradation products (FDP) in disorders associated with activation of the coagulation or fibrinolytic systems. Br J Haematol 60:715–722
Ishii H, Salem HH, Bell CE, Laposata EA, Majerus PW (1986) Thrombo-modulin, an endothelial anticoagulant protein, is absent from the human brain. Blood 67:362–365
Jaffe AS, Eisenberg PR, Wilner GD (1987) In vivo assessment of thrombosis and fibrinolysis during acute myocardial infarction. Prog Hematol 15:71–89
Jespersen J, Kluft C (1982) Decreased levels of histidine-rich glycoprotein (HRG) and increased levels of free plasminogen in women on oral contraceptives low in estrogen. Thromb Haemost 48:283–285
Juhan-Vague I, Valdier J, Alessi MC et al. (1987) Deficient t-PA release and elevated PA inhibitor levels in patients with spontaneous or recurrent deep venous thrombosis. Thromb Haemost 57:67–75
Kalendovsky Z, Austin J, Steele P (1975) Increased platelet aggregability in young patients with stroke. Arch Neurol 32:13–20
Kannel WB, Wolf PA, Castelli WP, D'Agostino RB (1987) Fibrinogen and risk of cardiovascular disease. JAMA 258:1183–1186
Kaplan KL, Owen J (1981) Plasma levels of β-thromboglobulin and platelet factor 4 as indices of platelet activation in vivo. Blood 57:199–202
Landi G, D'Angelo A, Boccardi E et al. (1987) Hypercoagulability in acute stroke: prognostic significance. Neurology 37:1667–1671
Lane DA, Gawel MJ, Wolff S et al. (1981) The effects of age and thrombotic stroke upon plasma concentrations of fibrinogen derivatives and platelet release. Thromb Haemost 46:437
Lane DA, Wolff S, Freland H, Gawel M, Foadi M (1983) Activation of coagulation and fibrinolytic systems following stroke. Br J Haematol 53:655–658
Lawson JA, Patrano C, Ciabattoni G, FitzGerald GA (1986) Long-lived enzymatic metabolites of thromboxane B_2 in the human circulation. Anal Biochem 155:198–205
Lijnen HR, Jacobs G, Collen D (1981) Histidine-rich glycoprotein in a normal and a clinical population. Thromb Res 22:519–523
Lindahl AK, Sandset PM, Abildgaard U (1990) Indices of hypercoagulation in cancer as compared with those in acute inflammation and acute infarction. Haemostasis 20:253–262
Meade TW, Mellows S, Brosovic M et al. (1986) Haemostatic function and ischaemic heart disease: principal results of the Northwick Park Heart Study. Lancet 2:533–537
Mehta J (1983) Platelets and prostaglandins in coronary artery disease. Rationale for use of platelet-suppressive drugs. JAMA 249:2818–2823
Mettinger KL, Nyman D, Kjellin KG, Sidén Å, Söderström CE (1979) Factor VIII related antigen, antithrombin III, spontaneous platelet aggregation and plasminogen activator in ischemic cerebrovascular disease. J Neurol Sci 41:31–38
Millenson MM, Bauer KA, Kistler JP et al. (1992) Monitoring 'mini-intensity' anticoagulation with warfarin: comparison of the prothrombin time using a sensitive thromboplastin with prothrombin fragment F_{1+2} levels. Blood 79:2034–2038
Mohr JP, Caplan LR, Melski JW et al (1978) The Harvard Cooperative Stroke Registry: a prospective registry. Neurology 28:754–762
Nossel HL (1981) Relative proteolysis of fibrin Bβ chain by thrombin and plasmin as a determinant of thrombosis. Nature 291:165–167
Oates JA, FitzGerald GA, Branch RA et al. (1988) Clinical implications of prostaglandin and thromboxane A_2 formation. New Engl J Med 319:689–698
Okada Y, Copeland BR, Tung M-M, del Zoppo GJ (1994) Fibrin forms in the perivasicular tissue during focal cerebral ischemia and reperfusion. (Abstract) Stroke 25:266
Otsuki Y, Kondo T, Shio H, Kameyama M, Koyama T (1983)

Platelet aggregability in cerebral thrombosis – analyzed for vessel stenosis. Stroke 14:368–371

Owen J, Kvam D, Nossel H, Kaplan KL, Kernoff PBA (1983) Thrombin and plasmin activity and platelet activation in the development of venous thrombosis. Blood 60:476–482

Peerschke EIB (1985) The platelet fibrinogen receptor. Semin Hematol 22:241–259

Pizzo SV, Petruska DB, Doman KA, Soong S, Fuchs HE (1985) Releasable vascular plasminogen activator and thrombotic strokes. Am J Med 79:407–411

Rick ME (1990) Protein C and protein S. Vitamin K-dependent inhibitors of blood coagulation. JAMA 263:701–703

Sacco RL, Owen J, Mohr JP, Tatemichi TK, Grossman BA (1989) Free protein S deficiency. A possible association with cerebrovascular occlusion. Stroke 20:1657–1661

Sandset PM, Andersson TR (1989) Coagulation inhibitor levels in pneumonia and stroke: changes due to consumption and acute phase reaction. J Intern Med 225:311–316

Shah AB, Beamer N, Coull BM (1985) Enhanced in vivo platelet activation in subtypes of ischemic stroke. Stroke 16:643–647

Sherman DG, Dyken ML, Fisher M, Harrison MJG, Hart RG (1989) Antithrombotic therapy for cerebrovascular disorders. Chest 95 (Suppl 1) 140S–155S

Sloan MA (1986) Thrombolysis and stroke – past and future. Arch Neurol 44:748–768

Solis OJ, Roberson GR, Taveras JM, Mohr JP, Pessin MS (1977) Cerebral angiography in acute cerebral infarction. Revist Interam Radiol 2:19–25

Sprengers DE, Akkerman JWN, Jansen BG (1986) Blood platelet plasminogen activator inhibitor: two different pools of endothelial cell type plasminogen activator inhibitor in human blood. Thromb Haemost 55:325–329

Stroke Prevention in Atrial Fibrillation Investigators (1991) The Stroke Prevention in Atrial Fibrillation Study. Final Results. Circulation 84:527–539

Takano K, Yamaguchi T, Okada Y et al. (1990) Hypercoagulability in acute ischemic stroke: analysis of the extrinsic coagulation reactions in plasma by a highly sensitive automated method. Thromb Res 58:481–491

Takano K, Yamaguchi T, Kato H, Omae T (1991) Activation of coagulation in acute cardioembolic stroke. Stroke 22:12–16

Takano K, Yamaguchi T, Uchida K (1992) Markers of a hypercoagulable state following acute ischemic stroke. Stroke 23:194–198

Taomoto K, Asada M, Kanazaua Y, Matsumoto S (1983) Usefulness of the measurement of plasma B-thromboglobulin (B-TG) in cerebrovascular disease. Stroke 14:518–524

Teitel JM, Bauer KA, Lau HK, Rosenberg RD (1982) Studies of the prothrombin activation pathway utilizing radioimmunoassays for the F_2/F_{1+2} fragment and thrombin–antithrombin complex. Blood 59:1086–1097

Tohgi H, Kawashima M, Tamura K, Suzuki H (1990) Coagulation–fibrinolysis abnormalities in acute and chronic phases of cerebral thrombosis and embolism. Stroke 21:1663–1667

Tohgi H, Konno S, Tamura K, Kimura B, Katsumi K (1992) Effects of low-to-high doses of aspirin on platelet aggregability and metabolites of thromboxane A_2 and prostacyclin. Stroke 23:1400–1403

Uchiyama S, Takeuchi M, Osawa M et al. (1983) Platelet function in thrombotic cerebrovascular disorders. Stroke 14:511–517

van Gijn J (1992) Aspirin: dose and indications in modern stroke prevention. Neurol Clin 10:193–207

Vermylen J, Verstraete M, Fuster V (1986) Role of platelet activation and fibrin formation in thrombogenesis. J Am Coll Cardiol 8:2B–9B

Wessler S, Gitel SN (1986) Pharmacology of warfarin and heparin. J Am Coll Cardiol 8:10B–20B

Whitaker AN, Elms JJ, Masci PP et al. (1984) Measurement of cross-linked fibrin derivatives in plasma: an immunoassay using monoclonal antibodies. J Clin Pathol 37:882–887

Wilhelmsen L, Svardsudd K, Korsan-Bengtsen K et al. (1984) Fibrinogen as a risk factor for stroke and myocardial infarction. New Engl J Med 311:501–505

Wong VL, Hofman FM, Ishii H, Fisher M (1991) Regional distribution of thrombomodulin in human brain. (Abstract) Stroke 22:152

Wu KK, Hoak JC (1974) A new method for the quantitative detection of platelet aggregates in patients with arterial insufficiency. Lancet 2:924–926

Yamazaki H, Takahashi T, Sano T (1975) Hyperaggregability of platelets in thromboembolic disorders. Thromb Diathes Haem 34:94–105

Yasaka M, Yamaguchi T, Miyashita T, Tsuchiya T (1990) Regression of intracardiac thrombus after embolic stroke. Stroke 21:1540–1544

8. Immunohematologic Mechanisms in Stroke

M. J. Fisher

Stroke investigators have made great progress in defining the neurochemical consequences of brain ischemia, as well as providing a scientific foundation for the medical and surgical therapies of secondary stroke prevention. Nevertheless, it is striking that relatively little is known of the initiating process of the clinical event constituting stroke. This process, lacking precise characterization, is typically referred to as 'hypoperfusion' or 'embolization'. Current neurodiagnostic technology has blurred, rather than clarified, this distinction. An example of this phenomenon comes from the landmark North American Symptomatic Carotid Endarterectomy Trial, in which the efficacy of surgical removal of symptomatic carotid plaques appeared to be directly related to the degree of hemodynamic compromise produced by these lesions (NASCET Collaborators 1991). Yet angiography performed in the first hours after stroke commonly shows possible embolic occlusion distal to carotid stenosis (del Zoppo et al. 1992). Transcranial Doppler studies of patients with symptomatic carotid plaques also suggest that intracranial embolism occurs in patients with these atheromatous lesions (Siebler et al. 1992). Another example is the pathophysiology of stroke associated with chronic non-valvular atrial fibrillation. Transesophageal echocardiographic studies have shown striking evidence of an embolic source in many of these patients (Lee et al. 1991). Nevertheless, both cerebral blood flow (Petersen et al. 1989) and transcranial Doppler (Ameriso et al. 1992) studies have demonstrated substantial intracranial hemodynamic effects of atrial fibrillation. It has therefore become increasingly appealing to identify mechanisms of stroke likely to bridge both embolic and flow-related processes.

Hemostasis Regulation

There is compelling, albeit indirect, evidence linking thrombosis and ischemic stroke pathophysiology.

1. Alterations of systemic hemostatic variables are prominent in association with both acute and chronic ischemic stroke (Fisher and Francis 1990).
2. The incidence of brain infarction can be significantly reduced by pharmacotherapy, i.e. aspirin, warfarin and ticlopidine, which interferes with the components of hemostasis pathways (Hass et al. 1989).
3. The importance of occlusive thrombus in acute brain infarction has been emphasized by the finding that thrombolytic therapies appear to enhance the recanalization of intracranial arteries following acute stroke (Hacke et al. 1988).
4. Brain infarction tends to occur in the early morning hours, when endogenous fibrinolytic activity is at its minimum (Marsh et al. 1990).

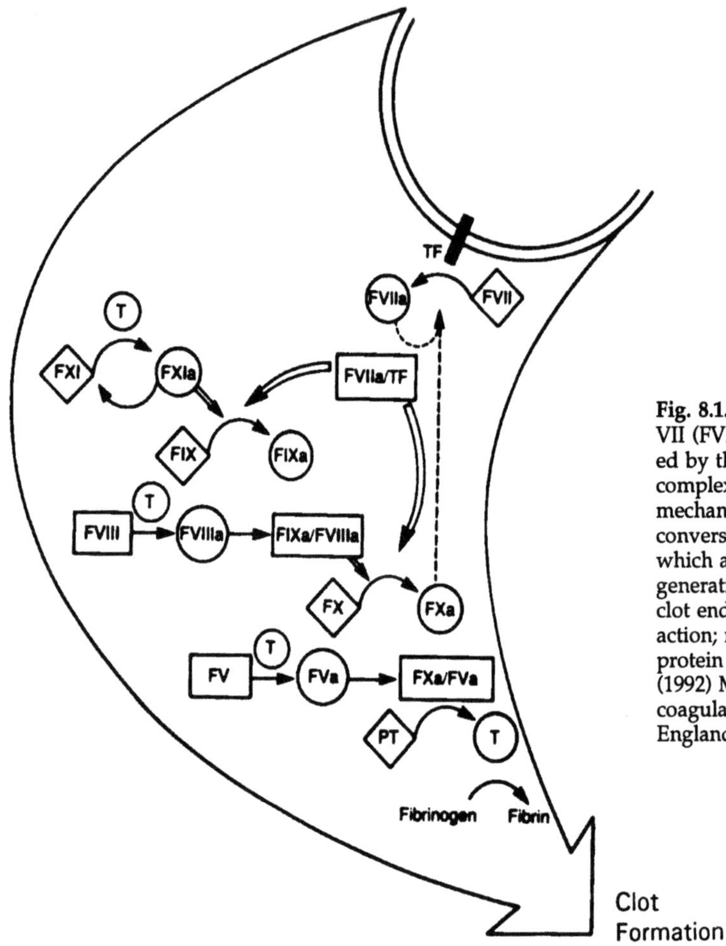

Fig. 8.1. Cell-surface tissue factor (TF) binds Factor VII (FVII). Factors IX (FIX) and X (FX) are activated by the tissue factor-activated Factor VIIa (FVIIa) complex. Blood clotting is accelerated by feedback mechanisms, e.g. prothrombin (PT) to thrombin (T) conversion or other enzymes (dashed arrows), which activate Factors VII, XI, VIII and V; fibrin generation and polymerization produce the fibrin clot end-product. Open arrows indicate enzymatic action; narrow solid arrows indicate conversion of protein functional state. From Furie B and Furie B (1992) Molecular and cellular biology of blood coagulation. Reproduced with permission of New England Journal of Medicine 326:800–806.

This association between thrombosis and brain infarction emphasizes the importance of understanding hemostasis regulation, both systemically and within the central nervous system. Nevertheless, hemostasis has been a relatively uncommon area of investigation for those interested in stroke. This is surprising, given the multidisciplinary emphasis of investigations of other neurological disorders, for example neuroimmunology. Hemostasis regulation may prove to be as relevant to stroke pathophysiology as immune regulation is to the etiology of multiple sclerosis; perhaps the future will bring us the emerging field of 'neurohemostasis' or 'neurobiology of coagulation'.

The traditional concepts of hemostasis consist of clearly distinct intrinsic and extrinsic pathways that are initiated by activation of circulating zymogens of the various coagulation factors. This view has given way to an often bewildering array of circulating and tissue-bound activators, inhibitors and zymogens, which interact in a manner sufficiently complex to discourage most non-experts in hemostasis. A snapshot of the newer version of the coagulation cascade is shown in Fig. 8.1; note the lack of notation of intrinsic and extrinsic pathways (pathways which are relevant mostly for in-vitro monitoring of hemostatic function), as well as the emphasis on tissue factor as the key initiator of clot formation (Furie and Furie 1992). Tissue factor is normally expressed by cells of the adventitia, but not the vascular endothelium (Drake et al. 1989). Interestingly tissue factor is also present in substantial quantities in brain, particularly in the cortex (Drake et al. 1989; Fleck et al. 1990), and tissue factor mRNA has been demonstrated in astrocytes (Eddleston et al. 1993).

The distribution of tissue factor suggests it has a function as a 'hemostatic envelope', ready to initiate the coagulation cascade when the integrity of the vasculature is disrupted (Drake et al. 1989). In contrast, cell culture studies have emphasized inducible tissue factor expression by vascular endothelium and monocytes (Conway et al. 1989). The usual stimuli for tissue factor induction are cytokines interleukin-1 (IL-1) and tumor necrosis factor (TNF), as well as lipopolysaccharide (LPS, a component of the wall of Gram-negative bacteria, also referred to as endotoxin). Both IL-1 and TNF are important participants in the inflammatory response; they are rapidly secreted by activated cells, usually macrophages but also astrocytes and microglia, have a short half-life (10–30 minutes), and bind to high-affinity specific cell-surface receptors (Benveniste 1992). IL-1 and TNF both exist in α and β forms, which have no more than 30% structural homology but bind to the same receptors; IL-1β and TNFα are the predominant forms in man. Among their many functions, both IL-1 and TNF enhance endothelial permeability and induce expression by endothelial cells of the leukocyte adhesion molecules P-selectin and intercellular adhesion molecule-1 (ICAM-1) (Benveniste 1992; Sloan et al. 1992). That both IL-1 and TNF can induce tissue factor expression demonstrates one of several critical linkages between the hemostatic and the immune systems.

There are many other ways in which the hemostatic system is regulated, including endothelial secretion of the platelet antiaggregant prostacylin (a prostaglandin derivative) and the action of tissue factor pathway inhibitor, which inactivates the tissue factor–Factor VIIa–Factor Xa complex (Sandset et al. 1992). Among the most important of the regulatory systems are the thrombolytic and the antithrombin systems. The thrombolytic system, best known by the use of exogenous tissue plasminogen activator (t-PA) for coronary as well as intracranial thrombosis, is endothelial-dependent and manifests its fibrin-degrading function by the balance of plasminogen activators (both t-PA and urokinase-type plasminogen activator, or u-PA) and inhibitors of plasminogen activation. Most important among the latter is plasminogen activator inhibitor-1 (PAI-1); the effects of t-PA predominate over PAI-1 in unstimulated endothelium. In addition to thrombolytic effects, plasminogen activators and inhibitors appear to have an important role in neuronal migration (Krystosek and Seeds 1981) and in the functional activity of immature astrocytes (Kalderon et al. 1990). As seen with tissue factor expression, cell culture studies have shown that the inflammatory mediators IL-1, TNF and LPS all appear capable of upregulating the synthesis of PAI-1, with reciprocal inhibition of t-PA (Scarpati and Sadler 1989). The effects of these mediators on thrombolysis is less well studied in vivo, but it appears that injection of LPS initially produces less than 3 hours of increased t-PA activity (probably due to the release of a pool of stored t-PA), which is followed by a longer-lived rise in plasminogen activator inhibitor activity; the latter probably reflects de novo protein synthesis (Suffredini et al. 1989).

Unstimulated endothelium also has important antithrombin properties, consisting of the heparin-like molecules that bind antithrombin III on the abluminal surface (De Agostini et al. 1990) and the thrombomodulin–protein C system. Whereas regulation of the former is not well understood, the latter has been studied in remarkable detail (Fig. 8.2). Thrombomodulin is an integral membrane protein that binds and alters the molecular configuration of thrombin. This process radically alters the function of thrombin, so that instead of its usual procoagulant activity thrombin activates the circulating zymogen protein C. Activated protein C, in the presence of its cofactor protein S, will in turn inactivate Factors Va and VIIIa as well as PAI-1. Protein S is present in free and bound forms; it is the free (unbound) protein S that can act as cofactor for activated protein C. A component of the complement pathway, C4b binding protein, reversibly binds protein S. The thrombomodulin–protein C system is strikingly sensitive to inflammatory processes. IL-1, TNF and LPS each downregulate endothelial thrombomodulin in vitro by inhibition of transcription, followed by endocytosis and degradation (Esmon et al. 1991). Levels of C4b binding protein may increase as part of the inflammatory response, resulting in an increased proportion of bound protein S, i.e. less cofactor available for activated protein C (Esmon et al. 1991). Human brain thrombomodulin has a heterogeneous distribution: it is found less frequently in blood vessels of the pons and putamen, two common sites of infarction (Wong et al. 1991).

There is, then, considerable evidence (mostly in vitro) demonstrating the procoagulant effects of inflammatory stimuli. Tissue factor expression, diminution of thrombolytic capacity and inhibition of the thrombomodulin–protein C systems can be seen in the presence of inflammatory mediators IL-1 and TNF. These responses can evolve in ways that are particularly complex. A good example of this is the coordinate expression of tissue factor with thrombomodulin downregulation. Following cytokine stimulation, endothelial cell tissue factor expression peaks by 6 hours, at which time loss of protein C activation is only half-maximal. The inhibitory

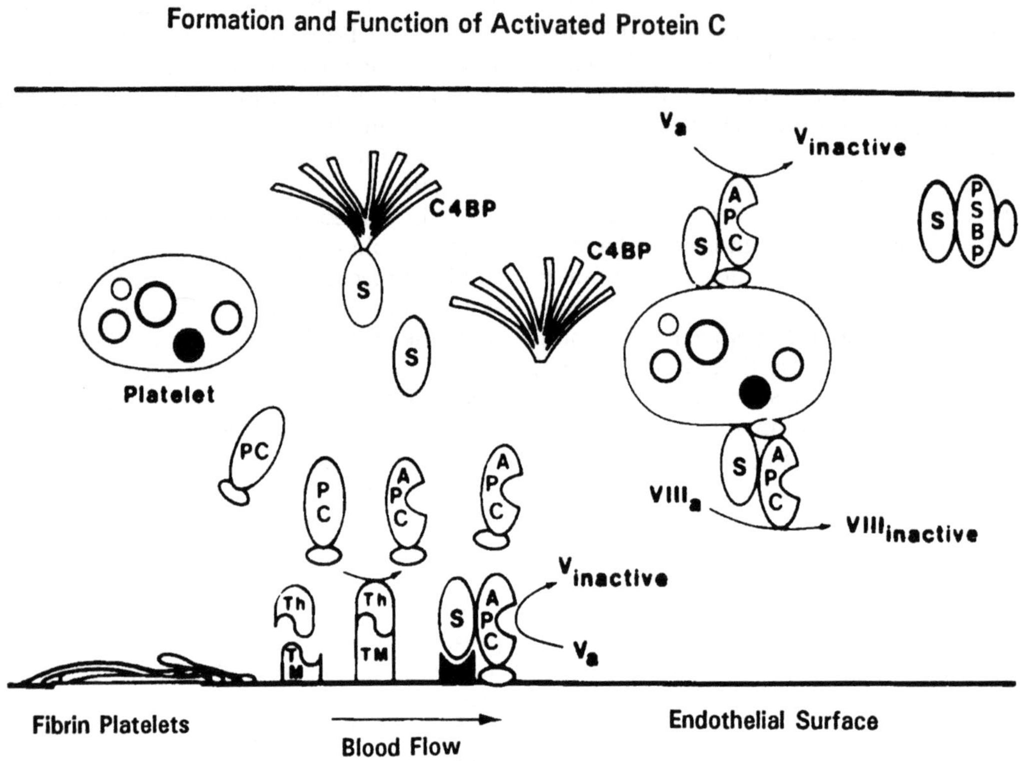

Fig. 8.2. Thrombin (Th) formed within the vasculature is carried by blood flow to thrombomodulin (TM). The TM–Th complex converts protein C (PC) to activated protein C (APC). The functional complex of APC with protein S (PS) inactivates Factors Va and VIIIa. PS reversibly complexes with C4b binding protein (C4BP); only the unbound PS can complex APC. From Esmon CT (1987) The regulation of natural anticoaglulant pathways. Reproduced with permission of Science 235:1348–1352.

effects on thrombomodulin peak at 24 hours, and a second inflammatory stimulus occurring at this point could have thrombogenic effects that are particularly profound (Esmon et al. 1991). It remains to be demonstrated whether such a scenario occurs in vivo.

Immunoregulation of Hemostasis and Vascular Disorders

Limited clinical and animal studies have added important information to the basic hemostatic regulatory processes outlined above. Bauer and colleagues investigated the effects of infusion of recombinant TNF in a group of cancer patients (Bauer et al 1989). They found a threshold dose, producing levels of approximately 30 pmol/l, at which there was enhanced thrombin generation and fibrinogen proteolysis; this indicated a net procoagulant effect of TNF.

The role of activated protein C in coronary occlusion has been investigated in the porcine model (Snow et al. 1991). A number of substantial findings have emerged from these studies:

1. Occlusion of the left anterior descending (LAD) coronary artery for only 30 seconds produced substantial activation of protein C.
2. This protein C activation was only demonstrable locally, i.e. in blood taken from the anterior interventricular vein, but not systemically.
3. Animals receiving a 2-minute occlusion of the LAD recovered less well than control animals if they were pretreated with a monoclonal antibody that blocks protein C activation.

These findings are of particular interest given the belief that most thrombomodulin resides in the capillaries. In the coronary circulation at least, the outcome following transient large-vessel occlusion appears to be substantially related to the hemostatic function of the microcirculation. A clinical observation consistent with these findings is an autopsy study of sudden cardiac death patients, showing microinfarcts in 38% of cases (Falk 1985).

The role of immunoregulation in stroke pathogenesis has received little attention. Most of the animal work has been produced by Hallenbeck and colleagues, whose findings can be summarized as follows:

1. Rats with risk factors for stroke (e.g. hypertension, diabetes, genetic predisposition and advanced age) responded to intracisternal injection of LPS by developing hemorrhagic infarction; this was not encountered in stroke risk factor-free animals (Hallenbeck et al. 1988).
2. Intravenous injection of LPS produced substantial increases in plasma TNF activity; this effect was greatly enhanced if the animal was hypertensive or had the combination of hypertension and genetic predisposition to stroke (Hallenbeck et al. 1991).
3. In addition to producing enhanced TNF activity, hypertensive rats responded to intravenous LPS with increases in prostacyclin, the platelet proaggregant thromboxane A_2, and the lipid mediator platelet-activating factor (Siren et al. 1992); the net effect of LPS in hypertensive animals may have been procoagulant.

Taken together, these studies clearly demonstrate that the presence of a key stroke risk factor (hypertension) enhances cytokine production in response to a standard stimulus (LPS). The source of TNF in these animals is uncertain. It is appealing to speculate that exaggerated increases in TNF contributed to the LPS-induced brain stem infarctions seen in animals with stroke risk factors in the initial study.

Immune-Mediated Hemostasis Regulation and Stroke

Clinical studies of immune mechanisms in stroke have concentrated on the role of anticardiolipin antibodies (aCL) and, to a lesser extent, lupus anticoagulants. Both are circulating immunoglobulins – IgG or IgM – that bind negatively charged or neutral phospholipids. Both aCL and lupus anticoagulants are antiphospholipid antibodies: they are identified by their ability to bind cardiolipin and prolong phospholipid-dependent clotting assays, respectively. Many aCLs have lupus anticoagulant activity and most lupus anticoagulants bind cardiolipin; the concordance between aCL and lupus anticoagulant appears to be approximately 45%–70% (Love and Santoro 1990).

Antiphospholipid antibodies were initially associated with systemic lupus erythematosus, but more recently their role in thrombotic disorders in non-lupus patients has been emphasized. The prevalence of high-titer aCL in the general population appears to be less than 3%; this is in contrast to the roughly 50% prevalence in lupus patients with thrombosis (Boey et al. 1983). Many groups of ischemic stroke patients have been surveyed, and the prevalence of aCL in these patients has been found to range from 7%–46%, with higher rates in younger populations (Coull et al. 1992).

The relationship between antiphospholipid antibodies and stroke seems to be based on the procoagulant activity of the antibodies. A number of in-vitro hemostatic abnormalities have been described with aCL, the most prominent being decreased synthesis or release of prostacyclin (Schorer et al. 1989) and interference with the thrombomodulin–protein C pathway (Freyssinet and Cazenave 1987). In-vivo analysis of hemostatic function in the presence of aCL will offer critical confirmation of these observations.

The role of cytokines in stroke patients has not been investigated systematically. One report describes 2 cancer patients who developed recurrent TIA-like episodes during treatment with interleukin-2; the latter activates endothelial cells in vivo, but has uncertain hemostatic effects (Bernard et al. 1990). More information is available concerning the relationship between systemic infections and ischemic stroke. Syrjanen and colleagues described a young (<50 years old) population of ischemic stroke patients, one-third of whom had signs or symptoms of febrile infection during the 1 month prior to the stroke. The majority were relatively mild upper respiratory tract infections, and the incidence was significantly higher than in the control population (Syrjanen et al. 1988). A later study of inner-city stroke patients of all ages showed that a similar proportion (one-third) of patients had febrile infections during the month prior to the stroke;

again, most of the infections were upper respiratory tract (Ameriso et al. 1991). In addition, high-titer aCLs were found more frequently in the infection-associated stroke patients: the latter had higher levels of the fibrin fragment D-dimer, and tended to have increased fibrinogen levels as well. These studies suggest that ischemic stroke is commonly preceded by a systemic inflammatory process which may play a role in stroke initiation. Confirmation of these observations would provide a definition of a clinically identifiable period of high stroke risk during which preventive therapies could be initiated or accentuated.

Future Directions

The Blood–Brain Barrier and Hemostasis

Although the blood–brain barrier has received enormous attention, particularly from the perspective of transport phenomena, its potential hemostatic consequences have not been explored. The configuration of capillary endothelium sharing basement membrane with astrocytes implies considerable possibilities for unique hemostatic regulation of the brain, i.e. brain-specific hemostasis. Astrocytes have the capacity to express tissue factor and t-PA as well as TNF and IL-1 (Chung et al. 1991). The effects of these latter two cytokines on endothelial cell function is profound, increasing permeability as well as upregulating leukocyte adhesion molecules. The hemostatic effects of these cytokines have already been discussed. Whether the blood–brain barrier does regulate hemostasis in vivo is only speculation at present.

Acute-Phase Reaction and Stroke Pathogenesis

The acute-phase response is an early and nonspecific response to injury, infection and inflammation that appears to be largely mediated by interleukin-6 (Gershenwald et al. 1990). Included in the response are increased levels of the specific acute-phase proteins C-reactive protein and serum amyloid P (Myers and Fleck 1988). A nonspecific component of the acute-phase reaction is increased levels of plasma fibrinogen. The latter is of particular interest in that several longitudinal studies have demonstrated that elevated fibrinogen is predictive of stroke, at least in men (Wilhelmsen et al. 1984; Kannel et al. 1987). The consequences of increased fibrinogen include reduced cerebral blood flow (Grotta et al. 1982), reduced middle cerebral artery blood flow velocity in the elderly (Ameriso et al. 1990), and enhanced progression of carotid artery atherosclerosis (Grotta et al. 1989). At present, it is again only speculation whether the transient increases in plasma fibrinogen, as seen with the acute-phase reaction, adversely affect stroke risk.

References

Ameriso SF, Paganini-Hill A, Meiselman HJ, Fisher M (1990) Correlates of middle cerebral artery blood flow velocity in the elderly. Stroke 21:1579–1583

Ameriso SF, Wong VLY, Quismorio FP, Fisher M (1991) Immunohematologic characteristics of infection-associated cerebral infarction. Stroke 22:1004–1009

Ameriso SF, Sager P, Fisher M (1992) Atrial fibrillation, congestive heart failure, and the middle cerebral artery: an ultrasound analysis. J Neuroimag 2:190–194

Bauer KA, ten Cate H, Barzegar S, Spriggs DR, Sherman ML, Rosenberg RD (1989) Tumor necrosis factor infusions have a procoagulant effect on the hemostatic mechanism of humans. Blood 74:165–172

Benveniste EN (1992) Inflammatory cytokines within the central nervous system: sources, function, and mechanisms of action. Am J Physiol 263 (Cell Physiol) 32:C1–C16

Bernard JT, Ameriso S, Kempf RA, Rosen P, Mitchell MS, Fisher M (1990) Transient focal neurologic deficits complicating interleukin-2 therapy. Neurology 40:154–155

Boey ML, Colaco CB, Gharavi AE, Elkon KB, Loizou S, Hughes GRV (1983) Thrombosis in SLE: striking association with the presence of circulating 'lupus' anticoagulant. Br Med J 287:1021–1023

Chung IY, Norris JG, Benveniste EN (1991) Differential tumor necrosis factor alpha expression by astrocytes from experimental allergic encephalomyelitis-susceptible and -resistant rat strains. J Exp Med 173:801–811

Conway EM, Bach R, Rosenberg RD, Konigsberg WH (1989) Tumor necrosis factor enhances expression of tissue factor mRNA in endothelial cells. Thromb Res 53:231–241

Coull BM, Levine SR, Brey RL (1992) The role of antiphospholipid antibodies in stroke. Neurol Clin 10:125–142

De Agostini AI, Watkins SC, Slayter HS, Youssoufian H, Rosenberg RD (1990) Localization of anticoagulantly active heparin sulfate proteoglycans in vascular endothelium: antithrombin binding on cultured endothelial cells and perfused rat aorta. J Cell Biol 111:1293–1304

del Zoppo G, Poeck K, Pessin MS et al (1992) Recombinant tissue plasminogen activator in acute thrombotic and embolic stroke. Ann Neurol 32:78–86

Drake TA, Morrissey JH, Edgington TS (1989) Selective cellular expression of tissue factor in human tissues. Am J Pathol 134:1087–1097

Eddleston M, de la Torre JC, Oldstone MBA, Loskutoff DJ, Edgington TS, Mackman N (1993) Astrocytes are the primary source of tissue factor in the murine central nervous system. J Clin Invest 92:349–358

Esmon CT, Taylor FB, Snow TR (1991) Inflammation and coagulation: linked processes potentially regulated through a common pathway mediated by protein C. Thromb Haemost 66:160–165

Falk E (1985) Unstable angina with fatal outcome: dynamic thrombosis leading to infarction and/or sudden death. Circulation 71:699–708

Fisher M, Francis R (1990) Altered coagulation in cerebral ischemia: platelet, thrombin, and plasmin activity. Arch Neurol 47:1075–1079

Fleck RA, Rao LVM, Rapaport SI, Varki N (1990) Localization of human tissue factor antigen by immunostaining with monospecific, polyclonal anti-human tissue factor antibody. Thromb Res 57:765–781

Freyssinet JM, Cazenave JP (1987) Lupus-like anticoagulants, modulation of the protein C pathway and thrombosis. Thromb Haemost 58:679–681

Furie B, Furie BC (1992) Molecular and cellular biology of blood coagulation. New Engl J Med 326:800–806

Gershenwald JE, Fong Y, Fahey TJ et al. (1990) Interleukin 1 receptor blockade attenuates the host inflammatory response. Proc Natl Acad Sci USA 87:4966–4970

Grotta J, Ackerman R, Correia J, Fallick G, Chang J (1982) Whole blood viscosity parameters and cerebral blood flow. Stroke 13:296–301

Grotta JC, Yatsu FM, Pettigrew LC (1989) Prediction of carotid stenosis by lipid and hematologic measurements. Neurology 39:1325–1331

Hacke W, Zeumer H, Ferbert A et al. (1988) Intra-arterial thrombolytic therapy improves outcome in patients with acute vertebro-basilar occlusive disease. Stroke 19:1216–1222

Hallenbeck JM, Dutka AJ, Kochanek PM, Siren A, Pezeshpour GH, Feuerstein G (1988) Stroke risk factors prepare brainstem tissues for modified local Shwartzman reaction. Stroke 19:863–869

Hallenbeck JM, Dutka AH, Vogel SN, Heldman E, Doron D, Feuerstein G (1991) Lipopolysaccharide-induced production of tumor necrosis factor activity in rats with and without risk factors for stroke. Brain Res 541:115–120

Hass WK, Easton JD, Adams HP et al. (1989) A randomized trial comparing ticlopidine hydrochloride with aspirin for the prevention of stroke in high-risk patients. New Engl J Med 321:501–507

Kalderon N, Ahonen K, Federoff S (1990) Developmental transition in plasticity properties of differentiating astrocytes: age-related biochemical profile of plasminogen activators in astroglial cultures. Glia 3:413–426

Kannel WB, Wolf PA, Castelli WP, D'Agostino RB (1987) Fibrinogen and risk of cardiovascular disease: the Framingham Study. JAMA 258:1183–1186

Krystosek A, Seeds NW (1981) Plasminogen activator release at the neuronal growth cone. Science 213:1532–1534

Lee RJ, Bartzokis T, Yeoh TK, Grogin H, Choi D, Schnittger I (1991) Enhanced detection of intracardiac sources of cerebral emboli by transesophageal echocardiography. Stroke 22:734–739

Love PE, Santoro SA (1990) Antiphospholipid antibodies: anticardiolipin and the lupus anticoagulant in systemic lupus erythematosus (SLE) and in non-SLE disorders. Ann Intern Med 112:682–698

Marsh E, Biller J, Adams H et al (1990) Circadian variation in onset of acute ischemic stroke. Arch Neurol 47:1178–1180

Myers MA, Fleck A (1988) Observations on the delay in onset of the acute phase protein response. Br J Exp Pathol 69:169–176

North American Symptomatic Carotid Endarterectomy Trial Collaborators (1991) Beneficial effect of carotid endarterectomy in symptomatic patients with high-grade carotid stenosis. New Engl J Med 325:445–453

Petersen P, Kastrup J, Videbaek R, Boysen G (1989) Cerebral blood flow before and after cardioversion of atrial fibrillation. J Cerebr Blood Flow Metab 9:422–425

Sandset PM, Warn-Cramer BJ, Maki SL, Rapaport SI (1992) Immunodepletion of extrinsic pathway inhibitor sensitizes rabbits to endotoxin-induced intravascular coagulation and the generalized Shwartzman reaction. Blood 78:1496–1502

Scarpati EM, Sadler JE (1989) Regulation of endothelial cell coagulant properties. J Biol Chem 264:20705–20713

Schorer AE, Wickham NWR, Watson KV (1989) Lupus anticoagulant induces a selective defect in thrombin-mediated endothelial prostacyclin release and platelet aggregation. Br J Haematol 71:399–407

Siebler M, Sitzer M, Steinmetz H (1992) Detection of intracranial emboli in patients with symptomatic extracranial carotid artery disease. Stroke 23:1652–1654

Siren AL, Heldman E, Doron D et al (1992) Release of proinflammatory and prothrombotic mediators in the brain and peripheral circulation in spontaneously hypertensive and normotensive Wistar–Kyoto rats. Stroke 23:1643–1651

Sloan DJ, Wood MJ, Charlton HM (1992) Leukocyte recruitment and inflammation in the CNS. TINS 15:276–280

Snow TR, Deal MT, Dickey DT, Esmon CT (1991) Protein C activation following coronary artery occlusion in the in situ porcine heart. Circulation 84:293–299

Suffredini AF, Harpel PC, Parrillo JE (1989) Promotion and subsequent inhibition of plasminogen activation after administration of intravenous endotoxin to normal subjects. New Engl J Med 320:1165–1172

Syrjanen J, Valtonen VV, Iivanainen M, Kaste M, Huttunen JK (1988) Preceding infection as an important risk factor for ischaemic brain infarction in young and middle aged patients. Br Med J 296:1156–1160

Wilhelmsen L, Svardsudd K, Korsan-Bengtsen K, Larsson B, Welin L, Tibblin G (1984) Fibrinogen as a risk factor for stroke and myocardial infarction. New Engl J Med 311:501–505

Wong VLY, Hofman F, Ishii H, Fisher M (1991) Regional distribution of thrombomodulin in human brain. Brain Res 556:1–5

9. Fibrinolysis and its Relevance to Acute Focal Cerebral Ischemia

G. J. del Zoppo

The use of thrombolytic agents in the treatment of acute cerebral ischemia is experimental. Approximately 80%–90% of focal cerebral ischemic events presenting as strokes within 8–24 hours of symptom onset (Solis et al. 1977) and 50%–70% of all strokes prospectively entered in contemporary stroke registries (Mohr et al. 1978; Kunitz et al. 1984) are due to atherothrombotic and embolic occlusive disease. It is on this basis that efforts to evaluate the efficacy of thrombus lysis in acute stroke have attracted renewed interest.

Two hypotheses are central to the study of thrombolysis: (1) that prompt restoration of the blood flow is essential for neuronal recovery and limitation of permanent injury; and (2) that neuronal recovery will be marked by measurable neurological improvement. The primary goal of thrombolytic intervention for stroke is the timely lysis of occluding arterial thrombus and the rapid restoration of flow to the ischemic territory. Because of the potential impact of fibrinolytic agents on thrombi, which may protect injured vessels, and their further effect on the coagulation system, a major concern about their use in cerebral ischemia is the risk of significant intracranial hemorrhage. In this chapter we will explore the recently published and anticipated outcomes of clinical experience with thrombolysis in acute focal cerebral ischemia from the perspective of the processes of thrombosis and the specific mechanisms of fibrin(ogen)olysis.

Thrombosis and Endogenous Fibrinolysis

Thrombus formation and the occlusion of a brain-supplying artery are central to focal cerebral ischemia. Thrombosis is an active complex process involving a breach of the antithrombotic features of the endothelia, coagulation system activation, platelet activation and recruitment, and thrombus fragmentation and embolism resulting from endogenous fibrinolysis. The complex molecular and cellular processes of thrombosis and atheroma formation are related (Bloom and Thomas 1987; Mustard et al. 1987). The central step in thrombosis is fibrin lattice formation, following the cleavage of fibrinogen by thrombin and cross-linking by Factor XIII of the forming fibrin matrix. Thrombus size may be limited by endogenous fibrinolysis, also triggered by thrombin generation (Gaffney 1987). Arterial flow characteristics in the vicinity of the thrombus, thrombus composition and vascular integrity are likely to contribute to thrombus

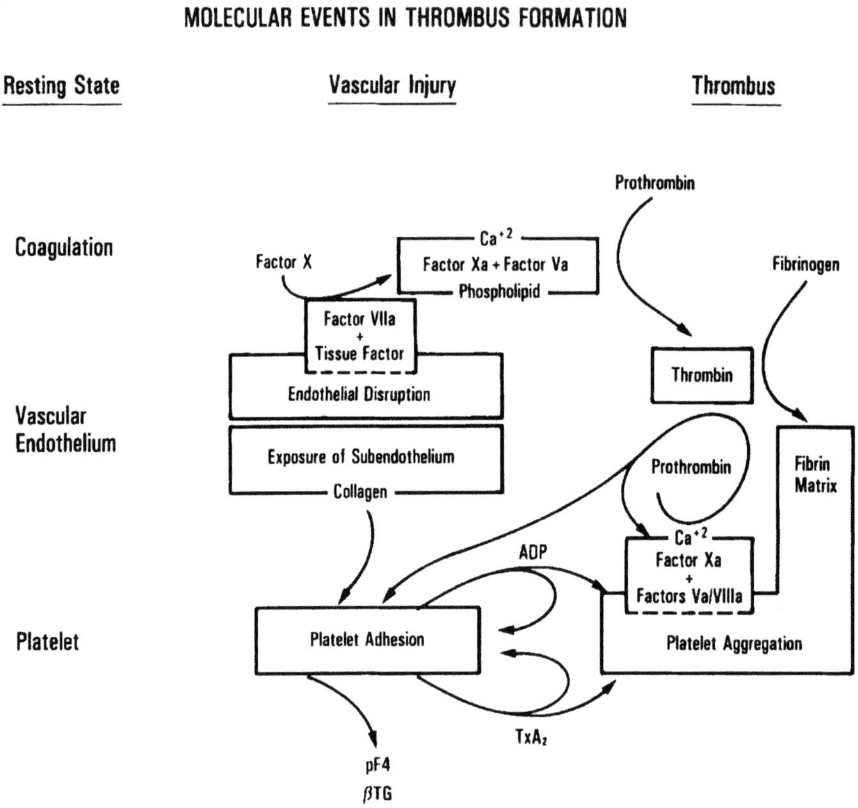

Fig. 9.1. Principal activation relationship of coagulation and platelet function in normal hemostasis.

susceptibility in the cerebrovasculature as elsewhere.

Thrombin is generated during coagulation system activation, and is itself the central figure in platelet activation and aggregation (Fig. 9.1). Tissue factor (TF)-mediated Factor VII activation, which results from exposure of Factor VII in the flowing plasma column to perivascular TF (Bach 1988; Morrissey et al. 1988), and TF on activated monocytes/macrophages (Niemetz 1972; Niemetz and Morrison 1977), as well as activation of coagulation by ischemia-related endothelial cell injury (Colucci et al. 1983; Drake et al. 1991) contribute to thrombin generation and thrombosis in situ. Similar mechanisms contribute to thrombosis upon subjacent atheromata. This is consistent with the observation that thrombi may be associated in situ with atheroma of the cervical internal carotid artery (ICA) at the flow divider, the carotid siphon, the middle cerebral artery (MCA) origin, vertebral artery origins, the vertebrobasilar junction and the posterior cerebral artery.

Thrombin stimulates platelet activation and release. Thrombin is locally generated via both extrinsic and intrinsic pathways of coagulation in a process which involves platelet membranes (Majerus et al. 1980). Platelets have a specific Factor XI receptor which, with high molecular weight kininogen (HMWK), promotes intrinsic coagulation. Specific receptors further promote the conversion of prothrombin to thrombin by binding Factors V and VIII which, together with membrane phospholipid, accelerate the conversion of Factor X to Xa. Conversely, Factor V is activated to Factor Va by thrombin (Owren 1944). On the platelet surface small quantities of thrombin are generated which promote α-granule secretion and the activation of Factor V; Factor Va serves as a high-affinity Factor Xa receptor. Thrombin may alter Factor V, which has a higher affinity for platelets. Platelets provide the phospholipid matrix for Factor X and Factor V, the so-called 'tenase complex'.

Platelet adherence and activation result in α and δ (dense) granule secretion of adenosine triphosphate (ATP), adenosine diphosphate (ADP), serotonin, β-thromboglobulin (βTG), platelet factor 4 (PF4), fibrinogen and von Willebrand factor (vWF) in a Ca^{2+}-dependent process. Factor V, indistinguishable from that in plasma, is also found in α granules (Breederveld et al. 1975). Released ADP promotes further aggregation (seen in the secondary wave of in vitro aggregometry) and secretion. Both circulating and platelet-derived vWF bind to the GPIIb/IIIa complex on the surface of stimulated platelets. vWF, through this process, is necessary for platelet–platelet interaction (Ruggeri et al. 1983; Berliner et al. 1988). This process may be interrupted by monoclonal antibodies (MoAb) against the peptide arg–gly–asp found in the vWF molecule (residues 1744–1746). GPIIb/IIIa also serves as the fibrinogen receptor on resting platelets (Savage and Ruggeri 1991), and allows platelet interaction in irreversible adhesion with vWF bound at the vascular surface.

Cross-linking of the fibrin matrix and stabilization of the thrombus is accomplished by Factor XIII bound to fibrinogen (Gaffney 1987). Factor XIIIa may also be bound to platelets. Apparently, Factor XIII cross-linking may confer resistance to plasmin (Schwartz et al. 1973; Gaffney and Whittaker 1979). This process involves the generation of interchain γ-glutamyl ϵ-lysyl linkage by Factor XIII-mediated transamidation (Curtis 1987).

Modeling of the thrombus is due to many forces, including shear stress, the underlying substrate (e.g. atheroma vs. subendothelium), polymorphonuclear (PMN) leukocyte/monocyte infiltration, and the relative fibrin content. High shear rates (e.g. up to 10 000/s), which characterize arterial flow, favor platelet deposition and are partly responsible for the platelet-rich character of arterial thrombi. Thrombus growth is limited by local molecular processes. Thrombin is directly inhibited by antithrombin III (AT III) in a stoichiometric fashion, and the thrombin–AT III complex is removed from the circulation (Adilgaard 1969). Free thrombin may bind rapidly to endothelial receptors, probably glycoaminoglycans, where interaction with AT III may occur (Lollar and Owen 1980; Bauer et al. 1983). AT III may also inhibit Factors IXa, Xa, XIa and XIIa (Damus et al. 1973). A complex but clinically important relationship exists between the circulating anticoagulant protein C and thrombin activity. Protein C is activated by thrombin on thrombomodulin, the endothelial receptor for protein C (Comp et al. 1982). The activated form (APC) cleaves both platelet-bound and plasma Factors Va and VIIIa, thereby producing irreversibly inactivated forms which downregulate local thrombin activity. Non-thrombin-related platelet activation may be inhibited by prostacyclin (PGI_2) secreted from the endothelium (Moncada and Vane 1979). In addition, proaggregatory ADP from platelets may be converted by an endothelial ADPase to adenosine monophosphate (AMP) and adenosine, which do not promote platelet aggregation.

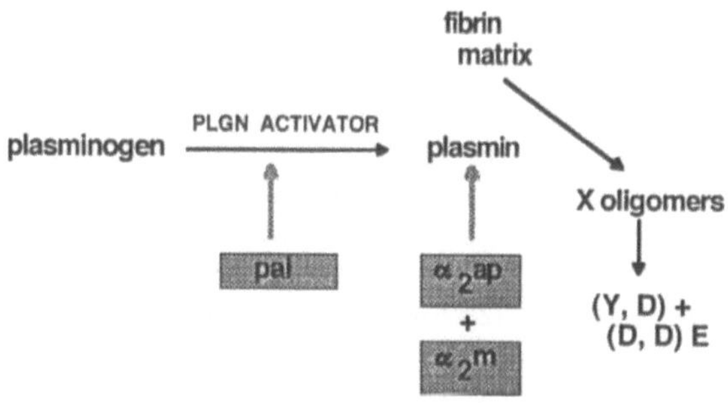

Fig. 9.2. General reactions of fibrinolysis. Plgn activator refers to either endogenous or exogenous plasminogen activators; α_2m is α_2-macroglobulin and $\alpha_2$ap is α_2-antiplasmin. Hatched fields and arrows refer to inhibition reactions. (Y,D) + (D,D) E are the primary fibrin degradation products of fibrinolysis.

Retardation of thrombus growth and the dissolution and fragmentation of thrombi are facilitated by the processes of endogenous fibrinolysis. Thrombolysis occurs pari passu with thrombus formation, but may be augmented when fibrin(ogen)olytic substances are infused. All thrombolytic substances are obligate plasminogen activators (PAs) which convert plasminogen to plasmin (Verstraete and Collen 1986) (Fig. 9.2). Endogenous PAs derived from vascular sources include tissue plasminogen activator (t-PA), single-chain urokinase plasminogen activator (scu-PA) and urokinase plasminogen activator (u-PA).

Plasminogen, a 92 kDa glycoprotein proenzyme of plasmin (791 amino acids), binds to fibrin by five kringle structures. Lysine-binding sites are also responsible for the binding of plasminogen to fibrin and to α_2-antiplasmin. Plasminogen may exist in several modified forms which differ from the native glu-plasminogen, by the presence of an NH_2-terminal lysine, methionine or valine (lys-plasminogen), the product of NH_2-terminal hydrolysis (Collen and Lijnen 1986). u-PA, one of the two physiologic plasminogen activators, generates the protease plasmin from plasminogen by cleavage of the arg^{560}–val^{561} linkage (Sjoholm et al. 1973; Rickli 1975). Endogenous tissue plasminogen activator and kallikrein may also generate plasmin from glu-plasminogen in vivo. Plasmin is a two-chain molecule consisting of a heavy (A) chain and a light (B) chain, which contains a trypsin-like active site (Groskopf et al. 1969). In addition to fibrin, fibrinogen and Factors II, Va and VIIIa may be cleaved by plasmin. When fibrin(ogen)olysis is not confined to the thrombus, a detectable 'systemic thrombolytic state' marked by decreased circulating plasminogen, fibrinogen and Factors V and VIII may occur. Circulating fibrin(ogen) degradation fragments appear which interfere with fibrin multimerization and thrombus stabilization; a transient anticoagulant state evolves secondary to the hypofibrinogenemia and factor depletion. Endogenous fibrinolysis does not elicit such an anticoagulant state in the absence of pathological conditions, such as disseminated intravascular coagulation (DIC).

The activities of t-PA and scu-PA are mostly confined to the thrombus. In vitro, the catalytic efficiency of single-chain and two-chain t-PA for the conversion of plasminogen to plasmin is stimulated to similar activity in the presence of fibrin (Ranby et al. 1982a,b; Rijken et al. 1982). The increased activity of the t-PA–plasminogen–fibrin ternary complex over t-PA–plasminogen confers apparent thrombus selectivity to t-PA, although the relative efficacy of t-PA, and perhaps of other plasminogen activators, is likely to depend upon the relative content of plasminogen in the thrombus. Indeed, a relative paucity of plasminogen may characterize aged thrombi, and may contribute to the relative resistance of thrombi to lysis by PAs. The appearance of t-PA in the circulation may be augmented by certain physiologic conditions, including exercise. Although the primary source of t-PA in pathophysiologic states is not yet defined, endothelial cells (umbilical vein) in vitro may be stimulated to secrete t-PA (Levin et al. 1986; Levin and Santell 1988) and its principal inhibitor, plasminogen activator inhibitor PAI-1 (Levin and Santell 1987). Little is yet known about the mechanism of action of scu-PA and its relative thrombus selectivity. Fragmentation of emboli in the cerebral vasculature occurs following cerebral thromboembolism (Bruckmann and Ferbert 1989). Although little is known about the embolic process per se, endogenous fibrinolysis, thrombus composition and local flow characteristics are important contributors.

Modulation of Fibrinolysis

Modulation of thrombolysis is achieved by two groups of endogenous inhibitors against plasmin and plasminogen activators, respectively. The requirement for activation of the proenzyme plasminogen to plasmin affords one level of regulation of fibrinolysis, which is most dependent upon the specific plasminogen activator. Within the circulating blood the inhibitors of free plasmin, α_2-antiplasmin (α_2AP) and α_2-macroglobulin (α_2M) effectively confine plasmin activity to the thrombus (Verstraete and Collen 1986). During thrombus dissolution, fibrin-bound plasmin is released into the circulation where it becomes complexed with α_2AP, the primary inhibitor of plasmin, to form the inhibited plasmin–α_2AP (PAP) complex. This protection may be overcome by infusion of PAs at pharmacologic concentrations which then produce detectable lysis of circulating fibrinogen and fibrin. Within the thrombus, fibrin-associated plasmin is protected from α_2M and α_2AP. Of the plasminogen activator inhibitors (PAIs), PAI-1, a 52 kDa glycoprotein, binds to and specifically inhibits both single and two-chain t-PA (Kruithoff 1988) and u-PA (Kruithof et al. 1984; Thorsen and Phillips 1984; Verheijen et al. 1984; Wiman et al. 1984; Kruithof et al. 1986). PAI-1 is found in the plasma and in platelet α-granules. PAI-2 and PAI-3 inhibit u-PA in the plasma. In addition, α_2M also inhibits/inactivates t-PA and the streptokinase–plasmin(ogen) complexes at low rates. The general effect of this multilayered system of inhibitors is to limit endogenous fibrinolysis to the thrombus. Infusion of plasminogen activators is intended to overwhelm these protective mechanisms.

Thrombus Composition

Differences in thrombus composition between atherothrombotic and cardiac-source thromboemboli may contribute to the relative resistance to endogenous fibrinolysis. Pelz and colleagues (1991) described 2 of 10 patients with persistent vertebrobasilar arterial (VBA) occlusion following systemic thrombolysis with thrombi of 'low fibrin content' set upon underlying atheromata at postmortem examination. It is known that the degree of fibrin cross-linking by Factor XIII activity may contribute to the resistance of thrombi to plasmin (Schwartz et al. 1973; Gaffney and Whittaker 1979). This implies that older thrombi may be relatively more resistant to the action of PAs. As noted earlier, the relative content of fibrin in thrombi and therefore their cellular constitution, may further contribute to plasmin resistance. In vitro studies indicate that an increased thrombus content in platelets may confer a resistance to fibrinolysis (Jacobson and Chandler 1965).

Whether thrombus age and organization may differ between atherothrombotic and cardiac-source emboli in either the carotid artery or the VBA territory is not known (Ito et al. 1983). It is possible that thrombus volume and location may influence resistance to fragmentation (del Zoppo et al. 1986b, 1988; Mori et al. 1988; Mori 1991). This is suggested by the observation that atheroma-based thrombi in the extracranial portion of the carotid artery appear less likely to recanalize than do distal branch occlusions (del Zoppo et al. 1992).

Spontaneous Recanalization

Spontaneous recanalization within the cerebrovasculature has been the subject of few prospective angiography-based studies and anecdotes (Dalal et al. 1965b; Irino et al. 1977; Ito et al. 1983; Taneda et al. 1985). The incidence of cerebrovascular patency increases with time from stroke onset to angiography. To what degree spontaneous recanalization contributes to patency is not known, as only one study has examined this issue with serial vascular procedures (Dalal et al. 1965a). Carotid territory occlusions seen 2–9 days after embolic stroke opened within 10 minutes to 8 days in this angiography-based study (Dalal et al. 1965b). Taken in aggregate, recent prospective angiographic studies have documented a decrease in carotid territory occlusion frequency from 76% at 6 hours (Fieschi et al. 1989) and 81% at 5.7 ± 1.4 hours (del Zoppo et al. 1992), to 58.7% (of 63 patients) at 24 hours (Solis et al. 1977), 57% (of 86 patients) at 3 days (Fieschi and Bozzao 1969), and 44.4% (of 18 patients) at 7 days (Irino et al. 1977) post ictus.

In a recent rt-PA dose-rate finding study, only 2 of 25 patients with ICA occlusion (8%) showed recanalization following a 60-minute rt-PA infusion within 8 hours of symptom onset (del Zoppo et al. 1992), in contrast to 37.5% (6 of 16) untreated patients with ICA occlusions at 7 days (Irino et al. 1977). Yamaguchi and colleagues (1984) have also noted that spontaneous recanalization is a relatively frequent occurrence following cerebral thromboembolism. Mori et al. (1992) have suggested that only 2 of 12 (16.7%) untreated patients had evidence of spontaneous recanalization within 4 hours of symptom onset. From the above limited information, it would seem unlikely that recanalization within hours of symptom onset would contribute more than 20% patency in patients presenting with carotid territory symptoms. Alternately, this may represent the incidence of non-thrombotic causes of focal ischemia.

Implementation of Fibrinolytic Agents in Cerebrovascular Ischemia

The potential value of fibrinolytic agents in focal cerebral ischemia is being evaluated in a series of prospective clinical trials (del Zoppo et al. 1993). Those trials rest upon (1) the observations that cerebral arterial occlusions occur very frequently in acute stroke; (2) an interval of at most 4–6 hours in humans may exist after cessation of blood flow, during which significant reversibility of neuronal function may occur (Astrup et al. 1977; Skyhoj-Olsen et al. 1983); (3) the possibility of an 'ischemic penumbra' of neuronal tissue peripheral to the ischemic core that is injured and 'stunned', but capable of at least partial functional recovery upon restoration of blood flow (Astrup et al. 1981); and (4) the presence of vascular collateral channels which may, in certain territories, protect neuronal tissue at risk. Although the contributions of (2)–(4) above may vary with the individual, the properties of specific fibrinolytic agents and their expected interaction with human thrombi in vivo are well characterized.

Agents with thrombolytic potential which have been applied to clinical thrombotic problems include the exogenous agents streptokinase (SK) and anisoylated plasminogen activator complex (APSAC), and the endogenous agents u-PA, rt-PA and scu-PA (or prourokinase).

Thrombolytic Agents

The unique and individual characteristics of the plasminogen activators which are used as thrombolytic agents define the limits of their pharmacologic efficacy. Whereas animal cerebral ischemia models have been used to evaluate the lytic efficacy of several plasminogen activators (del Zoppo 1990), their relative efficacy in human stroke is only now being systematically examined. It should be emphasized that post-fibrinolytic responses in one vascular bed (or specie) do not predict outcome in another vascular bed (or specie). Two general infusional approaches have been employed: 'front-loaded' infusions, in which a portion of the total dose of the agent is applied as an initial bolus, and non-'front-loaded' infusions, without bolus. The former is intended to achieve rapid steady-state plasma levels of the agent; the latter, a more prolonged exposure to the agent. Both infusional approaches seek to augment endogenous thrombolysis by overwhelming circulating plasmin and plasminogen activators. Plasminogen activators now used in clinical practice and relevant to the study of recanalization in acute stroke are summarized here. There is no reason to believe that one PA may be intrinsically more effective than another on the basis of thrombolytic activity alone. There is some evidence, from the relative incidence of coronary artery reocclusion, that agents (e.g. u-PA or SK) that can produce an anticoagulant state are more likely to prolong the patency of arteries with atheroma-based thrombi than agents without a direct effect on Factors V and VIII.

Tissue Plasminogen Activator (t-PA)

Tissue plasminogen activator (t-PA) is a single-chain 527 amino acid serine protease with an unglycosylated molecular weight of 58 kDa (Robbins et al. 1967; Rijken 1988). The native t-PA is synthesized and secreted by endothelial cells (Levin and Loskutoff 1982). The single-chain form is converted by plasmin to the two-chain form. This involves hydrolysis of the arg^{275}–$isoleu^{276}$ linkage. Both t-PA species are enzymatically active. The preferential cleavage of fibrin over fibrinogen in vivo suggests that both t-PAs have relative fibrin-selective properties. Fibrin-binding is attributed to the N-terminal finger domain and kringle-2 of t-PA (Verstraete and Collen 1986; Rijken 1988) because mutants devoid of these structures have markedly reduced fibrin-binding constants (Ehrlich et al. 1987). The interaction of kringle-2 with fibrin requires an inhibitable lysine-binding site not found in the finger domain. In normal human plasma t-PA antigen is present in 4–5 ng/ml. Infusion studies in humans of both species have demonstrated a half-life of 3–8 minutes (Verstraete and Collen 1986), although the biological half-life is believed to be somewhat longer. Recombinant t-PA (rt-PA) in single-chain or in two-chain form has been used in studies of acute ischemic stroke.

Urokinase (u-PA) and Single-Chain Urokinase Plasminogen Activator (scu-PA)

Urokinase (u-PA) exists in three forms which exhibit fibrin(ogen)olytic activity (Fletcher et al. 1965; Gunzler et al. 1982; Stump and Mann 1988).

Single-chain u-PA (scu-PA or pro-UK) is an endogenous plasminogen activator and a functional proenzyme of u-PA. It is unusual in possessing fibrin-selective activity (Lijnen et al. 1986; Bando et al. 1987). High molecular weight u-PA (54 kDa) is a fibrin-non-selective thrombolytic serine protease which activates plasminogen to plasmin directly by first-order kinetics (White et al. 1966). The relationship of scu-PA to u-PA is complex. scu-PA is a 411-amino acid 54 kDa glycoprotein with a single disulfide bridge between cys^{148} and cys^{279}. Cleavage of the lys^{158}–ile^{159} peptide bond, or removal of lys^{158}, produces the high molecular weight (54 kDa) two-chain u-PA consisting of an A-chain (157 amino acids) and a glycosylated B-chain (253 amino acids). Further cleavages at lys^{135} and arg^{156} produce low molecular weight (31 kDa) u-PA (Rijken 1988). Both high and low molecular weight species are enzymatically active in vivo and in vitro. Purified u-PA, whether isolated from urine or renal tissue (White et al. 1966; Bernick and Kwaan 1969), has a plasma half-life upon infusion of 9–12 minutes, and has measurable fibrin(ogen)olytic effects. scu-PA may be isolated from urine and kidney cell culture medium or prepared from *Escherichia coli* by recombinant techniques (Gunzler et al. 1982; Lijnen et al. 1986; Verstraete and Collen 1986). The precise mechanism underlying the fibrin-selectivity of scu-PA is not known (Lijnen et al. 1986).

Streptokinase (SK)

Streptokinase (SK) activates plasminogen to plasmin by complex kinetics. It is a 47 kDa glycoprotein derived from Group G *Streptococcus haemolyticus*. In addition to plasmin, [SK/plasminogen] and [SK/plasmin] species are capable of cleaving fibrinogen and fibrin (Castellino 1979). Free circulating plasmin and [SK/plasmin(ogen)] complexes degrade both fibrinogen and fibrin and inactivate prothrombin, Factor V and Factor VIII (Brogden et al. 1973), thereby producing a 'systemic lytic state' similar to that of u-PA. Non-stoichiometric generation of plasmin from SK infusions makes the effects of dose-rate adjustments unpredictable (Brogden et al. 1973). A unique situation is presented by the ubiquitous antistreptococcal and antistaphylococcal antibodies which inhibit the activity of streptokinase and APSAC, and staphylokinase, respectively. Generally, the presence of such antibodies requires an increased initial dose of the fibrinolytic agent.

Anisoylated Plasminogen Streptokinase Activator Complex (APSAC)

APSAC is a 131 kDa artificial complex of acylated human lys-plasminogen and streptokinase from group G *Streptococcus haemolyticus*. It has relative fibrin selectivity through the fibrin-binding properties of the lys-plasminogen kringles. Hydrolytic activation of the acyl-protected plasminogen-active site occurs by first-order kinetics in solution, generating plasmin by the exposed streptokinase (Smith et al. 1981). Activation may occur in fibrin-containing thrombi and on freely circulating fibrin(ogen). The biological activity of APSAC is expected to be longer than that of streptokinase because of the need for internal hydrolysis prior to fibrin cleavage. The thrombolytic potential of APSAC in experimental models is somewhat greater than that observed for streptokinase/plasminogen (Matsuo et al. 1981; Smith et al. 1982). The development of hypofibrinogenemia and the inactivation of Factors V and VIII are similar to those produced by SK alone, as are the presence and evidence of SK-related hypersensitivity reactions. One practical feature of APSAC to be noted is that, upon reconstitution, hydrolysis produces an increasing content of free SK with time.

Other Plasminogen Activators

The exogenous streptokinase–plasminogen complexes Fibrinolysin® and Thrombolysin® are no longer available and are of historical value. Staphylokinase has raised recent interest, but has not been used in clinical trails of brain ischemia to date. It is currently undergoing comparative studies in animal models (Collen et al. 1992). Plasminogen activators derived from the salivary glands of vampire bats are highly fibrin-selective agents. They are now being characterized in vivo in small-mammal non-cerebral thrombus models (Gandell 1992; Witt et al. 1992).

Novel Thrombolytic Agents

A number of artificial constructs of t-PA and scu-PA have been prepared with the intention of enhancing thrombolytic potency, shortening half-life, altering clearance and taking advantage of known synergistic actions. These include single-site t-PA mutants, t-PA/scu-PA and t-PA/u-PA chimerae (Pierard et al. 1987), u-PA/antifibrin monoclonal antibody (MoAb) and u-PA/antifibrin/antiplatelet MoAb conjugates (Runge et al. 1987), scu-PA deletion mutants (Gheysen et al. 1987), and other agents which are under development.

Thrombolysis: Hemorrhage

In addition to restoration of the flow in a thrombus-occluded vessel, the augmentation of endogenous fibrinolysis may dissolve thrombi which play a protective role. Hemorrhage may be the consequence. As thrombus formation is a natural result of denudation of vascular endothelia and exposure of the subendothelia and underlying tissue factor, plasminogen activators are unable to distinguish protective thrombi formed in consequence of vascular injury and disruption from thromboemboli. In this regard the therapeutic–hemorrhagic risk window may be quite narrow, and may vary with the organ system and the type of vascular injury. For instance, from aggregate studies of fibrinolytic agents (including SK, APSAC, u-PA, rt-PA and rscu-PA) in acute myocardial infarction, a rather constant incidence of symptomatic intracranial hemorrhage (ICH) ranging from 0% to 1.3% of treated patients has been observed (del Zoppo and Mori 1992). This is probably independent of the specific fibrinolytic agent: an overall incidence of 0.18% ICH with SK, 0.34% ICH with APSAC and 0.46% ICH with rt-PA at presently accepted doses, compared to an overall incidence of 0.02% ICH among untreated or control patients with acute myocardial infarction, has been recorded (del Zoppo and Mori 1992). The incidence and pathogenesis of ICH accompanying untreated acute myocardial infarction is often uncertain and has, erroneously, been compared with the incidence of hemorrhagic transformation following atherothrombotic and embolic stroke.

Hemorrhagic transformation, which consists of hemorrhagic infarction (HI) or parenchymatous hematoma (PH), is a common consequence of focal ischemia during thromboembolic stroke (Pessin et al. 1990, 1991), but not myocardial infarction. Among non-anticoagulated patients with acute cerebral infarction, HI has been documented in ‹10%–43% in prospective CT scan-based studies (Hart 1986; Ott et al. 1986; Lodder et al. 1988). Hemorrhagic transformation was observed in 46.4% of 140 patients with carotid territory embolic stroke, including 16 (11.4%) patients with PH (Okada et al. 1989), compared with 43.1% of 65 patients, 13.8% of whom had PH, in a separate study (Hornig et al. 1986). HI is probably more commonly associated with cardiogenic cerebral embolism than with thrombosis in situ, as noted by Yamaguchi et al. (1984) who, in an angiographic study, reported HI in 37.5% of 120 cases of cardiogenic cerebral embolism, but in only 1.9% of 105 cases of cerebral thrombosis in the carotid territory. It is of interest that PH, which is most often symptomatic (Pessin et al. 1990), seems to accompany acute carotid territory embolism in the setting of anticoagulant treatment (Meyer et al. 1963; Drake and Shin 1983; Babikian et al. 1989), rather than in the absence of anticoagulants (Furlan et al. 1982; Koller 1982). The long-standing concept that hemorrhagic transformation may result from arterial reperfusion is not supported by recent angiographic studies (del Zoppo et al. 1988; Mori et al. 1988, 1991; rt-PA Acute Stroke Study Group 1991). The findings by Ogata and co-workers (1989) of hemorrhagic transformation with persistent occlusion of the primary artery has suggested that hemorrhage may occur from other vascular sources (e.g. collateral channels) (Fisher and Adams 1951, 1987). When considered together, the presumed incidence of symptomatic hemorrhagic transformation is approximately 10%.

All thrombolytic agents carry the risk of intracranial hemorrhage (Aldrich et al. 1985; del Zoppo et al. 1986a). The possibility that exposure of the focal cerebral ischemic zone to a thrombolytic agent may increase the size or severity of naturally occurring hemorrhagic transformation is the main safety issue. Early studies, despite limitations in design (del Zoppo et al. 1986a), have suggested a significant incidence of symptomatic intracranial hemorrhage and death (Sussman and Fitch 1958; Clarke and Cliffton 1960; Herndon et al. 1960; Meyer et al. 1963, 1964; Fletcher et al. 1976; Hanaway et al. 1976; Abe et al. 1981a, b). The limitations of those studies urge caution in interpreting the incidence of ICH (del Zoppo et al 1986a). The more recent angiography- and CT scan-based acute thrombolysis studies have further defined the incidence of hemorrhagic transformation (Yamaguchi 1991; del Zoppo et al. 1992; Mori et al. 1992; von Kummer and Hacke 1992). In the carotid territory, the incidence of symptomatic hemorrhagic transformation, among three intra-arterial u-PA or streptokinase trials, was 9.3% (del Zoppo et al. 1988; Mori et al. 1988; Theron et al. 1989), compared with 10% from three prospective angiography-based trials of intravenous infusion of rt-PA (duteplase or alteplase) (del Zoppo et al. 1992; Mori et al. 1992; von Kummer and Hacke 1992) with similar entry criteria. The respective aggregate mean frequencies of arterial recanalization were 74.1% and 43.5%. Among the intra-arterial infusion trials, recanalization occurred in 5 of 9 patients with PH, and clinical deterioration was noted in 8 of the 9 patients.

In a non-angiographic systemic rt-PA (alteplase) infusion study, three (4.1%) symptomatic hemorrhages (all PHs) were documented by CT scan among 74 patients treated (Brott et al. 1990). The symptomatic episodes seemed to be related to the higher rt-PA doses (Brott et al. 1992), although there was no difference in incidence from that observed in the one angiography-based rt-PA dose-escalation study (del Zoppo et al. 1992) or one placebo-controlled trial (von Kummer and Hacke 1992) reported thus far. From these sources (del Zoppo et al. 1988; Mori et al. 1988: Theron et al. 1989; Brott et al. 1990; Yamaguchi 1991; del Zoppo et al. 1992; Mori et al. 1992; von Kummer and Hacke 1992) there was no relationship between intracerebral hemorrhage and recanalization; PH and HI were not associated with prior antiplatelet therapy; the incidence of hemorrhagic transformation increased significantly with time to treatment, particularly when rt-PA (duteplase) was initiated later than 6 hours from symptom onset; no difference in the incidence of symptomatic hemorrhage between rt-PA (duteplase) at two dose-rates and placebo (von Kummer and Hacke 1992) or among increasing dose-rates (del Zoppo et al. 1992) was observed; there was no relationship between the volume of infarction and the severity of hemorrhage; and the majority of hemorrhages occurred within the ischemic territory. This does not imply that in acute focal cerebral ischemia the incidence of hemorrhagic transformation is independent of rt-PA dose above those so far studied; it is indeed quite possible, as observed in certain individual cases (del Zoppo et al. 1992), that the timing and volume of hemorrhage were different from those observed in untreated patients.

Nonetheless, acute intervention trials reported to date with similar entry criteria have suggested a rather constant incidence of hemorrhage with associated clinical deterioration which is not substantially different from untreated patients described in the literature (Hart 1986; Ott et al. 1986; Lodder et al. 1988).

Thrombolysis: Restoration of Flow and Clinical Outcome

Studies of the clinical efficacy of fibrinolytic agents in cerebral ischemia have been directed at two goals, the facilitation of cerebral arterial recanalization, which is necessary for neurological recovery, and neurological recovery itself. Two different strategies have been employed: angiography-based studies which assess vascular outcome and related clinical benefits, and symptom-based studies which assess clinical outcome only, without the corroboration of vascular outcome. Here, either intravenous or intra-arterial administration of the agent has been employed. In general, the frequency of recanalization in intravenous studies remains below that of the intra-arterial studies in prospective studies (of different agents). In none of these situations has the agent or agent dose been optimized. Angiographic studies suggest that the delivery of an active agent may not occur if a stagnant blood column exists proximal to a fixed occlusion, unless delivery is directed at the thrombus (e.g. peripheral arterial infusions). The higher doses required by agent dilution and passive delivery to the thrombus may increase unwanted systemic side effects with the agents currently available. This is one rationale for the use of direct intra-arterial delivery of thrombolytic agents by catheter to the thrombus in cerebrovascular disease. Whereas trials of intravenous therapy have been either angiography-controlled or symptom-based, all trials of intra-arterial therapy are angiography-controlled. The intravenous non-angiographic studies have allowed very early interventions in acute stroke.

Angiography-Controlled Trials

Early limited angiography-based studies conducted with intravenous infusion fibrinolysin–plasmin (Sussman and Fitch 1958), Thrombolysin® (Meyer et al. 1964), streptokinase (Meyer et al. 1963) and u-PA (Araki et al. 1973) failed to demonstrate a significant clinical improvement in stroke patients. Delays in treatment, variable and prolonged treatment duration, inconsistent use of both pre- and post-treatment angiography, the prolonged interval between angiographies–too long to reasonably link the vascular outcome and the treatment–and the like treatment of 'stroke in progression' and completed stroke were some of the weaknesses of these exploratory studies.

More recent prospective angiography-based trials designed to assess the effect of intravenous rt-PA on recanalization have suggested that recanalization is feasible (Herderscheê et al. 1991; Mori et al. 1991; von Kummer 1991; del Zoppo et al. 1992; Yamaguchi et al. 1993) (Table 9.1). Treatment lasting 1–3 hours was initiated early (within 6 or 8 hours) after symptom onset. These studies attempted to optimize the rt-PA dose-rate relative to recanalization and hemorrhagic transformation (del Zoppo et al. 1992), or adopted one or more fixed dose-rates of rt-PA (either alteplase or duteplase) and evaluated recanalization and clinical outcome (Mori et al. 1991; von Kummer 1991; Yamaguchi 1991). Although an open prospective multicenter dose-escalation study demonstrated no recanalization dose-rate response with rt-PA (duteplase) (del Zoppo et al. 1992), four studies of patients with carotid territory occlusions with similar entry criteria indicated 34%–59% recanalization (Mori et al. 1991; von Kummer 1991; Yamaguchi 1991; del Zoppo et al. 1992), or a mean of 42% recanalization (74 of 176 patients in aggregate). Recanalization of internal carotid artery (ICA) occlusions (0–25%) was significantly less frequent at the dose-rates used than that of MCA occlusions (Mori et al. 1991; del Zoppo et al. 1992; Yamaguchi et al. 1993). A three-arm placebo-controlled, double-blind pilot trial of rt-PA (duteplase), comparing 20 MIU, 30 MIU and placebo, demonstrated a significantly better clinical improvement at 30 days in subjects receiving 30 MIU

Table 9.1. Prospective trials to assess the effect of intravenous rt-PA on recanalization

	Agent	No. of patients	Treatment		Recanalization		Hemorrhage		
		n	Δ(T-0) (h)	Duration (h)	No.	%	HI	PH	%
Carotid Territory									
del Zoppo et al. (1992)	rt-PA	93(104)	‹8	1	32	34.4	21	11	30.8
Mori et al. (1991)	rt-PA	19	‹6	1	9	47.4	8	2	52.6
	C	12			2	16.7	4	1	41.7
Yamaguchi et al. (1993)	rt-PA	47 (51)	‹6	1	10	21.3	20	4	47.1
	C	46 (47)			2	4.4	17	5	46.8
von Kummer (1991)	rt-PA	22	‹6	1.5	13	59.1	6	2	36.4
Vertebrobasilar Territory									
Herderscheê et al. (1991)	rt-PA	2	‹6	3	1	50.0	1	0	50.0
von Kummer (1991)	rt-PA	5	‹6	1.5	2	40.0	0	0	0
Yamaguchi (1991)	rt-PA	5	‹6	1	4	80.0	0	0	0

C, control (placebo); Δ(T-0), interval from onset to treatment; PH, parenchymatous hemorrhage; HI, hemorrhagic infarction.

(~0.5 MIU/kg/60 min) than those treated with placebo or 20 MIU (Mori et al. 1991). Patients undergoing recanalization had a better neurological outcome than non-recanalized patients (Mori et al. 1991). Von Kummer and associates (1992) have suggested that the presence of patent collateral channels contributed to a favourable neurological outcome. These studies have also suggested that the incidence of hemorrhagic transformation correlated neither with the use of rt-PA, with specific dose-rates, nor with the achievement of reperfusion (see below).

Among intra-arterial studies employing either regional infusion or local infusion techniques, cerebral arterial recanalization has been documented (Zeumer 1985; Zeumer et al. 1989) (Table 9.2). Although the clinical outcome results of single or small case studies have been difficult to interpret (Clarke and Cliffton 1960; Meyer et al. 1961; Atkin et al. 1964; Nenci et al. 1983; Miyakawa 1984; Zeumer et al. 1989), larger prospective open trials of regional or local infusion of u-PA or SK have demonstrated recanalization of acute carotid artery territory or vertebrobasilar artery territory occlusions (del Zoppo et al. 1988; Mori et al. 1988, 1991; Hacke et al. 1988; Theron et al. 1989; Zeumer et al. 1989; Matsumoto and Satoh 1991; Möbius et al. 1991). Recanalization of

Table 9.2. Intra-arterial studies employing regional or local infusion techniques

	Agent	No. of patients	Treatment		Infusion	Recanalization		Hemorrhage		
			Δ(T-0) (h)	Duration (h)		No.	%	HI	PH	%
Carotid Territory										
del Zoppo et al. (1988)	SK/u-PA	20	1–24	1.0–4.0	I/R	18	90.0	4	0	20.0
Mori et al. (1988)	u-PA	22	0.82–7	0.2–0.5	R	10	45.5	1	3	18.2
Theron et al. (1989)	SK/u-PA	12	2–504	‹1.0	I/R	12	100.0	1	2	25.0
Matsumoto and Satoh (1991)	u-PA	40	1–24	0.2–0.5	R	24	60.0	9	4	32.5
Vertebrobasilar Territory										
Nenci et al. (1983)	SK/u-PA	4	6–96	0.3/10–44	R	4	100.0	0	0	0
Hacke et al. (1988)	SK/u-PA	43	‹24	‹4/12–48	I	19	44.2	2	2	9.3
Zeumer et al. (1989)	u-PA	7	4–48	2	I	7	100.0	1	0	14.3
Möbius et al. (1991)	SK/u-PA	18	0.5–2.0	2	I	14	77.8	0	0	0
Matsumoto and Satoh (1991)	u-PA	10	3–24	0.2–0.7	R	4	40.0	0	1	10.0

Δ(T-0), interval from symptom onset to treatment; I, interventional; R, regional; PH, parenchymatous hemorrhage; HI, hemorrhagic infarction.

occluded arteries has been reported in 44%–100% of the patients entered in those trials, and a favorable clinical outcome associated with recanalization has been suggested by several series. Among 85 patients treated with u-PA or SK from the three largest studies, 47 (55.3%) underwent recanalization following intra-arterial infusion (del Zoppo et al. 1988; Hacke et al. 1988; Mori et al. 1988). A favorable clinical outcome was strongly associated with recanalization in 30 of these patients (63.8%). Mori and colleagues (1988, 1991) suggested resistance of ICA occlusions to reperfusion relative to more distal MCA obstructions. A limited pilot attempt with intra-arterial melanoma-derived t-PA in carotid territory occlusion had a mixed outcome (del Zoppo et al. 1988) and two single case reports of intra-arterial rt-PA infusion have appeared (Henze et al. 1987; Buteux et al. 1988).

Experience with fibrinolytic agents in vertebrobasilar ischemia has been restricted to a single retrospective comparison of recanalization–clinical outcome in 43 patients receiving intra-arterial u-PA or SK with 22 patients receiving other antithrombotic therapy. A survival benefit in those patients undergoing recanalization after fibrinolytic agent infusion was suggested (Hacke et al. 1988). This potentially significant result requires further study.

Symptom-Based (Clinical Outcome) Trials

In early studies employing intravenous plasmin, clinical results were equivocal (Clarke and Clifton 1960; Herndon et al. 1960). Angiography was performed in selected patients for diagnosis. Among 31 patients treated with intravenous u-PA in one study, no neurological improvement was noted; death occurred in 5 patients and intracranial hemorrhage occurred in 4 patients (Fletcher et al. 1976). Among non-angiographic systemic infusion studies using u-PA (Abe et al. 1981a,b; Atarashi et al. 1985; Otomo et al. 1985) or rt-PA (Otomo et al. 1988; Abe et al. 1990) two studies have shown equivocal clinical benefit (Abe et al. 1981a,b; Otomo et al. 1985). Clinical outcomes could not be linked to vascular reperfusion because of the very low u-PA dose-rates used, and the unusually long interval between symptom onset and treatment [Δ(T-0) ›3 days]. Koudstaal and colleagues (1988) described two patients with a fatal outcome following intravenous rt-PA (alteplase), in whom reperfusion injury was suspected but not proven.

In contrast, Brott and associates (1992) have examined the clinical effect of systemic infusion of rt-PA (alteplase) on neurological outcome within 90 minutes and in the interval 91–180 minutes (Haley et al. 1992) of focal cerebral ischemic symptoms. Thirty-four of 74 (46%) patients had a clinical improvement within 24 hours of rt-PA (alteplase) treatment, but no relationship to dose was observed. The persistence of neurological improvement was not commented upon. Symptomatic intracerebral hematomas occurred in 3 (4.1%) of the 74 patients. Despite the intensiveness of this study, the precise nature and number of vascular events and the contributions of spontaneous recanalization remain unknown. However, the feasibility of intervention very early after focal symptom onset has clearly been indicated. In a parallel open dose-escalation study, 20 patients received rt-PA (alteplase) within 91–180 minutes from symptom onset; 3 of these had an improvement by 24 hours later (Haley et al. 1992).

Summary

The molecular mechanisms of fibrinolysis and the pharmacologic characteristics of fibrin(ogen)olytic agents are relatively well defined, whereas an understanding of the processes of atherothrombosis and embolism in focal cerebral ischemia is still incomplete. Therefore, conclusions about the contributions of fibrinolysis to the management of acute thrombotic stroke must be provisional. At the time of writing we note that:

1. Recanalization of symptomatic occlusions within 4–6 hours of symptom onset by fibrinolytic agents is feasible.
2. A relationship between embolus location and recanalization efficacy may exist (Mori 1991; rt-PA Acute Stroke Study Group 1991). Occlusions of the cervical portion of the ICA by atheroma-based in situ thrombosis appear more resistant to systemic thrombolysis than do cardiogenic or artery-to-artery emboli of the intracranial portion of the ICA, or of the stem and major branches of the MCA.
3. Early partial or complete recanalization of symptomatic cerebral arteries correlate with a significant reduction in residual infarction volume on CT scan.

4. The incidence of cerebral arterial recanalization is apparently greater when thrombolytic agents are delivered by direct intra-arterial infusion than by systemic infusion.
5. The presence of (leptomeningeal) collateral channels may contribute to a favorable neurological outcome in patients with recanalization of the symptomatic cerebral artery.
6. Although the recanalization efficacy of interventional arterial infusion seems superior to regional or systemic infusion, no comparative study has been performed to test this notion directly (Zeumer 1985).
7. As yet only limited prospective evidence is available in patients with carotid territory ischemia to show that clinical improvement can occur following the infusion of a fibrinolytic agent (Mori et al. 1991; Brott et al. 1992).
8. The risks of intracranial hemorrhage with thrombolytic treatment is a troubling concern, and is the practical limitation of this approach. At present doses and with agents currently used, the incidence of hemorrhagic transformation probably does not exceed that suggested by literature reports of the untreated or anticoagulated stroke patients.
9. The most appropriate thrombolytic agent and the optimal dose and delivery system for early intervention in acute stroke have not yet been worked out.
10. The effects of thrombolytic treatment on stasis in the MCA microcirculation, a feature in some experimental models, remain unexplored at present.

The mechanisms of action of specific plasminogen activators, the known endogenous thrombolytic processes, and the little that is known about the antithrombotic characteristics of the cerebral vasculature do not yet suggest refinements of current approaches which may facilitate safe thrombolysis in acute stroke. Rather, more information about vascular events in cerebral ischemia is necessary.

References

Abe T, Kazama M, Naito I et al. (1981a) Clinical evaluation for efficacy of tissue culture urokinase (TCUK) on cerebral thrombosis by means of multicenter double-bind study. Blood Vessels 12:321–341

Abe T, Kazawa M, Naito I (1981b) Clinical effect of urokinase (60 000 units/day) on cerebral infarction: comparative study by means of multiple-center double-blind test. Blood Vessels 12:342–358

Abe T, Terashi A, Tohgi H, Sasoh S, Naito I (1990) Clinical efficacy of intravenous administration of SM-9527 (t-PA) in cerebral thrombosis. Clin Eval 18:39–69

Adilgaard U (1969) Binding of thrombin to antithrombin III. Scand J Clin Lab Invest 24:23–27

Aldrich MS, Sherman SA, Greenberg HS (1985) Cerebrovascular complications of streptokinase infusion. JAMA 253:1777–1779

Araki G, Minakami K, Mihara H (1973) Therapeutic effect of urokinase on cerebral infarction. Rinsho to Kenkyu 50:3317–3326

Astrup J, Symon L, Branston NM, Lassen N (1977) Cortical evoked potential and extracellular K^+ and H^+ at critical levels of brain ischemia. Stroke 8:51–57

Astrup J, Siesjo BK, Symon L (1981) thresholds in cerebral ischemia – the ischemic penumbra. Stroke 12:723–725

Atarashi J, Otomo E, Araki G, Itoh E, Togi H, Matsuda T (1985) Clinical utility of urokinase in the treatment of acute stage of cerebral thrombosis. Multicenter double-blind study in comparison with placebo. Clin Eval 13:659–709

Atkin N, Nitzberg S, Dorsey J (1964) Lysis of intracerebral thromboembolism with fibrinolysin. Report of a case. Angiology 15:346–439

Babikian VL, Kase CS, Pessin MS, Norrving B, Gorelick PB (1989) Intracerebral hemorrhage in stroke patients anticoagulated with heparin. Stroke 29:1500–1503

Bach RR (1988) Initiation of coagulation by tissue factor. CRC Crit Rev Biochem 23:339–368

Bando H, Okada K, Matsuo O (1987) Thrombolytic effect of prourokinase in vitro. J Fibrinol 1:169–176

Bauer PI, Machovich R, Aranyi P, Buki KG, Csonke E, Horvath I (1983) Mechanism of thrombin binding to endothelial cells. Blood 61:368–372

Berliner S, Niiya K, Roberts JR, Houghten RA, Ruggeri ZM (1988) Generation and characterization of peptide-specific antibodies that inhibit von Willebrand factor binding to glycoprotein IIb-IIIa without interacting with other adhesive molecules. J Biol Chem 263:7500–7505

Bernick MB, Kwaan HC (1969) Plasminogen activator activity in cultures from human tissues. An immunological and histochemical study. J Clin Invest 48:1740–1753

Bloom AL, Thomas DP (eds) (1987) Haemostasis and thrombosis. Churchill Livingstone, Edinburgh

Breederveld K, Giddings JC, ten Cate JW, Bloom AL (1975) The localization of factor V within normal human platelets and the demonstration of a platelet factor V in congenital factor V deficiency. Br J Haematol 29:405–412

Brogden RN, Speight TM, Avery GS (1973) Streptokinase: a review of its clinical pharmacology, mechanism of action, and therapeutic uses. Drugs 5:357–445

Brott T, Haley C, Levy D et al. (1990) Safety and potential efficacy of tissue plasminogen activator (t-PA) for stroke. Stroke 21:181

Brott TG, Haley EC, Levy DE et al. (1992) Urgent therapy for stroke. Part I. Pilot study of tissue plaminogen activator administered within 90 minutes. Stroke 23:632–640

Bruckmann H, Ferbert A (1989) Putaeminal hemorrhage after recanalization of an embolic MCA occlusion treated with tissue plasminogen activator. Neuroradiology 31:95–97

Buteux G, Jubault V, Suisse A (1988) Local recombinant tissue plasminogen activator to clear cerebral artery thrombosis developing soon after surgery. Lancet 1:1143–1144

Castellino FJ (1979) A unique enzyme–protein substrate modifier reaction: plasmin/streptokinase interaction. Trends Biochem Sci 4:1–5
Clarke RL, Cliffton EE (1960) The treatment of cerebrovascular thrombosis and embolism with fibrinolytic agents. Am J Cardiol 30:546–551
Collen D, Lijnen HR (1986) The fibrinolytic system in men. CRC Crit Rev Oncol Hematol 4:249
Collen D, Lijnen HR, Stassen JM (1992) Thrombolytic properties of staphylokinase and streptokinase in experimental animal models. Fibrinolysis 6 (Suppl 2):29 (Abstract)
Colucci M, Balconi G, Lorenzet R et al. (1983) Cultured human endothelial cells generate tissue factor in response to endotoxin. J Clin Invest 71:1893–1896
Comp PC, Jacocks RM, Ferrell GL, Esmon CT (1982) Activation of protein C in vivo. J Clin Invest 70:127–134
Curtis CG (1987) Plasma factor XIII. In: Bloom AL, Thomas DP (eds) Haemostasis and thrombosis. Churchill Livingstone, Edinburgh, p 217
Dalal PM, Shah PM, Aiyar RR (1965a) Arteriographic study of cerebral embolism. Lancet 2:358–361
Dalal PM, Shah PM, Sheth SC, Deshparde CK (1965b) Cerebral embolism: angiographic observations on spontaneous clot lysis. Lancet 1:61–64
Damus PS, Hicks M, Rosenberg RD (1973) Anticoagulant action of heparin. Nature 246:355–357
del Zoppo GJ (1988) Thrombolysis: new concepts in the treatment of stroke. In: Hennerici M, Sitzer G, Weger H-D (eds) Carotid artery plaques. Karger, Basel, pp 247–272
del Zoppo GJ (1990) Relevance of focal cerebral ischemia models. Experience with fibrinolytic agents. Stroke 21: IV155–IV160
del Zoppo GJ, Mori E (1992) Hematologic causes of intracerebral hemorrhage and their treatment. In: Batjer HH (ed) Spontaneous intracerebral hemorrhage. Neurosurg Clin North Am 3(3):637–658
del Zoppo GJ, Zeumer H, Harker LA (1986a) Thrombolytic therapy in acute stroke: possibilities and hazards. Stroke 17:595–607
del Zoppo GJ, Copeland BR, Waltz TZ, Tyroff J, Plow EF, Harker LA (1986b) The beneficial effect of intracarotid urokinase on acute stroke in a baboon model. Stroke 17:638–643
del Zoppo GJ, Ferbert A, Otis S et al. (1988) Local intraarterial fibrinolytic therapy in acute carotid territory stroke: a pilot study. Stroke 19:307–313
del Zoppo GJ, Poeck K, Pessin MS et al. (1992) Recombinant tissue plasminogen activator in acute thrombotic and embolic stroke. Ann Neurol 32:78–86
del Zoppo GJ, Mori E, Hacke W (1993) Thrombolytic therapy in acute ischemic stroke II. Springer-Verlag, Heidelberg
Drake ME, Shin C (1983) Conversion of ischemic to hemorrhagic infarction by anticoagulant administration. Report of two cases with evidence from serial computed tomographic brain scans. Arch Neurol 40:44–46
Drake TA, Hannani K, Fei H, Lavi S, Berliner JA (1991) Minimally oxidized low-density lipoprotein induces tissue factor expression in cultured human endothelial cells. Am J Pathol 138:601–607
Ehrlich HJ, Bang NW, Little SP et al. (1987) Biological properties of a kringleless tissue plasminogen activator (t-PA) mutant. Fibrinolysis 1:75–81
Fieschi C, Bozzao L (1969) Transient embolic occlusion of the middle cerebral and internal carotid arteries in cerebral apoplexy. J Neurol Neurosurg Psych 32:236–240
Fieschi C, Argentino C, Lenzi GL, Sacchetti ML, Toni D, Bozzao L (1989) Clinical and instrumental evaluation of patients with ischemic stroke within the first six hours. J Neurol Sci 91:311–321
Fisher CM, Adams RD (1951) Observations on brain embolism with special reference to the mechanism of hemorrhagic infarction. J Neuropathol Exp Neurol 10:92–94
Fisher CM, Adams RD (1987) Observations on brain embolism with special reference to hemorrhage infarction. In; Furlan AJ (ed) The heart and stroke. Exploring mutual cerebrovascular and cardiovascular issues. Springer-Verlag, New York, pp 17–36
Fletcher AP, Alkjaersig N, Sherry S, Genton E, Hirsh J, Bachmann F (1965) The development of urokinase as a thrombolytic agent. Maintenance of a sustained thrombolytic state in man by its intravenous infusion. J Lab Clin Med 65:713–731
Fletcher AP, Alkjaersig N, Lewis M et al. (1976) A pilot study of urokinase therapy in cerebral infarction. Stroke 7:135–142
Furlan AJ, Cavalier SJ, Hobbs RE, Weinstein MA, Modic MI (1982) Hemorrhage and anticoagulation after nonseptic embolic brain infarction. Neurology 32:280–282
Gaffney PJ (1987) Fibrinolysis. In: Bloom AL, Thomas DP (eds) Haemostasis and thrombosis. Churchill Livingstone, London, pp 223–244
Gaffney PJ, Whittaker AN (1979) Fibrin cross-links and lysis rates. Thromb Res 14:85–94
Gandell SJ (1992) Vampire bat salivary plasminogen activator (Bat-PA): exploration of efficacy and safety using animal models of arterial thrombosis and bleeding. Fibrinolysis 6 (Suppl 2):28 (Abstract)
Gheysen D, Lijnen HR, Pierard L et al. (1987) Characterization of a recombinant fusion protein of the kringle domain of tissue-type plasminogen activator with a truncated single chain urokinase type plasminogen activator. J Biol Chem 262:11779–11784
Groskopf WR, Summaria L, Robbins KC (1969) Studies on the active center of human plasmin. Partial amino acid sequence of a peptide containing the active center serine residue. J Biol Chem 224:3590
Gunzler WA, Steffens GJ, Otting F, Buse G, Flohe L (1982) Structural relationship between human high and low molecular mass urokinase. Hoppe Seylers Z Physiol Chem 563:133–141
Hacke W, Zeumer H, Ferbert A, Bruckmann H, del Zoppo GJ (1988) Intra-arterial thrombolytic therapy improves outcome in patients with acute vertebrobasilar oclusive disease. Stroke 19:1216–1222
Haley EC, Levy DE, Brott TG et al. (1992) Urgent therapy for stroke. Part II. Pilot study of plasminogen administered 91–180 minutes from onset. Stroke 23:641–645
Hanaway J, Torack R, Fletcher AP, Landau WM (1976) Intracranial bleeding associated with urokinase therapy for acute ischemic hemispheral stroke. Stroke 7:143–146
Hart RG (1986) Cerebral Embolism Study Group: timing of hemorrhagic transformation of cardioembolic stroke. In: Stober T (ed) Central nervous system control of the heart. Martinus Nijhoff Publishing, Boston, pp 229–232
Henze TH, Boeer A, Tebbe U, Romatowski J (1987) Lysis of basilar artery occlusion with tissue plasminogen activator. Lancet 1:1391
Herderscheê D, Limburg U, Hijdra A, Koster PA (1991) Recombinant tissue plasminogen activator in two patients with basilar artery occlusion. J Neurol Neurosurg Psych 54:71–73
Herndon RM, Meyer JS, Johnson JF, Landers J (1960) Treatment of cerebrovascular thrombosis with fibrinolysin. Preliminary report. Am J Cardiol 30:540–545
Hornig CR, Dorndorf W, Agnoli AL (1986) Hemorrhagic cerebral infarction: a prospective study. Stroke 17:179–185

Irino T, Taneda M, Minami T (1977) Angiographic manifestations in post-recanalized cerebral infarction. Neurology 27:471–475

Ito Z, Suzuki A, Hen R, Menura K (1983) Prognostic factors in spontaneous recanalization of middle cerebral artery occlusion. In: Ito Z, Kutsuzawa T, Yasui N (eds) Cerebral ischemia – an update. Exerpta Medica, Amsterdam, pp 159–166

Jacobson CD, Chandler AB (1965) Thrombolysis in vivo. I. Method, comparison of various thrombolytic agents and factors influencing thrombolysis. Scand J Clin Lab Invest 17 (Suppl 84): 209–224

Koller RL (1982) Recurrent embolic cerebral infarction and anticoagulation. Neurology 32:283–285

Koudstaal PJ, Stibbe J, Vermeulen M (1988) Fatal ischaemic brain oedema after early thrombolysis with tissue plasminogen activator in acute stroke. Br Med J 297: 1571–1574

Kruithoff EKO (1988) Inhibitors of plasminogen activators. In: Kluft C (ed) Tissue-type plasminogen activator (t-PA): physiological and clinical aspects. Volume 1. CRC Press, Boca Raton, pp 190–210

Kruithof EKO, Tran-Thang C, Bachmann F (1986) The fast-acting inhibitor of tissue-type plasminogen activator in plasma is also the primary plasma inhibitor of urokinase. Thromb Haemost 55:65–69

Kruithof EKO, Tran-Thang C, Ransijn A, Bachmann R (1984) Demonstration of a fast acting inhibitor of plasminogen activators in human plasma. Blood 68:907

Kunitz SC, Gross CR, Heyman A et al. (1984) The pilot stroke data bank: definition, design and data. Stroke 15:740–746

Levin EG, Loskutoff DJ (1982) Cultured bovine endothelial cells produce both urokinase and tissue-type plasminogen activators. J Cell Biol 94:631

Levin EG, Santell L (1987) Association of a plasminogen activator inhibitor (PAI-1) with the growth substratum and membrane of human endothelial cells. J Cell Biol 105:2543–2549

Levin EG, Santell L (1988) Stimulation and desensitization of tissue plasminogen activator release from human endothelial cells. J Biol Chem 263:6502–6507

Levin EG, Stern DM, Nawroth PP et al. (1986) Specificity of the thrombin-induced release of tissue plasminogen activator from cultured human endothelial cells. Thromb Haemost 56:115–119

Lijnen HR, Zamarron C, Blaber M, Winkler ME, Collen D (1986) Activation of plasminogen by pro-urokinase. I. Mechanism. J Biol Chem 261:1253–1258

Lodder J, Krijne-Kubat B, van der Lugt PJM (1988) Timing of autopsy-confirmed hemorrhagic infarction with reference to cardioembolic stroke. Stroke 19: 1482–1484

Lollar P, Owen WG (1980) Clearance of thrombin from circulation in rabbits by high-affinity binding sites on endothelium. J Clin Invest 66:1222–1230

Majerus PW, Miletich JP, Kane WP, Hoffmann SL, Stanford N, Jackson CM (1980) The formation of thrombin on platelet surface. In: Mann KG, Taylor FB (eds) The regulation of coagulation. Elsevier/North Holland, New York, p 215

Matsumoto K, Satoh K (1991) Topical intraarterial urokinase infusion for acute stroke. In: Hacke W, del Zoppo GJ, Hirschberg M (eds) Thrombolytic therapy in acute ischemic stroke. Springer-Verlag, Heidelberg, pp 207–212

Matsuo O, Collen D, Verstraete M (1981) On the fibrinolytic and thrombolytic properties of active-site *p*-anisoylated streptokinase–plasminogen complex (BRL 26921). Thromb Res 347:358

Meyer JS, Herndon RM, Gotoh F, Tazaki Y, Nelson JN, Johnson JF (1961) Therapeutic thrombolysis. In: Millikan CH, Siekert RG, Whisnant JP (eds) Cerebral vascular disease. Third Princeton Conference, Grune and Stratton, New York, pp 160–177

Meyer JS, Gilroy J, Barnhart MI, Johnson JF (1963) Therapeutic thrombolysis in cerebral thromboembolism. Neurology 13:927–937

Meyer JS, Gilroy J, Barnhart MI, Johnson JF (1964) Anticoagulants plus streptokinase therapy in progressive stroke. JAMA 189:373

Miyakawa T (1984) The cerebral vessels and thrombosis. Rinsho Ketsueki 25:1018–1026

Möbius E, Berg-Dammer E, Kühne D, Ahser HC (1991) Local thrombolytic therapy in acute basilar artery occlusion: experience with 18 patients. In: Hacke W, del Zoppo GJ, Hirschberg M (eds) Thrombolytic therapy in acute ischemic stroke. Springer-Verlag, Heidelberg, pp 213–215

Mohr JP, Caplan LR, Melski JW et al. (1978) The Harvard Cooperative Stroke Registry: a prospective registry of patients hospitalized with stroke. Neurology 28:754–762

Moncada S, Vane JR (1979) Arachidonic acid metabolites and the interactions between platelets and blood-vessel walls. New Engl J Med 300:1142–1147

Mori E (1991) Fibrinolytic recanalization therapy in acute cerebrovascular thromboembolism. In: Hacke W, del Zoppo GJ, Hirschberg M (eds) Thrombolytic therapy in acute ischemic stroke. Springer-Verlag, Heidelberg, pp 137–146

Mori E, Tabuchi M, Yoshida T, Yamadori A (1988) Intracarotid urokinase with thromboembolic occlusion of the middle cerebral artery. Stroke 19:802–812

Mori E, Yoneda Y, Ohkawa S et al. (1991) Double-blind placebo-controlled trial of intravenous recombinant tissue plasminogen activator (rt-PA) in acute carotid stroke. Neurology 41 (Suppl 1):347

Mori E, Yoneda Y, Tabuchi M et al. (1992) Intravenous recombinant tissue plasminogen activator in acute carotid artery territory stroke. Neurology 42:976–982

Morrissey JH, Revak D, Tejada P, Fair DS, Edgington TS (1988) Resolution of monomeric and heterodimeric forms of tissue factor, the high affinity cellular receptor for factor VII. Thromb Res 50:481–493

Mustard JF, Packham MA, Kinlough-Rathbone RL (1987) Mechanisms in thrombosis. In: Bloom AL, Thomas DP (eds) Haemostasis and thrombosis, 2nd edn. Churchill Livingstone, Edinburgh, pp 618–650

Nenci GG, Gresele P, Taramelli M, Agnelli G, Signorini E (1983) Thrombolytic therapy for thromboembolism of vertebrobasilar artery. Angiology 34:561–571

Niemetz J (1972) Coagulant activity of monocytes. J Clin Invest 51:307–313

Niemetz J, Morrison DC (1977) Lipid A as the biologically active moiety in bacterial endotoxin (LPS)-initiated generation of procoagulant activity by peripheral blood leukocytes. Blood 49:947–955

Ogata J, Yutani C, Imakita M et al. (1989) Hemorrhagic infarct of the brain without a reopening of the occluded arteries in cardioembolic stroke. Stroke 20:876–883

Okada Y, Yamaguchi T, Minematsu K et al. (1989) Hemorrhagic transformation in cerebral embolism. Stroke 20:598–603

Otomo E, Araki G, Itoh E, Tohgi H, Matsuda T, Atarashi J (1985) Clinical efficacy of urokinase in the treatment of cerebral thrombosis: multicenter double-blind study in comparison with placebo. Clin Eval 13:711–751

Otomo E, Tohgi H, Hirai S et al. (1988) Clinical efficacy of AK-124 (tissue plasminogen activator) in the treatment of cerebral thrombosis: study by means of multicenter double blind comparison with urokinase. Yakuri To Chiryo 16:3775–3821

Ott BR, Zamani A, Kleefield J, Funkenstein HH (1986) The

clinical spectrum of hemorrhagic infarction. Stroke 17:630–637
Owren P (1944) New investigations on the coagulation of blood. Proc Norwegian Acad Sci 21
Pelz P, Ladumer G, Giebritz E (1991) Neuropathological findings after thrombolytic therapy in acute ischemic stroke. In: Hacke W, del Zoppo GJ, Hirschberg M (eds) Thrombolytic therapy in acute ischemic stroke. Springer-Verlag, Heidelberg, pp 224–227
Pessin MS, del Zoppo GJ, Estol CJ (1990) Thrombolytic agents in the treatment of stroke. Clin Neuropharmacol 13:271–289
Pessin MS, Teal PA, Caplan LR (1991) Hemorrhagic infarction: guilt by association. AJNR 12:1123–1126
Pierard L, Jacobs P, Gheysen D et al. (1987) Mutant and chimeric recombinant plasminogen activators. J Biol Chem 262: 11771–11778
Ranby M, Bergsdorf N, Nilsson T (1982a) Enzymatic properties of the one- and two-chain form of tissue plasminogen activator. Thromb Res 27:175–183
Ranby M, Bergsdorf N, Norrman B, Svenson E, Wallen P (1982b) Tissue plasminogen activator kinetics. In: Davison JF, Bachmann F, Bouvier CA, Kruithof EKO (eds) Progress in fibrinolysis. Volume VI. Churchill-Livingstone, New York, p 182
Rickli EE (1975) The activation mechanism of human plasminogen. Thromb Diathes Haem 34:386–395
Rijken DC (1988) Structure/function relationships of t-PA. In: Kluft C (ed) Tissue type plasminogen activator (t-PA): physiological and clinical aspects. Volume 1. CRC Press, Boca Raton, pp 101–122
Rijken DC, Hoylaerts M, Collen D (1982) Fibrinolytic properties of one-chain and two-chain human extrinsic (tissue-type) plasminogen activator. J Biol Chem 257:2920–2925
Robbins KC, Summaria L, Hsieh B, Shah RJ (1967) The peptide chains of human plasmin. J Biochem 242:2333–2342
rt-PA Acute Stroke Study Group (1991) An open safety/efficacy trial of rt-PA in acute thromboembolic stroke: final report. Stroke 22:153
Ruggeri ZM, De Marco L, Gatti L, Bader R, Montgomery RR (1983) Platelets have more than one binding site for von Willebrand factor. J Clin Invest 72:1–12
Runge MS, Bode C, Matsueda GR, Haber E (1987) Antibody-enhanced thrombolysis: targeting of tissue plasminogen activator in vivo. Proc Natl Acad Sci USA 84:7659–7662
Savage B, Ruggeri ZM (1991) Selective recognition of adhesive sites in surface-bound fibrinogen by GPIIb-IIIa on nonactivated platelets. J Biol Chem 266:11227–11233
Schwartz ML, Pizzo SV, Hill RL, McKee PA (1973) Human factor XIII from plasma and platelets. Molecular weight, subunit structures, proteolytic activation and cross-linking of fibrinogen and fibrin. J Biol Chem 248:1395–1407
Sjoholm I, Wiman B, Wallen P (1973) Studies on the conformational changes of plasminogen induced during activation to plasmin and by 6-aminohexanoic acid. Eur J Biochem 39:471–479
Skyhoj-Olsen T, Larsen B, Herring M, Skawer EB, Lassen NA (1983) Blood flow and vascular reactivity in collateral perfused brain tissue: evidence of an ischemic penumbra. Stroke 14:332–341
Smith RAG, Dupe RJ, English PD, Green J (1981) Fibrinolysis with acylenzymes: a new approach to thrombolytic therapy. Nature 290:505–508
Smith RAG, Dupe FJ, English PD, Green J (1982) Acyl-enzymes as thrombolytic agents in a rabbit model of venous thrombosis. Thromb Haemost 47:269–274
Solis OJ, Roberson GR, Taveras JM, Neohr J, Pessin M (1977) Cerebral angiography in acute cerebral infarction. Revist Interam Radiol 2:19–25
Stump DC, Mann KG (1988) Mechanisms of thrombus formation and lysis. Ann Emerg Med 17:1138–1147
Sussman BJ, Fitch TSP (1958) Thrombolysis with fibrinolysin in cerebral arterial occlusion. JAMA 167: 1705–1709
Taneda M, Shimada N, Tsuchiya T (1985) Transient neurological deficits due to embolic occlusion and immediate reopening of the cerebral arteries. Stroke 16:522–524
Theron J, Courtheoux P, Casaseo A et al. (1989) Local intra-arterial fibrinolysis in the carotid territory. Am J Neuroradiol 10:753–765
Thorsen S, Phillips M (1984) Isolation of tissue-type plasminogen activator–inhibitor complexes from human plasma: evidence for a rapid plasminogen activator inhibitor. Biochim Biophys Acta 802:111
Verheijen JH, Chang GTG, Kluft C (1984) Evidence for the occurrence of a fast-acting inhibitor for tissue-type plasminogen activator in human plasma. Thromb Haemost 51:392
Verstraete M, Collen D (1986) Thrombolytic therapy in the eighties. Blood 67:1529–1541
von Kummer R (1991) Intravenous tissue plasminogen activator in acute stroke. In: Hacke W, del Zoppo GJ, Hirschberg M (eds) Thrombolytic therapy in acute ischemic stroke. Springer-Verlag, Heidelberg, pp 161–167
von Kummer R, Hacke W (1992) Safety and efficacy of intravenous tissue plasminogen activator and heparin in acute middle cerebral artery. Stroke 23:646–652
White FW, Barlow GH, Mozen MM (1966) The isolation and characterization of plasminogen activators (urokinase) from human urine. Biochemistry 5:2160–2169
Wiman B, Chmielewska J, Ranby M (1984) Inactivation of tissue plasminogen activator in plasma. Demonstration of a complex with a new rapid inhibitor. J Biol Chem 259:3644
Witt W, Baldus B, Bringmann P et al. (1992) The new plasminogen activators DSPAα1 and α2 from the vampire bat *Desmodus rotundus*. Fibrinolysis 6 (Suppl 2):28 (Abstract)
Yamaguchi T (1991) Intravenous rt-PA in acute embolic stroke. In: Hacke W, del Zoppo GJ, Hirschberg M (eds) Thrombolytic therapy in acute ischemic stroke. Springer-Verlag, Heidelberg, pp 168–174
Yamaguchi T, Minematsu K, Choki J-I, Ikeda M (1984) Clinical and neuroradiological analysis of thrombotic and embolic cerebral infarction. Jpn Circ J 48:50–58
Yamaguchi T, Hayaleawa T, Kileuchi H (1993) Intravenous tissue plasminogen activator in acute thromboembolic stroke: a placebo-controlled, double-blind trial. In: del Zoppo GJ, Mori E, Hacke W (eds) Thrombolytic therapy in acute ischemic stroke II. Springer-Verlag, Heidelberg, pp 59–65
Zeumer H (1985) Survey of progress: vascular recanalizing techniques in interventional neuroradiology. J Neurol 231:287–294
Zeumer H, Freitag HJ, Grzyka U, Neunzig HP (1989) Local intra-arterial fibrinolysis in acute vertebrobasilar occlusion. Technical developments and recent results. Neuroradiology 31:336–340

Concluding Comments

The authors of the chapters in this section on blood coagulability have reviewed in detail what is now known about clot formation and clot lysis. I will briefly elaborate herein only on selected aspects of this topic.

Utility of Measuring Coagulation Factors

Feinberg states the prevailing view that measurements of coagulation factors after a stroke are of dubious value since, as with acute-phase reactants, elevated levels after a stroke may be merely a post hoc epiphenomenon of thrombosis. Hart and Kanter (1990) also state in their review of hematologic disorders and ischemic stroke that 'it is uncertain whether elevated levels of coagulation factors predispose to thrombosis'. We have been measuring Factor VIII activity in patients with ischemic stroke for a number of years (Estol et al. 1989). The preliminary data show that there are probably three groups of patients with significant elevations of Factor VIII antigen and activity. The first is a group of patients with long-term, either congenital or acquired, increased Factor VIII activity. Such patients have been reported previously (Kosik and Furie 1980) and generally have a history of venous thrombosis, pulmonary embolism, frequent miscarriages and strokes. The second group are those who have increased Factor VIII activity due to intercurrent medical illness, such as acute inflammatory bowel disease. As Fisher has noted in his chapter, systemic illnesses, especially infection, cancer and inflammatory diseases, alter various cytokines and serum proteins. The erythrocyte sedimentation rate (ESR) has long been used as a non-specific marker of active body illness and probably most reflects changes in serum proteins. The increase in serine protein coagulation factors could at least partially explain hypercoagulability. The third group are patients in whom the Factor VIII elevation probably followed the stroke and soon returned to much lower levels. In this group other coagulation factors were also elevated acutely. Much work remains to be done, preferably prospectively before stroke, and after, to determine the relation of the levels of coagulation factors to any possible contributory hypercoagulability.

Red Blood Cell Disorders

Another important factor not mentioned in detail in this section is the erythrocyte. Both qualitative and quantitative disorders of red blood cells can promote thrombosis. Polycythemia is very important since blood flow is probably reduced by nearly half as the Hct rises from

Fig. 1

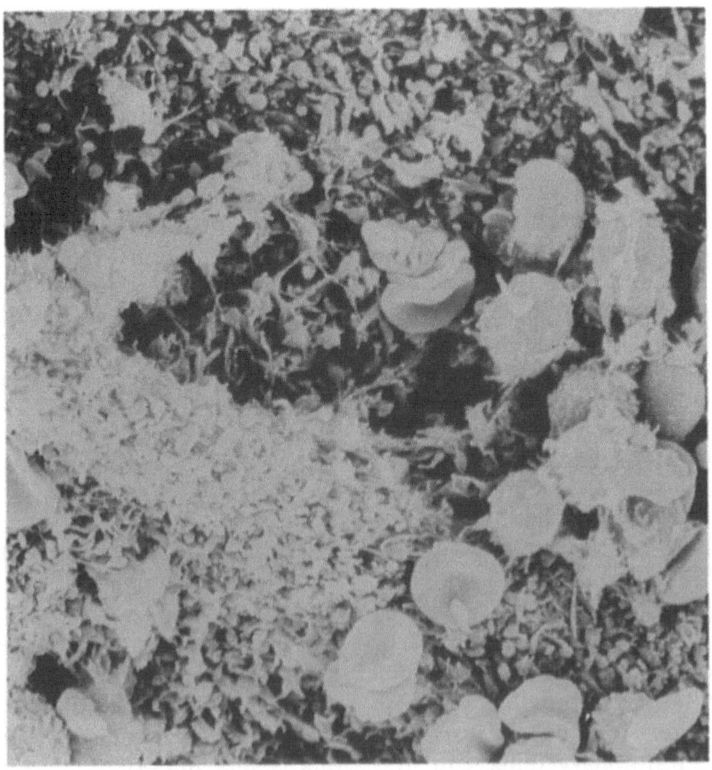

Fig. 2

Fig. 1. Scanning electron microscope image of a red thrombus composed of fibrin and erythrocytes, formed in a thrombogenic system, in a vessel with a low flow rate (courtesy of S. H. Hanson and Ch. Kessler, Emory University, Division of Hematology). From Caplan LR, Stroke: A Clinical Approach, 2nd edn, Butterworth-Heinemann, Boston, 1993, with permission.

Fig. 2. Scanning electron microscope image of a white fibrin–platelet thrombus formed in a high-flow system (courtesy of S. H. Hanson and Ch. Kessler, Emory University, Division of Hematology). From Caplan LR, Stroke: A Clinical Approach, 2nd edn, Butterworth-Heinemann, Boston, 1993, with permission.

42 to 52. The increased blood viscosity and reduced flow predisposes to coagulation, especially in narrowed atherosclerotic arteries. Sickle-cell disease, with its abnormally shaped erythrocytes, clearly also potentiates thrombosis in both small penetrating arteries and larger basal arteries (Adams et al. 1992). Sickle-cell disease also causes a proliferation of the connective tissue elements of the vascular intima and media, also promoting occlusive thrombosis. Sickle-cell–hemoglobin C disease and sickle-cell trait have also been complicated by ischemic strokes due to hypercoagulability. Paroxysmal nocturnal hemoglobinuria (PNH) is another disorder of red cell membranes that renders erythrocytes very sensitive to lysis by complement. There is an increased incidence of venous thrombosis in patients with PNH (Hart and Kanter 1990).

'Red Clots' versus 'White Clots'

Hematologists have for a long time distinguished between so-called 'red clots' and 'white clots' (Deyken 1967; Mustard et al. 1962). Red clots are composed mostly of red blood cells and fibrin, as illustrated in Fig. 1. These fibrin-rich clots are formed in slowly moving streams or areas of stagnation. Their formation does not require an abnormal blood vessel tissue thromboplastin. Clinically, red thrombi form mostly in veins and dural sinuses, in the heart in dilated atria, over infarcted hypokinetic or aneurysmal regions in the cardiac ventricles, and in very stenotic arteries.

In contrast, white clots are composed of platelets and fibrin and are relatively poor in red blood cells (Fig. 2). White clots form mostly in areas in which the vessel walls or endothelial surfaces are abnormal. An ideal situation for the formation of white platelet–fibrin nidi are regions of irregular plaque formation in large arteries and irregular lesions on cardiac valves. In many instances, a thrombus probably begins as a white platelet–fibrin clot and then a red thrombus is superimposed, having been laid down as a cap over the initial platelet mass (Deykin 1967; Mustard et al. 1962).

Recently, angioscopic examinations in patients with coronary artery disease and myocardial ischemia have shown differences in types of thrombi, depending on the type of coronary syndrome (Mizuno 1992; Mizuno et al. 1992). In patients with unstable angina, angioscopy showed grayish-white thrombi, whereas reddish thrombi were seen in all patients with acute myocardial infarction (Mizuno et al. 1992). In another study, after thrombolysis, white thrombi were seen to be older than red thrombi and had tighter fibrin networks (Uchida et al. 1986). Mizuno et al. (1992) postulated that white thrombi formed when blood flow was still present, but red thrombi were occlusive.

The concept of red versus white clots is of practical as well as theoretical interest, since the two types respond differently to therapeutic agents. Platelet–fibrin clots are inhibited by so-called antiplatelet aggregants such as aspirin, ticlopidine, sulfinpyrazone and others. Red clots are inhibited by standard anticoagulants such as heparin (including heparinoids and low molecular weight heparins) and warfarin compounds. In some circumstances, a mixture of aspirin and coumadin is probably most effective.

For years, I have used the idea of red and white clots to choose therapy for patients with brain ischemia and those with predisposing conditions that are risks for ischemia (Caplan 1993). I have used heparin followed by warfarin for acute, completely thrombosed arteries, continuing the treatment for 6–10 weeks only. During this period the occlusive thrombus organizes and adheres to the vessel wall and has little chance for subsequent embolization or propagation of clot. Collateral circulation also develops within this period. In situations which could promote red thrombus formation (dilated fibrillating cardiac atria and very stenotic intracranial arteries), I use long-term warfarin while monitoring the vascular lesions for changes. If the artery recanalizes or occludes, then the warfarin can probably be stopped (6 weeks after an occlusion). For patients with non-stenosing plaques I prescribe aspirin. Aspirin might also be effective prophylaxis for some cardiac lesions. In some patients, aspirin and warfarin might be most effective (Miller and Lees 1985).

Fibrinogen

Fibrinogen is a central factor in the formation of both platelet-rich and red blood cell-rich fibrin clots. In patients with rapid and diffuse increases in coagulation, e.g. disseminated intravascular coagulation (DIC), blood fibrinogen levels are rapidly decreased in addition to consumption of other coagulation factors and platelets. Fibrinogen is also a very important determinant of blood viscosity and blood flow. The hematocrit and fibrinogen levels are the two most important determinants of whole-blood viscosity (Grotta et al. 1982). Fibrinogen may also affect blood flow in the microcirculation. Recently, we have surveyed our patients with dementia due to severe white matter microvascular disease (Binswanger's) and found that the majority of such patients have abnormally high fibrinogen levels. The fibrinogen level seems to be an important prognostic factor in predicting which patients will have strokes, myocardial infarctions (Cook and Ubben 1990; Wilhelmsen et al. 1984) and recurrent brain ischemia (Coull et al. 1991). Physicians have attempted to reduce fibrinogen levels acutely using snake venom (Ancrod). Few, if any, studies have analyzed the determinants of the fibrinogen level, the relation to systemic illnesses and the sequential changes in fibrinogen levels before, during and after thrombotic events. Chronic reduction in the fibrinogen level may prove to be an important therapeutic strategy in the future, but clearly much work remains to be done about the role of fibrinogen in coagulation.

The Antiphospholipid Antibody Syndrome

Fisher, in his chapter, discusses the antiphospholipid antibody syndrome. The first-described antiphospholipid antibody was the so-called lupus anticoagulant (LA) (a misnomer, since this condition is characterized by increased coagulability, not bleeding). LA is an antiphospholipid antibody that interferes with the formation of the prothrombin activator, a complex of Ca^{2+}, Factors V and Xa and the phospholipid platelet Factor III (Simon 1990). The laboratory sign of the presence of LA is a prolonged APTT that does not correct when normal plasma is added, indicating the presence of a clotting inhibitor rather than deficiency of a needed component of the coagulation process (Hart and Kanter 1990). Although some patients with LA clearly have systemic lupus erythematosus, the majority do not. Thrombocytopenia and false-positive serological tests for syphilis (the VDRL uses a lipid antigen derived from heart tissue, a cardiolipin assay) are very common in patients with LA. The other phospholipid against which antibodies have been measured is cardiolipin. About 70% of patients with LA also have anticardiolipin antibodies and 70% of patients with anticardiolipin antibodies also have positive tests for LA (Hart and Kanter 1990).

Antiphospholipids are immunoglobulins from the IgG, IgM and IgA classes with specificities for anionic and neutral phospholipids (Pengo et al. 1987; Antiphospholipid Antibodies in Stroke Study Group 1990). IgG is probably the most common isotype but some patients have only IgM or IgA isotype immunoglobulins (Antiphospholipid Antibodies in Stroke Study Group 1990). Recurrent miscarriages, venous thrombosis and pulmonary embolism, cardiac valvular lesions, thrombocytopenia, myocardial infarction and strokes are undoubtedly more common in patients with antiphospholipid antibodies than in those without. It is still unclear how the antibodies cause thrombosis or stroke. Since many biological structures active in coagulation, including the endothelium of vessels, the heart and blood platelets, have phospholipid elements, antibodies against any of these components could be contributory. I and others (Coull, personal communication) have seen patients with all of the clinical features of the antiphospholipid antibody syndrome in whom neither LA nor anticardiolipin antibodies were detectable. The antibodies currently measured may simply be markers for an autoimmune process, and as yet we may not be measuring the antiphospholipid antibody(ies) responsible for the clinical syndrome. Knowledge about this syndrome has opened a potential door into the complex interactions of the immune and coagulation systems, as described by Fisher.

Coagulation Abnormalities and Strokes in Patients With Systemic Illnesses

Stroke seems to be an important complication of a variety of different acute and chronic disorders. Patients with cancer, especially mucinous adenocarcinomas, AIDS and inflammatory disorders, have an unexpectedly high incidence of stroke. Fisher, in his contribution, alludes to some of the possible mechanisms. We have described in an earlier report (Amico et al. 1989) how mucin might contribute to hypercoagulability in patients with adenocarcinoma. Of interest, these patients also have a high rate of marantic non-bacterial endocarditis, with lesions similar to those found in patients with systemic lupus erythematosus and in the antiphospholipid antibody syndrome.

Patients with strokes often develop myocardial infarcts (MIs) and phlebothrombosis and pulmonary embolism. Similarly, patients with MIs develop strokes. Customarily in the past, the concurrence of MI and stroke has usually been interpreted as the stroke being caused by cardiogenic embolism. When the MI followed the stroke, clinicians have suggested that some myocardial ischemia, in restrospect, probably preceded the stroke but was unrecognized. There may be another explanation. Systemic illnesses, including ischemia of any origin, may alter serum proteins and other chemical tissue and serological components that cause the coagulation system to be activated. We know that the ESR is elevated. When so-called acute-phase reactants are measured, they are often elevated. We have recently cared for 5 patients with Crohn's disease who had strokes, often associated with elevated Factor VIII activity. Others have pointed out the frequency of infections in adult patients with stroke, and in children before the first attack of Moya Moya syndrome (Suzuki and Kodama 1983). Clinicians and researchers are just beginning to clarify this interface between systemic disorders and stroke. Clearly, our clinical measurements are not adequate since patients with cancer and clinical hypercoagulability, including women who are pregnant, take oral contraceptives or are in the puerperium, have thrombotic lesions (especially dural sinus thrombosis) without important measurable changes in coagulation parameters.

Clinical Thrombolysis in Acute Stroke

Interest in the clinical use of fibrinolytic agents has clearly accelerated during the past 5 years. During this time there have been two international meetings in which the results to date of clinical fibrinolytic studies have been reported and discussed (Hacke et al. 1991; del Zoppo et al. 1993). del Zoppo presents some of this information in his chapter. We now know that there are many factors that are important in the success of fibrinolysis. These can be divided into those related to the disorder being treated and those more related to the particular drug and its delivery.

Disease-related factors are very critical. These include:

Anatomical factors: the size of the artery occluded and the width and length of the thrombus. Large, long thrombi in the internal carotid artery in the neck have proven very resistant to treatment, whereas small MCA branch emboli are more responsive.

Time factors: the longer the thrombus has been in place, the less likely the success of fibrinolytic treatment. This observation also relates to the decreased likelihood of reversing brain ischemia. Hemorrhagic complications also probably increase the longer treatment is delayed, since blood vessels that are damaged by ischemia are more likely to leak during reperfusion.

Disease factors: fresh cardioembolic occlusions are more likely to lyse than in-situ thrombosis. In-situ lesions usually have an admixture of arterial stenosing plaques and superimposed occlusive thrombosis. Of course, fibrinolysis does nothing to the stenosing plaque and so a recurrent occlusion is common. In artery-to-artery embolism the proximal lesion may limit the delivery of blood and fibrinolytic agents to the lesion site. In cardiac-origin lesions, some embolic material is composed of fibrous vegetations, calcium, valvular matter or organized clots that have been in the heart chambers for some time. These lesions will obviously be less responsive to lysis than fresh red thromboemboli.

The delivery of the drug is also very important. We still do not know whether intravenous or direct intra-arterial introduction is preferred. Clearly, intravenous use is easier and does not require rapid, skilled intervention by an experienced neuroradiologist. For this reason, intravenous use would be more generally applicable clinically. The optimal agent has also not been clarified. del Zoppo reviews data on the various agents used and their clot specificity and systemic actions. The dose, speed and duration of delivery are also not as yet determined, nor are the risk/benefit features of each agent known. The next decade should allow clarification of these practical clinical points. Undoubtedly, the clinical introduction of fibrinolytic agents at the bedside has greatly increased our knowledge about fibrinolysis and the body's natural fibrinolytic systems and their physiology and pathophysiology.

References

Adams R, McKie,V, Nichols F et al. (1992) The use of transcranial ultrasonography to predict stroke in sickle cell disease. New Engl J Med 326:605–610

Amico L, Caplan LR, Thomas C (1989) Cerebrovascular complications of mucinous cancers. Neurology 39:522–526

Antiphospholipid Antibodies in Stroke Study Group (1990) Clinical and laboratory findings in patients with antiphospholipid antibodies and cerebral ischemia. Stroke 21:1268–1273

Caplan LR (1993) Stroke: a clinical approach, 2nd edn. Butterworths, Boston

Cook NS, Ubben D (1990) Fibrinogen as a major risk factor in cardiovascular disease. Therapy Pharm Stud 11:444–451

Coull BM, Beamer NB, de Garmo PL et al. (1991) Chronic blood hyperviscosity in subjects with acute stroke, transient ischemic attacks, and risk factors in stroke. Stroke 22: 162–168

del Zoppo GJ, Mori E, Hacke W (1993) Thrombolytic therapy in acute ischemic stroke II. Springer-Verlag, Heidelberg (in press)

Deykin D (1967) Thrombogenesis. New Engl J Med 276:622–628

Estol C, Pessin MS, DeWitt LD et al. (1989) Stroke and increased factor VIII activity. Neurology 39 (Suppl 1):159

Grotta J, Ackerman R, Correia J et al. (1982) Whole-blood viscosity parameters and cerebral blood flow. Stroke 13:296–298

Hacke W, del Zoppo GJ, Hirschberg M (1991) Thrombolytic therapy in acute ischemic stroke. Springer-Verlag, Heidelberg

Hart RG, Kanter MC (1990) Hematologic disorders and ischemic stroke: a selective review. Stroke 21:1111–1121

Kosik K, Furie B (1980) Thrombotic strokes associated with elevated plasmin factor VIII. Ann Neurol 8:435–437

Miller A, Lees RS (1985) Simultaneous therapy with antiplatelet and anticoagulant drugs in symptomatic cardiovascular disease. Stroke 16:668–675

Mizuno K (1992) Angioscopic examination of the coronary arteries: what have we learned? Heart Dis Stroke 1:320–324

Mizuno K, Satomura K, Miyamoto A et al. (1992) Angioscopic evaluation of coronary-artery thrombi in acute coronary syndromes. New Engl J Med 326:287–291

Mustard JF, Murphy EA, Rousell HC et al. (1962) Factors influencing thrombus formations in vivo. Am J Med 33:621–627

Pengo V, Thiagarajan P, Shapiro SS, Heine MJ (1987) Immunologic specificity and mechanism of action of IgG lupus anticoagulant. Blood 70:69–76

Simon L (1990) Case records of the Massachusetts General Hospital. Case 11–1990. New Engl J Med 322:754–769

Suzuki J, Kodama N (1983) Moya Moya disease – a review. Stroke 14:104–109

Uchida Y, Masuo M, Tomaru T et al. (1986) Fiberoptic observation of thrombosis and thrombolysis in isolated human coronary arteries. Am Heart J 112:691–696

Wilhelmsen L, Swardsudd K, Korsan-Bengtsen K et al. (1984) Fibrinogen as a risk factor for stroke and myocardial infarction. New Engl J Med 311:501–505

Section IV

The Blood Vessels

Prefatory Comments

Stroke is a term that refers to brain dysfunction and brain injury caused by abnormalities of the blood vessels supplying the brain. Vascular lesions are diverse and can affect the brain in a variety of ways. These include mechanical obstruction of the arteries, causing decreased perfusion distal to the blockage; roughening of the intimal and luminal surfaces, with break-off of materials causing distal embolization; functional constriction of the arteries, causing decreased perfusion often with thrombosis; abnormal dilation of arteries (often called dolichoectasia or dilatative arteriopathy), with altered blood flow; rupture of the arteries, causing brain and/or subarachnoid hemorrhage; and obstruction and occlusion of major veins and dural sinuses, causing decreased drainage of blood from brain regions and resultant edema, ischemia and hemorrhage. Although I have separated in this text sections on the blood vessels, blood flow and the blood, in reality these are intimately interwoven, interrelated and interdependent. Changes in the blood vessels often lead to activation of platelets and activation of the coagulation cascade. The endothelium has a variety of secretory functions that affect blood flow, blood vessel contractility and blood coagulation.

Atherosclerosis

This term is very widely used, yet I find it very difficult to define. Does atherosclerosis differ from arteriosclerosis? Do these entities include all degenerative changes in arteries and arterioles or only those related to lipid-containing atheromas? A 60-year-old man has a history of smoking, hypercholesterolemia with very high LDL and low HDL levels, diabetes, angina pectoris and claudication in both legs. Ultrasound shows an 85% narrowing of his left internal carotid artery and the endarterectomy specimen shows a very thick lipid-rich plaque with ulceration. Contrast this patient with a 19-year-old African–American man with sickle-cell disease who has a left cerebral infarct due to severe stenosis of the left middle cerebral artery (MCA). The arterial lesion consists of severe fibrous and smooth muscle cell hypertrophy with a 95% narrowing of the MCA lumen. Each patient has severe vascular obstruction and luminal narrowing. Does each have atherosclerosis?

Do the terms atherosclerosis and arteriosclerosis also pertain to degenerative changes in small penetrating intracranial arteries? The abnormalities in intracranial arteries in patients with hypertension ordinarily do not involve lipid deposition. The media is thickened with fibrinoid deposition and fibromuscular hypertrophy with increased connective tissue. Herein, I will separate the discussions of degenerative changes in the large extracranial and intracranial arteries from abnormalities found in the penetrating small intracranial arteries, since these vessels are quite different morphologically and the pathological changes are so dissimilar.

Large extracranial arteries and the aorta in humans begin to show abnormalities quite early in life: fatty streaks are present in teenagers in these locales. Studies of the distribution and severity of atherosclerotic lesions in young patients dying of other problems show a kaleidoscope of successive changes during the ensuing years, leading to plaque formation and luminal narrowing (McGill et al. 1968). Intimal hypertrophy, subintimal lipid deposition, proliferation of smooth muscle cells and calcification of the arterial wall have all been described (Ross 1986; Desmouliere and Gabbiana 1992). Fisher, in his chapter in this section, reviews knowledge about the pathogenesis of these large-artery lesions. A number of different provocative and inciting factors could lead to similar end results. The role of various cytokines and endothelial factors is now being recognized and studied. Fisher will also discuss the important observation that the sites and severity of atherosclerotic lesions vary greatly, depending on the race and sex of the patient. This topic is reviewed elsewhere and will be discussed in various sections of this book (Caplan et al. 1986). The study of atherosclerotic lesions in vivo has been greatly aided by the emergence of effective ultrasound imaging of the arteries. This capability has made it possible to accurately measure the thickness of the arterial walls (Touboul et al. 1992) and to sequentially follow atherosclerotic plaques for ulceration and progression and regression of stenosing lesions (Hennerici et al. 1985; Furst et al. 1992).

Small-Artery Disease

Knowledge about the morphology and progression of large-artery extracranial disease has come from careful microscopic studies of carotid endarterectomy specimens (Fisher and Ojemann 1986) and from correlation of ultrasound with histology and prognosis (O'Donnell et al. 1985; Sterpetti et al. 1988). These studies of carotid specimens involve modern histologic techniques. In contrast, the studies of the morphology of intracranial artery lesions are quite old.

Studies by Fisher (1965, 1969) in the middle of the 20th century showed a high incidence of a lesion that he called variously lipohyalinosis, fibrinoid degeneration and segmental arterial disorganization. This arterial lesion involved some lipid changes in the media, fibrinoid changes and disorganization of the arterial wall. When these histologic studies were carried out the incidence of severe uncontrolled hypertension was high, and there was less effective pharmacologic therapy of hypertension. So-called malignant hypertension was very common. Fisher's studies involved meticulous analysis of thousands of serial brain sections. Fisher and colleagues also described different vascular pathologies that could also lead to obstructive changes in penetrating deep brain arteries (Fisher and Caplan 1971; Fisher 1977). This disorder, which I call 'intracranial branch atheromatous disease' (Caplan 1989), is much more akin to atheromatous changes in large arteries. Atheromas in parent arteries, such as the MCA or the basilar artery, either partially block the origin of branches or extend into the orifices of the branches. In other patients, a small 'microatheroma' arises within the origin of the branch itself. Unfortunately, there have been no recent studies analyzing the spectrum and frequency of the microvascular changes that underlie deep brain infarcts and other forms of microvasculopathy. Hommel and Gray, in their chapter, review the various changes that occur in the brain microvasculature. Experimental scanning electron microscope studies of spontaneously hypertensive rats show early changes in endothelial cells and platelet–endothelial interactions (Hazama et al. 1979). Recent experimental studies have raised the question of whether small penetrating artery occlusions are often due to embolization of materials from the proximal arterial system (Futrell et al. 1989) Clearly, the neuropathology of microvascular disease needs much work.

Physical Aspects of Arterial Lesions

The physics and physiology of flow within blood vessels are very important in understanding vessel wall disease. The best-studied vessel is the internal carotid artery (ICA). Transparent life-sized elastic models, constructed to replicate the human vessel (Kerber and Heilman 1992), and transparent segments of the artery studied after necropsy with cinemicrographic techniques (Karino and Motomiya 1983; Motomiya and Karino 1984) have yielded considerable insight into flow dynamics. The shape and configuration and supporting structures of the human ICA are characteristic and unique. A bulbous dilatation of the artery at its origin is followed by a straight segment in the pharynx in which the vessel is free and readily movable. The ICA then makes a

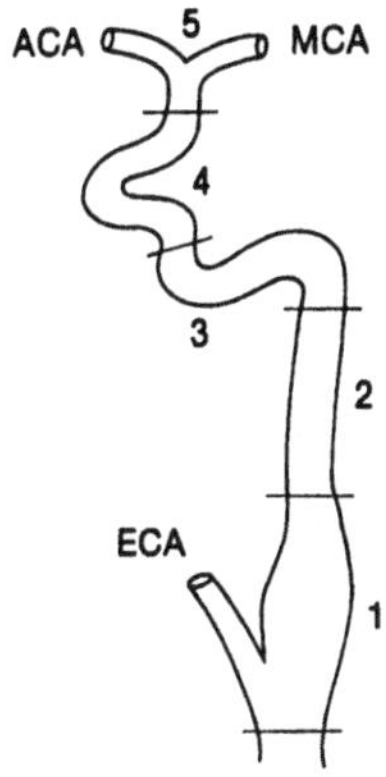

Fig. 1. The divisions of the internal carotid artery used for constructing a life-sized transparent elastic model. From Kerber and Heilman (1992) Flow dynamics in the human arotid artery: preliminary observations using a transparent elastic model. Reproduced with permission of AJNR 13:173–180.

right-angled bend in its petrous portion and enters the cavernous segment after another right-angled bend, terminating in a T-shaped bifurcation (Kerber and Heilman 1992). Fig. 1 shows the ICA model. Flow through this oddly pretzel-shaped tube is laminar or pulsatile (Grady 1984). Shear stress, defined as the force or stress that one lamina exerts on adjacent laminae or surfaces (Grady 1984), varies within the lumen but is most pronounced at the vascular wall. The viscosity of the blood also strongly influences shear stress. At times, flow is not streamlined or laminar but becomes irregular in motion, with circling and tumbling (Grady 1984). Flow within the very thin 'boundary layer' at the very furthest point from the middle of the stream is slowest and nearly stagnant (Grady 1984). It is this thin film boundary layer, which lies adjacent to the vascular wall, in which interactions between the blood column and the wall occur. When velocity of flow is high (as will occur in areas of stenosis), the shear stress on the wall is high and endothelial desquamation and adherence of platelets to the vascular wall occur. Turbulence and endothelial injury both activate platelets and the components of the coagulation cascade.

Recently, it has become feasible and practical to study these physical and rheological aspects of blood flow in vivo using ultrasound. Study of pulsed Doppler waveforms and velocities and color-flow Doppler imaging of the ICAs has yielded very important information about flow in normal and diseased arteries (Nicholls et al. 1989; Steinke et al. 1990a,b). Flow separation is the rule (79% of normal arteries) but there are varying spatial and temporal distributions of the flow separation zones (Steinke et al. 1990a). Flow separation can be limited to the proximal ICA, the external carotid artery (ECA) or both. Secondary flow up the artery may extend from the ICA to the ECA around the flow divider between these two arteries. Plaques usually start to form on the posterior wall of the common carotid artery (CCA) and ICA just opposite the flow divider. Flow near the flow divider is usually laminar or anterograde, but flow in the boundary layer zone on the posterolateral aspect of the vessel often shows transient flow reversal (Nicholls et al. 1989). The presence of flow separation in the ICA seems to predict the absence of serious disease of the vessel wall (Nicholls et al. 1989). Further analysis of velocities and flow promises to yield new insights into the pathophysiology of atherosclerotic wall lesions. Unfortunately, cerebral angiography of the cervicocranial arteries, usually considered as the gold standard for vascular lesions, is documented by a series of sequential films which are reviewed later. Cine-angiography, the rule in the coronary circulation, is seldom performed in the brain circulation. Blood flow is a dynamic complex process not able to be recreated by an analysis of the morphology and measurements of the lesion. Ultrasound is a better reflection of the dynamics. Schmid-Schönbein and Perktold, in their contribution, review in detail the physics and physiology of flow in the arteries and the various factors that lead to vessel wall injury. They also emphasize the interaction between the vessel wall, the blood flow and the blood.

The Endothelium

One of the most important advances in our understanding of disease of vessel walls is the realization of the role played by the endothelium. Furchgott and Zawadzki (1980) reported that an intact endothelium was required for acetylcholine to produce relaxation of vascular smooth muscle. This work suggested the presence of an endothelium-derived relaxing factor, which was subsequently identified as nitric oxide (Palmer et al. 1987; Greenberg et al. 1992). Other investigators found that endothelial cells also release vasoconstrictor peptides, which are usually now referred to as 'endothelins' (Greenberg et al. 1992; Vane et al. 1990). Ryan and Trachtenberg, in their contribution, review our present knowledge of the endothelium and its biochemistry and functions.

'Vasospasm'

Perhaps the most controversial topic that will be discussed in this section is that of functional changes in the blood vessels – vasoconstriction. No-one doubts that vasoconstriction can and does occur, but its role in human disease is still uncertain. During the first half of the 20th century TIAs were often explained as 'slight artery spasms'. Research into the pathology of atherosclerosis led observers to the interpretation that ischemia was explained by diminished blood flow caused by mechanical luminal obstruction and emboli arising from atherosclerotic lesions. Spasm became a discredited explanation given little credence.

New interest in vasospasm arose from studies of the coronary circulation in patients with cardiac ischemia. Spasm was shown in the coronary arteries by cine-angiography during the instillation of dye, and could be reversed pharmacologically in some patients. Coronary artery spasm occurred with and without coexistent atherosclerosis. Vasoconstriction was recognized in the cerebral circulation, when it was appreciated that delayed brain ischemia after subarachnoid hemorrhage (SAH) was often accompanied by narrowing of the basal arteries bathed in bloody spinal fluid. Vasoconstriction had also long been known to be a feature of migraine. These observations have led to a general acknowledgment that vasospasm (reversible vasoconstriction) can and does occur in the craniocerebral circulation. Still unsettled and much debated was whether vasospasm (with or without structural vascular lesions) underlies a significant number of TIAs and brain infarcts. Recently, Faraci et al. (1991), working in Heistad's laboratory, showed that atherosclerotic lesions in experimental animals (monkeys) facilitated the ease of provocation of vasoconstriction. Might vasoconstriction explain a number of cases of ischemia of unknown cause – a frequent final diagnosis still in many patients with brain ischemia? Sloan, in his detailed review, discusses the most recent experimental and clinical data about vasoconstriction.

References

Caplan LR (1989) Intracranial branch atheromatous disease: a neglected, understudied, and underused concept. Neurology 39:1246–1250

Caplan LR, Gorelick PB, Hier DB (1986) Race, sex and occlusive cerebrovascular disease: a review. Stroke 17:648–655

Desmouliere A, Gabbiani G (1992) The role of arterial smooth muscle cells in the pathogenesis of atherosclerosis. Cerebrovasc Dis 2:63–71

Faraci FM, Lopez AG, Breese K et al. (1991) Effect of atherosclerosis on cerebral vascular responses to activation of leukocytes and platelets in monkey. Stroke 22:790–796

Fisher CM (1965) Lacunes: small deep cerebral infarcts. Neurology 15:774–784

Fisher CM (1969) The arterial lesions underlying lacunes. Acta Neuropathol 12:1–15

Fisher CM (1977) Bilateral occlusion of basilar branches. J Neurol Neurosurg Psych 40:1182–1189

Fisher CM, Caplan LR (1971) Basilar artery branch occlusion: a cause of pontine infarction. Neurology 21:900–905

Fisher CM, Ojemann R (1986) A clinico-pathological study of carotid endarterectomy plaques. Rev Neurol 42:573–589

Furchgott RF, Zawadzki JV (1980) The obligatory role of endothelial cells in the relaxation of arterial smooth muscle by acetylcholine. Nature 288:373–376

Furst H, Hartl WH, Jansen I et al. (1992) Color-flow Doppler sonography in the identification of ulcerative plaques in patients with high-grade carotid artery stenosis. AJNR 13:1581–1587

Futrell N, Millikan C, Watson BD et al. (1989) Embolic stroke from a carotid arterial source in the rat. Pathology and clinical implications. Neurology 39:1050–1056

Grady PA (1984) Pathophysiology of extracranial cerebral arterial stenosis – a critical review. Stroke 15:224–236

Greenberg DA, Chan J, Sampson HA (1992) Endothelias and the nervous system. Neurology 42:25–31

Hazama F, Ozaki T, Amano S (1979) Scanning electron microscopic study of endothelial cells of cerebral arteries from spontaneously hypertensive rats. Stroke 10:245–252

Hennerici M, Trockel U, Rautenberg W, Kladetzky RG (1985) Spontaneous progression and regression of small carotid atheroma. Lancet 1: 1415–1419

Karino T, Motomiya M (1983) Flow visualization in isolated transparent natural blood vessels. Biorheology 20:119–127

Kerber CW, Heilman CB (1992) Flow dynamics in the human carotid artery: preliminary observations using a transparent elastic model. AJNR 13:173–180

McGill H, Arias-Stella J, Carbonell L et al. (1968) General findings of the International Atherosclerosis Project. Lab Invest 18:498–502

Motomiya M, Karino T (1984) Flow patterns in the human carotid artery bifurcation. Stroke 15:50–56

Nicholls SC, Phillips DJ, Primozich JF et al. (1989) Diagnostic significance of flow separation in the carotid bulb. Stroke 20:175–182

O'Donnell TF, Erdoes L, Mackey WC et al. (1985) Correlation of B-mode ultrasound imaging and arteriography with pathologic findings at carotid endarterectomy. Arch Surg 120:443–449

Palmer RMJ, Ferrige AG, Moncada S (1987) Nitric-oxide release accounts for the biological activity of endothelium-

derived relaxing factor. Nature 327:524–526
Ross R (1986) The pathogenesis of atherosclerosis – an update. New Engl J Med 314:488–500
Steinke W, Kloetzsch C, Hennerici M (1990a) Variability of flow patterns in the normal bifurcation. Atherosclerosis 84:121–127
Steinke W, Kloetzsch C, Hennerici M (1990b) Carotid artery disease assessed by color Doppler flow imaging: corelation with standard Doppler sonography and angiography. AJNR 11:259–266
Sterpetti AV, Schultz RD, Feldhaus RJ et al. (1988) Ultrasonic features of carotid plaque and the risk of subsequent neurologic deficits. Surgery 104:652–660
Touboul PJ, Prati P, Scarabin P-Y et al. (1992) Use of monitoring software to improve the measurement of carotid wall thickness by B-mode imaging. J Hypertension 10 (Suppl 5):488–500
Vane JR, Anggard EE, Botting RM (1990) Regulatory functions of the vascular endothelium. New Engl J Med 323:27–36

10. Atherosclerosis

M. Fisher

Atherosclerosis of large extra- and intracranial arteries is an important substrate for many ischemic strokes (Bogousslavsky et al. 1988). The related arteriopathies observed in small intracranial penetrating arteries, lipohyalinosis and microatheroma are associated with most subcortical lacunar infarcts. Coronary artery atherosclerosis is associated with sources of cardioembolic stroke such as ischemic atrial fibrillation, myocardial infarction and ventricular aneurysms. The majority of ischemic strokes are directly or indirectly related to atherosclerosis at diverse locations and the clinician interested in stroke pathophysiology must also be cognizant of the atherosclerotic process. Understanding the epidemiology, pathologic basis and cellular mechanisms related to atherosclerosis will lead to an enhanced appreciation of interventions directed at inhibiting development and promoting regression of atherosclerotic lesions at multiple arterial sites.

Atherosclerosis at multiple arterial locations and its potential pathophysiologic importance was recognized by many 19th century pathologists, as exemplified by Virchow (Virchow 1856), and the relationship of coronary atherosclerosis to symptomatic myocardial ischemia was appreciated by Herrick in 1912. Over the ensuing decades, the relevance of coronary atherosclerosis to the increasing epidemic of myocardial ischemia became more apparent, and risk factors such as hypertension, cigarette smoking and hyperlipidemia were noted (Manon et al. 199 The seminal studies of Fisher and colleagues in th 1950s and 1960s extended these observations t the cerebrovasculature. In a series of detailed clir icopathologic studies, these investigators descr bed the relationship between large-vessel extra ranial atherosclerosis in the carotid an vertebrobasilar arteries with ischemic stroke an transient ischemic attacks (TIA) (Fisher 1954; Fisl er et al. 1965). In the 1960s and 1970s Fishe performed similar clinicopathologic studies lacunar stroke syndromes related to small intr cranial vascular lesions (Fisher 1965; 1979). Thes studies paved the way to our current classificatio of ischemic stroke subtypes and helped to sugge therapeutic interventions, both medical and surg cal, which can reduce stroke incidence.

Epidemiology

Large- and small-vessel atherosclerosis can b directly related to 70%–80% of all ischem strokes. Age, hypertension, cigarette smokir and diabetes mellitus are well established ris factors for the development of atheroscleros and ischemic stroke (Dyken 1991). Hyperlipid mia, an important risk factor for coronary athe osclerosis and myocardial ischemia, is less clea ly linked to cerebral arterial atheroscleros

Recent studies do support the relationship between carotid atherosclerosis and elevations of total and low-density lipoprotein (LDL) cholesterol (Tell et al. 1988). Patients with underlying symptomatic coronary artery atherosclerosis have a substantially increased stroke risk, as might be anticipated if the pathogenesis of atherosclerosis in these two arterial territories is similar.

The relationship of race to cerebral atherosclerosis is intriguing. The early studies of Fisher previously alluded to described atherosclerotic lesions and related atherothrombosis primarily at the origin of the internal carotid artery (ICA), but the ischemic stroke patients studied were overwhelmingly Caucasian. Angiographic studies in symptomatic Caucasian patients confirmed that proximal ICA lesions predominated, and that isolated intracranial atherosclerosis was distinctly uncommon (Caplan et al. 1986). In studies where black stroke and TIA patients were also evaluated, a different picture began to emerge. In blacks significantly more supraclinoid ICA and middle cerebral artery (MCA) stem lesions and fewer proximal ICA lesions were detected than in whites. These observations were confirmed by the Stroke Data Bank. Interestingly the relationship of race to asymptomatic cerebral atherosclerosis may be somewhat different. McGarry et al. performed an extensive autopsy series on blacks and whites aged 10–69 years dying of causes not associated with clinically symptomatic atherosclerosis (McGarry et al. 1988). They observed no significant racial differences in raised carotid artery atherosclerotic lesions in men of either race, and that black women had a greater prevalence of such lesions that white women. Intracranial advanced atherosclerotic lesions were more common in blacks of both sexes at all ages. Regarding risk factors, blacks consistently have higher prevalence rates of hypertension than whites, and higher blood pressure levels (Initazari et al. 1990). These observations suggest that the location for the development of symptomatically important atherosclerotic plaques differs between blacks and whites, and that hypertension or other unidentified factors may have an impact on this disparity. A speculation would be that greater degrees of blood pressure elevation enhance intracranial plaque development, perhaps related to changes in flow dynamics, and also affect where plaques destabilize, leading to symptom development.

Asians also demonstrate differences in atherosclerotic plaque locations compared to whites. Symptomatic Japanese patients demonstrated significantly more stenosis of intracranial arteries than white TIA patients (Nishimaru et al. 1984). Small-vessel intraparenchymal arterial sclerosis was observed more frequently at autopsy in Japanese living in Japan than among Japanese living in Hawaii, associated with an increased risk for cerebral infarction (Fig. 10.1) (Mitsuyama et al. 1979). This observation suggests that environmental factors, perhaps primarily diet, can affect the location of atherogenesis in cerebral arteries. A similar study in Chinese living in the USA also showed that TIA and ischemic stroke in this population was associated with significantly more severe intracranial atherosclerosis and with less extracranial atherosclerosis than in a comparable white study group (Feldmann et al. 1990). There were no differences in any risk factors between the two groups. A recent report from the EC/IC Bypass Trial directly compared lesion sites and risk factors among whites, Asians and blacks with TIA or minor strokes. The prevalence of isolated MCA or tandem ICA and MCA lesions was significantly greater in Asians than in whites, whereas ICA lesions were more frequent among whites (Table 10.1) (Initazari et al. 1990). For all of these lesion locations blacks had an intermediate frequency falling between the rates for orientals and whites. Hypertension, diabetes and cigarette smoking were significantly more common among blacks than the other two groups. The group with isolated MCA disease tended to have lower levels for most risk factors.

Table 10.1. Location of atherosclerotic lesions by race (reproduced with permission of Archives of Neurology 1990; 47:1081)

	Oriental (n=223)	Black (n=86	White (n=1058)
MCA*	20	10	5
ICA	46	64	82
TANDEM	34	26	13
Carotid bruit	2	12	21
Ischemic heart disease	10	15	22
Intermittent claudication	1	12	14

*P value of all entries is less than 0.001.
MCA, middle cerebral artery; ICA, internal carotid artery; TANDEM, lesions on both MCA and ICA.

Taken in aggregate, these studies confirm differences in the patterns of atherosclerotic lesion

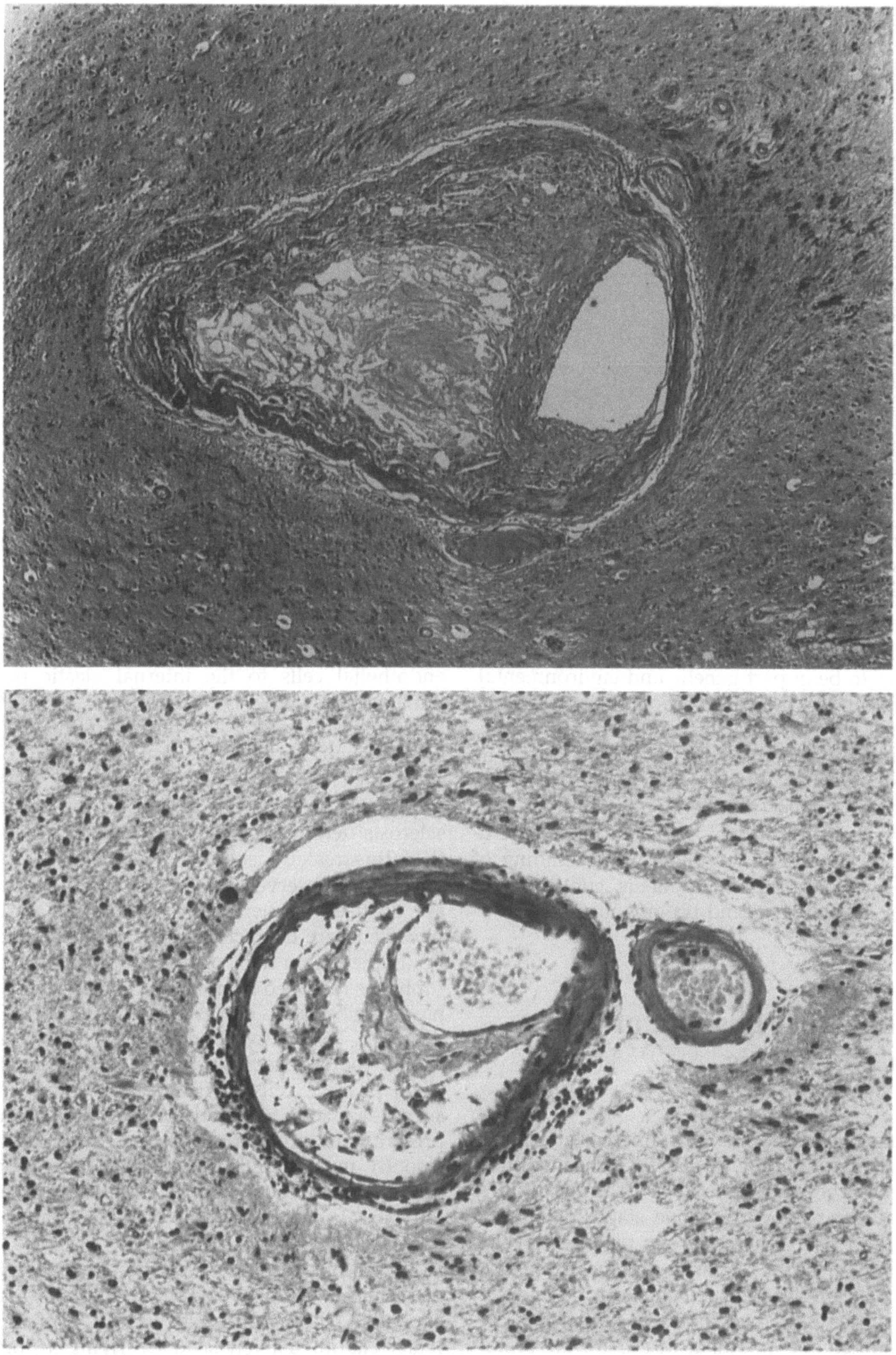

a

b

Fig. 10.1. (a) and (b) Pathological specimens demonstrating lipohyalinosis and microatherometous changes in patients who had suffered lacunar infarctions. (Courtesy Dr J. Ogata, Osaka, Japan)

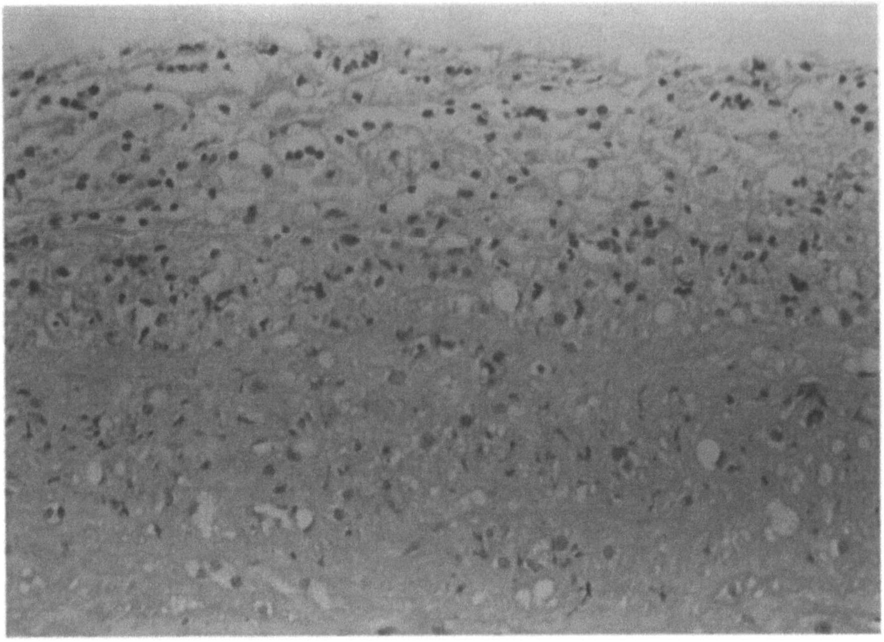

Fig. 10.2. A fatty streak with numerous lipid-filled foam cells. (Reproduced with permission of Cerebrovascular Brain Metabolism Reviews 1991; 3:116)

development among the races. These differences appear to be in part genetic and environmental. Dissecting out the contribution of these various factors, and perhaps differences in the pathological substrate for lesion development, will be important. Such differences will help to guide therapy, because intervention regarding both risk factors and cellular contributors to atherosclerosis requires a complete understanding of how these various components contribute to lesion development and progression.

Pathologic Substrate of Atherosclerosis

During the development of atherosclerosis several distinct pathological stages can be identified, and the process appears to be a chronically progressive transition from one stage to the next (McGill 1988). Atherosclerotic plaques develop within the intima of large and medium-sized muscular arteries. The proximal carotid and vertebral arteries, as well as the basilar artery, carotid siphon and proximal MCA, are the preferential sites for plaque growth among the extra- and intracranial arteries. The normal arterial intima extends from the luminal surface, covered by endothelial cells to the internal elastic membrane, which marks the border with the adventitia (Stary et al. 1992). The normal intima varies, and is composed of two distinct regions, the subintimal proteoglycan layer and the deeper musculoelastic layer. The earliest lesion of atherosclerosis detectable within the arterial intima is the fatty streak (Masuda and Ross 1990a,b). Fatty streaks are observed as early as childhood, and are diffusely scattered throughout the arterial tree. They are characterized by the presence of lipid-laden foam cells, primarily formed by macrophages recruited from blood monocytes which have imbibed LDL-cholesterol (Fig. 10.2) (Tsukada et al. 1986). Lymphocytes are also found within the lesions, and appear to be primarily CD-8 positive cells, which have cytotoxic and suppressor activity. Smooth muscle cells are also present and are derived from native intimal smooth muscle cells and those which have migrated from the media. When viewed grossly, fatty streaks appear as yellowish raised areas on the intima (Walker et al. 1986).

The fibrous plaque is a more advanced lesion which can encroach upon the lumen as it expands. Fibrous plaques usually contain an intact endothelial cell layer overlying a fibrous cap (Fig. 10.3). Within the fibrous plaque are abundant foam cells, transformed smooth mus-

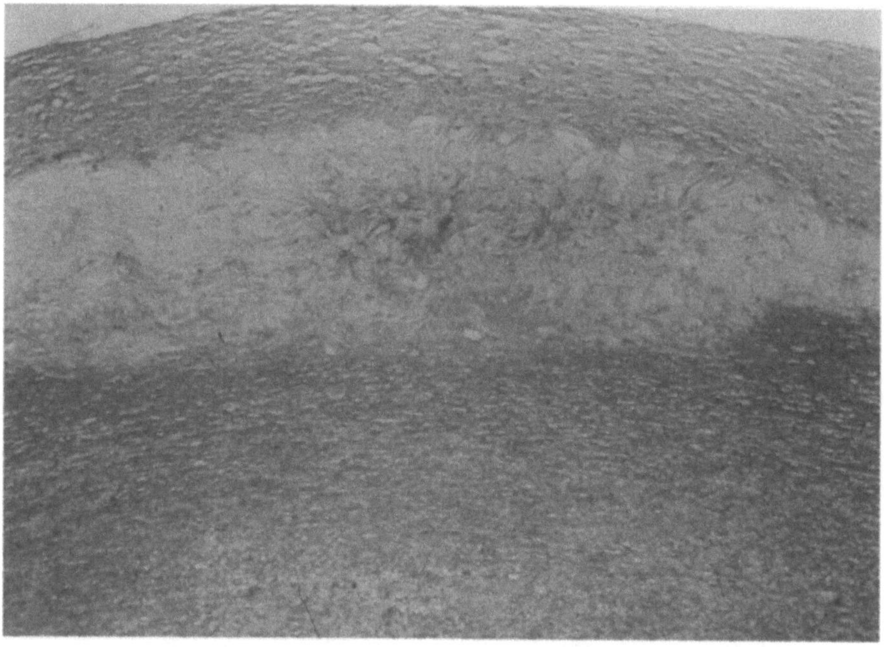

Fig. 10.3. A fibrous plaque from a human aorta. (Reproduced with permission of Cerebrovascular Brain Metabolism Reviews 1991; 3:117)

cle cells, lymphocytes and macrophages and a connective tissue matrix. As the lesion expands a central necrotic core develops, containing cellular debris, intracellular lipid droplets and cholesterol crystals. The extent of free lipid deposition versus connective tissue proliferation will determine the consistency of the plaque, i.e. soft or hard (McGill 1988). Small arterioles are present primarily at the base, deep within the plaque. Fibrous plaques can begin to develop in late adolescence or early adulthood and then progress over decades. They are usually observed initially at arterial branch-points and proximal segments of large and medium-sized arteries (Cornhill et al. 1985). Fatty streaks appear to be the precursors of fibrous plaques, but clearly not all fatty streaks progress to more advanced lesions. An intermediate lesion between fatty streaks and fibrous plaques has been identified, supporting this concept (Geer et al. 1960). The evolutionary nature of atherogenesis is also demonstrable in animal models, where serial sacrifice of animals at different time points after initiation of atherogenic diet clearly shows lesion progression at the same arterial site.

The last pathological lesion of atherosclerosis is the complicated plaque. This lesion is similar to the fibrous plaque, but also shows evidence of calcification and hemosiderin deposition (McGill 1988). The luminal surface may show evidence of ulceration or endothelial cell disruption, cracks or crevices. As complicated plaques enlarge, compensatory luminal dilatation begins to fail and hemodynamically significant stenosis can develop (Glagov et al. 1988).

The progressive increase in plaque burden within the carotid or vertebrobasilar arterial systems is usually a slow insidious process, although abrupt enlargement related to the incorporation of luminal thrombi may occur (Fuster et al. 1992). Lesions can progress to complete occlusion without symptom development. However, in many patients clinical events such as TIAs, strokes or myocardial infarction occur abruptly, presumably related to acute plaque destabilization. In the coronary arteries, mural thrombi associated with disruption of the luminal plaque surface are very frequently observed (Falk 1989). The luminal surface disruption may be an ulceration or just a minor fissuring of the endothelial cell lining, leading to thrombus formation on the luminal surface and arterial compromise. In coronary ischemia many of these events occur in plaques of moderate size, i.e. those causing less than 70% stenosis (Hackett et al 1988). Lipid-rich soft plaques may be more prone to such luminal surface disruptions than highly fibrotic plaques. Shear forces and abrupt changes of intraluminal pressure may induce plaque surface disruption (Richardson et al.

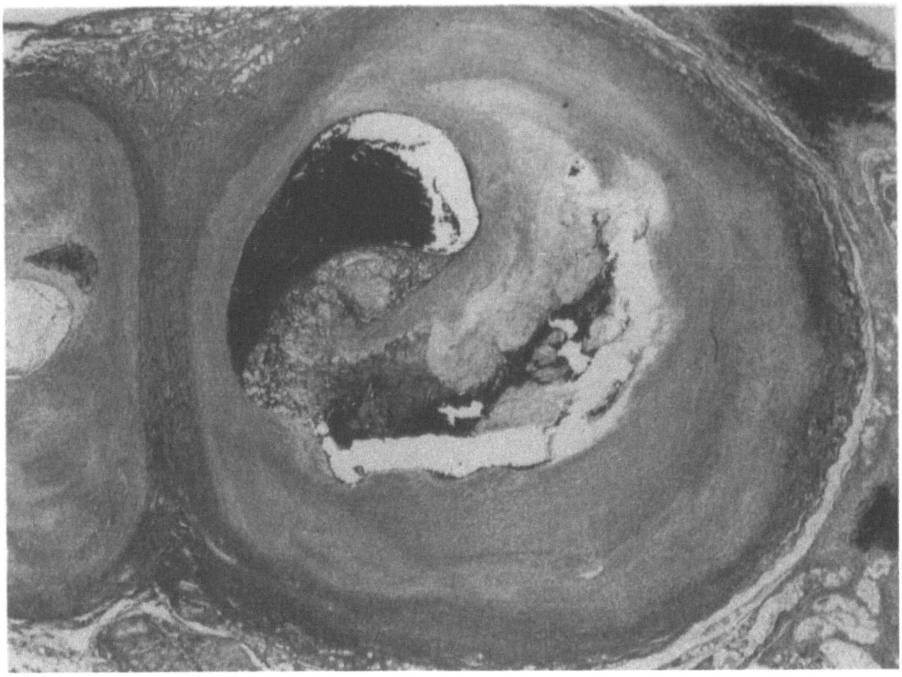

Fig. 10.4. An example of luminal thrombus superimposed upon a disrupted carotid artery plaque in a patient who died of massive carotid territory stroke. (Courtesy Dr J. Ogata, Osaka, Japan)

1989). Primary intraplaque hemmorrhage has been proposed as another mechanism for acute plaque disruption, but this mechanism is not widely accepted as a common cause for symptom development.

Acute plaque disruption must also be an important mechanism for symptom development in the cerebral arteries. Initially the importance of primary intraplaque hemorrhage for acute symptom development was proposed based upon the study of carotid endarterectomy specimens (Imparato et al. 1983). Very few detailed autopsy studies of acutely symptomatic carotid or vertebrobasilar arterial specimens have been described. Most studies of symptomatic arteries have come from endarterectomy specimens, which have not been evaluated at the close intersegment intervals (100–200 μm), performed in the coronary studies. Several recent studies of carotid arteries from symptomatic TIA patients demonstrate that intraplaque hemorrhage is common, but a similar frequency of intraplaque hemorrhage has also been observed in asymptomatic carotid arteries (Bassiouny 1989). There also appears to be no obvious relationship between the age of the hemorrhages and the time when the symptomatic events occurred (Bernstein et al. 1990). Fisher and Ojemann analyzed carotid endarterectomy specimens at 8–10 μm intervals and observed intraplaque hemorrhages in only 39% of the specimens obtained from TIA patients (Fisher and Ojemann 1986). In over 90% of cases the intraplaque hemorrhage could be traced to an area of surface disruption, implying that the intraplaque blood was secondary to luminal disruption. Two recent autopsy studies evaluated carotid arteries from patients who died acutely after carotid occlusion (Masawa et al 1990; Ogata et al. 1990). In both studies luminal surface disruption of the plaques was observed, related to the arterial thrombus (Fig. 10.4). An important analysis of carotid artery specimens from large symptomatic and asymptomatic carotid artery trial is under way. Preliminary results suggest that plaque ulceration is common (83%) and that intraplaque hemorrhage occurs with relatively equal frequency in both symptomatic and asymptomatic carotid plaques (Fisher et al. 1992). These studies suggest that acute plaque destabilization is related to luminal surface disruption, and that intraplaque hemorrhage is usually a secondary event which does not initiate symptom development. It is likely that the sequence of events in the carotid and vertebrobasilar arterial systems is similar to that in the coronary arteries. This scenario is supported by observations that antiplatelet and anticoagulant

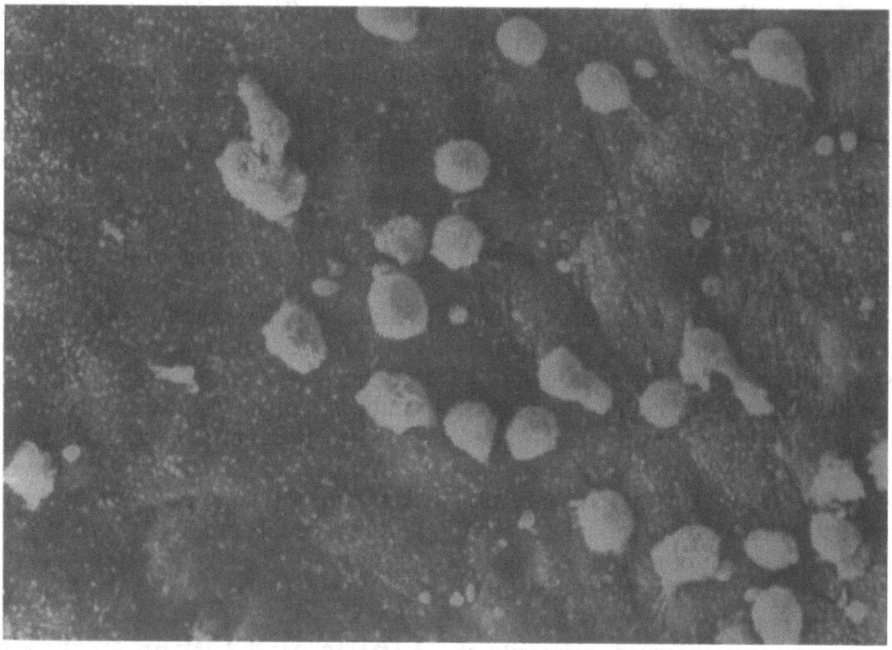

Fig. 10.5. Monocyte adherence to the vascular surface in rats shortly after initiation of a hypercholesterolemic diet. (Courtesy J. Nunnari)

therapy can reduce the development of TIA/ stroke in patients with substantial cerebral artery stenosis, an unlikely occurrence if primary intraplaque hemorrhage was an important cause of symptom development (Sherman et al. 1989).

Cellular Aspects of Atherogenesis

Studying the pathologic lesions of atherosclerosis provides a freeze-frame at varying time points and has provided suggestions regarding the cell types involved and the nature of the progressive process. Animal models of atherogenesis, primarily using dietary hypercholesterolemia in addition to studying human atherosclerotic plaques from populations of various ages, provided much information about plaque evolution and the cellular aspects of atherogenesis. The three major cellular contributors to atherogenesis appear to be monocytes/macrophages, smooth muscle cells (SMC) and endothelial cells (Masuda and Ross 1990a,b). Platelets and lymphocytes are also involved, but to a lesser degree. The interrelationship of the cellular contributors to atherogenesis and risk factors was incorporated into the 'response to injury hypothesis of atherosclerosis' proposed and updated by Ross and colleagues (Ross et al. 1984). This hypothesis is a modern refinement of concepts originally suggested by Virchow in the 1800s. (Virchow 1856).

From animal studies it is clear that the earliest cellular event after the start of an atherogenic diet is the adherence to the arterial wall of circulating leukocytes, primarily mononuclear, and their migration into the intima (Fig. 10.5) (Joris et al. 1983). Fatty streak formation ensues with abundant foam-cell formation, primarily enlarged macrophages which imbibe LDL-cholesterol. SMCs also take up lipid to become foam cells and transform from a contractile to a synthetic state (Manderson et al. 1989). It has been hypothesized that functional or morphologic endothelial cell injury due to a variety of factors such as hyperlipidemia, hypertension, cigarette smoking, radiation, homocysteine, viruses, physical forces and other factors, can initiate endothelial injury (Ross et al. 1984). In response to the endothelial injury, monocytes adhere and enter the intima to begin the process of foam-cell formation. The monocytes/macrophages, endothelial cells and SMCs release a variety of growth factors, cytokines and lymphokines which promote intercellular signaling, cellular proliferation and transformation (Mitchinson and Ball 1987;

Jaffee 1987). Two well studied promoters are platelet-derived growth factor (PDGF) and interleukin-1 (IL-1) (Ross 1986; Raines et al. 1989). Other potentially important growth factors are fibroblast growth factor, transforming growth factor and epidermal growth factor (Klagsbrun and Edelman 1989). γ-Interferon, colony-stimulating factor, tumor necrosis factor and endothelial-derived growth factor may also be important (Munro and Cotran 1988). All three cell types, which are important contributors to atherogenesis, produce oxygen free radical species and these toxic oxygen metabolites may be important contributors to the cellular aspects of atherogenesis; this relationship will be discussed in detail later.

As fatty streaks transform into fibrous plaques, endothelial surface disruption is observed, especially in regions of endothelial cell thinning (Masuda and Ross 1990a,b). Platelet microthrombi can be detected attached to exposed foam cells. T lymphocytes are observed, primarily at the periphery, in both early and more advanced lesions associated with macrophages. Their precise contribution remains uncertain, but the presence of lymphocytes supports the inflammatory concept of atherogenesis (Emerson and Robertson 1988). In fibrous plaques, SMCs which have migrated from the media are apparent. These transformed synthetic SMCs probably contribute to the formation of the extracellular matrix and fibrous tissue proliferation. As the fibrous plaque continues to expand, luminal compromise ensues and endothelial surface disruption may develop, leading to clinical symptoms.

This complex cascade of cellular events and interactions implies that atherogenesis is a dynamic, but not inexorable, process. Understanding how the primary cellular constituents of atherosclerosis promote the process and the role of growth factors and cytokines can suggest intervention strategies at a cellular level. Risk factor modification, such as lowering elevated cholesterol, reducing hypertension or cessation of smoking, can also be interpreted at a cellular level. The response-to-injury hypothesis suggests that atherosclerosis is not inevitable, but that various types of intervention can impede or perhaps reverse atherosclerosis at a fundamental level (Fig. 10.6).

The slow progression of atherosclerotic plaques implied by the response-to-injury hypothesis may be supervened upon by rapid plaque expansion. This probably occurs by the incorporation of thrombus into the lesion, as was originally described by Von Rokitansky 140 years ago (Von Rokitansky 1852). An updated version of this hypothesis suggests that mural thrombi at sites of endothelial surface disruption can be re-endothelialized and included in the plaque (Fuster et al. 1992). In coronary artery studies, thrombi of various ages overlying fissures are observed (Bini et al. 1989). Digested thrombi may explain some of the hemosiderin detected in advanced atherosclerotic lesions. Thrombin within plaques may also contribute to cellular proliferation by enhancing C-*fos* oncogene expression, a promoter of SMC protein synthesis and hypertrophy (Berk et al. 1990). Inhibiting platelet-induced arterial thrombi or primary thrombus formation may not only reduce the frequency of clinical symptoms, but also impede plaque progression related to incorporation of thrombi.

Lipids and Other Risk Factors

Analysis of the risk factors associated with ischemic stroke and vascular occlusive disease is important in understanding the development of cerebral atherosclerosis. The relationship between elevated total and LDL-cholesterol levels and coronary artery atherosclerosis and ischemia is well established (Expert Panel 1988). Hypercholesterolemia is used in isolation to induce the rapid formation of atherosclerosis in many animal species, including primates. In humans, a direct correlation between elevated total and LDL-cholesterol levels and ischemic heart disease has been demonstrated. It is also clear that reducing total and LDL-cholesterol levels confers a significant benefit in reducing the incidence of subsequent coronary ischemic events, and also the progression of coronary atherosclerosis (La Rosa et al. 1990). The relationship of elevated total and LDL-cholesterol levels to extra- and intracranial atherosclerosis and ischemic stroke is not as obvious. However, a recent review showed that only three of 26 studies failed to document a relationship between hyperlipidemia and cerebrovascular atherosclerosis, despite many methodological differences (Tell et al. 1988). Two surveys of asymptomatic healthy populations, both men and women, showed that total and LDL-cholesterol levels were independent risk factors for the presence of intimal–medial carotid thickening and plaque formation, as demonstrated by ultrasonography (Salonen et al. 1988; Bonithon-Kopp et al. 1991).

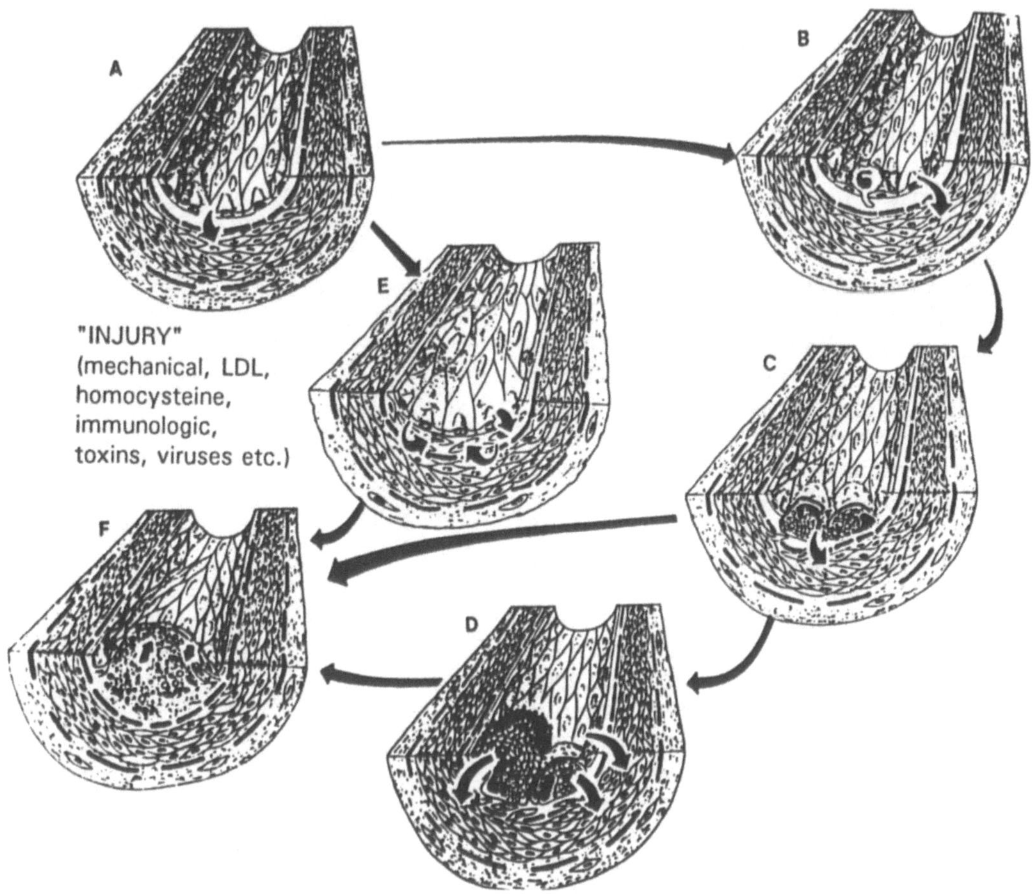

Fig. 10.6. The response to injury cascade, as outlined by Dr Russell Ross. (Reproduced with permission of New England Journal of Medicine 1986; 314:496)

Elevated LDL-cholesterol has also been observed to be a predictor for the progression of carotid artery thickening. Elevated total and LDL-cholesterol levels appear to be an independent risk factor for ischemic stroke, although a less potent one than hypertension or diabetes (Qizibash et al. 1992).

High-density lipoprotein (HDL)-cholesterol is involved with reverse cholesterol transport, i.e. the removal of cholesterol from atherosclerotic plaques. Serum HDL-cholesterol levels are inversely related to the risk of ischemic heart disease (Castelli et al. 1986). Reduced HDL-cholesterol levels are an independent risk factor for ischemic stroke and carotid atherosclerosis. Lipoprotein (a) (Lp(a)) is another lipoprotein which, when elevated, is linked to the risk for developing atherosclerosis and ischemic heart disease (Armstrong et al. (1986); Berk et al. 1990; Zenker et al. (1986)). Interestingly, Lp(a) contains apoprotein (a), which features a marked structural similarity to plasminogen. This suggests that Lp(a) may interfere with the fibrinolytic system (Miles and Plow 1990). Lp(a) is also an independent risk factor for ischemic stroke and carotid atherosclerosis.

The relationship of other risk factors to atherosclerosis and ischemic vascular disease, both coronary and cerebral, has been well established. Hypertension has been recognized as a risk factor for stroke and myocardial infarction for several decades, and reducing elevated blood pressure definitely reduces stroke risk (Collins et al. 1990). Cigarette smoking has recently been demonstrated to be an independent stroke risk factor, and also a risk factor for cerebrovascular atherosclerosis (Whisnant et al. 1990). The number of years of smoking correlates with the severity of carotid stenosis: cessation of smoking may limit further progression. Regression of coronary or carotid atherosclerosis has not been documented with smoking cessation, but cessation does appear to reduce subsequent stroke risk. Diabetes mellitus is another independent risk factor for cerebrovascular atherosclerosis and ischemic stroke (Dyken 1991), although it is unclear whether strict diabetic control reduces subsequent stroke risk or the progression of established atherosclerosis. Interestingly, low to moderate levels of alcohol consumption confer a reduction in risk for carotid atherosclerosis in patients with TIAs or strokes, and may also be associated with reduced ischemic stroke risk (Bogousslavsky et al. 1990). This potential benefit is counterbalanced by an enhanced risk for hemorrhagic stroke with increasing alcohol consumption. Two recent well designed studies from the Mayo Clinic assessed the impact of risk factors on the presence and severity of extracranial and intracranial carotid atherosclerosis, evaluated by angiography (Whisnant et al. 1990; Ingall et al. 1991). At both sites the duration of cigarette smoking was the strongest predictor of stenosis. Hypertension was an independent risk factor at both sites, while diabetes and age were independent risk factors for extracranial carotid atherosclerosis and apolipoprotein A-1 (the apolipoprotein associated with HDL-cholesterol) was inversely associated with intracranial atherosclerosis. Risk factor data provide a strong basis for identifying people at increased risk for plaque formation and ischemic vascular disease. Intervention can then be initiated to reduce risk factors by behavioral changes, dietary modification and pharmacologic therapy.

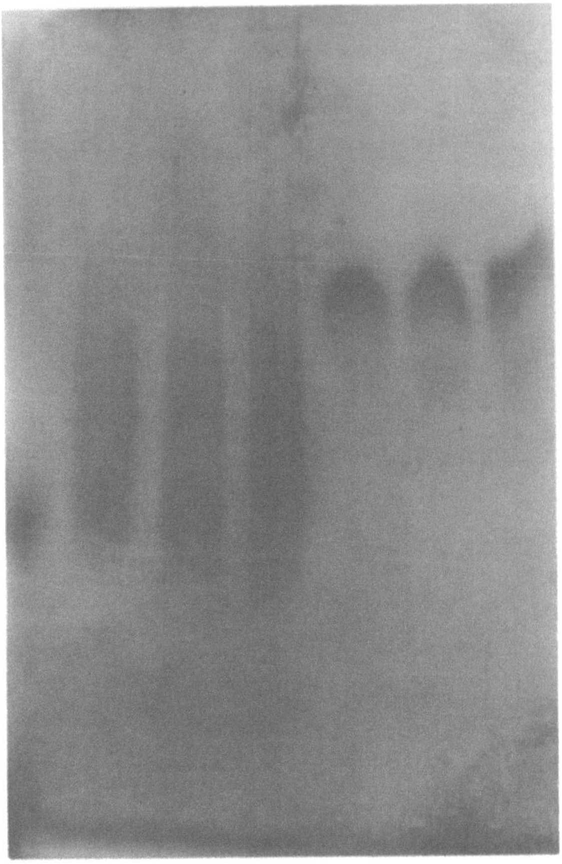

Fig. 10.7. An electrophoretic gel of LDL-cholesterol demonstrating non-oxidized LDL in the three left lanes and oxidized LDL in the three right lanes.

A Unifying Hypothesis: Lipids and Cellular Aspects of Atherosclerosis

The relationship of elevated total and LDL-cholesterol levels to atherosclerosis and symptomatic cardiovascular diseases is clear. The lipid hypothesis of atherosclerosis states that excess cholesterol accumulates in cells of the arterial wall, provoking the development of atherosclerosis (Yatsu and Fisher 1989). The interaction of cholesterol with arterial wall cells, previously discussed to be important in the cellular events of atherosclerosis, is an important current research area. LDL-cholesterol is transported by endothelial cells into the subintima, especially when serum LDL-cholesterol levels are elevated (Jaffee 1987). This LDL-cholesterol can then be imbibed by macrophages and SMC which have accumulated in response to prior endothelial cell injury, as outlined in the response-to-injury hypothesis. Native unmodified LDL-cholesterol is taken up relatively slowly by these cells via its receptor. The scavenger receptor, described by Brown and Goldstein, present on macrophages takes up modified LDL-cholesterol at a much faster rate than the native LDL-cholesterol receptor (Brown et al. 1981). Chemical modification of LDL-cholesterol, which masks the epsilon amino groups of lysine residues, converts it to a form recognized by the scavenger receptor, leading to enhanced LDL-cholesterol uptake and foam-cell formation (Brown et al. 1981). Oxidation of LDL appears to be the most important mechanism for its modification; oxygen free radicals produced by macrophages, SMC and endothelial cells can induce this transformation (Fig. 10.7). Oxidatively modified LDL has been isolated from atherosclerotic plaques in both animals and humans, and autoantibodies against oxidized LDL have been observed in the plasma of atherosclerotic rabbits and humans (Carew et al. 1987;

Yla-Herttuala et al. 1989; Parums et al. 1990). Autoantibody titers to oxidized LDL have recently been demonstrated to be an independent predictor for the progression of carotid atherosclerosis, as assessed by ultrasonography after a 2-year follow-up (Salonon et al. 1992). These studies suggest that oxidized LDL may be a key step in early atherogenesis, and that oxidized LDL is not just an in-vitro phenomenon.

Oxidized LDL has several additional properties which could help to potentiate the atherosclerotic process (Table 10.2). It has cytotoxic effects and can induce or continue endothelial cell injury, an important early step in atherogenesis (Davies and Thomas 1985). Oxidized LDL has chemoattractant properties for circulating monocytes and can directly enhance their accumulation in the arterial wall (Quinn et al. 1987). Production of other monocyte chemoattractants released by endothelial cells is also stimulated by oxidized LDL, increasing the vessel wall macrophage population. Paradoxically, oxidized LDL appears to inhibit the motility of vessel wall-resident macrophages, reducing their egress. Oxidized LDL stimulates endothelial-cell release of colony-stimulating factor, a cell mitogen. Interleukin-1 production may also be enhanced, and this cytokine can induce SMC transformation and proliferation. Thus, oxidized LDL can potentiate and modify the cellular contributors to atherosclerosis.

Table 10.2. Potential atherogenic effects of oxidized LDL-cholesterol

Enhances foam-cell formation
Promotes endothelial injury
Chemoattractant for circulating monocyte
Inhibits egress of vessel wall macrophages
Induces cytokine production

The importance of oxidized LDL in the initiation and continuing development of atherosclerotic plaques remains unclear, despite the multitude of potential contributions (Steinberg 1991). Further study of the basic science related to oxidized LDL may illuminate its role. The concept of relating lipids to the cellular aspects of atherogenesis is intriguing, and oxidized LDL appears to be an initial attempt to explain this interaction. Other risk factors for atherosclerosis also have interactions with the cellular contributors to atherosclerosis. Cigarette smoke contains potential cellular toxins, which can induce endothelial cell injury and initiate the cellular injury cascade (Seiffert et al. 1981). Platelet aggregation and fibrinogen levels are enhanced by cigarette smoking, and these factors may be initiators of thrombus formation and clinical symptoms (Fitzgerald et al. 1981). Hypertension influences the hemodynamic environment within the arterial system and can affect arterial wall cells mechanically (Glagov et al. 1988). Additionally, the interaction between blood cells and the arterial wall cells is affected by high blood pressure, leading to enhanced contact between 'noxious' blood elements and the arterial wall. Diabetes may affect the metabolic activity of arterial wall cells and promote cellular injury. The final common pathway effects of these and other less well defined risk factors should be considered at a cellular level, if therapies beyond risk factor modification are being considered.

Lipid-Lowering Therapy

The strong relationship between elevated total and LDL-cholesterol and ischemic heart disease related to coronary atherosclerosis has stimulated many therapeutic trials designed to reduce elevated levels of these lipids. The first approach to lipid-lowering therapy should always be dietary modification (Manon et al. 1992). Patients with total cholesterol levels above 240 mg/dl, or LDL-cholesterol levels above 160 mg/dl, should be advised to reduce total cholesterol intake to below 300 mg/day, and fat should comprise less than 30% of total caloric intake (Consensus Conference 1985). If dietary modification does not reduce elevated total and/or LDL-cholesterol to a satisfactory level – ideally a total cholesterol less than 220 mg/dl and LDL-cholesterol less than 160 mg/dl – then drug intervention may be advisable.

Several types of lipid-lowering drug have been evaluated in clinical trials and shown to reduce clinical end-points and the progression of coronary artery atherosclerosis. The bile-acid sequestrant cholestyramine was compared to placebo in the Lipid Research Clinics Study: for every 1% reduction in LDL-cholesterol level a 2% reduction in symptomatic ischemic heart disease rate was observed and the drug was more effective than placebo in reducing cholesterol levels and symptomatic events (Lipid Research Clinics 1984). Gemfibrozil, a fibric acid derivative, was studied in the Helsinki Heart Study and was found to reduce mean LDL-cholesterol levels by 11% and increase HDL-cholesterol levels by 11%

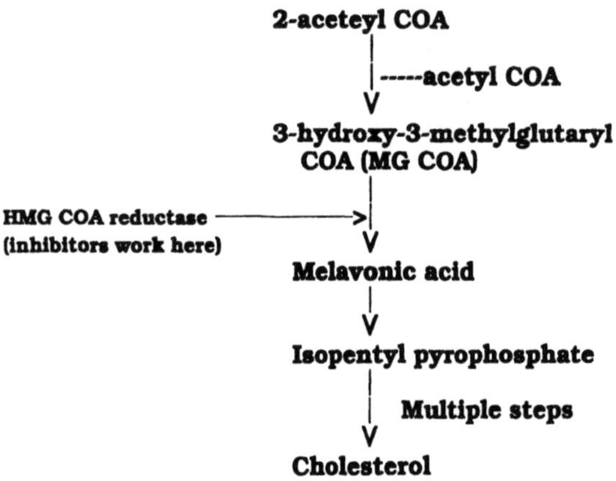

Fig. 10.8. The HMGCoA pathway.

(Frick et al. 1987). This combination resulted in a 34% reduction in ischemic heart disease, compared to dietary controls. Niacin also reduces hyperlipidemia, and this drug in combination with another bile-acid sequestrant, colestipol, was evaluated in men after coronary artery bypass concerning its effects on coronary atherosclerosis (Blankenhorn et al. 1987). The group receiving both drugs had a 43% reduction in LDL-cholesterol levels and a 37% increase in HDL-cholesterol levels compared to a 5% decrease and 2% increase in dietary controls receiving placebo. Angiographic follow-up at 2 years showed lesion progression in non-grafted coronary vessels in 61% of controls and 39% of drug-treated patients ($P < 0.01$). Lesion regression was observed in 16% of the treated group and only 2% of controls ($P < 0.002$). These results conclusively demonstrated that lipid-lowering drug therapy can reduce symptoms and impede plaque development, but side-effects such as bloating, diarrhoea, skin discomfort etc. limit patient compliance with bile-acid sequestrants and niacin.

A new approach to reducing elevated levels of total and LDL-cholesterol is to inhibit intrinsic synthesis by blocking hydroxymethyl glutaryl coenzyme A (HMGCoA) reductase (Fig. 10.8) (Grundy 1988). Several HMGCoA reductase inhibitors have been developed and evaluated in clinical trials. Lovastatin has been available for several years and two new HMGCoA reductase inhibitors, simvastatin and pravastatin, have recently been introduced. All three drugs markedly reduce total and LDL-cholesterol levels in a dose-related fashion, but simvastatin appears to be more effective than the other two (Bocuzzi et al. 1991). Both simvastatin and lovastatin have been shown to significantly reduce early coronary atherosclerosis in a rabbit model (Kobayashi et al. 1989). In a human trial, lovastatin plus colestipol reduced LDL-cholesterol by 46% and increased HDL-cholesterol by 15% (Brown et al. 1990). Angiographic follow-up demonstrated a significant reduction in lesion progression and an increase in regression in the drug-treated group, compared to dietary controls. Definitive proof that HMGCoA reductase inhibitors reduce clinical events is not yet available, but several large studies addressing this issue are in progress. Based upon prior studies with other lipid-lowering drugs, such proof should become available soon. HMGCoA reductase inhibitors do have adverse effects and the most important ones are elevated liver enzymes, myopathy, lens opacification and rash. The side-effect profile for the newer drugs may be better than that for lovastatin.

The role of lipid-lowering drugs in preventing ischemic stroke is uncertain. If elevated total and LDL-cholesterol is a risk factor we can anticipate a benefit, although it has not yet been established in the coronary artery trials. Trials to evaluate the effects of HMGCoA reductase inhibitors on the progression of carotid artery lesions are in progress. Documenting the inhibition of lesion progression in asymptomatic hyperlipidemic patients would imply a potential role for these drugs in primary stroke prevention, because stroke risk positively correlates with increasing degrees of carotid artery narrowing.

Potential Cellular Therapies for Atherosclerosis

Epidemiologic studies have shown that populations that consume large quantities of fish and marine mammals, rich in n-3 fatty acids, have a reduced incidence of myocardial infarction and probably ischemic stroke (Hirai et al. 1980). These observations led to studies of n-3 fatty acids in animal atherosclerosis models. Dietary n-3 fatty acid supplementation was observed to reduce primary development of coronary atherosclerosis in swine models (Goodnight et al. 1992). Similar results were obtained in primates with dietary substitution of n-3 fatty acids, and Davis et al. also showed a reduction in carotid atherosclerosis (Davis et al. 1987). These beneficial effects on atherosclerosis were accompanied by modest or non-significant changes in lipid levels. Secondary inhibition of lesion progression has not been demonstrated, although n-3 fatty acids may help atherosclerotic arteries to maintain their luminal diameter, despite plaque enlargement.

The effects of a small increase in dietary fish consumption was evaluated after myocardial infarction in a number of patients. Eating fish at least twice a week was found to reduce the relative risk of subsequent ischemic heart disease mortality by 33% ($P < 0.01$) in comparison to a group advised to reduce saturated fat intake (Burr et al. 1989). Several small human studies with n-3 fatty acids have been reported in patients undergoing percutaneous transluminal coronary angioplasty (PTCA) to try and reduce subsequent restenosis (Goodnight et al. 1992). Important variables in the trials include the dose of n-3 fatty acids used; when they are started prior to PTCA; and the angiographic analysis used to evaluate treatment efficacy. Variable results have been reported, with some trials reporting significant reductions in retenosis and others showing no effects. A meta-analysis of these trials demonstrated an odds reduction of 24% (CI 2 to 45%, $P < 0.05$). Two large trials of n-3 fatty acids after PTCA are in progress and should provide definitive results. No trials with n-3 fatty acids in native coronary or carotid atherosclerosis have been reported, but such trials to evaluate the effects of dietary n-3 fatty acid supplementation on atherosclerosis progression are indicated.

Dietary n-3 fatty acids have no significant lipid-lowering effects on hyperlipidemia in humans and in most animal atherosclerosis models (Witt et al. 1989). It is likely that n-3 fatty acids exert their positive impact on atherosclerosis at a cellular level. A variety of potentially beneficial effects on cytokines, growth factors and cellular mediators of atherogenesis have been described, and others can be anticipated (Fisher et al. 1989). These substances can also lower serum fibrinogen.

Calcium-Channel Antagonists

Many animal studies have evaluated the effects of calcium-channel antagonists on atherosclerosis. Dihydropyridines (i.e. nicardipine, isradipine), benzothiazepines (i.e. diltiazem) and phenylalkylamines (i.e. verapamil), all of which block voltage-activated slow calcium channels, have been studied in a large variety of atherosclerosis models (Jackson and Bowyer 1989). The majority of studies demonstrated protective effects when the drugs were begun before or at the same time as the atherogenic stimulus. When calcium antagonists are started later, after atherosclerosis has begun, beneficial effects are not observed. Two large human trials in patients with coronary atherosclerosis have been reported, one using nifedipine and the other nicardipine (Lichtlen et al. 1990; Waters et al. 1990). In both studies, pretreatment coronary angiograms were compared to follow-up angiograms obtained 2 or 3 years later in a placebo-controlled randomized double-blind design. The overall results of both studies were unimpressive. However, nifedipine significantly reduced the appearance of new lesions and nicardipine significantly inhibited the progression of minimal lesions (initially less than 20% stenosis). In neither study were more advanced lesions inhibited from progressing. Regarding carotid atherosclerosis, a large multicenter trial is in progress comparing isradipine and hydrochlorothiazide (Furberg et al. 1989). The rate of carotid lesion progression in hypertensive patients is being assessed with ultrasound techniques during a 3-year period.

Calcium metabolism may influence atherogenesis at many different points. A variety of potentially beneficial cellular effects of calcium antagonists have been described, as well as effects on blood pressure (Schmitz et al. 1991). However, in animals calcium-channel antagonists inhibit atherosclerosis independent of blood pressure reduction. Calcium antagonists reduce cholesterol accumulation in the arterial wall and also reduce foam-cell formation. Calcium antagonists inhibit cholesterol esterification and receptor uptake of LDL-cholesterol. The most important cellular effect may be the inhibition of SMC migration and proliferation, as well as SMC production of collagen and extracellular matrix proteins (Jackson and Bowyer 1989).

Angiotensin-Converting Enzyme Inhibitors, Heparin Derivatives and Other Medical Therapies

SMC proliferation is an important process in atherosclerosis, and it has been suggested that local angiotensin release regulates SMC proliferative responses (Naftilan et al. 1989). Angiotensin-converting enzyme (ACE) inhibitors have been investigated as inhibitors of myointimal proliferation and early atherosclerosis independent of blood pressure effects. In rats, the ACE inhibitor cilazapril reduced SMC proliferation and migration in the carotid artery after denuding endothelial injury (Powell et al. 1989). In hyperlipidemic rabbits another ACE inhibitor had significant antiatherogenic effects. Cilazapril did not inhibit myofibrotic changes in swine carotid arteries after catheter-induced vessel-wall injury, similar to that induced during coronary angioplasty (Lam et al. 1992). Despite these mixed results, human trials with ACE inhibitors are proceeding.

Heparin and heparin derivatives have been considered as antiatherogenic agents because they also inhibit SMC proliferative responses (Clones and Karnovsky 1977). Native anticoagulant-inducing heparin can probably not be used for long-term suppression of atherosclerosis, because of the potential for bleeding side-effects. Fragments of the heparin molecule without substantial anticoagulant effects have been derived and continue to demonstrate antiproliferative effects (Edelman 1990). Local delivery of a heparin derivative to the site of arterial wall proliferation very effectively inhibits arterial occlusion. Heparin derivatives will soon be evaluated in patients to try and reduce restenosis after PTCA.

A variety of other cellular interventions for atherosclerosis are at various stages of evaluation. Inhibition of acylcoenzyme A/cholesterol *o*-acyltransferase (ACAT) provides another way to reduce cellular lipid accumulation by inhibiting cholesterol esterification and the accumulation of cholesterol in macrophages. An ACAT inhibitor was shown to reduce fatty streak and foam-cell formation in a hypercholesterolemic rabbit model after endothelial cell injury without cholesterol lowering (Bocan et al. 1991). Antiproliferative drugs such as colchicine are another potential approach to cellular intervention for atherosclerosis. Growth factors are important mediators of atherogenesis, and monoclonal antibodies which inhibit growth factors such as PDGF, or block its receptor, are being developed. Clearly, the development of potential therapies to inhibit atherosclerosis has been enhanced as knowledge about the cellular mechanisms of atherosclerosis has advanced. We can anticipate new and innovative cellular therapies in the near future which can be combined with risk factor modifiers such as lipid-lowering therapy.

References

Armstrong VW, Cremer P, Eberle E et al. (1986) The association between serum Lp(a) concentrations and angiographically assessed coronary atherosclerosis. Atherosclerosis 6:249–257

Bassiouny HS, Davis H, Massawa N, Gewerts BL, Glagov S, Zarins CK (1989) Critical carotid stenosis: morphologic and chemical similarity between symptomatic and asymptomatic plaques. J Vasc Surg 9:202–212

Berk BC, Taubman MB, Cragoe EJ, Fenton JW, Griedling KK (1990) Thrombin signal transduction mechanisms in rat vascular smooth muscle cells. J Biol Chem 265:17334–17990

Bernstein NM, Krajewski A, Lewis AJ, Norris JW (1990) Clinical significance of carotid plaque hemorrhage. Arch Neurol 47:958–959

Bini A, Fenoglia JJ, Mesa-Tejada R, Kudryk B, Kaplan KL (1989) Identification and distribution of fibrinogen, fibin and fibrin(ogen) degradation products in atherosclerosis: use of monoclonal antibody. Arteriosclerosis 9:109–121

Blankenhorn DH, Nessim SA, Johnson RA et al (1987) Beneficial effects of combined colestipol/niacin therapy on coronary atherosclerosis and coronary venous bypass grafts. JAMA 257:3233–3240

Bocan TMA, Mueller SB, Uhlendorf PD, Newton RS, Krause BR (1991) Comparison of CI-976, an ACAT inhibitor, and selected lipid-lowering agents for antiatherosclerotic activity in iliac–femoral and thoracic aortic lesions. Arteriosc Thromb 11:1830–1843

Bocuzzi SJ, Bocangera TS, Walker JF, Shapiro DR, Keegan ME (1991) Longterm safety and efficacy profile of simvastatin. Am J Cardiol 68:1127–1131

Bogousslavsky J, Van Melle G, Regli F (1988) The Lausanne Stroke Registry. Stroke 19:1083–1092

Bogousslavsky J, Van Melle G, Despland PA, Regli F (1990) Alcohol consumption and carotid atherosclerosis in the Lausanne Stroke Registry. Stroke 21:715–720

Bonithon-Kopp C, Scarabin PY, Taquet A, Toubol PJ, Malmejae A, Guize L (1991) Risk factors for early carotid atherosclerosis in middle-aged French women. Arteriosc Throm 11:960–972

Brown G, Albers JJ, Fisher LD et al. (1990) Regression of coronary artery disease as a result of intensive lipid-lowering therapy in men with high levels of apolipoprotein. New Engl J Med 323:1289–1298

Brown MS, Kovanen PT, Goldstein JL (1981) Regulation of plasma cholesterol by lipoprotein receptors. Science 212:628–635

Burr ML, Gilbert JF, Holliday RM et al. (1989) Effects of changes in fat, fish and fibre intakes on death and myocardial infarction: diet and reinfarction trial (DART). Lancet 2:757–701

Caplan LR, Gorelick PB, Hier DB (1986) Race, sex and occlusive cerebrovascular disease: a review. Stroke 17:64849–64855

Carew TE, Schwenke DC, Steinberg D (1987) Antiatherogenic effect of probucol unrelated to its hypocholesterolemic effect. Proc Natl Acad Sci USA 84:7725–7729

Castelli WP, Garrison RJ, Wilson PWF, Abbot RD, Kalousdian S, Kannol WB (1986). Incidence of coronary heart disease and lipoprotein cholesterol levels: the Framingham Study. JAMA 256:2835–2838

Clones AW, Karnovsky MJ (1977) Supression by heparin of smooth muscle cell proliferation in injured arteries. Nature 256:625–626

Collins R, Peto R, Maemahon S et al. (1990) Blood pressure, stroke and coronary heart disease: part 2, short-term reductions in blood pressure: overview of randomized trials in their epidemiologic context. Lancet 335:827–838
Consensus Conference (1985) Lowering blood cholesterol to prevent heart disease. JAMA 253:2080–2086
Cornhill JF, Barrett WA, Herderick EE, Mahley RW, Fry DL (1985) Topographic study of sudanophilic lesions in cholesterol-fed minpigs by image analysis. Arteriosclerosis 5:415–426
Davies MJ, Thomas AC (1985) Plaque fissuring: the cause of acute myocardial infarction, sudden ischemic death, and crescendo angina. Br Heart J 53:363–373
Davis HR, Bridenstine RT, Vesselinovitch D, Wissler RW (1987) Fish oil inhibits development of atherosclerosis in rhesus monkeys. Arteriosclerosis 7:441–449
Dyken ML (1991) Stroke risk factors. In: Norris JW, Hachinski VC (eds) Prevention of stroke. Springer-Verlag, New York, pp 83–102
Edelman ER (1990) Effect of controlled adventitial heparin delivery on smooth muscle cell proliferation following endothelial injury. Proc Natl Acad Sci USA 87:3773–3778
Emerson EE, Robertson AL (1988) T-lymphocytes in aortic and coronary intimas. Their potential role in atherogenesis. Am J Pathol 130:369–376
Expert Panel (1988) Report of the National Cholesterol Education Program Expert Panel on detection, evaluation and treatment of high blood cholesterol in adults. Arch Intern Med 148:36–69
Falk E (1989) Morphologic features of unstable atherothrombotic plaques underlying acute coronary syndromes. Am J Cardiol 63:114E–120E
Feldmann E, Daneault N, Kwan E et al. (1990) Chinese–white differences in the distribution of occlusive cerebrovascular disease. Neurology 40:1541–1545
Fisher CM (1954) Occlusion of the carotid arteries: further experiences. Arch Neurol 72:187–204
Fisher CM (1965) Lucunes – small deep infarcts. Neurology 15:774–778
Fisher CM (1979) Capsular infarcts: the underlying vascular lesion. Arch Neurol 36:65–73
Fisher CM, Ojemann RC (1986) A clinico-pathologic study of endarterectomy plaques. Rev Neurol 142:573–589
Fisher M, Leaf A, Levine PH (1989) N-3 fatty acids and cellular aspects of atherosclerosis. Arch Intern Med 149:1726–1728
Fisher CM, Gore I, Okabe N, White PD (1965) Atherosclerosis of the carotid and vertebral arteries: extracranial and intracranial. J Neuropathol Exp Neurol 24:455–407
Fisher M, Martin A, Cosgrove M, Pirsi A, Norris J (1992) Carotid artery plaques in the NASCET and ACAS projects. Neurology 42 (Suppl 3): 204
Fitzgerald GA, Oates JA, Nowak J (1981) Cigarette smoking and hemostatic function. Am Heart J 115:267–271
Frick MH, Elo D, Haapa K et al. (1987) Helsinki Heart Study: primary prevention trial with gemfibrozil in middle-aged men with dyslipidemia: safety of treatment, changes in risk factors and incidence of coronary heart disease. New Engl J Med 317:1237–1245
Furberg CD, Byington RP, Borhani NA (1989) Multicenter isradipine diuretic atherosclerosis study. Am J Med 86 (Suppl 4A):37–39
Fuster V, Badimon L, Badimon JJ, Cheseboro JH (1992) The pathogenesis of coronary artery disease and the acute coronary syndromes. New Engl J Med 326:292–250
Geer JC, McGill HC, Robertson WB et al (1960) Histologic characteristics of coronary artery fatty streaks. Lab Invest 18:565–570
Glagov S, Zarins C, Giddens DP, Ku DN (1988) Hemodynamics and atherosclerosis. Arch Pathol Lab Med 122:1018–1031
Goodnight SH, Cairns JA, Fisher M, Fitzgerald GA (1992) Assesment of the therapeutic use of N-3 fatty acids in vascular disease and thrombosis. Chest 102:3745–3845
Grundy SM (1988) HMG-CoA reductase inhibitors for treatment of hypercholesterolemia. New Engl J Med 319:24–33
Hackett D, Davies G, Maseri A (1988) Pre-existing coronary stenosis in patients with first myocardial infarction is not necessarily severe. Eur Heart J 9:1313–1317
Herrick JB (1912) Clinical features of sudden obstruction of the coronary arteries. JAMA 58:2015–2020
Hirai A, Hamazaki T, Terano T (1980) Eicosapentaenoic acid and platelet function in Japanese. Lancet 2:1132–1133
Imparato AM, Riles RS, Mintzer R et al. (1983) The importance of hemorrhage in the relationship between gross morphologic and cerebral symptoms in 376 carotid artery plaques. Ann Surg 197:195–203
Ingall TJ, Homer D, Baker HL, Kottke BA, O'Fallon WM, Whisnant JP (1991) Predictors of intracranial carotid artery atherosclerosis. Arch Neurol 48:687–691
Initazari D, Hachinski VC, Taylor W, Barnet HJM (1990) Racial differences in the anterior circulation in cerebrovascular disease. How much can be explained by risk factors? Arch Neurol 47: 1080–1084
Jackson CL, Bowyer DE (1989) Mechanism of antiatherogenic action of calcium antagonists. Atherosclerosis 90:17–26
Jaffee EA (1987) Cell biology of endothelial cells. Hum Pathol 18:234–239
Joris I, Zand T, Nunnari JJ, Krolikowski FJ, Magno G (1983) Studies on the pathogenesis of atherosclerosis. 1. Adhesion and immigration of mononuclear cells in the aorta of hypercholesterolemic rats. Am J Pathol 116:56–68
Klagsbrun M, Edelman ER (1989) Biological and biochemical properties of fibroblast growth factors. Arteriosclerosis 9:269–278
Kobayashi M, Ishida F, Takahashi T et al. (1989) Preventive effects of MK-733 (simvastatin), an inhibitor of HMG-CoA reductase on hypercholesterolemia and atherosclerosis induced by cholesterol feeding in rabbits. Jpn J Pharmacol 49:125–133
Lam JYT, Lacoste L, Bourassa MG (1992) Cilazapril and early atherosclerotic changes after balloon injury of porcine carotid arteries. Circulation 85:1542–1547
La Rosa JC, Hunning Lake D, Bush D et al. (1990) The cholesterol facts: a summary relating dietary fats, serum cholesterol and coronary heart disease. Circulation 81:1721–1733
Lichtlen PR, Hugenholtz PG, Rafflebeul W, Hecker H, Jost S, Deckers JW (1990) Retardation of angiographic progression of coronary artery disease. Lancet 335:1109–1113
Lipid Research Clinics (1984) Coronary Primary Prevention Trial Results. I. Reduction in incidence of coronary heart disease. JAMA 251:351–364
McGarry P, Solberg LA, Guzman MA, Strong JP (1988) Cerebral atherosclerosis in New Orleans: comparison of lesions by age, sex and race. Lab Invest 52:533–539
McGill HC (1988) The pathogenesis of atherosclerosis. Clin Chem 34:833–839
Manderson JA, Mosse PRL, Safstrom JA, Young SB, Campbell GR (1989) Balloon catheter injury to rabbit carotid artery. 1. Changes in smooth muscle phenotype. Atherosclerosis 9:289–298
Manon JE, Tosteson H, Ridker PM et al. (1992) The primary prevention of myocardial infarction. New Engl J Med 326:1406–1416

Masawa N, Hoshida Y, Joshita T, Ishihara N, Yamada T (1990) Three-dimensional morphologic analysis of thrombotic occlusive arteries in autopsies of atherosclerotic cerebral infarction. Stroke 21:1–33

Masuda J, Ross R (1990a) Altherogenesis during low level hypercholesterolemia in the non human primate. 1 Fatty streak formation. Arteriosclerosis 10:164–177

Masuda J, Ross R (1990b) Atherogenesis during low level hypercholesterolemia in the non human primate. II Fatty streak conversion to fibrous plaque. Arteriosclerosis 10:178–187

Miles LA, Plow EF (1990) Lp(a): an interloper into the fibrinolytic system. Thromb Haemost 63:331–335

Mitchinson MJ, Ball RY (1987) Macrophages and atherogenesis. Lancet 2:146–149

Mitsuyama Y, Thompson LR, Hayashi T et al. (1979) Autopsy study of cerebrovascular disease in Japanese men who lived in Kiroshima, Japan, and Honolulu, Hawaii. Stroke 10:389–395

Munro MJ, Cotran RJ (1988) Biology of disease. The pathogenesis of atherosclerosis: atherogenesis and inflammation. Lab Invest 58:249–261

Naftilan AJ, Pratt RE, Dzan VJ (1989) Induction of PDGF-A chain and C-*myc* gene expression by angiotensin II in cultured rat vascular smooth muscle cells. J Clin Invest 83:1419–1424

Nishimaru K, McHenry LC, Toole JF (1984) Cerebral angiographic and clinical differences in carotid system transient ischemic attacks between American Caucasians and Japanese patients. Stroke 15:56–59

Ogata J, Masuda J, Yutani C et al. (1990) Rupture of atheromatous plaque as a cause of thrombotic occlusion of stenotic internal carotid artery. Stroke 21:1740–1745

Parums DV, Brown DL, Mitchinson DJ (1990) Serum antibodies to oxidized low density lipoprotein and ceroid in chronic periaortitis. Arch Pathol Lab Med 119:383–387

Powell JS, Clozel JP, Muller RKM et al. (1989) Inhibitors of angiotensin-converting enzyme prevent myointimal proliferation after vascular injury. Science 245:186–188

Qizibash N, Duffy SW, Warlow C, Mann J (1992) Lipids are risk factors for ischemic stroke. Cerebrovasc Dis 2:127–136

Quinn MT, Parthasarathy S, Fong LG, Steinberg D (1987) Oxidatively modified low density lipoproteins: a potential role in recruitment and retention of monocyte/macrophages during atherogenesis. Proc Natl Acad Sci USA 84:2995–2998

Raines EW, Dower SK, Ross R (1989) Interleukin-1 mitogenic activity for fibroblasts and smooth muscle cells is due to PDGF-AA. Science 243:393–395

Richardson PD, Davies MJ, Born GVR (1989) Influence of plaque configuration and stress distribution on fissuring of coronary atherosclerotic plaques. Lancet 2:941–944

Ross R (1986) The pathogenesis of atherosclerosis: an update. New Engl J Med 314:488–500

Ross R, Wight TN, Strandess E, Thiele B (1984) Human atherosclerosis I. Cell constitution and characteristics of advanced lesions of the superficial femoral artery. Am J Pathol 114:79–93

Salonen R, Seppanen K, Rauramaa R, Salonen JT (1988) Prevalence of carotid atherosclerosis and serum cholesterol levels in Eastern Finland. Atherosclerosis 8:788–792

Salonon JT, Yla-Hertulla, S, Yamamoto R et al. (1992) Autoantibody against oxidized LDL and progression of carotid atherosclerosis. Lancet 339:883–887

Schmitz G, Hankowitz J, Kovacs EM (1991) Cellular processes in atherogenesis: potential targets of Ca^{2+} channel blockers. Atherosclerosis 11:109–132

Seiffert GF, Keown K, Moore SW (1981) Pathologic effects of tobacco smoke inhalation on arterial intima. Surg Forum 32:353–359

Sherman DG, Dyken ML, Fisher M, Harrison MJB, Hart RB (1989) Antithrombotic therapy for cerebrovascular disease. Chest 95 (Suppl): 140S–155S

Stary HC, Blankenhorn DH, Chandler AB et al. (1992) A definition of the intima of human arteries and of its atherosclerosis prone regions. Arterio Throm 12:120–134

Steinberg D (1991) Antioxidants and atherosclerosis: a current assessment. Circulation 84:1920–1925

Tell GS, Gouse J, Furberg GD (1988) Relation between blood lipid, lipoproteins and cerebrovascular atherosclerosis: a review. Stroke 19:423–430

Tsukada T, Rosenfeld MR, Ross R, Gown AM (1986) Immunocytochemical analysis of cellular components in lesions of atherosclerosis in Watanabe and fat-fed rabbits using monoclonal antibodies. Arteriosclerosis 6:601–613

Virchow R (1856) Plogose und Thrombose in Gefassytem, Gesammelte Abhandlungen zur wissenschaftlichen Medezin. Meidinger Sohn, Frankfurt-am-Main, p 458

Von Rokitansky C (1852) A manual of pathologic anatomy, Volume 4. Sydenham Society, London, p 202

Walker LN, Reidy MA, Bowyer DE (1986) Morphology and cell kinetics of fatty streak lesion formation in the hypercholesterolemic rabbit. Am J Pathol 125:450–459

Waters D, Lesperance J, Franletich M et al. (1990) A controlled clinical trial to assess the effect of a calcium channel blocker on the progression of coronary atherosclerosis. Circulation 82:1990–1953

Whisnant JP, Homer D, Ingall TJ, Baker HL, O'Fallon W, Wiebers DO (1990) Duration of cigarette smoking is the strongest predictor of severe extracranial carotid artery atherosclerosis. Stroke 21: 707–714

Witt TJ, Lofgren RP, Nichol KL et al. (1989) Fish oil supplementation does not lower plasma cholesterol in men with hypercholesterolemia. Ann Intern Med 111:900–905

Yatsu FM, Fisher M (1989) Atherosclerosis: current concepts on pathogenesis and interventional therapies. Ann Neurol 26:3–12

Yla-Herttuala S, Palinski W, Rosenfeld MR et al. (1989) Evidence for the presence of oxidatively modified low density lipoprotein in atherosclerotic lesions of rabbit and man. J Clin Invest 84:1086–1095

Zenker G, Koltringer P, Boneb, Niederkorn K, Pfeifer K, Jurgens G (1986) Lipoprotein (a) as a strong predicator for cerebrovascular disease. Stroke 17:942–945

11. Cerebral Vasoconstriction: Physiology, Pathophysiology and Occurrence in Selected Cerebrovascular Disorders

M. A. Sloan

As clinical and instrumental capabilities become more sophisticated, the number of recognized mechanisms by which cerebral ischemia may occur has grown. In recent years, it has been recognized that cerebral ischemia may be due to occlusive disease of large or small intracranial arteries or large extracranial arteries, cardioembolic (including transcardiac) or intra-arterial embolism, inflammatory diseases, arterial anomalies etc. In some patients hemorrhagic cerebrovascular disease may be complicated by ischemia as well. Increasing attention has refocused on the old concept that cerebral vasoconstriction or 'vasospasm' may be an important cause of cerebral ischemia. Basic neuroanatomical and neuropharmacological investigations have defined the physiological aspects of the aminergic and peptidergic pathways and mechanisms involving the cerebral circulation. In addition, a number of reports discuss the exploding field of endothelium-derived factors which may affect cerebral vascular tone.

Cerebral vasoconstriction may occur when an imbalance develops between factors which produce vasoconstriction and vasodilation of large or small cerebral vessels. The literature on this topic, as it pertains to cerebrovascular disorders, is large and continues to expand rapidly. In this chapter, the occurrence and mechanisms of cerebral vasoconstriction in subarachnoid hemorrhage, Call's syndrome, illicit and licit drug abuse, and eclampsia and postpartum cerebral angiopathy will be discussed on the basis of representative publications in each subject area. It is hoped that this synthesis of a diverse topic will foster further investigations which will shed light on an intriguing clinical phenomenon, thus leading to improved patient care and outcome.

Factors Affecting Cerebral Vascular Tone

Innervation of the Cerebral Circulation

The types, origin, pathways, physiology and constituent neurotransmitters and neuropeptides of the sympathetic, parasympathetic and sensory systems innervating or modulating the cerebral circulation have recently been reviewed (Edvinsson and MacKenzie 1977; Walters et al 1986; Edvinsson 1987; Uddman and Edvinsson 1989; Macfarlane and Moskowitz 1994; Suzuki and Hardebo 1993) and are summarized in Table 11.1. Such pathways have been found in a variety of animal species, including humans. However, studies in small numbers of animals show regional (Edvinsson 1987; Moskowitz et al. 1983),

Table 11.1 Aminergic and peptidergic innervation of the cerebral circulation

System	Neuromodulator	Origin	Distribution	Function	Relation to endothelium	Receptor-mediated	Inhibitors
Sympathetic	Norepinephrine (NE)	SCG	forebrain basal arteries, arterioles, veins	VC VD	– –	α_1 β_1	α β
		SG	VA, BA	VC VD	– –	α_1 β_1	α β
	Neuropeptide Y (NPY)	SCG	forebrain › hindbrain	VC	–		Ca^{2+} free, Ca^{2+} blocker
Parasympathetic	Acetylcholine (ACh)	SPG OG MG	forebrain	VD	+	muscarinic	atropine
	Vasoactive intestinal polypeptide (VIP)	SPG	forebrain and hindbrain basal arteries, arterioles, veins	VD	–	+	
	Peptide histidine isoleucine (PHI)	SPG OG	forebrain basal arteries, arterioles, veins	VD			
Sensory	*Tachykinins* Substance P (SP)	TG (V_1, V_2)	forebrain arteries, arterioles, veins	VD	+	+	+
		SP	hindbrain arteries, arterioles, veins	VD		+	+
	Neurokinin A (NKA)	TG	forebrain and hindbrain arteries, arterioles, veins	VD	+	+	+
	Calcitonin gene-related peptide (CGRP)	TG	forebrain and hindbrain arteries, arterioles, veins	VD	–	+	+
Other	Vasopressin (VP)	PV, SO	?	VC, VD dose related	?	?	?
	Natriuretic peptide (NP)	HT septum	?	VC: intra-parenchymal VD: pial	?	?	?

SCG, superior cervical ganglion; SG, stellate ganglion; VA, vertebral artery; BA, basilar artery SP, spinal ganglia; SPG, sphenopalatine ganglion; OG, otic ganglion; +, dependent; –, independent; MG, local microganglia; HT, Hypothalamus; TG, Trigeminal ganglion; V_1, ophthalmic division V_2, maxillary division PV, paraventricular nucleus; SO, supraoptic nucleus; VC, vasoconstriction VD, vasodilation.

interspecies (Liu-chen et al. 1983), and inter-individual (Liu-chen et al. 1983) differences in nerve content (Edvinsson 1987; Moskowitz et al. 1983; Liu-chen et al. 1983) or reactivity (Edvinsson 1987) of the compounds, such as acetylcholine (ACh) (Edvinsson 1987) and substance P (SP) (Liu-chen et al. 1983). Several of the peptides are synthesized from mRNA as large precursor molecules which are then processed to smaller fragments by proteolysis and post-translational modifications (sulfation and/or amidation). For example, the tachykinins (substance P (SP), and neurokinin A (NKA)) derive from the preprotachykinins (Uddman and Edvinsson 1989). These neuromodulators reside in various ganglia and perivascular nerve fibers. Pathways for these various neuromodulators are mostly unilateral, although noradrenergic fibers may have a small bilateral distribution and midline vessels have a bilateral sensory system innervation (Edvinsson 1987; Macfarlane and Moskowitz 1994). Interestingly, many of these compounds are colocalized, such as norepinephrine (NE) with neuropeptide Y (NPY) (Jansen et al. 1986), vasoactive intestinal peptide (VIP) with peptide histidine isoleucine (PHI), VIP with NPY, SP with NKA, SP with calcitonin gene-related peptide (CGRP), and SP with NKA and CGRP (Edvinsson 1987; Macfarlane and Moskowitz 1994). These arrangements may provide for balanced vasoconstrictor and vasodilator effects (VIP + NPY) or additive or synergistic vasoconstrictor and vasodilator effects (CGRP + SP), respectively. VIP acts upon more peripheral arterioles (Brayden and Bevan 1986; Ballon et al. 1986), whereas SP and CGRP may act upon vessels in the circle of Willis (Moskowitz 1984; Edvinsson 1987; Uddman and Edvinsson 1989; Macfarlane and Moskowitz 1994). Vasodilator effects may be mediated by both cAMP and cGMP second-messenger systems (Edvinsson and MacKenzie 1977; Edvinsson et al. 1985; Wei et al. 1992b).

The sympathetic (Edvinsson and MacKenzie 1977; Edvinsson 1987; MacFarlane and Moskowitz 1994), parasympathetic (Van Riper and Bevan 1992) and sensory (Goadsby et al. 1988; Linnik et al. 1989) systems may become activated by various stimulatory phenomena (Edvinsson and MacKenzie 1977; Edvinsson 1987; Van Riper and Bevan 1992; MacFarlane and Moskowitz 1994) and disease states (Goadsby et al. 1988; Linnik et al. 1989). Limited data exist on the relation between cerebral innervation and cerebral infarction. Postganglionic parasympathetic fibers innervating the circle of Willis may serve a protective role against focal cerebral ischemia. In a rat model of MCA occlusion and temporary (45 minutes) bilateral common carotid occlusion (Kano et al. 1991), unilateral and ipsilateral selective parasympathectomy or combined parasympathectomy and sensory fiber ablation resulted in a significantly larger infarct volume. Immunohistochemistry studies demonstrated a significant reduction in VIP- and CGRP-containing fibers.

Endothelium-Derived Factors and Vasomotor Function

It was only 13 years ago that Furchgott and Zawadzki (1980) reported that relaxation of isolated rabbit aorta and other arteries induced by ACh and other agonists for muscarinic receptors depended on the presence of endothelial cells; in denuded arteries, ACh caused contraction. Basic physiology, pharmacology and the role in disease states played by endothelium-derived relaxing factor(s) (EDRF) and endothelium-derived contracting factor(s) (EDCF) have been the subject of numerous scientific papers and review articles. In cerebral arteries the endothelium may, in part, act to sense changes in pressure and flow, thus leading to contraction and dilation, respectively (Furchgott and Vanhoutte 1989; Bevan and Gaw 1992; Rosenblum 1992). In fact, EDRF may modulate noradrenergic constriction of cerebral arterioles (Bauknight et al. 1992). A number of agents will produce endothelium-dependent relaxation or contraction of cerebral arteries. These results may be due to the use of differing experimental conditions and methods, different animal species, different vessels, or other reasons (Katusic et al. 1988; Onoue et al. 1988; Faraci et al. 1991a,b). Selected agents are summarized in Table 11.2.

Table 11.2. Agents that produce endothelium-dependent vasomotor activity

Vasoconstriction	Vasodilation
Acetylcholine	Acetylcholine
A23187 (calcium ionophore)	A23187 (calcium ionophore)
Substance P	ADP
Endothelin	ATP
Anoxia	Bradykinin
Thrombin	Kallikrein
Vasopressin	Thrombin
	Vasopressin
	Substance P
	CGRP

The mechanism(s) of endothelium-dependent relaxation and the identification of EDRFs in the cerebral vascular bed have been the subject of intensive investigation and controversy. Some investigators have hypothesized that, since L-arginine is apparently required for the synthesis of EDRF, EDRF is nitric oxide (NO) (Faraci 1990; Moncada et al. 1991). Recently, the presence of a rich unilateral innervation of the circle of Willis by nerve fibers containing NO synthase immunoreactivity (NOS-IR) was demonstrated in rats, probably originating in the sphenopalatine ganglion (Nozaki et al. 1992b). In cynomolgous monkeys (Thompson et al. 1992), continuous release of NO may be necessary to maintain homeostatic cerebral vasodilation. Inhibition of NO synthase by N^G-monomethyl-L-arginine (L-NAME) leads to moderate pial artery constriction, reduced cerebral blood flow, enlarged perivascular spaces and cortical leaky sites with constricted arterioles and endothelial pinocytotic vesicles and microvilli (Prado et al. 1992). Others have suggested that there is probably more than one EDRF (Marshall et al. 1988; Furchgott and Vanhoutte 1989; Rosenblum 1992; Kontos 1993) or that EDRF in the cerebral circulation is NO or an NO-containing substance, such as a nitrosothiol (Rosenblum 1992; Wei et al. 1992a; Kontos 1993).

Briefly, ACh acts on an endothelial muscarinic receptor to induce influx of Ca^{2+} into cells, which in the presence of calmodulin and NADPH leads to activation of NO synthase, thus generating NO from L-arginine (Furchgott and Vanhoutte 1989; Kontos 1993). NO then diffuses into the extracellular space as 'EDRF', combines with the iron of the heme component of soluble guanylate cyclase, induces a conformational change to activate the enzyme and generates cGMP from GTP, thus leading to smooth muscle cell relaxation and vasodilation. A related mechanism also occurs. Once NO is generated, it may combine with endogenous thiols to produce a nitrosothiol which, after release, directly activates soluble guanylate cyclase in a fashion independent of NO release. This is supported by the observation that a nitrosothiol, *S*-nitro-L-cysteine, resembled the EDRF from ACh much more closely than did NO. Moreover, NO itself, nitroprusside and nitroglycerin activate soluble guanylate cyclase indirectly via release of CGRP from sensory nerve fibers (Marshall et al. 1988; Rosenblum 1992; Wei et al. 1992b).

Recently, one of the endothelium-derived contraction factors, endothelin (Shigeno and Mima 1990), was discovered. This is the most potent vasoconstrictor known to date ($EC_{50} = 4 \times 10^{-10}$ M) and its action is long-lasting. At present three isopeptides, endothelin-1, endothelin-2 and endothelin-3, have been identified, with endothelin-1 being the best characterized. Endothelin's action upon cerebral vessels may be from the adventitial side, and at times may be endothelium-independent (Shigeno and Mima 1990; Suzuki et al. 1990). Each endothelin isopeptide acts via a specific vascular smooth muscle cell receptor, with the potency of contraction in the order of endothelin-1 › -2» -3. Endothelin rapidly increases intracellular Ca^{2+} (early phase) followed by a slowly progressive increase (delayed phase). Other effects of endothelin include stimulation of DNA synthesis (Shigeno and Mima 1990), a contribution to cellular proliferation and growth, increased production of the vasoconstrictor prostaglandins PGF_2 and thromboxane (Spatz et al. 1992; Stanimirovic et al. 1993) and reduced microvascular perfusion (Willette et al. 1990).

Vascular smooth muscle contraction occurs as the result of a complicated process (Somlyo and Himpens 1989). A variety of external stimuli cause the elevation of cytosolic Ca^{2+} concentration to a critical level, partly as a result of influx of extracellular Ca^{2+} and also by mobilization from intracellular Ca^{2+} storage sites (Sasaki et al. 1986). Binding of an agonist, such as a noradrenergic agent, to a membrane receptor leads to the activation of G protein and phospholipase C, which converts phosphatidyl inositol to inositol triphosphate and diacyglycerol. Inositol triphosphate may contribute to the rise in intracellular Ca^{2+}, which promotes the formation of a Ca^{2+}–calmodulin complex. This complex stimulates myosin light-chain kinase, thus triggering the phosphorylation of myosin. The generation of cross-linkages between myosin and actin leads to contraction and muscle tension. Diacylglycerol may activate the protein kinase C system, leading to prolonged contraction (Somlyo and Himpens 1989; Shigeno and Mima 1990; Takanashi et al. 1992). The early response to endothelin requires the presence of extracellular Ca^{2+} and is attenuated by dihydropyridine Ca^{2+} channel antagonists. The slowly progressive and sustained contraction seems to be related to the delayed phase of increase in intracellular Ca^{2+}, perhaps resulting from the operation of protein kinase C (Shigeno and Mima 1990).

Other Vasoactive Compounds

A variety of factors also affect endothelial responses. Increasing age (Hongo et al. 1988a; Hatake et al. 1990) or hypertension (Hongo et al. 1988a) may reduce relaxation responses. The presence of atherosclerosis may potentiate constrictor responses of cerebral and ocular blood vessels to thromboxane or serotonin (5-HT) (Faraci et al. 1989), and is amenable to treatment (Faraci et al. 1991a). Subtle endothelial injury may reversibly impair endothelial-derived responses, but not smooth muscle function (Rosenblum et al. 1987; Ellison et al. 1989). A number of other substances affect cerebral vascular tone. High concentrations of KCl favor vasoconstriction (Furchgott 1983; Furchgott and Vanhoutte

1989; White et al. 1991). Products of platelet aggregation also tend to favor vasoconstriction. These include 5-HT (Hongo et al. 1988b) and vasoactive prostanoids, such as PGD_2, $PGF_{2\alpha}$ and thromboxanes (Kanamaru et al. 1987a; Onoue et al. 1988; Katusic et al. 1988; Faraci et al. 1991b).

Some components of blood non-specifically reduce cerebral vascular tone in isolated canine basilar arteries (White 1986). Antithrombin III may produce endothelium-independent inhibition of contractile responses to cumulative doses of KCl, 5-HT, uridine triphosphate (UTP) and thrombin. α-2 macroglobulin may inhibit contractile responses to high concentrations of KCl, 5-HT and thrombin. Kallikrein, through an endothelium-dependent mechanism, may inhibit contractile responses to high KCl concentrations and 5-HT and irreversibly block the response to thrombin. Heparin may relax vessels previously contracted by plasmin and thrombin by enhancing the action of antithrombin III (Kapp et al. 1985, 1987; Chimowitz and Pessin 1987). Prostacyclin (PGI_2) is a well known eicosanoid released from endothelial cells and is a powerful platelet antiaggregant and vasodilator. Finally, capric acid, a C10 saturated fatty acid, and monounsaturated fatty acids are potent vasodilators (White et al. 1991).

Vasoconstriction in Disease States

Vasospasm After Subarachnoid Hemorrhage

The pathogenesis, pathophysiology, detection and management of cerebral vasospasm following aneurysmal subarachnoid hemorrhage (SAH) have recently been reviewed. Diverse experimental (Heros et al. 1983; Chyatte and Sundt 1984; Kassell et al. 1985; Duff et al. 1988; Mayberg et al. 1990a; Findlay et al. 1991a; Macdonald et al. 1991) and clinical (Fisher et al. 1980; Kistler et al. 1983; Sloan 1993a) studies convincingly show that the amount and duration of vessel exposure to blood in the basal cisterns and the subarachnoid space determine the occurrence of delayed vasoconstriction of cerebral arteries, often referred to as vasospasm (VSP). Two major theories explain the mechanism(s) of VSP. The structural theory rests on the observation that vasospasm leads to smooth muscle cell damage/myonecrosis and induces a change in smooth muscle cell phenotype. Subsequent migration of these cells (myofibroblasts) into the tunica intima is followed by their proliferation (Kapp et al. 1985; Findlay et al. 1991a; Pluta et al. 1992b; Yamamoto et al. 1992; Iwasa et al. 1992; Terai et al. 1993). The spasmogen theory suggests that VSP is prolonged vasoconstriction caused by or associated with prolonged exposure to a variety of chemical mediators, such as blood breakdown products [hemoglobin (Fujiwara et al. 1986; Kanamaru et al. 1987b; Hongo et al. 1988a; Macdonald and Weir 1991; Duff et al. 1988; Mayberg et al. 1990a), oxyhemoglobin (Findlay et al. 1991a; Tanaka and Chiba 1988; Onoue et al. 1989; Macdonald et al. 1991; Macdonald and Weir 1991; Ohlstein and Storer 1992), and bilirubin (Tanaka et al. 1992)], eicosanoids (Yokota et al. 1987, 1989; Nosko et al. 1988; Rodriguez y Baena et al. 1989; Chehrazi et al. 1989; Rosenblum et al. 1990; Gaetani et al. 1990; Juvela et al. 1990a,b, 1991; Ohkuma et al. 1991; O'Neill et al. 1992), free radicals (Sasaki et al. 1986; Yokota et al. 1987; Steinke et al. 1989; Findlay et al. 1991a; Macdonald and Weir 1991; Harada and Mayberg 1992), inflammation (Chyatte and Sundt 1984; Kassell et al. 1985; Chyatte 1989; Gaetani et al. 1991), immunologic/complement system derangements (Kassell et al. 1985; Kasuya et al. 1988; Kasuya and Shimizu 1989; Yanamoto et al. 1992a), vasoconstrictor neuropeptides (Suzuki et al. 1989; Ide et al. 1990; Juul et al. 1990a; Ohlstein and Storer 1992; Pluta et al. 1992a; Spatz et al. 1992; Yamaura et al. 1992) and disturbed endothelial function (Fujiwara et al. 1986; Nakagomi et al. 1987a,b, 1988; Kanamaru et al. 1987b; Tanaka and Chiba 1988; Hongo et al. 1988b; Kim et al. 1989a; Edwards et al. 1992; Hatake et al. 1992) with impaired vasodilator activity (Hongo et al. 1989).

These theories are based upon experiments performed in a wide variety of animal species (including humans), using different (one- or two-SAH) models and various in-vitro and in-vivo conditions, with assessments at different time points after SAH by various methodologies. As such, the experimental studies are difficult to compare and reconcile, although many findings are consistent. The two theories are not mutually exclusive but are interdependent, thus resulting in multifactorial biochemical mechanism(s) for the production of VSP. The occurrence and severity of clinically important vasospasm will depend upon the intensity of the stimulus (blood and its byproducts), individual differences in responses to spasmogens and structural concomitants, and the success of therapeutic interventions.

A number of structural changes occur in cerebral arteries after SAH. Several studies have shown that arterial denervation (Hara et al. 1986; Duff et al. 1986, 1987; Svendgaard et al. 1986; Tsukahara et al. 1986, 1988; Rivilla et al. 1989) and changes in noradrenergic receptors (Tsukahara et al. 1986, 1988) occur after SAH. However, these changes do not correlate with the development of angiographic VSP (Hara et al. 1986; Svendgaard et al. 1986) or affect responses to vasoconstrictor substances (5-HT, NE or blood) (Duff et al. 1987). Morphological arterial changes include intraluminal platelet adhesion and aggregation with white-cell adhesion and thrombus formation; increased endothelial pinocytotic activity or channel formation; blood–arterial wall barrier disruption with increased permeability; intimal swelling, proliferation, degeneration and desquamation; unfolding and corrugation of the endothelium; corrugation of the internal elastic lamina; myonecrosis; migration and proliferation of smooth muscle cells; presence of myofibroblasts; medial infiltration of lymphocytes, plasma cells and macrophages; increased collagen in the extracellular matrix; adventitial collections of inflammatory cells; and deposition of IgG and complement (Kassell et al. 1985; Seifert et al 1989b; Nakagomi et al. 1989a,b; Mayberg et al. 1990b; Ohta et al. 1992; Yamamoto et al. 1992). Recent studies report variable levels of α-actin (Macdonald et al. 1992; Yamamoto et al. 1992; Terai et al. 1993) and desmin (Macdonald et al. 1992; Terai et al. 1993), decreased levels of calmodulin and calcium pyriantimonate, increased fibrinogen and fibronectin levels (Macdonald et al. 1992) and elevated DNA synthesis (Somlyo and Himpens 1989) in arterial walls after SAH. Some investigators believe that some pathological changes do not directly contribute to vessel narrowing (Mayberg et al. 1990b). On the other hand, some of these changes, such as increased extracellular collagen (Mayberg et al. 1990b), myofibroblasts which compact the collagen lattice (Yamamoto et al. 1992) and platelet-derived growth factors (Iwasa et al. 1992) may act to augment prolonged vasoconstriction or inhibit the maintenance of normal vascular tone (Sasaki et al. 1989; Mayberg et al. 1990b).

Chronically vasospastic arteries also have a number of functional abnormalities. These include increased thickness and stiffness (Bevan et al. 1987; Kim et al. 1989b), decreased distensibility (Bevan et al. 1987; Kim et al. 1989b), abnormal spontaneous myogenic tone (Bevan et al. 1987; Bevan et al. 1988), reduced endothelium-dependent relaxation to histamine and calcium ionophore A23187 (Kanamaru et al. 1989), some increased sensitivity to 5-HT (Bevan et al. 1987), increase in resting tension (Kim et al. 1989b), shortening of the optimum length for maximum contraction (L_{max}) (Kim et al. 1989b) and reduced constrictor responses to KCl (Kim et al. 1989b; Kanamaru et al. 1989), $PGF_{2\alpha}$ (Kanamaru et al., 1989) and UTP (Kanamaru et al. 1989). The decreased distensibility, increase in resting tension and shortening of L_{max} all favor a smaller arterial diameter after SAH (Kim et al. 1989b). The situation is different during the acute post-SAH period. In the early period after SAH, elevations in cytosolic Ca^{2+} levels have been demonstrated (Somlyo and Himpens 1989; Takenaka et al. 1991; Takanashi et al. 1992). In one study, there was a biphasic course (peaks at day 2 and day 11 after SAH) (Takenaka et al. 1991). Early (<24 hours after SAH) accumulation of cytosolic Ca^{2+} has been associated with increased vascular reactivity to 5-HT and UTP (Debdi et al. 1992).

Blood byproducts, such as hemoglobin (Hb), oxyhemoglobin (OxyHb) and (to a lesser extent) bilirubin play an important role in cerebral vasoconstriction after SAH (Mayberg et al. 1990a; Findlay et al. 1991a; Macdonald and Weir 1991; Macdonald et al. 1991). Both Hb and OxyHb cause ultrastructural changes consistent with VSP. The time course of appearance of OxyHb in cerebrospinal fluid is similar to the time course of VSP. Adequate amounts of OxyHb are present near vasospastic arteries for long enough to produce maximum contraction (Macdonald and Weir 1991). Hb and OxyHb may contribute to VSP by the following mechanisms: (a) impaired endothelium-dependent relaxation (Fujiwara et al. 1986; Nakagomi et al. 1987a,b; Kanamaru et al. 1987b; Hongo et al. 1988b; Macdonald and Weir 1991; Hatake et al. 1992; Katusic et al. 1993) due to endothelial damage (Macdonald and Weir 1991), decreased EDRF synthesis (Nakagomi et al. 1987a; Hongo et al. 1988b; Mayberg et al. 1990a; Macdonald and Weir 1991), inactivation of EDRF (Tanaka and Chiba 1988; Hongo et al. 1988b; Edwards et al. 1992), or decreased transfer of EDRF to smooth muscle cells (Kim et al. 1989a); (b) increased intracellular free Ca^{2+} (Takenaka et al. 1991, 1992); (c) increased concentration of inositol triphosphate (Macdonald and Weir 1991); (d) increased release of vasoconstrictor eicosanoids or reduced PGI_2 (Macdonald and Weir 1991); (e) release of superoxide anion radical (O_2) and propagation of lipid peroxidation

(Macdonald and Weir 1991); (f) enhanced release of endothelin (MacDonald and Weir 1991; Ohlstein and Storer 1992); (g) inhibition of vasodilatory nerves and potentiation of vasoconstricting nerves (Macdonald and Weir 1991); and (h) accentuation of the contractile responses to hypoxia, 5-HT, KCl and vasoactive prostanoids (Macdonald and Weir 1991).

Analysis of diverse experimental and clinical investigations indicates that early removal of subarachnoid blood offers the best chance to prevent VSP (Findlay et al. 1991a; Macdonald and Weir 1991). The rationale for intrathecal thrombolytic therapy is to augment the limited fibrinolytic activity of the CSF, promote rapid clot lysis and thereby promote clearance of entrapped erythrocytes prior to hemolysis and release of spasmogens near vessels in the subarachnoid space (Yoshida et al. 1985; Alksne et al. 1986, 1988; Findlay et al. 1988, 1989b; Sloan 1994). Experimental studies with small numbers of pigs (Alksne et al. 1986, 1988), cynomolgus monkeys (Findlay et al. 1988, 1989a,b, 1990), dogs (Seifert et al. 1989a) and cats (Yamamoto et al. 1991b; Brinker et al. 1991) using plasmin (Alksne et al. 1986, 1988) or various doses of recombinant tissue plasminogen activator (rt-PA) have shown several potentially valuable effects of intrathecal thrombolytic therapy after SAH. In general, early [30 minutes (Brinker et al. 1991) to 54 hours (Seifert et al. 1989a) after first induced SAH] application of single (Seifert et al. 1989a; Brinker et al. 1991) or multiple (Findlay et al. 1988) bolus injections through an Ommaya reservoir (Findlay et al. 1988) or the use of sustained-release gel rt-PA (Findlay et al. 1989a,b, 1990) has resulted in the near-complete disappearance of subarachnoid clot in treated animals. Angiographic studies (Findlay et al. 1988, 1989a,b, 1990) have shown significant reductions in the occurrence of VSP, especially severe VSP, in treated animals compared with controls. Brain infarction occurred only in an occasional placebo-treated animal (Findlay et al. 1988, 1989a). In pathological studies (Alksne et al. 1986, 1988; Seifert et al. 1989a), intimal proliferation (Alksne et al. 1986, 1988) and proliferative vasculopathy (luminal narrowing, corrugation of the elastic lamina and subendothelial thickening and proliferation) (Seifert et al. 1989a) were significantly less likely to be seen in treated animals. In one study (Yamamoto et al. 1991b), intimal platelet accumulation was inhibited if tissue plasminogen activator was given within 10 minutes of SAH. In one study (Brinker et al. 1991), intrathecal rt-PA abolished the elevation in CSF outflow resistance by permitting near-normal CSF absorption, thus facilitating the rapid clearance of blood breakdown products. In these studies, no histological evidence of brain inflammation was seen (Findlay et al. 1988, 1990; Seifert et al. 1989a). In one study (Findlay et al. 1988), no significant change in coagulation status occurred in either treatment or placebo groups. In two studies (Findlay et al. 1988, 1989a), incisional bleeding occurred in one animal in each group. In one study (Findlay et al. 1989a), petechial hemorrhages occurred in one animal in each group. No intracerebral, intraventricular, epidural, or subdural hematomas occurred in these studies (Findlay et al. 1988, 1989a,b, 1990; Seifert et al. 1989a; Brinker et al. 1991). Recent promising small clinical studies (Shiobara et al. 1985; Mizoi et al. 1991; Findlay et al. 1991b; Zabramski et al. 1991; Öhman et al. 1991; Stolke and Seifert 1992) have suggested the presence of partial to total clot removal and a reduced incidence of VSP on angiography (Findlay et al. 1991b; Zabramski et al. 1991; Öhman et al. 1991) or transcranial Doppler (TCD) (Stolke and Siefert 1992) following intrathecal rt-PA therapy.

Disturbances of calcium homeostasis are believed to be important in the pathogenesis of VSP after SAH (Bevan et al. 1988; Shibuya et al. 1988; Tsuji and Cook 1989; Haley et al. 1990, 1991; Vorkapic et al. 1991; Findlay et al. 1991a; Kassell et al. 1991; Marzatico et al. 1991; Ram et al. 1991; Szabo et al. 1991; Toshima et al. 1992), since disruption in intracellular calcium regulation and calcium overloading can cause smooth muscle cell injury (Takanashi et al. 1992; Debdi et al. 1992). Some authors (Bevan et al. 1988; Takanashi et al. 1992), believe that vasospasm occurs in two phases: acute vasospasm may relate to excessive Ca^{2+} entry through receptor-operated channels and mobilization of calcium from intracellular stores. Chronic vasospasm may result from the effects of Ca^{2+}-induced sustained muscle contraction, leading to cell damage, inflammation, fibrosis and functional derangements (Bevan et al. 1988).

As a result, a number of calcium-channel antagonists, such as magnesium (Ram et al. 1991), diltiazem (Bevan et al. 1988), clentiazem (Vorkapic et al. 1991), nimodipine (Tsuji and Cook 1989; Findlay et al. 1991a), nicardipine (Haley et al. 1990, 1991; Kassell et al. 1991), and AT-877 (Shibuya et al. 1988, 1992) have been used to prevent or reverse VSP. In a rat SAH model, magnesium sulfate, a natural Ca^{2+} antagonist, caused dramatic dilation of the basilar artery (Ram et al. 1991). When given before SAH, diltiazem (Bevan et al. 1988) and clentiazem (Vorkapic et al. 1991) protect against functional arterial changes (Bevan et al. 1988) and chronic cerebral VSP (Vorkapic et al. 1991), respectively. Nimodipine pretreatment may lessen acute vasoconstriction caused by KCl and 5-HT (Tsuji and Cook 1989). Nimodipine has not been uniformly effective in preventing or reversing angiographic VSP, but may either have a direct cytoprotective effect or dilate small collateral channels (Findlay et al. 1991a). The lack of effect on angiographic VSP may reflect the inhibitory effect of hemoglobin (Toshima et al. 1992) and other factors. Nicardipine may reduce the capacity of cerebral lipid peroxidation (Marzatico et al. 1991). Nicardipine has been shown to be effective in preventing both symptomatic and angiographic VSP (Haley et al. 1990; Kassell et al. 1991), as indicated by a significantly lower proportion of patients with mean flow velocities greater than 200 cm/s monitored by TCD (Haley et al. 1991). A new calcium antagonist, AT-877 or HA1077 (hexahydro-1-(5-isoquinoline sulfonyl)-1H-1,4-diazepine hydrochloride, or fasudil hydrochloride), inhibits the action of free intracellular calcium ion, protein kinases A, G and C and myosin light-chain kinase as well as antagonizing the actions of endothelin (Shibuya et al. 1988, 1992). AT-877 dilates spastic cerebral arteries (Shibuya et al. 1988, 1992) and reduces symptomatic VSP, VSP-associated low densities on CT scan, and the number of patients with a poor outcome at 1 month after SAH (Shibuya et al. 1992). Recent studies also indicate that the protein kinase C system is activated during chronic VSP (Matsui et al. 1991; Germann et al. 1991; Minami et al. 1992; Nishizawa et al. 1992); which may be inhibited by the specific inhibitors H-7 (Matsui et al. 1991; Minami et al. 1992) and calphostin C (Minami et al. 1992), but not W-7 (calmodulin inhibitor) (Matsui et al. 1991), calpeptin (inhibitor of calpain, an activator of protein kinase C) (Minami et al. 1992), or nicardipine (Matsui et al. 1991).

Qualitative and quantitative platelet abnormalities are associated with VSP after SAH (Chyatte 1989; Gaetani et al. 1990; 1991; Juvela et al. 1990a; Rosenblum et al. 1990; Findlay et al. 1991a; Macdonald and Weir 1991; Ohkuma et al. 1991; O'Neill et al. 1992; Spatz et al. 1992). Increased platelet count (Juvela et al. 1990a), increased platelet aggregability (Juvela et al. 1990a; Ohkuma et al. 1991), increased thromboxane release (Juvela et al. 1990b, 1991; Ohkuma et al. 1991; O'Neill et al. 1992), increased prostaglandin release (PGD_2, PGE_2, $PGF_{2\alpha}$) (Rodriguez y Baena et al. 1989; Chehrazi et al. 1989; Steinke et al. 1989; Gaetani et al. 1990; Kanamaru et al. 1991; O'Neill et al. 1992), decreased PGI_2 (Nosko et al. 1988), increased leukotrienes (Yokota et al. 1989; Gaetani et al. 1990; Kobayaski et al. 1992), increased β-thromboglobulin (Ohkuma et al. 1991) and increased 6-oxo-prostaglandin $F_{1\alpha}$ (a PGI_2 metabolite) have been reported to occur after SAH. Diffuse (Juvela et al. 1991), severe (Chehrazi et al. 1989; Juvela et al. 1991) or symptomatic (Chehrazi et al. 1989) VSP has been associated with increased levels of thromboxane B_2 (Juvela et al. 1990a, 1991; Ohkuma et al. 1991), increased levels of $PGF_{2\alpha}$ (Chehrazi et al. 1989), increased levels of β-thromboglobulin (Ohkuma et al. 1991) and lower PGI_2 levels (Nosko et al. 1988). Prostaglandin $F_{2\alpha}$ and the stable thromboxane A_2 analogue SQ-26,655 each appear to constrict cerebral arteries by two mechanisms, first by promoting calcium uptake from low-affinity binding sites through receptor-operated channels sensitive to calcium antagonists, and secondly by releasing calcium from depletable internal stores (Wendling and Harakal 1991). Nimodipine (Juvela et al. 1990b) and nicardipine (Rodriguez y Baena et al. 1989) may reduce the release of thromboxane (Juvela et al. 1990b), PGD_2 and PGE_2 (Rodriguez y Baena et al. 1989) and leukotriene C_4 (Rodriguez y Baena et al. 1989).

The effects of inhibition of prostaglandin and thromboxane synthesis on experimental and human VSP are equivocal at best (Macdonald and Weir 1991; Tokiyoshi et al. 1991; O'Neill et al. 1992). Inhibitors of cyclo-oxygenase (Macdonald and Weir 1991) or the leukotriene antagonist ONO-1078 (Kobayaski et al. 1992) have had no effect (Macdonald and Weir 1991) or only partly prevent VSP (Macdonald and Weir 1991; Kobayaski et al. 1992). In one study (Chyatte 1989), treatment with ibuprofen and high-dose methylprednisolone significantly reduced the increase in PGE_2 release. In another study (Gaetani et al. 1991), high-dose methylprednisolone or the 21-aminosteroid U74006F significantly reduced the ex vivo synthesis of PGD_2, PGE_2 and leukotriene C_4, whereas U74006F increased the early synthesis of 6-keto-$PGF_{1\alpha}$. In one study (O'Neill et al. 1992), no significant correlation was found between CSF levels of prostaglandin E_2, thromboxane B_2 and 6-oxo-prostaglandin $F_{1\alpha}$ with clinical grade on admission, occurrence of symptomatic VSP, clinical outcome at 3 months or incidence of ischemia as a cause of death or disability. A selective inhibitor of 5-lipoxygenase, AA-861 [2-(12-hydroxydodeca-5,10-diynyl)-3,5,6-trimethyl-1,4-benzoquinone], may reduce angiographic VSP and improve the contractile function of arteries (Yokota et al. 1987).

Free radicals may also contribute to the occurrence of VSP (Macdonald and Weir 1991). In extracellular fluid, (ferrous) Hb is rapidly autooxidized to (ferric) methemoglobin and undergoes degradation to bilirubin, globin and elemental iron. OxyHb and ferric iron catalyze the conversion of superoxide to hydroxy radicals and singlet oxygen (Harada and Mayberg 1992). In conjunction with the iron in Hb, O_2 has been postulated to initiate and propagate lipid peroxidation by the Haber–Weiss reaction and Fenton chemistry. Lipid peroxides can constrict and damage cerebral arteries (Macdonald and Weir 1991) and contribute to functional impairment of the endothelium (Steinke et al. 1989). Concentrations of lipid peroxides in the CSF correlate with VSP (Sasaki et al. 1986; Macdonald and Weir 1991). Superoxide dismutase and catalase are poor antagonists of OxyHb-induced cerebral artery contraction (Macdonald and Weir 1991). Perivascular application of deferoxamine, a ferric ion chelator and antioxidant, may reduce arterial narrowing (Harada and Mayberg 1992), perhaps by preventing the disruption of cytoskeletal components such as F-actin and vimentin (Comair et al. 1993). U74006F may scavenge superoxide anions and possibly act as a membrane-localized iron chelator (Steinke et al. 1989). In one study (Kanamaru et al. 1991), a significant reduction in both VSP and subarachnoid clot malondialdehyde (a byproduct of lipid peroxidation) was shown in U74006F-treated monkeys.

Immunologic/inflammatory derangements may also occur after SAH (Kassell et al. 1985). In one study (Ryba et al. 1991) the immunostimulants thymostimuline and inosine dimethylamino-isopropanol-*p*-acetamido-benzoate aggravated angiopathy following experimental SAH. These changes were prevented by the use of cyclosporine A and azathioprine. In other studies (Kasuya et al. 1988; Kasuya and Shimizu 1989), the coagulation, complement, fibrinolytic and kallikrein–kinin systems in the CSF of patients with SAH were activated. In one experimental study (Yanamoto et al. 1992a) and an accompanying clinical investigation (Yanamoto et al. 1992b), the synthetic multiserine protease (Clr, Cl esterase, thrombin, plasmin, kallikrein and trypsin) inhibitor FUT-175 significantly reduced arterial narrowing when given early after SAH.

Finally, vasoactive neuropeptides may play a role in the occurrence of VSP. In two studies (Ide et al. 1990; Yamaura et al. 1992), endothelin-1 levels were increased on day 1 (Ide et al. 1990) or day 2 (Yamaura et al. 1992) after SAH and were associated with vasoconstriction. This vasoconstriction may be inhibited by the presence of a Ca^{2+}-free bath (Yamaura et al. 1992), nicardipine (Ide et al. 1990; Yamaura et al. 1992), the monoclonal antibody AwETN40 (Yamaura et al. 1992) or papaverine (Ide et al. 1990). More recent studies (Suzuki et al. 1992; Hamann et al. 1993) show conflicting results. In one study (Suzuki et al. 1992), CSF endothelin-like immunoreactivity levels were elevated on days 5–7 after SAH and remained high until the end of the second week. In another study (Hamann et al. 1993) there was no elevation in plasma or CSF levels of big endothelin in patients with or without VSP. These discrepancies may be explained by small numbers of patients, different times of study, methodological differences and other factors. Reports of the levels of NPY in the CSF and plasma are also conflicting. One experimental study (Pluta et al. 1992a) found no changes in CSF or plasma levels of NPY after SAH, even in the presence of VSP. In one clinical study (Suzuki et al. 1989), increased levels of NPY were found between days 6 and 11 after SAH. In another clinical study (Juul et al. 1990a), a close relationship between external jugular vein levels of NPY and flow velocity acceleration detected by TCD was observed, with the highest levels in patients with significant cerebral hemodynamic changes (MCA flow velocities >120 cm/s and increases in the hemodynamic index or V_{mca}/V_{ica}). Cerebral innervation by CGRP fibers (Edvinsson et al. 1990) and CGRP immunoreactivity (Edvinsson et al. 1991) are reduced after SAH. In one study (Juul et al. 1990b), CGRP concentrations in the external jugular vein were higher than normal in patients with cerebral VSP after SAH. After experimental SAH, the vasodilator response to CGRP is unaffected or increased (Johnston et al. 1990). CGRP may reduce elevated MCA flow velocities detected by TCD (Naylor et al. 1991). The significance of these findings remains to be determined (European CGRP in Subarachnoid Haemorrhage Study Group 1992).

Migraine

The theories of pathogenesis (Moskowitz 1984, 1991; Welch 1987; Edmeads 1991; Moskowitz and Buzzi 1991; Humphrey 1991), nature of blood flow changes (Olesen 1991; Friberg 1991) and other vascular disturbances (Olesen 1991; Babikian 1993) and clinical aspects pertaining to stroke (Broderick and Swanson 1987; Bogousslavsky et al. 1988; Moen et al. 1988; Rothrock et al. 1988; Caplan 1991) in migraine have recently been reviewed. Three major theories explain the occurrence of migraine. The neurogenic hypothesis maintains that migraine attacks originate in the brain after being triggered by one of many variables that alter the threshold of a neuronal circuit composed of orbitofrontal–brain-stem projections to the intrinsic noradrenergic system. It was found that during attacks of migraine, a wave of oligemia begins in the occipital lobe and spreads forward across the hemisphere, transgressing vascular boundaries at a rate of 2–3 mm/min (Olesen 1991; Edmeads 1991). The trigeminovascular system may release SP into the walls of cerebral vessels, dilate pial arteries, increase permeability and activate cells that participate in the inflammatory process, thus leading to vascular head pain (Moskowitz 1984; 1991; Edmeads 1991; Macfarlane and Moskowitz 1993). The occurrence of head pain with balloon inflation in the MCA (Nichols et al. 1990) supports this hypothesis.

The vascular hypothesis holds that a vascular process is responsible for the reduction in cerebral blood flow seen in some migraine attacks (Edmeads 1991; Humphrey 1991). Although the observation of spreading depression tends to weaken this theory, some investigators suggest that if the Compton scatter is properly considered, then the blood flow data may also be explained on the basis of a fixed area of deepening ischemia, perhaps related to vasospasm (Edmeads 1991; Friberg 1991). The finding of interictal regional CBF asymmetry suggests the occurrence of either abnormal vascular regulation between attacks or, alternatively, that repeated transient short-lasting ischemia might cause selective neuronal damage (Friberg 1991).

The systemic hypothesis holds that altered serotonergic neurotransmission affects the brain, platelet functions and the gastrointestinal tract (Edmeads 1991). There is abundant circumstantial but inconclusive evidence of serotonergic aberrations in migraine, such as disturbances in brain 5-HT levels and the observation of headache relief with serotonergic agonists (Humphrey 1991). Platelet membrane abnormalities, platelet hyperaggregability and coagulopathy are thought to play a secondary role, but may potentiate brain ischemia (Welch 1987; Moen et al. 1988; Welch and Levine 1990).

The trigeminovascular system (Moskowitz 1984, 1991; Edmeads 1991; Moskowitz and Buzzi 1991) links the neurogenic and vascular theories and may be the 'final common pathway' which is activated to generate the migraine syndrome. Antidromic activation of the trigeminal nerve leads to plasma protein extravasation, release of the vasodilators SP and CGRP, and produces headache pain. Various vasoconstrictors, such as NE (Humphrey 1991), 5-HT (Humphrey 1991), ergots (Humphrey 1991; Krootila et al. 1992) and sumatriptan (Buzzi et al. 1991; Nozaki et al. 1992a) block *c-fos* gene expression (Buzzi et al. 1991; Nozaki et al. 1992a), plasma protein extra-vasation (Buzzi and Moskowitz 1990), and the vasodilatory effects of CGRP (Buzzi et al. 1991; Krootila et al. 1992) in experimental animals. Sumatriptan, a 5-HT agonist, relieves headache due to MCA dilation (Friberg et al. 1991) and is a very effective fast-acting and well tolerated treatment for the pain of migraine attacks (Subcutaneous Sumatriptan International Study Group 1991).

The relationship between migraine, vasoconstriction of intracranial arteries and cerebral ischemia is complex. Recent clinical studies (Broderick and Swanson 1987; Bogousslavsky et al. 1988; Rothrock et al. 1988b; Moen et al. 1988; Caplan 1991; Ferbert et al. 1991) and review articles (Welch and Levine 1990; Babikian 1993) provide conclusive evidence that stroke occurs in migraineurs. Stroke may be related to or induced by migraine (Welch and Levine 1990). By an unknown mechanism, transient or prolonged vasoconstriction of cerebral arteries occurs, at times accompanied by activation of the platelet and coagulation systems, thus leading to vascular stenosis or occlusion and transient or permanent cerebral ischemia (Moen et al. 1988; Thie et al. 1988; Welch and Levine 1990; McLaughlin et al. 1990; Caplan 1991). Cerebral ischemia often (but not always) occurs in the setting of a prolonged typical migraine attack (Broderick and Swanson 1987; Bogousslavsky et al. 1988; Caplan 1991). Patients may develop migrainous symptoms after a stroke as well (Broderick and Swanson 1987; Caplan 1991). Stroke risk factors are often not present (Welch and Levine 1990; Bogousslavsky et al. 1988; Caplan 1991), but in some patients contribute to the pathogenesis. Atherosclerosis leads to arterial rigidity and may potentiate vasoconstrictor responses to vasoactive agents (Faraci et al. 1989).

Transcranial Doppler (TCD) has been used to investigate the occurrence of abnormal vascular reactivity in migraine (Thie et al. 1988; Gomez et al. 1988; Abernathy et al. 1990; McLaughlin et al. 1990; Diener et al. 1991; Caplan 1991; Ferbert et al. 1991; Rieke et al. 1992; Zanette et al. 1992). Studies of migraine patients in the headache-free phase have shown increased (Thie et al 1988; Abernathy et al. 1990; Rieke et al. 1992), decreased (Friberg et al. 1991) or normal (Gomez et al. 1988; Ferbert et al. 1991; Zanette et al. 1992) intracranial flow velocities (FV) and increased (Gomez et al. 1988; McLaughlin et al. 1990) or decreased reactivity to arterial P_{CO_2} changes. When compared with values obtained during headache-free intervals, FVs may increase (Thie et al. 1988; Zanette et al. 1992), decrease (Friberg et al. 1991; Diener et al. 1991; Zanette et al. 1992) or not change (Rieke et al. 1992) during attacks. Patients with aura who are studied within 6 hours of a migraine attack may have reduced FVs and increased pulsatility (suggestive of vasoconstriction) whereas patients without aura, studied somewhat later, may have increased diastolic FVs and reduced pulsatility (suggestive of vasodilation) (Zanette et al. 1992). Therapeutic agents such as flunarizine, ergotamine and sumatriptan have no effect on FVs during attacks (Diener et al. 1991), whereas nitroglycerin may decrease FVs (Babikian 1993). These results are difficult to reconcile. Discrepancies may be the result of studying non-homogeneous populations of migraine sufferers, lack of available data on continuous FV values during the aura and headache phases of attacks, small numbers of patients, methodological differences, spontaneous variations in vessel caliber and resultant FVs, and other factors (Babikian 1993).

Call's Syndrome

In 1988, Call et al. described 4 patients with prolonged reversible segmental cerebral vasoconstriction and reviewed 15 other patient reports. Symptoms were frequently abrupt in onset and consisted of headache, altered consciousness, fluctuating and multifocal signs and seizures. Cerebral angiography showed segmental proximal and distal narrowing and dilation with intervening segments of normal caliber. These vascular changes were diffuse, bilateral and severe, and resolved in 3 of 4 patients with or without specific therapy. Two were treated for cerebral vasculitis (1 with a negative biopsy after 6 months of immunosuppressive therapy and 1 with possible hypersensitivity angiitis), 1 had a 10-year history of vascular headaches with visual prodrome, and 1 was clinically improved at 8 months after onset but had no follow-up angiography. The authors also mentioned that similar cases occurred following carotid endarterectomy, during Guillain–Barré neuropathy and postpartum. The disorder, although of heterogeneous nature and various nosologies, appears to be more common in young women, suggesting a hormonal basis. This syndrome may represent a more severe form of migraine (Call et al. 1988). However, we do not now know if these patients represent an atypical form of cerebral vasculitis (Nadeau and Watson 1983), a severe form of migraine-induced vasoconstriction (Welch and Levine 1990), vasoconstriction associated with licit or illicit drug use/abuse (Sloan 1993b), or other disorders.

Licit and Illicit Drug Use/Abuse

The occurrence, pathophysiology and mechanisms of stroke associated with licit and illicit drug use/abuse have recently been reviewed (Gorelick 1990; Sloan 1993b). Recent data on the vascular effects, as well as types and mechanisms of cerebrovascular disease associated with licit and illicit drugs derive from experimental studies (Rumbaugh et al. 1971, 1976; Alksne and Greenhoot 1974; Fayad et al. 1992; Vitullo et al. 1989; Trouvé et al. 1990; Huang et al. 1990; Wang et al. 1990; Dohi et al. 1990; Kreupfer and Branch 1990; Powers and Madden 1990; Mena et al. 1990; Billman 1990; Egashira et al. 1991), case reports (Armstrong and Hayes 1961; Wooten et al. 1983; Boyko et al. 1987; Glick et al. 1987; Rothrock et al. 1988a; Krendel et al. 1990; Fredericks et al. 1991), case series (Kase et al. 1987; Levine et al. 1990; Daras et al. 1991; Dhuna et al. 1991; Petersen et al. 1991; Sloan et al. 1991) and review articles (Gorelick 1990; Sloan 1993b). The list of drugs associated with acute cerebrovascular disease includes ethanol (Gorelick 1990; Sloan 1993b), heroin (Citron et al. 1970; Gorelick 1990; Sloan et al. 1991; Sloan 1993b), cocaine (including hydrochloride, 'crack', and 'freebase') (Citron et al. 1970; Vitullo et al. 1989; Billman 1990; Dohi et al. 1990; Gorelick 1990; Krendel et al. 1990; Kreupfer and Branch 1990; Mena et al. 1990; Powers and Madden 1990; Levine et al. 1990; Petersen et al. 1991; Daras et al. 1991; Fredericks et al. 1991; Sloan et al. 1991; Huang et al. 1990; Sloan 1993b; Wang et al. 1990; Egashira et al. 1991), amphetamines (dextroamphetamine, methamphetamine and methylphenidate) (Citron et al. 1970; Rumbaugh et al. 1971, 1976; Rothrock et al. 1988a; Gorelick 1990; Wang et al. 1990; Sloan et al. 1991; Sloan 1993b), lysergic acid diethylamide (LSD) (Gorelick 1990; Sloan 1993b), phencyclidine (PCP) (Boyko et al. 1987; Sloan et al. 1991; Gorelick 1990; Sloan 1993b), over-the-counter (OTC) sympathomimetic drugs (phenylpropanolamine (PPA), ephedrine (EPH), pseudoephedrine (PS-EPH), and congeners) (Wooten et al. 1983; Glick et al. 1987; Kase et al. 1987; Gorelick 1990; Sloan et al. 1991; Sloan 1993b) T's and blues (tripelennamine and pentazocine) (Caplan et al. 1982) and perhaps marijuana (Sloan 1993b). All subtypes of transient and permanent ischemic events in the retinal, spinal, anterior and posterior circulations, as well as all subtypes of hemorrhagic stroke, have been reported.

The mechanism(s) by which the various drugs produce acute cardiovascular and cerebrovascular disease may be explained in part by their pharmacologic effects on neurotransmitters and physiologic actions (Sloan 1993b). Most of these compounds (heroin, cocaine, amphetamines, OTC sympathomimetics, PCP and LSD) gain direct entry into the central nervous system (CNS). For the OTC sympathomimetics, compounds devoid of —OH groups tend to cross the blood–brain barrier and have enhanced CNS activity. Substitution at the α-carbon blocks oxidation by monoamine oxidase (MAO), while β-carbon substitution increases agonist activity at α- and β -receptors and reduces stimulant action. These structure–activity relationships explain why some compounds are likely to have clinically important CNS effects. Many drugs either increase catecholamine release from central noradrenergic nerve terminals (amphetamines, PPA, EPH), block reuptake of catecholamines (cocaine, PCP) or increase plasma catecholamine levels (ethanol, cocaine, T's and blues). Sometimes, similar effects on central serotonergic and dopaminergic neurotransmission (amphetamine, cocaine and PCP) are observed. In cats (Edvinsson et al. 1975), cocaine treatment may lead to prolonged increased prejunctional sensitivity of the cerebral arteries. T's and blues may antagonize central cholinergic function. With the exception of heroin and LSD, all drugs of interest may produce mild to profound blood pressure elevation, at times with bradycardia or tachycardia. Peripheral vasoconstriction may occur with cocaine, amphetamines and ephedrine.

Many drugs have effects upon the cerebral vasculature. Vasospasm may be caused by ethanol (Gorelick 1990), cocaine (Gorelick 1990; Levine et al. 1990; Sloan 1993b), amphetamines (Rumbaugh et al. 1971, 1976; Wang et al. 1990) and PPA (Kase et al. 1987; Sloan 1993b). Evidence from in-vitro and in-vivo experiments as well as human subjects suggests that cocaine not only potentiates the physiologic response to catecholamines but also may act independently to produce vasoconstriction of vascular smooth muscle. In isolated ferret aorta, cocaine concentrations $<10^{-4}$ M led to an endothelium-independent but dose-dependent contractile response which could be antagonized by the α_2-antagonist prazosin (Egashira et al. 1991). Pathologic changes consistent with disturbed Ca^{2+} homeostasis (e.g. myocardial contraction bands) are found more frequently in the hearts of individuals who died from acute cocaine toxicity than in those dying from other drug-related causes (Billman 1990). Cocaine causes an increase in intracellular Ca^{2+} levels and enhances spontaneous release of Ca^{2+} from the sarcoplasmic reticulum (Billman 1990). In coronary arteries and arterioles <65 μm in diameter (Vitullo et al. 1989; Billman 1990), cocaine and its norcocaine metabolite cause an endothelium-independent contractile response, with a resultant decrease in coronary blood flow. These effects are antagonized by the Ca^{2+} antagonists diltiazem and nitrendipine, but not the α_1-antagonist phentolamine (Vitullo et al. 1989; Billman 1990).

Recent experiments in rats (Kreupfer and Branch 1990) and cats (Powers and Madden 1990) and evaluation of acute cocaine users with single photon emission computed tomography (SPECT) (Mena et al. 1990) suggest that cocaine and its metabolites benzoylecgonine (Powers and Madden 1990) and norcocaine (Powers and Madden 1990) may cause transient (Kreupfer and Branch 1990) or prolonged (Mena et al. 1990) cerebral vasoconstriction. Early experiments in rats showed that cortical pial arteries, arterioles, veins and venules could be affected. Increasing concentrations of cocaine hydrochloride led to capillary rupture. More recently, venular vasospasm and postcapillary microhemorrhages were also demonstrated (Huang et al. 1990). Magnesium, a natural Ca^{2+} antagonist, produces dose-dependent inhibition (20%–85%) of cocaine -induced arteriolar spasms and attenuation (up to 85%–90%) of venular vasospasm and microhemorrhages (Huang et al. 1990). In a recent angiographic study in rabbits (Wang et al. 1990), low doses of intravenous cocaine, methamphetamine or both led to mild vasodilation. High doses of cocaine or methamphetamine alone led to little basilar artery VSP. Coadministration of high doses of cocaine and methamphetamine led to severe basilar artery VSP. However, in anesthetized cats (Dohi et al. 1990), topical application of cocaine in concentrations of 10^{-6} M and 10^{-5} M led to pial arteriolar dilation, whereas a concentration of 10^{-4} M led to dilation of all arterioles. This action was eliminated by intravenous propranolol. These discrepant findings may be explained in part by species differences, varying methodologies and dose regimens and other factors. The vasoconstrictor effect of cocaine may be due to a mechanism similar to that seen in peripheral vascular smooth muscle (Vitullo et al. 1989; Billman 1990; Egashira et al. 1991) and cardiac muscle (Billman 1990).

Other drugs have significant vasoactive effects. Phencyclidine, LSD and mescaline each produce a contractile response when directly applied to isolated canine basilar and middle cerebral arteries (Gorelick 1990). Experimental studies in monkeys suggest that intrathecal norepinephrine may induce biphasic VSP that is morphologically similar to the vasculopathy seen after SAH (Alksne and Greenhoot 1974). In monkeys, intravenous methamphetamine produces immediate (<10 minutes) decrease in vessel caliber in some animals, whereas others develop SAH with generalized VSP, multifocal infarction, edema, petechial hemorrhages and microaneurysms (Rumbaugh et al. 1971) similar to those seen in drug addicts with 'necrotizing angiitis' (Citron et al. 1970). In monkeys and rats, exposure to methamphetamine and methylphenidate for periods ranging from 1 week to 1 year produces vascular occlusion, reduced vessel caliber, poor vascular filling, thrombosis, acute to subacute neuronal changes and microhemorrhages (Rumbaugh et al. 1976).

Patients may develop symptoms after chronic abuse, 'binge' use, re-exposure after prolonged abstinence, or even first exposure to the offending agent. Presenting symptoms include vascular headache, depressed level of consciousness and focal neurological signs, such as hemiparesis and seizures (Sloan 1993b). The association between drug use and headaches has received increased attention in substance abuse treatment programs (El-Mallakh et al. 1991; Dhopesh et al. 1991) and the acute hospital setting (Dhuna et al. 1991): 89.5% of patients had headaches at some point in their lives in one study (El-Mallakh et al. 1991), and 49.6% of patients had headaches in another study (Dhopesh et al. 1991). In one study (El-Mallakh et al. 1991), 34% had a history of migraine headaches without aura. However, patients who abuse drugs and have no personal or family history of migraine may develop headache with migrainous features. Patterns of headache associated with cocaine use include severe vascular headaches without neurological complications immediately after drug ingestion, frequent headaches of increasing severity during 'binge' use, and increasingly severe headaches with acute drug withdrawal (Dhuna et al. 1991). Detailed assessment of migraine precipitants is lacking in these studies (El-Mallakh et al. 1991; Dhopesh et al. 1991; Dhuna et al. 1991) and the nature and significance of the association between drug use, migraine and stroke remains to be determined (Sloan 1993b).

Transient or chronic arterial vasoconstriction has been suspected or demonstrated in normal subjects and patients with stroke following drug use. A patient with pheochromocytoma and elevated plasma catecholamine levels had segmental constriction in all secondary and tertiary vessels on cerebral angiography (Armstrong and Hayes 1961). In one study (Fayad et al. 1992), a 15% increase in pulsatility index was found during TCD monitoring of the effects of acute cocaine administration in normal subjects, suggesting an increase in cerebrovascular resistance. Extracranial and intracranial large-vessel vasculopathy, either stenosis or occlusion, has been observed in patients using heroin (Gorelick 1990; Sloan 1993b), cocaine (Levine et al. 1990; Daras et al. 1991; Petersen et al. 1991), amphetamines (Rothrock et al. 1988a), LSD (Gorelick 1990; Sloan 1993b), PPA (Kase et al. 1987; Glick et al. 1987), EPH (Wooten et al. 1983), PS-EPH (Sloan 1993b) and T's and blues (Caplan et al. 1982). In patients with chronic vasculopathy, segmental and/or multifocal narrowing of the vessels might be due to either a primary toxic effect or perhaps a complication of unrecognized SAH (Sloan 1993b). Beading of vessels or 'arteritis' has been described in patients who abuse heroin (Gorelick 1990; Sloan 1993b), various amphetamines (Citron et al. 1970; Rothrock et al. 1988a), PPA (Kase et al. 1987; Glick et al. 1987), EPH (Wooten et al. 1983), PS-EPH (Sloan 1993b) and T's and blues (Caplan et al. 1982). Cocaine-associated ischemic stroke might be due to large-vessel (extracranial and intracranial) stenotic or occlusive vascular disease (Levine et al. 1990; Daras et al. 1991; Petersen et al. 1991), superficial pial vessel occlusion (Levine et al. 1990; Sloan et al. 1991; Daras et al. 1991; Petersen et al. 1991), cardiogenic embolism (Sloan 1993b), perforating vessel occlusion (Levine et al. 1990; Sloan et al. 1991; Peterson et al. 1991), vasculitis (Krendel et al. 1990; Fredericks et al. 1991) and triggering expression of meningovascular syphilis (Sloan 1993b). Interestingly, subcortical (Levine et al. 1990; Sloan et al. 1991; Petersen et al. 1991) and lacunar (Levine et al. 1990; Sloan et al. 1991) lesions were found in as many as 50%–63% of infarcts in recent case

Whether this latter finding reflects intracranial small-vessel vasospasm (Vitullo et al. 1989) is unknown.

There are few studies of the vascular pathology in drug-induced cerebrovascular disease. Levine et al. (1990) reported a patient who died following hemorrhagic infarctions in the left anterior and middle cerebral artery territories. The left intracranial carotid, middle cerebral and anterior cerebral arteries were smaller than their right-sided counterparts. Histopathological examination of these vessels showed normal endothelium, intima, media, elastic lamina and adventitia. Whether these findings reflect vasoconstriction or reduced vessel caliber related to decreased demand for blood flow is unknown. Krendel et al. (1990) reported two cocaine users who developed cerebral vasculitis. The 'crack' user had a biphasic course, with dysarthria, confusion and right hemiparesis. A right temporal lobe biopsy showed acute vasculitis involving small cortical vessels and multinucleated giant cells in an area of infarction. The intravenous cocaine user developed a non-dominant syndrome in the setting of Westergren ESR = 108 mm/h and positive Hb_sAg and core antibody. A right frontal cortical and leptomeningeal biopsy showed vasculitis involving small cortical vessels. Fredericks et al. (1991) described a patient who used intranasal and intravenous cocaine and ethanol who developed encephalopathy, aphasia and bilateral corticospinal tract abnormalities associated with CSF protein = 185 mg/dl. Right front biopsy revealed lymphocytic infiltration in perivascular collections and within the walls of arterioles, endothelial swelling of small arterioles and scattered interstitial edema. It appears that cocaine may be associated with a non-necrotizing non-leukocytoclastic small-vessel arteritis (Fredericks et al. 1991) or trigger expression of viral infection-associated vasculitis (Nadeau and Watson 1983; Krendel et al. 1990).

It is also possible that a vasculopathy is induced by direct toxic or immunologic mechanisms (Sloan 1993b). Glick et al. (1987) reported a patient with intracerebral and subarachnoid hemorrhage following PPA use. Brain and leptomeningeal biopsy revealed necrotizing vasculitis of small arteries and veins, with intimal infiltration of polymorphonuclear leukocytes, vessel wall thickening and luminal reduction, fragmentation of the elastic lamina and vessel occlusion. Wooten et al. (1983) reported a patient with intracerebral hemorrhage who had leukocytoclastic vasculitis present on a skin biopsy. Shibata et al. (1991) reported a patient who developed fatal massive subarachnoid and intracerebral hemorrhage after intravenous methamphetamine abuse. Angiography showed non-filling of intracranial vessels and extravasation of contrast from the right pericallosal artery. Cross-sections of many arteries showed various degrees of medial necrosis but no inflammatory cells. Finally, Boyko et al. (1987) described a patient with fatal subarachnoid hemorrhage following PCP abuse. Autopsy showed a 1 mm perforation in the basilar artery surrounded by a hyperemic border, with rare inflammatory cells but no vasculitis.

Frequent abuse of multiple drugs and/or ethanol by individual patients tends to confound efforts to assign cause and effect to specific agents (Sloan 1993b). More pathological data from larger numbers of patients are needed to clarify the nature of the vasculopathy that occurs in drug abusers.

Eclampsia and Postpartum Cerebral Angiopathy

The historical, pathophysiological, clinical, radiological, pathological and therapeutic aspects of toxemia of pregnancy (eclampsia) have recently been reviewed (Donaldson 1989). Eclampsia is diagnosed in a pregnant or peripartum woman when blood pressure is >160 mmHg systolic or >110 mmHg diastolic on two occasions 6 hours apart; proteinuria is 4+ on dipstick or >5 g/24 h; there is oliguria <400 ml/24 h; there are stupor, headache, focal deficits, scotomata, blurred vision or seizures; there is pulmonary edema or cyanosis; there is unexplained hepatic dysfunction or right upper quadrant pain; and there is thrombocytopenia. Neuropathological findings include cortical petechiae, microinfarcts, subcortical hemorrhages, cerebral edema and small amounts of subarachnoid hemorrhage. The rapid increases in blood pressure contribute to hypertensive encephalopathy with arteriolar vasoconstriction which, if the eclampsia is untreated, permits precapillary thrombosis and fibrinoid thrombi, and leads to brain microinfarcts and central retinal artery occlusion. There is also enhanced vascular sensitivity to endogenous pressor peptides and amines and increased cerebrovascular resistance (Will et al. 1987).

Large cerebral artery vasoconstriction can also occur in eclampsia (Manelfe et al. 1983; Will et al. 1987; Brick 1988a; Trommer et al. 1988). Will et al. (1987) described 3 patients with toxemia who developed acute neurological deterioration in the immediate postpartum period. Patients 1 and 3 had subarachnoid hemorrhage on CSF examination and CT scan, respectively. Cerebral angiography revealed marked diffuse cerebral vasoconstriction of all intracranial vessels on day 1 and days 4–5, respectively. Patient 1 had multiple infarcts in all vascular territories, and a follow-up cerebral angiogram 10 months later was normal; symptoms and signs resolved completely within 3 days. Patient 2 developed symptoms 10 hours after delivery and CT scan showed decreased attenuation in the cerebral hemispheric border zones bilaterally. Lumbar puncture was not performed. Cerebral angiography showed delayed and poor filling of terminal pial artery branches. Seven days later, she was normal. Brick (1988a) described a patient with a cardiac murmur who developed transient ischemic attacks in the final days of pregnancy, associated with a 70% stenosis of the left MCA. Symptoms resolved after delivery and systemic anticoagulation. Repeat angiography 3 months later was normal. Trommer et al. (1988) described a patient who developed hypertension, decreased consciousness and gaze palsy postpartum. Initial CT scan was negative, but follow-up CT scan on day 2 showed lucency in the upper pons and midbrain. Cerebral angiography on day 4 showed diffuse cerebral vasoconstriction in small and medium-sized vessels of the anterior and posterior circulations. Lumbar puncture was not performed. Five days later, she had normal blood pressure and a normal neurological examination.

The diagnosis and pathophysiology of cerebral vasoconstriction complicating toxemia are controversial. Some authors do not believe that cerebral vasoconstriction occurs in this setting (Goodlin 1988). However, the angiographic data are convincing. The etiology may well be accompanying SAH (Raroque 1989), since it was documented in cases 1 and 3 of Will et al. (1987). However, lumbar puncture was not performed in case 2 of Will et al. (1987), the case of Brick (1988a) and the case of Trommer et al. (1988). As shown in case 1 of Will et al. (1987), SAH may be present even if the CT scan does not demonstrate its presence. Interestingly, the vasoconstriction seems to have occurred very early after onset – days 1 or 2 – which is unusual after aneurysmal SAH. This might reflect the combined effect of SAH and enhanced reaction to vasoactive peptides and amines (Will et al 1987; Findlay et al. 1991a; Macdonald and Weir 1991). The case of Brick (1988a) and the series of Manelfe et al. (1983) suggest that the vasoconstriction may be due to female reproductive steroids which lead to reversible intimal proliferation (Brick 1986b). However, in the case of Brick (1988a), the angiogram shows a filling defect which might be an embolus. Negative transthoracic echocardiography does not rule out the presence of occult sources of cardiogenic emboli. With regard to therapeutics, the cases of Will et al. (1987) and Trommer et al. (1988) developed symptoms despite significant Mg^{2+} levels. This suggests that either the Mg^{2+} level was inadequate, Mg^{2+} was ineffective because of endothelial impairment (Szabo et al. 1991), or that other factors are involved. Steroids were used in the cases of Will et al. (1987), but the role of these agents in this setting is unclear.

Barinagarrementeria et al. (1992) described a patient without toxemia who, after cesarean section and delivery of the placenta, received intravenous ergonovine. Within seconds the patient developed severe headache, confusion, blindness and right hemisensory deficit. MRI scan on day 4 showed bilateral temporo-occipital and left thalamic infarctions. Cerebral angiograms showed vasoconstriction of the basilar and right posterior inferior cerebellar arteries and non-filling of the posterior cerebral arteries. Three months later the vasoconstriction had partially resolved and there was partial filling of the left posterior cerebral artery. Similar cases have been reported by Rascol et al. (1980), Henry et al. (1984) and Bogousslavsky et al. (1989). In the case of Bogousslavsky et al. (1989), resolution of the vasoconstriction was suggested by serial TCD examinations.

These cases confirm that ergonovine use may cause arterial vasoconstriction and cerebral infarction. The contribution of other vasoconstrictor drugs, such as cocaine and OTC sympathomimetics, to the occurrence of cerebral constriction in this setting should be investigated. Ergonovine and other ergot agents should be given with caution to pregnant patients with migraine, Raynaud's phenomenon or other conditions which may predispose to VSP (Barinagarrementeria et al. 1992). The effectiveness of Ca^{2+}-channel antagonists or Mg^{2+} in this setting remains to be established.

References

Abernathy M, Wieneke J, Ramos M et al. (1990) Transcranial Doppler: intracranial blood flow velocities in headache-

free migraineurs and non-headache-prone volunteers. Neurology 40:213
Alksne JF, Greenhoot JH (1974) Experimental catecholamine-induced chronic cerebral vasospasm. J Neurosurg 41:440–445
Alksne JF, Branson PJ, Bailey M (1986) Modification of experimental post-subarachnoid hemorrhage vasculopathy with intracisternal plasmin. Neurosurgery 19:20–25
Alksne JF, Branson PJ, Bailey M (1988) Modification of experimental post-subarachnoid vasculopathy with intracisternal plasmin. Neurosurgery 23:335–337
Armstrong FS, Hayes GJ (1961) Segmental cerebral arterial constriction associated with pheochromocytoma. J Neurosurg 18:843–846
Babikian VL (1993) Transcranial Doppler evaluation of patients with ischemic cerebrovascular disease. In: Babikian VL, Wechsler LR (eds) Transcranial Doppler ultrasonography. BC Decker/Mosby, St. Louis, pp 87–104
Ballon BJ, Wei EP, Kontos HA (1986) Superoxide anion radical does not mediate vasodilation of cerebral arterioles by vasoactive intestinal polypeptide. Stroke 17:1287–1290
Barinagarrementeria F, Cantú C, Balderrama J (1992) Postpartum cerebral angiography with cerebral infarction due to ergonovine use. Stroke 23:1364–1366
Bauknight GC, Faraci FM, Heistad DD (1992) Endothelium-derived relaxing factor modulates noradrenergic constriction of cerebral arterioles in rabbits. Stroke 23:1522–1526
Bevan JA, Gaw AJ (1992) Flow-dilation of rabbit cerebral arteries is associated with a muscle-derived relaxing factor. Stroke 23:155
Bevan JA, Bevan RD, Frazee JG (1987) Functional arterial changes in chronic cerebrovasospasm in monkeys. An in vitro assessment of the contribution to arterial narrowing. Stroke 18:472–481
Bevan RD, Bevan JA, Frazee JG (1988) Diltiazem protects against functional changes in chronic cerebrovasospasm in monkeys. Stroke 19:73–79
Billman GE (1990) Mechanisms responsible for the cardiotoxic effects of cocaine. FASEB J 4:2469–2475
Bogousslavsky J, Regli F, Van Melle G, Payot M, Uske A (1988) Migraine stroke. Neurology 38:223–227
Bogousslavsky J, Despland PA, Regli F, Dubuis PY (1989) Postpartum cerebral angiopathy: reversible vasoconstriction assessed by transcranial Doppler ultrasound. Eur Neurol 29:102–105
Boyko OB, Burger PC, Heinz ER (1987) Pathological and radiological correlation of subarachnoid hemorrhage in phencyclidine abuse. J Neurosurg 67:446–448
Brayden JE, Bevan JA (1986) Evidence that vasoactive intestinal polypeptide (VIP) mediates neurogenic vasodilation of feline cerebral arteries. Stroke 17:1189–1192
Brick JF (1988a) Vanishing cerebrovascular disease of pregnancy. Neurology 38:804–806
Brick JF (1988b) Cerebrovascular disease in the postpartum period. Stroke 19:1572
Brinker T, Seifert V, Stolke D (1991) Effect of intrathecal fibrinolysis on cerebrospinal fluid absorption after experimental subarachnoid hemorrhage. J Neurosurg 74:789–793
Broderick JP, Swanson JW (1987) Migraine-related strokes. Clinical profile and prognosis in 20 patients. Arch Neurol 44:868–871
Buzzi MG, Moskowitz MA (1990) The antimigraine drug, sumatriptan (GR43175), selectively blocks neurogenic plasma extravasation from blood vessels in dura mater. Br J Pharmacol 99: 202–206
Buzzi MG, Carter WB, Shimizu T, Heath H, Moskowitz MA (1991) Dihydroergotamine and sumatriptan attenuate levels of CGRP in plasma in rat superior sagittal sinus during electrical stimulation of the trigeminal ganglion. Neuropharmacology 30:1193–1200
Call GK, Fleming MC, Sealfon S, Levine H, Kistler JP, Fisher CM (1988) Reversible cerebral segmental vasoconstriction. Stroke 19:1159–1170
Caplan LR (1991) Migraine and vertebrobasilar ischemia. Neurology 41:55–61
Caplan LR, Thomas C, Banks G (1982) Central nervous system complications of addiction to 'T's and Blues'. Neurology 32:623–628
Chehrazi BB, Giri S, Joy RM (1989) Prostaglandins and vasoactive amines in cerebral vasospasm after aneurysmal subarachnoid hemorrhage. Stroke 20:217–224
Chimowitz MI, Pessin MS (1987) Is there a role for heparin in the management of complications of subarachnoid hemorrhage? Stroke 18:1169–1172
Chyatte D (1989) Prevention of chronic cerebral vasospasm in dogs with ibuprofen and high-dose methylprednisolone. Stroke 20:1021–1026
Chyatte D, Sundt TM (1984) Cerebral vasospasm after subarachnoid hemorrhage. Mayo Clin Proc 59:498–505
Citron BP, Halpern M, McCarron M et al. (1970) Necrotizing angiitis associated with drug abuse. New Engl J Med 283:1003–1111
Comair YG, Schipper HM, Brem S (1993) The prevention of oxyhemoglobin-induced endothelial and smooth muscle cytoskeletal injury by deferoxamine. Neurosurgery 32:58–65
Daras M, Tuchman AJ, Marks S (1991) Central nervous system infarction due to cocaine use. Stroke 22:1320–1325
Debdi M, Seylaz J, Sercombe R (1992) Early changes in rabbit cerebral artery reactivity after subarachnoid hemorrhage. Stroke 23:1154–1162
Dhopesh V, Maany I, Herring C (1991) The relationship of cocaine to headache in polysubstance abusers. Headache 31:17–19
Dhuna A, Pascual-Leone A, Belgrade M (1991) Cocaine-related vascular headaches. J Neurol Neurosurg Psych 64:803–806
Diener HC, Peters C, Rudzio M et al. (1991) Ergotamine, flunarizine, and sumatriptan do not change cerebral blood flow velocity in normal subjects and migraineurs. J Neurol 238:245–250
Dohi S, Jones MD, Hudak ML, Traystman RJ (1990) Effects of cocaine on pial arterioles in cats. Stroke 21:1710–1714
Donaldson JO (1989) Eclampsia. In: Donaldson JO (ed) Neurology of pregnancy. W.B. Saunders, Philadelphia, pp 269–310
Duff TA, Scott G, Feilbach JA (1986) Ultrastructural evidence of arterial denervation following experimental subarachnoid hemorrhage. J Neurosurg 64:292–297
Duff TA, Feilbach JA, Scott G (1987) Does cerebral vasospasm result from denervation supersensitivity? Stroke 18:85–91
Duff TA, Louie J, Feilbach JA, Scott G (1988) Erythrocytes are essential for development of cerebral vasculopathy resulting from subarachnoid hemorrhage in cats. Stroke 19:68–72
Edmeans J (1991) What is migraine? controversy and stalemate in migraine pathophysiology. J Neurol 238:S2–S5
Edvinsson L (1987) Innervation of the cerebral circulation. Ann NY Acad Sci 519:334–348
Edvinsson L, MacKenzie ET (1977) Amine mechanisms in the cerebral circulation. Pharmacol Rev 28:275–348
Edvinsson L, Aubineau P, Owman C, Sercombe R, Seylaz J (1975) Sympathetic innervation of cerebral arteries: prejunctional supersensitivity to norepinephrine after sympathectomy or cocaine treatment. Stroke 6:525–530

Edvinsson L, Fredholm BB, Hamel E, Jansen I, Verrecchia C (1985) Perivascular peptides relax cerebral arteries concomitant with stimulation of cyclic adenosine monophosphate accumulation or release of an endothelium-derived relaxing factor in the cat. Neurosci Lett 58:213–217

Edvinsson L, Delgado-Zygmunt T, Ekman R, Jansen I, Svendgaard N-Aa, Uddman R (1990) Involvement of perivascular sensory fibers in the pathophysiology of cerebral vasospasm following subarachnoid haemorrhage. J Cerebr Blood Flow Metab 10:602–607

Edvinsson L, Ekman R, Jansen I, Kingman TA, McCulloch J, Uddman R (1991) Reduced levels of calcitonin gene-related peptide-like immunoreactivity in human blood vessels after subarachnoid haemorrhage. Neurosci Lett 121:151–154

Edwards DH, Byrne JV, Griffith TM (1992) The effect of chronic subarachnoid hemorrhage on basal endothelium-derived relaxing factor activity in intrathecal cerebral arteries. Neurosurgery 76:830–837

Egashira K, Morgan KG, Morgan JP (1991) Effects of cocaine on excitation–contraction coupling of aortic smooth muscle from the ferret. J Clin Invest 87:1322–1328

Ellison MD, Erb DE, Kontos HA, Povlishock JT (1989) Recovery of impaired endothelium-dependent relaxation after fluid-percussion brain injury in cats. Stroke 20:911–917

El-Mallakh RS, Kranzler HR, Kamanitz JR (1991) Headaches and psychoactive substance use. Headache 31:584–587

European CGRP in Subarachnoid Haemorrhage Study Group (1992) Effect of calcitonin-gene-related peptide in patients with delayed postoperative cerebral ischemia after aneurysmal subarachnoid haemorrhage. Lancet 339:831–834

Faraci FM (1990) Role of nitric oxide in regulation of basilar artery tone in vivo. Am J Physiol 259:H1216–H1221

Faraci FM, Williams JK, Breese KR, Armstrong ML, Heistad DD (1989) Atherosclerosis potentiates constrictor responses of cerebral and ocular blood vessels to thromboxane in monkeys. Stroke 20:242–247

Faraci FM, Armstrong ML, Heistad DD (1991a) Dietary treatment of atherosclerosis abolishes hyperresponsiveness of retinal blood vessels to serotonin in monkeys. Stroke 22:1405–1408

Faraci FM, Mayhan WG, Heistad DD (1991b) Responses of rat basilar artery to acetylcholine and platelet products in vivo. Stroke 22:56–60

Fayad PB, Price LH, McDougle CJ, Pavalkis FJ, Brass LM (1992) Acute hemodynamic effects of intranasal cocaine on the cerebral circulation. Stroke 23:457

Ferbert A, Busse D, Thron A (1991) Microinfarction in classic migraine? a study with magnetic resonance imaging findings. Stroke 22:1010–1014

Findlay JM, Weir BKA, Steinke D et al. (1988) Effect of intrathecal thrombolytic therapy on subarachnoid clot and chronic vasospasm in a primate model of SAH. J Neurosurg 69:723–735

Findlay JM, Weir BKA, Gordon P et al. (1989a) Safety and efficacy of intrathecal thrombolytic therapy in a primate model of cerebral vasospasm. Neurosurgery 24:491–498

Findlay JM, Weir BKA, Kanamaru K et al. (1989b) Intrathecal fibrinolytic therapy after subarachnoid hemorrhage: dosage study in a primate model and review of the literature. Can J Neurol Sci 16:28–40

Findlay JM, Weir BKA, Kanamaru K et al. (1990) The effect of timing of intrathecal fibrinolytic therapy on cerebral vasospasm in a primate model of subarachnoid hemorrhage. Neurosurgery 26:201–206

Findlay JM, Macdonald RL, Weir BKA (1991a) Current concepts of pathophysiology and management of cerebral vasospasm following aneurysmal subarachnoid hemorrhage. Cerebrovasc Brain Metab Rev 3:336–361

Findlay JM, Weir BKA, Kassell NF et al. (1991b) Intracisternal recombinant tissue plasminogen activator after aneurysmal subarachnoid hemorrhage. J Neurosurg 75:181–188

Fisher CM, Kistler JP, Davis JM (1980) Relation of cerebral vasospasm to subarachnoid hemorrhage visualized by computed tomographic scanning. Neurosurgery 6:1–9

Fredericks RK, Lefkowitz DS, Challa VR, Troost BT (1991) Cerebral vasculitis associated with cocaine abuse. Stroke 22:1437–1439

Friberg L (1991) Cerebral blood flow changes in migraine: methods, observations and hypotheses. J Neurol 238:S12–S17

Friberg L, Olesen J, Iversen H et al. (1991) Migraine pain associated with middle cerebral artery dilation: reversal by sumatriptan. Lancet 338:13–17

Fujiwara S, Kassell NF, Sasaki T, Nakagomi T, Lehman RM (1986) Selective hemoglobin inhibition of endothelium-dependent vasodilation of rabbit basilar artery. J Neurosurg 64:445–452

Furchgott RF (1983) Role of endothelium in responses of vascular smooth muscle. Circ Res 53:557–573

Furchgott RF, Vanhoutte PM (1989) Endothelium-derived relaxing and contracting factors. FASEB J 3:2007–2018

Furchgott RF, Zawadzki JV (1980) The obligatory role of endothelial cells in the relaxation of arterial smooth muscle by acetylcholine. Nature 288:373–376

Gaetani P, Marzatico F, Rodriguez y Baena R et al. (1990) Arachidonic acid metabolism and pathophysiologic aspects of subarachnoid hemorrhage in rats. Stroke 21:328–332

Gaetani P, Marzatico F, Lombardi D, Adinolfi D, Rodriguez y Baena R (1991) Effect of high dose methylprednisolone and U74006F on eicosanoid synthesis after subarachnoid hemorrhage in rats. Stroke 22:215–220

Germann P, Laher I, Bevan JA (1991) Platelets augment rabbit cerebral artery constriction by activating protein kinase C. Stroke 22:1534–1540

Glick R, Hoying J, Cerullo L, Perlman S (1987) Phenylpropanolamine: an over-the-counter drug causing central nervous system vasculitis and intracerebral hemorrhage. Neurosurgery 20:969–974

Goadsby PJ, Edvinsson L, Ekman R (1988) Release of vasoactive peptides in the extracerebral circulation of humans and the cat during activation of the trigeminovascular system. Ann Neurol 23:193–196

Gomez CR, Gomez SM, Horenstein S (1988) Abnormal cerebral vasoreactivity in twins with migraine and stroke. Ann Neurol 24:158

Goodlin RC (1988) Cerebral vasospasm and eclampsia. Stroke 19:1446

Gorelick PB (1990) Stroke from alcohol and drug abuse. Postgrad Med 88:171–178

Haley EC, Torner JC, Kassell NF et al. (1990) Cooperative randomized study of nicardipine in subarachnoid hemorrhage: preliminary report. In: Sano K, Takakura K, Kassell NF, Sasaki T (eds) Cerebral vasospasm. Tokyo University Press, Tokyo, pp 519–525

Haley EC, Kassell NF, Torner JC, Kongable G (1991) Nicardipine ameliorates angiographic vasospasm following aneurysmal subarachnoid hemorrhage (SAH). Neurology 41(Suppl I):I–346

Hamann G, Isenberg E, Strittmatter M, Schimrigk K (1993) Absence of big endothelin in subarachnoid hemorrhage. Stroke 24:383–386

Hara H, Nosko M, Weir B (1986) Cerebral perivascular nerves

in subarachnoid hemorrhage. A histochemical and immunohistochemical study. J Neurosurg 65:531–539

Harada T, Mayberg MR (1992) Inhibition of delayed arterial narrowing by the iron-chelating agent deferoxamine. J Neurosurg 77:763–767

Hatake K, Kakishita E, Wakabayashi I, Sakiyama N, Hishida S (1990) Effect of aging on endothelium-dependent vascular relaxation of isolated human basilar artery to thrombin and bradykinin. Stroke 21:1039–1043

Hatake K, Wakabayashi I, Kakishita E, Hishida S (1992) Impairment of endothelium-dependent relaxation in human basilar arteries after subarachnoid hemorrhage. Stroke 23:1111–1117

Henry PY, Larre P, Aupy M, Lafforgue JR, Orgogozo JM (1984) Reversible cerebral arteriopathy associated with administration of ergot derivatives. Cephalalgia 4:171–178

Heros RC, Zervas NT, Varsos V (1983) Cerebral vasospasm after subarachnoid hemorrhage. Ann Neurol 14:599–608

Hongo K, Nakagomi T, Kassell NF et al. (1988a) Effects of aging and hypertension on endothelium-dependent vascular relaxation in rat carotid artery. Stroke 19:892–897

Hongo K, Ogawa H, Kassell NF et al. (1988b) Comparison of intraluminal and extraluminal inhibitory effects of hemoglobin on endothelium-dependent relaxation of rabbit basilar artery. Stroke 19:1550–1555

Hongo K, Tsukahara T, Kassell NF, Ogawa H (1989) Effect of subarachnoid hemorrhage on calcitonin gene-related peptide-induced relaxation in rabbit basilar artery. Stroke 20:100–104

Huang QF, Gebrewold A, Altura BT, Altura BM (1990) Cocaine-induced cerebral vascular damage can be ameliorated by Mg^{2+} in rat brains. Neurosci Lett 109:113–116

Humphrey PPA (1991) 5-Hydroxytryptamine and the pathophysiology of migraine. J Neurol 238:S38–S44

Ide K, Yamakawa K, Sasaki T, Nakagomi T, Saito I, Takakura K (1990) The role of endothelin in the pathogenesis of vasospasm following subarachnoid hemorrhage. Stroke 21(Suppl I):I-68

Iwasa K, Smith RR, Bernanke DH (1992) Non-muscle vascular constriction after subarachnoid hemorrhage. Role of growth factors derived from platelets. Stroke 23:153

Jansen I, Uddman R, Hocherman M et al. (1986) Localization and effects of neuropeptide Y, vasoactive intestinal polypeptide, substance P, and calcitonin gene-related peptide in human temporal arteries. Ann Neurol 20:496–501

Johnston FG, Bell BA, Robertson IJA et al. (1990) Effect of calcitonin-gene-related peptide on postoperative neurological deficits after subarachnoid haemorrhage. Lancet 335:869–872

Juul R, Edvinsson L, Fredriksen TA, Ekman R, Brubakk AO, Gisvold SE (1990a) Changes in the levels of neuropeptide Y-LI in the external jugular vein in connection with vasoconstriction following subarachnoid haemorrhage in man. Involvement of sympathetic neuropeptide Y in cerebral vasospasm. Acta Neurochir 107:75–81

Juul R, Edvinsson L, Gisvold SE, Ekman R, Brubakk AO, Fredriksen TA (1990b) Calcitonin gene-related peptide-LI in subarachnoid hemorrhage in man: signs of activation of the trigemino-cerebrovascular system. Br J Neurosurg 4:171–180

Juvela S, Kaste M, Hillbom M (1990a) Platelet thromboxane release after subarachnoid hemorrhage and surgery. Stroke 21:566–571

Juvela S, Kaste M, Hillbom M (1990b) Effect of nimodipine on platelet function in patients with subarachnoid hemorrhage. Stroke 21:1283–1288

Juvela S, Öhman J, Servo A, Heiskanen O, Kaste M (1991) Angiographic vasospasm and release of platelet thromboxane after subarachnoid hemorrhage. Stroke 22:451–455

Kanamaru K, Waga S, Kojima T, Fujimoto K, Itoh H (1987a) Endothelium-dependent relaxation of canine basilar arteries. Part I: difference between acetylcholine- and A23187-induced relaxation and involvement of lipoxygenase metabolite(s). Stroke 18:932–937

Kanamaru K, Waga S, Kojima T, Fujimoto K, Niwa S (1987b) Endothelium-dependent relaxation of canine basilar arteries. Part II: inhibition by hemoglobin and cerebrospinal fluid from patients with aneurysmal subarachnoid hemorrhage. Stroke 18:938–943

Kanamaru K, Weir BKA, Findlay JM, Krueger CA, Cook DA (1989) Pharmacologic studies on relaxation of spastic primate cerebral arteries in subarachnoid hemorrhage. J Neurosurg 71:909–915

Kanamaru K, Weir BKA, Simpson I, Witbeck T, Grace M (1991) Effect of 21-aminosteroid U-74006F on lipid peroxidation in subarachnoid clot. J Neurosurg 74:454–459

Kano M, Moskowitz MA, Yokota M (1991) Parasympathetic denervation of rat pial vessels significantly increases infarction volume following middle cerebral artery occlusion. J Cerebr Blood Flow Metab 11:628–637

Kapp JP, Clower BR, Azar FM, Yabuno N, Smith RR (1985) Heparin reduces proliferative angiopathy following subarachnoid hemorrhage in cats. J Neurosurg 62:570–575

Kapp J, Neill WR, Satter JE, Barnes TY (1987) Systemic heparin in the early management of ruptured intracranial aneurysms: review of 104 consecutive cases and comparison with concurrent controls. Neurosurgery 20:564–570

Kase CS, Foster TE, Reed JE, Spatz EL, Girgis GN (1987) Intracranial hemorrhage and phenylpropanolamine use. Neurology 37:399–404

Kassell NF, Sasaki T, Colohan ART, Nazar G (1985) Cerebral vasospasm following aneurysmal subarachnoid hemorrhage. Stroke 16:562–572

Kassell NF, Haley EC, Torner JC, Kongable G (1991) Nicardipine and angiographic vasospasm. J Neurosurg 74:341A

Kasuya H, Shimizu T (1989) Activated complement components C3a and C4a in cerebrospinal fluid and plasma following subarachnoid hemorrhage. J Neurosurg 71:741–746

Kasuya H, Shimizu T, Okada T, Takahashi T, Summerville T, Kitamura K (1988) Activation of the coagulation system in the subarachnoid space after subarachnoid haemorrhage: serial measurements of fibrinopeptide A and bradykinin of cerebrospinal fluid and plasma in patients with subarachnoid haemorrhage. Acta Neurochir 91:120–125

Katusic ZS, Shepard JT, Vanhoutte PM (1988) Endothelium-dependent contractions to calcium ionophore A23187, arachidonic acid, and acetylcholine in canine basilar arteries. Stroke 19:476–479

Katusic ZS, Milde JH, Cosentino F, Mitrovic BS (1993) Subarachnoid hemorrhage and endothelial L-arginine pathway in small brain stem arteries in dogs. Stroke 24:392–399

Kim P, Lorenz RR, Sundt TM, Vanhoutte PM (1989a) Release of endothelium-derived relaxing factor after subarachnoid hemorrhage. J Neurosurg 70:108–114

Kim P, Sundt TM, Vanhoutte PM (1989b) Alterations of mechanical properties in canine basilar arteries after subarachnoid hemorrhage. J Neurosurg 71:430–436

Kistler JP, Crowell RM, Davis KR et al. (1983) The relation of cerebral vasospasm to the extent and location of subarachnoid blood visualized by CT scan. A prospective study. Neurology 33:424–436

Kobayaski H, Ide H, Handa Y, Aradachi H, Arai Y, Kubota T (1992) Effect of leukotriene antagonist on experimental delayed cerebral vasospasm. Neurosurgery 31:550–556

Kontos HA (1993) Nitric oxide and nitrosothiols in cerebrovascular and neuronal regulation. Stroke (Suppl):I-155–I-158
Krendel DA, Ditter SM, Frankel MR, Ross WK (1990) Biopsy-proven cerebral vasculitis associated with cocaine abuse. Neurology 40:1092–1094
Kreupfer MM, Branch CA (1990) Effects of cocaine on renal sympathetic nerve activity and cerebral blood flow. FASEB J 4:A321
Krootila K, Oksala O, Zschauer A, Palkama A, Uusitalo H (1992) Inhibitory effect of methysergide on calcitonin gene-related peptide-induced vasodilation and ocular irritative changes in the rabbit. Br J Pharmacol 106:404–418
Levine SR, Brust JCM, Futrell N et al. (1990) Cerebrovascular complications of the use of the 'crack' form of alkaloidal cocaine. New Engl J Med 323:699–704
Linnik MD, Sakas DE, Uhl GR, Moskowitz MA (1989) Subarachnoid blood and headache: altered trigeminal tachykinin gene expression. Ann Neurol 25:179–184
Liu-chen L-Y, Han DH, Moskowitz MA (1983) Pia arachnoid contains substance P originating from trigeminal neurons. Neuroscience 9:803–808
Macdonald RL, Weir BKA (1991) A review of hemoglobin and the pathogenesis of cerebral vasospasm. Stroke 22:971–982
Macdonald RL, Weir BKA, Runzer TD et al. (1991) Etiology of vasospasm in primates. J Neurosurg 75:415–424
Macdonald RL, Weir BKA, Young JD, Grace MGA (1992) Cytoskeletal and extracellular matrix proteins in cerebral arteries following subarachnoid hemorrhage in monkeys. J Neurosurg 76:81–90
Macfarlane R, Moskowitz MA (1994) The innervation of pial blood vessels and their role in cerebrovascular regulation. In: Caplan LR (ed) Brain ischemia: basic concepts and their clinical relevance. (Clinical Medicine and the Nervous System Series) Springer-Verlag, London (in press)
McLaughlin JR, Gomez GR, Gomez SM et al. (1990) Exaggerated cerebral vasoconstriction as an interictal marker of migraine. Ann Neurol 28:330
Manelfe C, Guiraud A, Rascol M, Clanet M, Bonafe A (1983) Postpartum cerebral angiopathy – angiographic study. Report of 6 cases. Am J Neuroradiol 4:1149
Marshall JJ, Wei EP, Kontos HA (1988) Independent blockade of cerebral vasodilation from acetylcholine and nitric oxide. Am J Physiol 255:H847–H854
Marzatico F, Gaetani P, Spanu G, Buratti E, Rodriguez y Baena R (1991) Effects of nicardipine treatment on Na^{+}–K^{+} ATPase and lipid peroxidation after experimental subarachnoid haemorrhage. Acta Neurochir 108:128–133
Matsui T, Sugawa M, Johshita H, Takuwa Y, Asano T (1991) Activation of the protein kinase C-mediated contractile system in canine basilar artery undergoing chronic vasospasm. Stroke 22:1183–1187
Mayberg MR, Okada T, Bark DH (1990a) The role of hemoglobin in arterial narrowing after subarachnoid hemorrhage. J Neurosurg 72:634–640
Mayberg MR, Okada T, Bark DH (1990b) The significance of morphological changes in cerebral arteries after subarachnoid hemorrhage. Neurosurgery 72:626–633
Mena I, Giombetti R, Mody CK et al. (1990) Acute cerebral blood flow changes with cocaine intoxication. Neurology 40(Suppl I):I-179
Minami N, Tani E, Maeda Y, Yamaura I, Fukami M (1992) Effects of inhibitors of protein kinase C and calpain in experimental delayed cerebral vasospasm. J Neurosurg 76:111–118
Mizoi K, Yoshimoto T, Fujiwara S et al. (1991) Prevention of vasospasm by clot removal and intrathecal bolus injection of tissue-type plasminogen activator. Preliminary report. Neurosurgery 28:807–813
Moen M, Levine SR, Newman DS, Dull-Baird A, Brown GG, Welch KMA (1988) Bilateral posterior cerebral artery strokes in a young migraine sufferer. Stroke 19:525–528
Moncada S, Palmer RMJ, Higgs EA (1991) Nitric oxide: physiology, pathophysiology, and pharmacology. Pharmocol Rev 43:109–142
Moskowitz MA (1984) The neurobiology of vascular head pain. Ann Neurol 16:157–168
Moskowitz MA (1991) The visceral organ brain: implications for the pathophysiology of vascular head pain. Neurology 41:182–186
Moskowitz MA, Buzzi MG (1991) Neuroeffector functions of sensory fibers: implications for headache mechanisms and drug actions. J Neurol 238:S18–S22
Moskowitz MA, Brody M, Liu-chen L-Y (1983) In vitro release of immunoreactive substance P from putative afferent nerve endings in bovine pia arachnoid. Neuroscience 9:809–814
Nadeau SE, Watson RT (1983) Neurologic manifestations of vasculitis and collagen vascular syndromes. In: Joynt RJ (ed) Clinical neurology. J.B. Lippincott Company, Philadelphia, Vol. 4, Chapter 59, pp 1–133
Nakagomi T, Kassell NF, Sasaki T et al. (1987a) Effect of subarachnoid hemorrhage on endothelium-dependent vasodilation. J Neurosurg 66:915–923
Nakagomi T, Kassell NF, Sasaki T, Fujiwara S, Lehman RM, Torner JC (1987b) Impairment of endothelium-dependent vasodilation induced by acetylcholine and adenosine triphosphate following experimental subarachnoid hemorrhage. Stroke 18:482–489
Nakagomi T, Kassell NF, Sasaki T et al. (1988) Effect of removal of the endothelium on vasoconstriction in canine and rabbit basilar arteries. J Neurosurg 68:757–766
Nakagomi T, Kassell NF, Sasaki T et al. (1989a) Time course of the blood arterial wall barrier disruption following experimental subarachnoid haemorrhage. Acta Neurochir 98:176–183
Nakagomi T, Kassell NF, Johshita H, Lehman RM, Fujiwara S, Sasaki T (1989b) Blood arterial wall barrier disruption to various sized tracers following subarachnoid haemorrhage. Acta Neurochir 99:76–84
Naylor AR, Robertson IJA, Edwards CRW et al. (1991) Cerebral vasospasm following subarachnoid haemorrhage: effect of calcitonin gene-related peptide on middle cerebral artery velocities using transcranial Doppler. Surg Neurol 36:278–280
Nichols FT, Mawad M, Mohr JP, Stein BM, Hilal S, Michelsen WJ (1990) Focal headache during balloon inflation in the internal carotid and middle cerebral arteries. Stroke 21:555–559
Nishizawa S, Nezu N, Uemura K (1992) Direct evidence for a key role of protein kinase C in the development of vasospasm after subarachnoid hemorrhage. J Neurosurg 76:635–639
Nosko M, Schulz R, Weir B, Cook DA, Grace M (1988) Effects of vasospasm on levels of prostacyclin and thromboxane A_2 in cerebral arteries of the monkey. Neurosurgery 22:45–50
Nozaki K, Moskowitz MA, Boccalini P (1992a) CP-93,129, sumatriptan, dihydroergotamine block *c-fos* expression within rat trigeminal nucleus caudalis caused by chemical stimulation of the meninges. Br J Pharmacol 106:409–415
Nozaki K, Moskowitz MA, Maynard KI et al. (1992b) Origin and distribution of nitric oxide synthase–immunoreactive nerve fibers in rat cerebral arteries. Stroke 23:154
Ohkuma H, Suzuki S, Kimura M, Sobata E (1991) Role of

platelet function in symptomatic cerebral vasospasm following aneurysmal subarachnoid hemorrhage. Stroke 22:854–859
Ohlstein EH, Storer BL (1992) Oxyhemoglobin stimulation of endothelin production in cultured endothelial cells. J Neurosurg 77:274–278
Öhman J, Servo A, Heiskanen O (1991) Effect of intrathecal fibrinolytic therapy on clot lysis and vasospasm in patients with aneurysmal subarachnoid hemorrhage. J Neurosurg 75:197–201
Ohta T, Satoh G, Kuroiwa T (1992) The permeability change of major cerebral arteries in experimental vasospasm. Neurosurgery 30:331–336
Olesen J (1991) Cerebral and extracranial circulatory disturbances in migraine: pathophysiological implications. Cerebrovasc Brain Metab Rev 3:1–28
O'Neill P, Walton S, Foy PM, Shaw MDM (1992) Role of prostaglandins in delayed cerebral ischemia after subarachnoid hemorrhage. Neurosurgery 30:17–27
Onoue H, Nakamura N, Toda N (1988) Endothelium-dependent and -independent responses to vasodilators of isolated dog cerebral arteries. Stroke 19:1388–1394
Onoue H, Nakamura N, Toda N (1989) Prolonged exposure to oxyhemoglobin modifies the response of isolated dog middle cerebral arteries to vasoactive substances. Stroke 20:657–663
Petersen PL, Roszler M, Jacobs I, Wilner HI (1991) Neurovascular complications of cocaine abuse. J Neuropsych Clin Neurosci 3:143–149
Pluta RM, Deka-Starosta A, Zauner A, Morgan JK, Muraszko KM, Oldfield EH (1992a) Neuropeptide Y in the primate model of subarachnoid hemorrhage. J Neurosurg 77:417–423
Pluta RM, Zauner A, Morgan JK, Muraszko KM, Oldfield EH (1992b) Is vasospasm related to proliferative arteriopathy? J Neurosurg 77:740–748
Powers RH, Madden JA (1990) Vasoconstrictive effects of cocaine metabolites and structural analogues on cat cerebral arteries. FASEB J 4:A1095
Prado R, Watson BP, Kuluz J, Dietrich WD (1992) Endothelium-derived nitric oxide synthase inhibition. Effects on cerebral blood flow, pial artery diameter, and vascular morphology in rats. Stroke 23:1118–1124
Ram Z, Sadeh M, Shacked I, Sahar A, Hadani M (1991) Magnesium sulfate reverses experimental delayed cerebral vasospasm after subarachnoid hemorrhage in rats. Stroke 22:922–927
Raroque HG (1989) Cerebral vasospasm in eclampsia. Stroke 20:826
Rascol A, Guiraud B, Manelfe C, Clanet M (1980) Accidents vasculaires cérébraux de la grossesse et du post partum. 2èmes Conferences de la Salpétrière sur Le Maladies Vasculaires Cérébrales. JB Baillière, Paris, pp 84–127
Rieke K, Baker L, Dalessio D et al. (1992) Transcranial Doppler flow velocity measurements and vasomotor response in migraine patients. J Neuroimag 2:58
Rivilla F, Marin J, Sánchez-Ferrer CF, Salaices M, Ramos PG (1989) Ultrastructural changes induced by experimental subarachnoid haemorrhage and 6-hydroxydopamine in cat cerebral arteries. Acta Neurochir 100:158–163
Rodriguez y Baena R, Gaetani P, Marzatico F, Benzi G, Pacchiarini L, Paoletti P (1989) Effects of nicardipine on the ex-vivo release of eicosanoids after experimental subarachnoid hemorrhage. J Neurosurg 71:903–908
Rosenblum WI (1992) Endothelium-derived relaxing factor in brain blood vessels is not nitric oxide. Stroke 23:1527–1532
Rosenblum WI, Povlishock JT, Wei EP, Kontos HA, Nelson GH (1987) Ultrastructural studies of pial vascular endothelium following damage resulting in loss of endothelium-dependent relaxation. Stroke 18: 927–931
Rosenblum WI, Nelson GH, Nishimura H (1990) Leukotriene constriction of mouse pial arterioles in vivo is endothelium-dependent and receptor mediated. Stroke 21: 1618–1620
Rothrock JF, Rubenstein R, Lyden PD (1988a) Ischemic stroke associated with methamphetamine inhalation. Neurology 38: 589–592
Rothrock JF, Walicke P, Swenson M, Lyden P, Logan W (1988b) Migrainous stroke. Arch Neurol 45:63–67
Rumbaugh CL, Bergeron RT, Scanlon RL et al. (1971) Cerebral vascular changes secondary to amphetamine abuse in the experimental animal. Radiology 101:345–351
Rumbaugh CL, Fang HCH, Higgins RE, Bergeron RT, Segall HD, Teal JS (1976) Cerebral microvascular injury in experimental drug abuse. Invest Radiol 11:282–294
Ryba M, Iwanska K, Walski M, Patuszko M (1991) Immunomodulators interfere with angiopathy but not vasospasm after subarachnoid hemorrhage in rats. Acta Neurochir 108: 81–84
Sakaki T, Kassell NF, Zaccarello M (1986) Dependence of cerebral arterial contractions on intracellularly stored Ca^{2+}. Stroke 17:95–97
Sasaki S, Kuwabara H, Ohta S (1986) Biological defence mechanism in the pathogenesis of prolonged vasospasm in the patients with ruptured intracranial aneurysms. Stroke 17:196–202
Sasaki S, Ohue S, Kohno K, Takeda S (1989) Impairment of vascular reactivity and changes in intracellular calcium and calmodulin levels of smooth muscle cells in canine basilar arteries after subarachnoid hemorrhage. Neurosurgery 25:753–761
Seifert V, Stolke D, Reale E (1989a) Ultrastructural changes of the basilar artery following experimental subarachnoid haemorrhage. A morphological study on the pathogenesis of delayed cerebral vasospasm. Acta Neurochir 100:164–171
Seifert V, Eisert WG, Stolke D et al. (1989b) Efficacy of single intracisternal injection of recombinant tissue plasminogen activator to prevent delayed cerebral vasospasm after experimental subarachnoid hemorrhage. Neurosurgery 25:590–598
Shibata S, Mori K, Sekine I, Suyama H (1991) Subarachnoid and intracerebral hemorrhage associated with necrotizing angiitis due to methamphetamine abuse. Neurol Med Chir (Tokyo) 31:49–52
Shibuya M, Suzuki Y, Takayasu M et al. (1988) The effects of an intracellular calcium antagonist HA1077 on delayed cerebral vasospasm in dogs. Acta Neurochir 90:53–59
Shibuya M, Suzuki Y, Suzita K et al. (1992) Effect of AT877 on cerebral vasospasm after aneurysmal subarachnoid hemorrhage. Results of a prospective placebo-controlled double-blind trial. J Neurosurg 76:571–577
Shigeno T, Mima T (1990) A new vasoconstrictor peptide, endothelin: profiles as vasoconstrictor and neuropeptide. Cerebrovasc Brain Metab Rev 2:227–239
Shiobara R, Kawase T, Toya S et al. (1985) 'Scavenger surgery' for subarachnoid hemorrhage. II. Continuous ventriculocisternal perfusion using artificial cerebrospinal fluid with urokinase. In: Auer LM (ed) Timing of aneurysm surgery. Walter de Gruyter, Berlin, p 365
Sloan MA (1993a) Detection of vasospasm after subarachnoid hemorrhage. In: Babikian VL, Wechsler LR (eds) Transcranial Doppler ultrasonography. BC Decker/Mosby, St. Louis, pp 105–127

Sloan MA (1993b) Cerebrovascular disorders associated with licit and illicit drugs. In: Fisher M, Bogousslavsky J (eds) Current review of cerebrovascular disease. Current Science Ltd, Philadelphia, pp 48–62

Sloan MA (1994) Thrombolysis and intracranial hemorrhage. In: Feldmann E (ed) Intracerebral hemorrhage. Futura Publishing Company Inc (in press).

Sloan MA, Kittner SJ, Rigamonti D, Price TR (1991) Occurrence of stroke associated with use/abuse of drugs. Neurology 41:1358–1364

Somlyo A, Himpens B (1989) Cell calcium and its regulation in smooth muscle. FASEB J 3:2266–2276

Spatz M, Bacic F, Stanimirovic D, Uematsu S, McCarron RM, Bembry J (1992) Endothelin 1 and PGD_2 stimulates the production of PGE_2 and thromboxane B_2 in human cerebrovascular endothelium. Stroke 23:156

Stanimirovic DB, Yamamoto T, Uematsu S, Spatz M (1993) Receptor mediated induction of vasoconstrictive prostaglandins by endothelin-1 in human brain endothelial cells. Stroke 24:168

Steinke DE, Weir BKA, Findlay JM, Tanabe T, Grace M, Krushelnycky BW (1989) A trial of the 21-aminosteroid U74006F in a primate model of chronic cerebral vasospasm. Neurosurgery 24:179–186

Stolke D, Seifert V (1992) Single intracisternal bolus of recombinant tissue plasminogen activator in patients with aneurysmal subarachnoid hemorrhage: preliminary assessment of efficacy and safety in an open clinical study. Neurosurgery 30:877–881

Subcutaneous Sumatriptan International Study Group (1991) Treatment of migraine with sumatriptan. New Engl J Med 325:316–321

Suzuki N, Hardebo JE (1993) The cerebrovascular parasympathetic innervation. Cerebrovasc Brain Rev 5:33–46

Suzuki R, Masaoka H, Hirata Y et al. (1992) The role of endothelin-1 in the origin of cerebral vasospasm in patients with aneurysmal subarachnoid hemorrhage. J Neurosurg 77:96–100

Suzuki Y, Sato S, Suzuki H et al. (1989) Increased neuropeptide Y concentrations in cerebrospinal fluid from patients with aneurysmal subarachnoid hemorrhage. Stroke 20:1680–1684

Suzuki Y, Satoh S, Ikegaki I et al. (1990) Endothelin causes contraction of canine and bovine arterial smooth muscle in vitro and in vivo. Acta Neurochir 104:42–47

Svendgaard NA, Brismar J, Delgado TJ, Rosengren E, Stenevi U (1986) Subarachnoid hemorrhage in the rat: effect on the development of cerebral vasospasm of lesions in the central serotonergic and dopaminergic systems. Stroke 17:86–90

Szabo C, Faragó M, Dóra E, Horváth I, Kovách AGB (1991) Endothelium-dependent influence of small changes in extracellular magnesium concentration on the tone of feline middle cerebral arteries. Stroke 22:785–789

Takanashi Y, Weir BKA, Vollrath B, Kasuga H, Macdonald RL, Cook D (1992) Time course of changes in contraction of intracellular free calcium in cultured cerebrovascular smooth muscle cells exposed to oxyhemoglobin. Neurosurgery 30:346–350

Takenaka K, Yamada H, Sakai N, Ando T, Nakashima T, Nishimura Y (1991) Induction of cytosolic free calcium elevation in rat vascular smooth-muscle cells by cerebrospinal fluid from patients after subarachnoid hemorrhage. J Neurosurg 75:452–457

Takenaka K, Kassell NF, Ardo T, Sakai N, Yamoda H (1992) Cerebrospinal fluid from subarachnoid hemorrhage patients elevates DNA synthesis and intracellular calcium in cerebral smooth muscle cells. Stroke 23:156

Tanaka Y, Chiba S (1988) Relationship between extraluminal oxyhemoglobin and endothelium-dependent vasodilation in isolated perfused canine internal carotid arteries. Neurosurgery 23:158–161

Tanaka Y, Kassell NF, Machi T, Toshima M, Dougherty DA (1992) Effect of bilirubin on rabbit cerebral arteries in vivo and in vitro. Neurosurgery 30:195–201

Terai Y, Smith RR, Bernanke DH, Iwasa K (1993) Myofibroblasts in human cerebral vasospasm – immunohistological characterization. Stroke 24:166

Thie A, Spitzer K, Lachenmayer L et al. (1988) Prolonged vasospasm in migraine detected by noninvasive transcranial Doppler ultrasound. Headache 28:183–186

Thompson BG, Pluta RM, Girton M, Oldfield EH (1992) Nitric oxide (NO) mediates chemoregulation of CBF in primates. Stroke 23:154

Tokiyoshi K, Ohnishi T, Nii Y (1991) Efficacy and toxicity of thromboxane synthetase inhibitor for cerebral vasospasm after subarachnoid hemorrhage. Surg Neurol 36:112–118

Toshima M, Kassell NF, Sasaki T, Tanaka Y, Machi T (1992) The effect of hemoglobin on vasodilatory effect of calcium antagonists in the isolated rabbit basilar artery. J Neurosurg 76:670–678

Trommer BL, Homer D, Mikhael MA (1988) Cerebral vasospasm and eclampsia. Stroke 19:326–329

Trouvé R, Nahas GG, Manger WM, Vinyard C, Goldberg S (1990) Interactions of nimodipine and cocaine on endogenous catecholamines in the squirrel monkey. Proc Soc Exp Biol Med 193:171–175

Tsuji T, Cook DA (1989) Effect of nimodipine on canine cerebrovascular responses to 5-hydroxytryptamine and potassium chloride after exposure to blood. Stroke 20:105–111

Tsukahara T, Taniguchi T, Shimohama S, Fujiwara M, Handa H (1986) Characterization of beta adrenergic receptors in human cerebral arteries and alterations of the receptors after subarachnoid hemorrhage. Stroke 17:202–207

Tsukahara T, Taniguchi T, Miwa S et al. (1988) Presynaptic and postsynaptic α_2-adrenergic receptors in human cerebral arteries and their alteration after subarachnoid hemorrhage. Stroke 19:80–83

Uddman R, Edvinsson L (1989) Neuropeptides in the cerebral circulation. Cerebrovasc Brain Metab Rev 1:230–252

Van Riper DA, Bevan JA (1992) Electrical field stimulation-mediated relaxation of rabbit middle cerebral artery. Evidence of a cholinergic endothelium-dependent component. Circ Res 70:1104–1112

Vitullo JC, Karam R, Mekhail N, Wicker P, Engelmann GL, Khairallah PA (1989) Cocaine-induced small vessel spasm in isolated rat hearts. Am J Pathol 135:85–91

Vorkapic P, Bevan JA, Bevan RD (1991) Clentiazem protects against chronic cerebral vasospasm in rabbit basilar artery. Stroke 22:1409–1413

Walters BB, Gillespie SA, Moskowitz MA (1986) Cerebrovascular projections from the sphenopalatine and otic ganglia to the middle cerebral artery of the cat. Stroke 17:488–494

Wang A-M, Suojanen JN, Colucci VW, Rumbaugh CL, Hollenberg NK (1990) Cocaine and methamphetamine-induced acute cerebral vasospasm: an angiographic study in rabbits. Am J Neuroradiol 11:1141–1146

Wei EP, Kukreja R, Kontos HA (1992a) Effect in cats of inhibition of nitric oxide synthesis on cerebral vasodilation and endothelium-derived relaxing factor from acetylcholine. Stroke 23:1623–1629

Wei EP, Moskowitz MA, Boccalini P, Kontos HA (1992b) Calcitonin gene-related peptide mediates nitroglycerine and sodium nitroprusside-induced vasodilation in feline cerebral arterioles. Circ Res 70:1313–1319

Welch KMA (1987) Migraine. A biobehavioral disorder. Arch Neurol 44:323–327
Welch KMA, Levine SR (1990) Migraine-related stroke in the context of the International Headache Society classification of head pain. Arch Neurol 47:458–462
Wendling WW, Harakal C (1991) Effects of prostaglandin $F_{2\alpha}$ and thromboxane A_2 analogue on bovine cerebral arterial tone and calcium fluxes. Stroke 22:66–72
White RP (1986) Vasodilator proteins: role in delayed cerebral vasospasm. Stroke 17:207–213
White RP, Ricca GF, El-Bauomy AM, Robertson JT (1991) Identification of capric acid as a potent vasorelaxant of human basilar arteries. Stroke 22:469–476
Will AD, Lewis KL, Hinshaw DB et al. (1987) Cerebral vasoconstriction in toxemia. Neurology 37:1555–1557
Willette RN, Sauermelch C, Ezekiel M, Feuerstein G, Ohlstein EH (1990) Effect of endothelin on cortical microvascular perfusion in rats. Stroke 21:451–458
Wooten MR, Khangure MS, Murphy MJ (1983) Intracerebral hemorrhage and vasculitis related to ephedrine abuse. Ann Neurol 13:337–340
Yamamoto Y, Clower BR, Haining JL, Asari S, Smith RR (1991a) Adventitial red blood cells produce intimal platelet accumulation in cerebral arteries of cats following subarachnoid hemorrhage. Stroke 22:373–377
Yamamoto Y, Clower BR, Haining JL, Smith RR (1991b) Effect of tissue plasminogen activator on intimal platelet accumulation in cerebral arteries after subarachnoid hemorrhage in cats. Stroke 22:780–784
Yamamoto Y, Smith RR, Bernanke DH (1992) Accelerated non-muscle contraction after subarachnoid hemorrhage. Culture and characterization of myofibroblasts from human cerebral arteries in vasospasm. Neurosurgery 30:337–345
Yamaura I, Tani E, Maeda Y, Minami N, Shindo H (1992) Endothelin-1 of canine basilar artery in vasospasm. J Neurosurg 76:99–105
Yanamoto H, Kikuchi H, Okamoto S, Nozaki K (1992a) Preventive effect of synthetic serine protease inhibitor, FUT-175, on cerebral vasospasm in rabbits. Neurosurgery 30:351–357
Yanamoto H, Kikuchi H, Sato M, Shimizu Y, Yoneda S, Okamoto S (1992b) Therapeutic trial of cerebral vasospasm with the serine protease inhibitor, FUT-175, administered in the acute stage after subarachnoid hemorrhage. Neurosurgery 30:358–363
Yokota M, Tani E, Maeda Y, Kokubu K (1987) Effect of 5-lipoxygenase inhibitor on experimental delayed cerebral vasospasm. Stroke 18:512–518
Yokota M, Tani E, Maeda Y (1989) Biosynthesis of leukotrienes in canine cerebral vasospasm. Stroke 20:527–533
Yoshida Y, Ueki S, Takahashi A (1985) Intrathecal irrigation with urokinase in ruptured cerebral aneurysm cases. Basic studies and clinical application. Neurol Med Chir 24:987–995
Zabramski JM, Spetzler RF, Lee KS et al. (1991) Phase I trial of tissue plasminogen activator for the prevention of vasospasm in patients with aneurysmal subarachnoid hemorrhage. J Neurosurg 75:189–196
Zanette EM, Agnoli A, Roberti C, Chiarotti F, Cerbo R, Fieschi C (1992) Transcranial Doppler in spontaneous attacks of migraine. Stroke 23:680–685

12. Endothelial Cells and Cerebrovascular Disease

J. D. Trachtenberg and U. S. Ryan

The endothelium is a complex, unique organ with an aggregate mass equal to that of the liver. It is structurally unique in its strategic interposition between the blood and all other tissues in the body, and interacts directly with cells and molecules arriving via the blood or from tissues (Ryan 1986). In the quiescent state the endothelium acts to maintain the blood's fluidity and resist thrombosis, but thrombotic functions take over when the cells are perturbed. The endothelium also acts as a selective barrier between the elements in the blood and the extravascular space, and serves to convey signals between the tissues and the circulating elements. During infection, or when activated by various cytokines, the endothelium functions to attract leukocytes, allowing their emigration to sites of inflammation. The endothelium also has local regulatory functions, being a key regulator of vascular tone as well as an important modulator of vascular remodeling through its influence on the growth of the underlying vascular smooth muscle cells and by secretion of extracellular matrix components. Toward all these functions, the endothelium synthesizes a diverse armament of hormones and vasoactive mediators, some of which are expressed constitutively and others which are only induced in response to various stimuli. However, the same anatomic location which makes the endothelium so unique and diverse in its biologic functions makes it both an immediate target of vascular injury in conditions such as stroke and an ideal candidate to initiate events leading to cerebrovascular damage.

Endothelial Cells and the Blood–Brain Barrier

The interactions between neighboring endothelial cells (ECs) serve to maintain vascular integrity, regulate vascular permeability and control leukocyte traffic. The junctions between ECs vary in different parts of the circulation and in different organs. In much of the vasculature, where ECs form a continuous monolayer, gap junctions predominate (Larson and Sheridan 1982). A decrease in both gap junctions and tight junctions is seen in areas of regenerating endothelium (Spagnoli et al. 1982). In contrast, in the brain and retina, where the development of edema can be dangerous, tight (occluding) junctions predominate. The blood–brain barrier is recognized to be a highly selective anatomic and physiological barrier which regulates the entry and exit of cerebral nutrients and biologically important substances necessary for the maintenance of cerebral metabolism and neuronal activity (Brightman et al. 1970). Although the single most important feature is the presence of EC high-resistance tight junctions, there are also other distinguishing features of the ECs. These specialized features include a continuous capillary basement membrane, a paucity of endothelial caveolae, abundant EC mitochondria and a lack of endothelial fenestrations (Oldendorf et al. 1977; Bowman et al. 1979). Amino acids, glucose and biogenic amines as well as other nutrients gain entry via a complex system of membrane transporters.

Endothelial Hemostatic Functions

ECs have both thrombotic and thromboresistant properties. In the quiescent state they are antithrombotic, but can be induced to become procoagulant. ECs inhibit thrombus formation by interfering with the coagulation cascade, inhibiting platelet adhesion/aggregation and activating fibrinolytic pathways. The luminal surface of the EC contains anionic heparin-like glycosaminoglycans which are synthesized by the ECs and have the capacity to bind thrombin, the key enzyme of blood coagulation, and the serum protein antithrombin III, a serum protein which binds and inactivates thrombin (Marcum and Rosenberg 1984; Hatton and Moar 1985; Shimada and Ozawa 1985). In the presence of ECs thrombin reacts primarily with antithombin III, resulting in rapid inactivation of the enzyme (Marcum et al. 1984). Thrombin also interacts with thrombomodulin on the EC surface. Thrombomodulin binds both thrombin and protein C, and the thrombin–thrombomodulin–protein C complex markedly accelerates the activation of protein C (Esmon et al. 1982). Protein C, with the intermediary protein S, inhibits Factor Va and Factor XIII in the coagulation cascade (Walker et al. 1979 ; Stern et al. 1986). ECs are also thought to produce a lipoprotein-associated inhibitor of coagulation known as TFPI (tissue factor pathway inhibitor) or LACI (lipoprotein-associated coagulation inhibitor) (Broze 1992). This factor is a potent inhibitor of Factor Xa and the Factor VIIa/TF-complex (Broze et al. 1988). Although there is no direct evidence of its being made in ECs, it is currently hypothesized that the heparin-releasable form is associated with endothelial glycosaminoglycans (Sandset et al. 1988). Following extensive vascular injury with vessel occlusion, TFPI has been shown to prevent reocclusion of the injured arterial segment opened with thrombolytic agents (Haskel et al. 1991). ECs also synthesize prostacyclin (PGI_2), a potent inhibitor of platelet aggregation (Weksler et al. 1977). Other products released by ECs which prevent platelet adhesion/aggregation include adenosine as well as nitric oxide (NO) (Alheid et al. 1987; Radomski et al. 1987b; MacDonald et al. 1988). Moreover, there is clear synergism between the antiaggregatory effect of PGI_2 and the effect of NO at subthreshold levels (Radomski et al. 1987a).

Clot dissolution through the enhancement of the fibrinolytic system is another EC anticoagulant defense mechanism (Gross et al. 1982; Erickson et al. 1985). ECs synthesize both tissue-type plasminogen activators (t-PA) and urokinase-type plasminogen (u-PA) activators (Levin and Loskutoff 1982). The synthesis and release of plasminogen activators by ECs is stimulated by thrombin, activated protein C, epinephrine, vasopressin and bradykinin (Levin et al. 1984; Sakata et al. 1985; Smith et al. 1985). In addition, t-PA appears to be more specific for fresh clot, since fibrin but not fibrinogen (at physiologic concentrations) is required for full enzymatic activity, localizing the fibrinolytic response to areas of ongoing clot formation (Loskutoff and Mussoni 1983).

In contrast to the anticoagulant mechanisms described above, the EC, when perturbed mechanically or exposed to bacterial endotoxin, thrombin, tumor necrosis factor or interleukin-1, can interact with blood components to promote coagulation (Colucci et al. 1983; Bevilacqua et al. 1986a; Nawroth and Stern 1986). The procoagulant activity is the result of increased production of tissue factor (thromboplastin), von Willebrand factor (Factor VIII_{vw}), plasminogen activator inhibitor, platelet-activating factor and extracellular matrix components (Rodgers et al. 1983; Wagner and Marder 1984). Tissue factor, released by damaged cells, functions with Factor VII in activating Factor X, initiating the 'extrinsic pathway' of blood coagulation and potentiating the cleavage of Factor IX (activating the 'intrinsic pathway' as well). In the presence of Factor VIII_{vw}, activated Factor IX activates Factor X. The activated Factor X, formed via either pathway, then converts prothrombin to thrombin in a reaction which requires Factor V. Thrombin converts soluble fibrinogen to insoluble fibrin, activates platelets and inhibits vessel-wall fibrinolytic activity (Gelehrter and Sznycer-Laszuk 1986). The Factor VIII_{vw} synthesized by ECs is also bound by collagen types I, III, IV and V, and helps mediate platelet adhesion to the subendothelium through a specific glycoprotein receptor on the platelet surface, glycoprotein IIb/IIIa (which also serves as a receptor for fibrinogen) (Sakariassen et al. 1986). The importance of F-VIII_{vw} in platelet adhesion to the subendothelium and subsequent thrombosis is evidenced by the fact that in von Willebrand's disease there is a marked decrease in platelet adhesion, which can be reversed by administration of F-VIII_{vw} (Weiss et al. 1978). ECs also synthesize other proteins which stimulate the coagulation cascade (although this may not be their primary function). These include thrombospondin, collagens and fibronectin (Jaffe and Mosher 1978). Procoagulant activity also involves alterations in fibrinolysis through the regulated expression of t-PA and specific plasminogen activator inhibitors (Bevilacqua et al. 1986b; Nachman et al. 1986; Emeis and Kooistra 1986). In fact, the activity of t-PA in plasma is based on the amount of t-PA released as well as on the level of plasminogen activator inhibitor (PAI) present (Brommer et al. 1984; Nilsson et al. 1985).

Thus, the hemostatic potential of the endothelium results from a complex balance between active factors which have opposing biological actions and between active factors and specific inhibitors. Upset of this balance could lead to conditions predisposing to both embolic and thrombotic stroke.

Endothelial Control of Vascular Wall Function

ECs play a key role in modulating vasomotor tone. Many substances produced by ECs effect local vasomotor tone and a complex interaction exists between those elements which control hemostasis, vascular modeling and vasomotor reactivity. Angiotensin-converting enzyme (ACE) localized on ECs has established a role for endothelium in the control of blood pressure (Ryan et al. 1976). The renin–angiotensin system is also known to play a role in vascular remodeling (Powell et al. 1991; Hermans et al. 1992). Of those substances secreted by ECs having vasoreactive properties, PGI_2, endothelial-derived relaxing factor (EDRF) or NO, and endothelin are currently under extensive study.

PGI_2 is a potent vasodilator and antiaggregant. Discovered in 1976, it is a member of the prostaglandin family synthesized from arachidonic acid (Moncada et al. 1976). PGI_2's release by ECs can be stimulated by pulsatile pressure as well as by the endogenous mediators thrombin, bradykinin, angiotensin II, histamine, platelet-derived growth factor (PDGF), IL-1 and adenine nucleotides (Baenziger et al. 1981; Crutchley et al. 1983; Bhagyalakshmi and Frangos 1989). The effects of PGI_2s are opposed by thromboxane, a prostanoid produced by activated platelets which induces platelet aggregation and vasoconstriction (Hamberg et al. 1975). Aspirin, an irreversible inhibitor of cyclo-oxygenase, has been shown to be effective in the prevention or treatment of cerebrovascular disease (Steering Commitee of the Physician's Health Study Research Group 1989). Its efficacy relates to the EC's ability to synthesize cyclo-oxygenase continuously, whereas the platelet cannot. This leads to a cumulative inhibition of thromboxane A_2 formation while the ECs continue to produce PGI_2 (Fitzgerald et al. 1983).

EDRF was first described by Furchgott and Zawadzki, who demonstrated that the relaxation of the rabbit aorta by acetylcholine depended on the presence of an endothelial lining and was the result of a non-prostanoid diffusible substance (Furchgott and Zawadzki 1980). Subsequently, Furchgott and Ignarro independently suggested that EDRF may be NO or a related species (Khan and Furchgott 1987; Ignarro et al. 1987). This has since been confirmed by Moncada and colleagues (Palmer et al. 1987; Hutchinson et al. 1987).

NO is synthesized from the amino acid L-arginine by the enzyme NO synthase (NOS) and its actions are mediated by increases in cellular cGMP (Rapoport and Murad 1983; Palmer et al. 1988; Moncada et al. 1991). Arginine analogs (such as L-NG-monomethyl arginine) are potent inhibitors of NO production and their effects can be reversed by arginine. Blockade of NO sythesis in animals results in marked increases in blood pressure, belying the importance of tonic NO release in maintaining resting blood pressure (Aisaka et al. 1989; Rees et al. 1989). Several isozymes of NOS exist. The constitutive enzyme is Ca^{2+}/calmodulin dependent and releases picomolar concentrations of NO. In several cell types, a second form of Ca^{2+}-independent NOS can be induced by inflammatory mediators (Busse and Mulch 1990; Knowles et al. 1990). The inducible

NOS produces NO at the nanomolar level; and its induction can be prevented by pretreatment with glucocorticoids (Radomski et al. 1990). NO, in addition to mediating vasodilation, inhibits platelet adhesion and aggregation as previously described. It also limits SMC proliferation and inhibits leukocyte–EC interactions (Garg and Hassid 1989; Assender et al. 1991; Kubes et al. 1991). In vitro, monocyte adherence to ECs is inhibited by NO (Bath et al. 1991). After reperfusion injury, sodium nitrite prevents leukocytes from adhering and infiltrating into the vessel wall (Johnson et al. 1990). This may be mediated by a direct effect of NO on EC adhesion molecule expression.

More than 30 substances have been shown to elicit endothelium-dependent relaxations of isolated blood vessels, including acetylcholine, the calcium ionophore A23187, bradykinin, thrombin and endothelin, indicating the importance of this factor in local autoregulation of vascular tone. Interestingly, there appear to be some minor differences between cerebral and peripheral arteries, at least with respect to the ability of acetylcholine and A23187 to induce endothelium-dependent relaxation (Kanamaru et al. 1987).

Defects in EC production of EDRF/NO have been seen in a number of disease states, including atherosclerosis, hypertension and diabetes, and may contribute to their pathogenesis (Gryglewski et al. 1988; Flavahan 1992). There is mounting evidence to suggest that mild trauma to microvessels is sufficient to impair the dilating responses of cerebral (pial) microvessels temporarily without irreversible injury to the vessels (Rosenblum et al. 1987). Defects in endothelium-dependent relaxation may also play a role in the pathogenesis of cerebral vasospasm following subarachnoid hemorrhage. Hemoglobin, fibrin degradation products and plasmin, released by clot present in the cerebrospinal fluid after hemorrhage, can interfere with either the release and/or the action of NO and prostacyclin (Kassel et al. 1985). For example, in the canine basilar artery, endothelium-dependent relaxations to vasopressin and thrombin are markedly attenuated after two successive intrathecal injections of blood, whereas the vasodilatory response to papaverine is preserved (Kim et al. 1989).

In addition to the vasodilators PGI_2 and NO, the endothelium also releases vasoconstrictor substances in response to a variety of stimuli (Rubanyi and Vanhoutte 1985). In 1988, Yanagisawa et al. identified endothelin, a linear 21-amino acid peptide secreted by ECs (Yanagisawa et al. 1988). Endothelin is the most potent vasoconstrictor substance yet discovered, with a potency 10 times that of angiotensin II. Intense vasoconstriction and deceased blood flow have been seen in many species in which ET-1 has been infused, although it is rapidly removed from the bloodstream, suggesting a local vasoregulatory role (De Nucci et al. 1988; Goetz et al. 1988). Three pharmacologically separate endothelin isopeptides have been identified in mammalian species, named endothelin-1 (ET-1), endothelin-2 and endothelin-3. ET-1 is the only one known to be made by ECs.

Many substances, including thrombin, epinephrine, transforming growth factor-β, and the Ca^{2+} ionophore A23187, increase preproendothelin mRNA as well as the release of vasoactive ET-1 from ECs (Yanagisawa et al. 1988). The release of ET-1 is slow, consistent with the fact that it requires new synthesis. ET-1 binds to a specific membrane receptor, leading to the release of calcium from intracellular stores activating protein kinase C (Simonson et al. 1989a). The fact that ET-1 is secreted by numerous cell types and that receptors are widely distributed among many tissues, including blood vessels, brain, lungs, kidneys, adrenal glands, spleen and intestines, suggests that endothelin has multiple biologic roles (Davenport et al. 1989).

In addition to its long-acting vasoconstrictor and pressor actions ET-1 has numerous biologic activities, including mitogenicity for mesangial cells, Swiss 3T3 cells, capillary ECs and VSMCs (Yanagisawa and Masaki 1989; Simonson et al. 1989b; Takuwa et al. 1989; Hirata et al. 1989; Vigne et al. 1990; Bobik et al. 1990). Elevated levels of ET-1 have been detected in patients who have suffered a myocardial infarction as well as in acute renal failure and hypertensive states, but it is unclear as to whether the elevated ET-1 levels actually contribute to the pathogenesis of these diseases or are a result of concurrent disease processes (Miyauchi et al. 1989; Tomita et al. 1989). The authors have demonstrated, in a rat model of arterial injury, that high levels of ET-1 worsen postangioplasty restenosis and may act through a direct mitogenic effect on the underlying VSMCs (Trachtenberg et al. 1993). Markedly elevated plasma endothelin levels have also been seen in acute ischemic stroke and may reflect enhanced synthesis by damaged ECs within the infarcted tissue (Ziv et al. 1992). In the brain, the local leakage of ET-1 may induce severe and prolonged constriction of collateral vessels, and may significantly worsen the outcome following an ischemic event.

Leukocyte–Endothelia-Cell Adhesive Interactions

Interaction of leukocytes with the endothelium is a routine physiologic function. Under normal circumstances over 75% of granulocytes are adherent to the endothelium, where they remain ready for release by specific stimuli to join the circulating pool (Cartwright et al. 1964). Lymphocytes, on the other hand, circulate through the plasma and as part of their normal course emigrate through specialized postcapillary venules in lymphoid tissues, returning to the plasma after a few hours (Ford 1975). When this delicate system is upset by an inflammatory process, changes occur in the endothelium which allow not only adhesion but also the emigration of leukocytes to the site. These interactions of leukocytes with the endothelium are carefully controlled by specific adhesive receptors. At present, three different groups of adhesive receptors/ligands are known to participate in leukocyte adhesion. These include proteins of the integrin family or Leu-CAMs (leukocyte cellular adhesion molecules) (Hynes 1986; Ruoslahti 1991), members of the immunoglobulin-related molecules or ICAMs (intercellular adhesion molecules) (Williams and Barclay 1988) and carbohydrate-binding proteins called selectins or LEC-CAMs (lectin-epidermal growth factor-complement cell adhesion molecule) (Brandley et al. 1990; Springer and Lasky 1991).

The most important integrins present on leukocytes belong to the β1(CD11/CD18) and β2 (VLA-4, CD49d/CD29) subfamilies (Arnout 1990; Larson and Springer 1990). Although integrins are constitutively expressed on the surface of leukocytes, they are normally in a functionally inactive state. Conversion to the active state is rapid in response to chemotactic factors, cytokines, antigens or mitogens (Buyon et al. 1990; Valmu et al. 1991). The importance of leukocyte integrins to immunologic defense against inflammatory processes is demonstrated in the human disease leukocyte adhesion deficiency (LAD). Neutrophils from patients with LAD lack expression of leukocyte integrins and exhibit defective leukocyte–EC adhesion in in vitro assays, correlating with the absence of these cells at sites of inflammation (Kishimoto et al. 1991).

ECs possess at least three adhesive receptors in the immunoglobulin family, namely ICAM-1, ICAM-2 and VCAM (INCAM-110), which serve as counter-receptors for the leukocyte integrins (Marlin and Springer 1987; Springer 1990; Gahmberg et al. 1991). They are single-chain n-glycosylated polypeptides. ICAM-1 contains five immunoglobulin domains. It is present not only on ECs but also on a variety of other cells, including leukocytes (Dustin et al. 1986; Marlin and Springer 1987; Simmons et al. 1988; Springer 1990). ECs constitutively express ICAM-1 in low amounts, but increased levels are easily induced by a variety of cytokines, including interferon-Y, IL-1 and TNF-α (Dustin et al. 1986). Following stimulation, increased expression can be detected within 4 hours and expression is maintained for over 24 hours. ICAM-2 is similar to ICAM-1 but contains only two immunoglobulin domains. ICAM-2 is constitutively expressed by ECs and is not increased after cytokine activation (Nortamo et al. 1991). This may indicate more of a role for ICAM-2 in routine EC–leukocyte interactions, whereas ICAM-1 becomes more important in activated ECs. VCAM-1 is also expressed following endothelial activation by cytokines (Carlos et al. 1990). VCAM-1 selectively binds mononuclear cells such as lymphocytes and monocytes, and is the counter-receptor for the β2 integrin VLA4 (Elices et al. 1990). VCAM-1 has been found at sites of chronic inflammation but its expression is not restricted to ECs. Macrophages and fibroblast-like cells also appear to express VCAM-1 (Koch et al. 1991).

Three members of the selectin family of adhesive receptors currently exist: L-selectin (LEC-CAM-1, LAM-1, Leu 8 and TQ-1), E-selectin (LEC-CAM-2, ELAM-1, endothelial–leukocyte adhesion molecule) and P-selectin (LEC-CAM-3, PAGEM, platelet activation-dependent granulocyte external-membrane protein, GMP-140, CD62) (Bevilacqua et al. 1989; Bevilacqua et al. 1991; Michi et al. 1991; McEver 1991; Lasky 1991; Kansas 1992). L-selectin is confined to leukocytes and is involved in leukocyte homing. E-selectin is found on ECs and P-selectin is found on both ECs and platelets. The term selectins was chosen as these molecules mediate selective cell–cell contacts through lectin-like mechanisms. These molecules are transmembrane glycoproteins which contain an amino-terminal lectin domain followed by an epidermal growth factor-like domain, and a varying number of complement-like consensus repeats.

P-and E-selectin bind sialylated Lewis x (SLeX) carbohydrates, which are commonly found on glycoproteins and glycolipids of myeloid cells (Phillips et al. 1990; Walz et al. 1990; Berg et al. 1991; Polley et al. 1991; Tiemeyer et al. 1991; Zhou et al. 1991). The natural ligand for L-selectin remains elusive.

E-selectin is an adhesion protein for monocytes and neutrophils, but not lymphocytes. It is expressed on the EC surface 4–6 hours after stimulation by cytokines, and is downregulated by 24 hours (Pober et al. 1986). It is not contained within the cytoplasm of cells and requires new protein synthesis for expression (Bevilacqua et al. 1989). E-selectin has been shown to be expressed on microvascular endothelium transiently in certain pathologic settings, particularly those acute and chronic inflammatory processes in which cytokine generation is thought to occur (Cotran et al. 1986). P-selectin is present in both platelets and ECs, although the endothelial protein is primarily found in postcapillary venules (McEver 1991). Endothelial P-selectin is expressed constitutively and is located in the membranes of Weibel-Palade bodies, the secretory granules of endothelium in which large multimers of F-VIII$_{vw}$ are stored (McEver et al. 1989; Hattori et al. 1989; Bonfanti et al. 1989). Following stimulation with agonists such as thrombin or histamine, P-selectin is rapidly distributed to the cell surface, where it binds neutrophils, monocytes and platelets (Geng et al. 1990). In stimulated cultured endothelium, surface expression of P-selectin is rapid: within 10 minutes. Downregulation then occurs over the next 30 minutes. Exposure of ECs to oxidants results in prolonged expression of P-selectin on their surface, and thus may provide a direct role for oxygen radicals in neutrophil adhesion (Patel et al. 1991). P-selectin is an excellent candidate for directing the initial adherence of unstimulated neutrophils and monocytes to sites of inflammation. Progression of the inflammatory stimulus can then activate other adhesive receptors on both cells, thus stimulating leukocyte emigration to the site of tissue damage (Lo et al. 1989). P-selectin's presence on platelets and its localization with F-VIII$_{vw}$ also suggest a role for this molecule in hemostasis. Stimulated platelets expressing P-selectin then may serve to recruit neutrophils and monocytes to the thrombus. The activated monocyte with its procoagulant surface could then serve to amplify thrombin generation and clot formation.

Clearly, the identification and characterization of these adhesive molecules has provided new insights into the increasingly complex interactions of ECs, platelets and leukocytes, and current investigation has focused on the exploitation of this knowledge to identify paths for therapeutic intervention in cerebrovascular disorders.

Endothelial Cells in Ischemia/Reperfusion

The pathology of ischemic insults to the heart or brain can be viewed as a 'one-two' punch. The initial insult results in tissue hypoxia with its associated cellular alterations, whereas the second insult occurs after reperfusion, with the generation of toxic oxygen radicals. It is this 'reperfusion' phenomenon which has hampered the treatment of ischemic events, as we have learned that the simple restoration of flow into ischemic tissue is clearly not curative and may have serious pathologic consequences.

The initial hypoxic episode results in specific alteration in cellular metabolism as the oxygen-deprived cell switches from aerobic metabolism to anerobic metabolism, decreasing the cellular pH concurrent with a buildup in metabolic waste products such as lactate. Cells starved of oxygen for too long will eventually die; however, a number of cells can survive the initial hypoxic episode quite well. A hypoxic environment is not necessarily toxic to ECs, but the hypoxic state does induce multiple reversible changes in endothelial pathology (Ogawa et al. 1990). One of the most important changes is a switch from the normal anticoagulant state of the EC to one which promotes clot formation. Hypoxia induces thrombomodulin activity on the EC cell surface, resulting in the activation of Factor X as well as the inhibition of EC fibrinolytic activity. Hypoxia also results in an inhibition of the barrier function of the endothelium, and since the production of EDRF requires oxygen the hypoxia and ischemia result in vasospasm. Clearly, the initial hypoxic insult sets the stage for continued thrombosis and contributes to the pathology seen in cerebrovascular injury. For instance, the loss of the blood–brain barrier in areas of ischemic vascular injury is a direct result of injury to the endothelium.

Neutrophils also appear to play a role in the ischemic insult. As mentioned above, oxygen radicals lead to and result in the activation of leukocytes which bind to endothelium. Obstruction of capillaries with neutrophils is one of the causes of the 'no-reflow phenomenon' which leads to further regional perfusion defects (Kloner et al. 1974; Engler et al. 1983). We have demonstrated in in-vitro studies that the mechanism of endothelial killing by neutrophils is a result of the conversion of hydrogen peroxide to the hydroxyl radical, a reaction catalyzed by metal ions via the Fenton reaction (Varani et al. 1985. In our in-vitro work it became clear that the source of the metal ions was the ECs themselves (Gannon et al. 1987). As certain areas of the brain are rich in iron, and as the CSF has no iron-binding capacity, cerebral injury results in release of large amounts of metal ions which could stimulate lipid peroxidation and contribute to cell death.

With reperfusion, the stage is thus set for the generation of oxygen radicals. Superoxide generation has been demonstrated experimentally on the surface of the brain in cats following reperfusion after a period of complete ischemia. The oxygen radicals thus generated have profound effects, both on the local vascular system and on the tissues. With respect to the vascular system, oxygen radical generation further stimulates neutrophil binding, increases vascular permeability and destroys NO, thus generating the toxic peroxynitrite which can lead to hydroxyl radical formation. The loss of the NO contributes to further vasoconstriction as well as platelet adhesion and aggregation. The oxygen radicals generated are also directly damaging to the cells themselves, causing lipid peroxidation and membrane disruptions, DNA damage, mitochondrial destruction and destruction of the cytoskeleton. The importance of the radicals in the ultimate pathogenesis of the ischemic insult is belied by the fact that oxygen radical scavengers such as superoxide dismutase, as well as non-specific antioxidants and metal ion scavengers, appear to attenuate the severity of the ischemic insult (Harper and Castleden 1990; Imaizumi et al. 1990; Chan 1992).

Conclusion

Despite their deceptive thinness in transverse section, ECs are highly active and responsive cells. It is clear from the above discussion that ECs perform a multitude of functions, some constitutively and some which are induced following activation. It is not yet known whether all aspects of endothelial activation occur concomitantly; nevertheless, they share some common mechanisms of cellular signal transduction.

The pathogenesis of stroke involves a number of situations that would be expected to alter the normal interactions of ECs with blood-borne vasoactive substances and to alter the communication systems between endothelium and other cell types. Thus, conditions that do not necessarily involve the loss or death of ECs, and fall short of monolayer disruption, may set the stage for occlusive or ischemic episodes which later result in irreversible endothelial damage. Studies of the mechanisms underlying the regulation of EC activation and endothelial–leukocyte interaction should lead to an improved understanding of modulations of EC structure, function and behavior during a variety of cerebrovascular diseases, including stroke.

References

Aisaka K, Gross SS, Griffith OW, Levi R (1989) NG-methyl-arginine, an inhibitor of endothelium-derived nitric oxide synthesis, is a potent pressor agent in the guinea pig: does nitric oxide regulate blood pressure in vivo? Biochim Biophys Acta 160:881

Alheid U, Frolich JC, Fostermann U (1987) Endothelium-derived relaxing factor from cultured human endothelial cells inhibits aggregation of human platelets. Thromb Res 47:561–571

Arnout MA (1990) The structure and function of the leukocyte adhesion molecules CD11/CD18. Blood 75: 1037–1050

Assender JW, Southgate KM, Newby AC (1991) Does nitric oxide inhibit smooth muscle proliferation? J Cardiovasc Pharmacol 31:S104–S107

Baenziger NL, Fogerty FJ, Metz LF, Chernuta LF (1981) Regulation of histamine-mediated prostacyclin synthesis in cultured human vascular endothelial cells. Cell 24:915

Bath PM, Hassall DG, Gladwin AM, Palmer RM, Martin JF (1991) Nitric oxide and prostacyclin. Divergence of inhibitory effects on monocyte chemotaxis and adhesion to endothelium in vitro. Arterioscler Thromb 11:254–260

Berg EL, Robinson MK, Mansson O, Butcher EC, Magnani JL (1991) A carbohydrate domain common to both sialyl Le (a) and sialyl Le (X) is recognized by the endothelial cell leukocyte adhesion molecule ELAM-1. J Biol Chem 266:14869–14872

Bevliacqua MP, Pober JS, Majeau GR, Fiers W, Cotran RS, Gimbrone MA Jr (1986a) Recombinant tumor necrosis factor induces procoagulant activity in cultured human vascular endothelium: characterization and comparison with the actions of interleukin 1. Proc Natl Acad Sci USA 83:4533

Bevliacqua MP, Schleef RR, Gimbrone MA Jr, Loskutoff DJ (1986b) Regulation of the fibrinolytic system of cultured human vascular endothelial cells by interleukin-1. J Clin Invest 78:578

Bevliacqua MP, Stengelin S, Gimbrone MA Jr, Seed B (1989) Endothelial leukocyte adhesion molecule 1: an inducible receptor for neutrophils related to complement regulatory proteins and lectins. Science 243:1160–1165

Bevilacqua M, Butcher E, Furie B et al. (1991) Selectins: a family of adhesion receptors. Cell 67:223

Bhagyalakshmi A, Frangos JA (1989) Mechanism of shear-induced prostacyclin production in endothelial cells. Biochem Biophys Res Commun 158:31–37

Bobik A, Grooms A, Millar JA, Mitchell A, Grinpukel S (1990)

Growth factor activity of endothelin on vascular smooth muscle. Am J Physiol 258:C408–C415
Bonfanti R, Furie BC, Furie B, Wagner DD (1989) PADGEM (GMP140) is a component of Weibel-Palade bodies of human endothelial cells. Blood 73:1109-1112
Bowman PD, Betz AL, Goldstein GW (1979) Characteristics of cultured brain capillaries. J Cell Biol 83:95a
Brandley BK, Swiedler SJ, Robbins PW (1990) Carbohydrate ligands of the LEC cell adhesion molecules. Cell 63:861-863
Brightman BW, Klatzo I, Olsson Y, Reese TS (1970) The blood-brain barrier to proteins under normal and pathologic conditions. J Neurol Sci 10:215–239
Brommer EJP, Verheijen JH, Chang GTG, Rijken DC (1984) Masking of fibrinolytic response to stimulation by an inhibitor of tissue-type plasminogen activator in plasma. Thromb Haemost 52:154
Broze GJ Jr, Warren LA, Novotny WF, Higuchi DA, Girard JJ, Miletich JP (1988) The lipoprotein-associated coagulation inhibitor that inhibits the factor VIII-tissue factor complex also inhibits factor Xa: insight into its possible mechanism of action. Blood 71:335–433
Broze GJ Jr, (1992) The role of tissue factor pathway inhibitor in a revised coagulation cascade. Semin Hematol 29:159–169
Busse R, Mulch A (1990) Induction of nitric oxide synthase by cytokines in vascular smooth muscle cells. FEBS Lett 275:87–90
Buyon JP, Slade SG, Reibman J et al. (1990) Constitutive and induced phosphorylation of the alpha- and beta- chains of the CD11/CD18 leukocyte integrin family. Relationship to adhesion-dependent functions. J Immunol 144:191–197
Carlos TM, Schwartz BR, Kovach NL et al. (1990) Vascular cell adhesion molecule-1 mediates lymphocyte adherence to cytokine-activated cultured human endothelial cells. Blood 76:965–970
Cartwright GE, Athens JW, Wintrobe MM (1964) Kinetics of granulopoiesis in normal man. Blood 24:780–803
Chan PH (1992) Antioxidant-dependent amelioration of brain injury: role of CuZn-superoxide dismutase. J Neurotrauma 9:S417–S423
Colucci M, Balcon GI, Lorenzet R et al. (1983) Cultured human endothelial cells generate tissue factor in response to endotoxin. J Clin Invest 71:1893
Cotran RS, Gimbrone MA Jr, Bevilacqua MP, Mendrick DL, Pober JS (1986) Induction and detection of a human endothelial activation antigen in vivo. J Exp Med 164:661–666
Crutchley DJ, Ryan US, Fischer GH (1983) Bradykinin-induced release of prostacyclin and thromboxanes from bovine pulmonary artery endothelial cells. Biochem Biophys Acta 751:91
Davenport AP, Nunez DJ, Hall JA, Kaumann AJ, Brown MJ (1989) Autoradiographical localization of binding sites for porcine 125I endothelin-1 in humans, pigs, and rats: functional relevance in humans. J Cariovasc Pharmacol 13 (Suppl 5): S166–S170
De Nucci G, Thomas R, D'Orleans-Juste P et al. (1988) Pressor effects of circulating endothelin are limited by its removal in the pulmonary circulation and by the release of prostacyclin and endothelium-derived relaxing factor. Proc Natl Acad Sci USA 85:9797–9800
Dustin ML, Rothlein R, Bhan AK, Dinarello CA, Springer TA (1986) Induction by IL-1 and interferon-gamma: tissue distribution, biochemistry, and function of a natural adherence molecule (ICAM-1). J Immunol 137:245–254
Elices MJ, Osborn L, Takada Y et al. (1990) VCAM-1 on activated endothelium interacts with the leukocyte integrin VLA4 at a site distinct from the VLA4/fibronectin binding site. Cell 60:577–584
Emeis JJ, Kooistra T (1986) Interleukin-1 and lipopolysaccharide induce an inhibitor of tissue-type plasminogen activator in vivo and in cultured endothelial cells. J Exp Med 163:1260
Engler RL, Schmid-Schönbein GW, Pavelec RS (1983) Leukocyte capillary plugging in myocardial ischemia and reperfusion in dogs. Am J Pathol 111:98–111
Erickson LA, Schleef RR, Ny T, Loskutoff DJ (1985) The fibrinolytic system and the vascular wall. Clin Haematol 14:513
Esmon NL, Owen WG, Esmon CT (1982) Isolation of a membrane bound cofactor for thrombin-catalyzed activation of protein C. J Biol Chem 257:859
Fitzgerald GA, Oates JA, Hawiger J et al. (1983) Endogenous biosynthesis of prostacyclin and thromboxane and platelet function during chronic administration of asprin in man. J Clin Invest 71:676–688
Flavahan NA (1992) Atherosclerosis or lipoprotein-induced endothlial dysfunction. Potential mechanisms underlying reduction in EDRF/nitric oxide activity. Circulation 85:1927–1938
Ford WL (1975) Lymphocyte migration and immune responses. Prog Allergy 19:1–59
Furchgott RF, Zawadzki JV (1980) The obligatory role of endothelial cells in the relaxation of arterial smooth muscle by acetylcholine. Nature 373:299
Gahmberg CG, Nortamo P, Zimmermann D, Ruoslahti E (1991) The human leukocyte-adhesion ligand, intercellular-adhesion molecule 2. Expression and characterization of the protein. Eur J Biochem 195:177–182
Gannon DE, Varani J, Phan SH et al. (1987) Source of iron in neutrophil-mediated killing of endothelial cells. Lab Invest 57:37–44
Garg UC, Hassid A (1989) Nitric oxide-generating vasodilators and 8-bromo-cyclic guanosine monophosphate inhibit mitogenesis and proliferation of cultured rat vascular smooth muscle cells. J Clin Invest 83:1774–1777
Gelehrter TD, Sznycer-Laszuk R (1986) Thrombin induction of plasminogen activator-inhibitor in cultured human endothelial cells. J Clin Invest 77:165–169
Geng JG, Bevilacqua MP, Moore KL et al. (1990) Rapid neutrophil adhesion to activated endothelium mediated by GMP-140. Nature 343:757–760
Goetz KL, Wang BC, Madwed JB, Zhu JL, Leadley RJ Jr (1988) Cardiovascular, renal, and endocrine responses to intravenous endothelin in conscious dogs. Am J Physiol 255:R1604–1608
Gross JL, Moscatelli D, Jaffe EA, Rifkin DB (1982) Plasminogen activator and endothelial cells. Surgery 91:550
Gryglewski RJ, Botting RM, Vane JR (1988) Mediators produced by the endothelial cell. Hypertension 12:530–548
Hamberg M, Svensson J, Samuelson B (1975) Thromboxanes: a new group of biologically active compounds derived from prostaglandin endoperoxides. Proc Natl Acad Sci USA 72:2994–2998
Harper GD, Castleden CM (1990) Drug therapy in patients with recent stroke. Br Med Bull 46:181–201
Haskel EJ, Torr SR, Day KC et al. (1991) Prevention of arterial reocclusion after thrombolysis with recombinant lipoprotein-associated coagulation inhibitor. Circulation 84:821–827
Hatton MWC, Moar SL (1985) A role for percellular proteoglycan in the binding of thrombin or antithrombin III? The effects of proteoglycan-degrading enzymes and glycosaminoglycan-binding proteins on I-thrombin binding by the rabbit thoracic aorta in vitro. Thromb Haemost 53:228
Hattori R, Hamilton KK, Fugate RD, McEver RP, Sims PJ

(1989) Stimulated secretion of endothelial von Willebrand factor is accompanied by rapid redistribution to the cell surface of the intracellular granule membrane protein GMP-140. J Biol Chem 264:7768–7771

Hermans WR, Rensing BJ, Foley DP et al. (1992) Therapeutic dissection after successful coronary balloon angioplasty: no influence on restenosis or on clinical outcome in 693 patients. The MERCATOR Study Group (Multicenter European Research Trial with Cilazapril after Angioplasty to prevent Transluminal Coronary Obstruction and Restenosis). J Am Coll Cardiol 20:767–780

Hirata Y, Takagi Y, Fukuda Y, Marumo F (1989) Endothelin is a potent mitogen for rat vascular smooth muscle cells. Atherosclerosis 78:225–228

Hutchinson PJ, Palmer RM, Moncada S (1987) Comparative pharmacology of EDRF and nitric oxide on vascular strips. Eur J Pharmacol 141:445–451

Hynes RO (1986) Integrins: a family of cell surface receptors. Cell 48:549–554

Ignarro LJ, Byrns RE, Buga GM, Wood KS (1987) Endothelium-derived relaxing factor (EDRF) released from artery and vein appears to be nitric oxide (NO) or a closely related radical species. Fed Proc 46:644 (Abstract)

Imaizumi S, Woolworth V, Fishman RA, Chan PH (1990) Liposome-entrapped superoxide dismutase reduces cerebral infarction in cerebral ischemia in rats. Stroke 21:1312–1317

Jaffe EA, Mosher DF (1978) Synthesis of fibronectin by cultured human endothelial cells. J Exp Med 147:1779

Johnson Gps, Tsao D, Malloy D, Leffer AM (1990) Cardioprotective effects of acidified sodium nitrite in myocardial ischemia with reperfusion. J Pharmacol Exp Ther 252:35–41

Kanamaru K, Waga S, Kojima T, Fujimoto K, Itoh H (1987) Endothelium-dependent relaxation of canine basilar arteries. Part 1: Difference between acetylcholine- and A23187-induced relaxation and involvement of lipoxygenase metabolite(s). Stroke 18:932–937

Kansas GS (1992) Stucture and function of L-selectin. APMIS 100:287–293

Kassel NF, Sasaki T, Colohan ART, Nazer GB (1985) Cerebral vasospasm following aneurysmal subarachnoid hemorrhage. Stroke 16:562–572

Khan MT, Furchgott RF (1987) Similarities of behavior of (NO) and endothelium-derived relaxing factor in a perfusion cascade bioassay system. Fed Proc 46:385 (Abstract)

Kim P, Sundt TM, Vanhoutte PM (1989) Alterations in endothelium-dependent responsiveness of the canine basilar artery. J Neurosurg 69:239–246

Kishimoto TK, Larson RS, Corbi AL, Dustin ML, Staunton DE, Springer TA (1991) The leukocyte integrins. Adv Immunol 46:149–182

Kloner RA, Ganote CE, Jennings RB (1974) The 'no-reflow' phenomenon after temporary coronary occlusion in the dog. J Clin Invest 54:1496–1502

Knowles RG, Merrett M, Salter M, Moncada S (1990) Differential induction of brain, lung and liver nitric oxide synthase by endotoxin in the rat. Biochem J 270:833–836

Koch AE, Burrows JC, Haines GK, Carlos TM, Harlan JM, Leibovich SJ (1991) Immunolocalization of endothelial and leukocyte adhesion molecules in human rheumatoid and osteoarthritic synovial tissues. Lab Invest 64:313–320

Kubes P, Suzuki M, Granger DN (1991) Nitric oxide an endogenous modulator of leukocyte adhesion. Proc Natl Acad Sci USA 88:4651–4655

Larson DM, Sheridan JD (1982) Intercellular junctions and transfer of small molecules in primary endothelial cell cultures. Am J Pathol 92:183–191

Larson RS, Springer TA (1990) Structure and function of leukocyte integrins. Immunol Rev 114:181–217

Lasky LA (1991) Lectin cell adhesion molecules (LEC-CAMs): a new family of cell adhesion proteins involved with inflammation. J Cell Biochem 45:139–146

Levin E, Loskutoff DJ (1982) Cultured bovine endothelial cells produce both urokinase and tissue-type plasminogen activators. J Cell Biol 94:631

Levin EG, Marzec U, Anderson J, Harker LA (1984) Thrombin stimulates tissue plasminogen activator release from cultured human endothelial cells. J Clin Invest 74:188

Lo SK, Van Seventer GA, Levin SM, Wright SD (1989) Two leukocyte receptors (CD11a/CD18 and CD11b/CD18) mediate transient adhesion to endothelium by binding to different ligands. J Immunol 143:3325–3329

Loskutoff DJ, Mussoni L (1983) Interactions between fibrin and the plasminogen activators produced by cultured endothelial cells. Blood 62:62

MacDonald PS, Read MA, Dusting GJ (1988) Synergistic inhibition of platelet aggregation by endothelium-derived relaxing factor and prostacyclin. Thromb Res 49:437

McEver RP, Beckstead JH, Moore KL, Marshall-Carlson L, Bainton df (1989) GMP-140, a platelet alpha-granule membrane protein, is also synthesized by vascular endothelial cells and is localized in Weiber-Palade bodies. J Clin Invest 84:92–99

McEver RP (1991) GMP-140: a receptor for neutrophils and monocytes on activated platelets and endothelium. J Cell Biochem 45:156–161

Marcum JA, McKenney JB, Rosenberg RD (1984) Acceleration of thrombin–antithrombin complex formation in rat hind quarters via heparin-like molecules bound to the endothelium. J Clin Invest 74:341

Marcum JA, Rosenberg RD (1984) Heparin-like molecules are synthesized by cultured endothelial cells. Biochem Biophys Res Commun 126:365

Marlin SD, Springer TA (1987) Purified intercellular adhesion molecule-1 (ICAM-1) is a ligand for lymphocyte function-associated antigen 1 (LFA-1). Cell 51:813–819

Michi J, Qiu QY, Kuerer HM (1991) Homing receptors and addressins. Curr Opin Immunol 3:373–382

Miyauchi T, Yanagisawa M, Tomizawa T et al. (1989) Increased plasma concentrations of endothelin-1 and big endothelin-1 in acute myocardial infarction. [Letter] Lancet 2:53–54

Moncada S, Palmer RMJ, Higgs EA (1991) Biosynthesis and endogenous roles of nitric oxide. Pharmacol Rev 43:109

Moncada S, Gryglewski RJ, Bunting S, Vane JR (1976) An emzyme isolated from arteries transforms prostaglandin endoperoxides to an unstable substance that inhibits platelet aggregation. Nature 263:663–665

Nachman RL, Hajjar KA, Silverstein RL, Dinarello CA (1986) Interleukin-1 induces endothelial cell synthesis of plasminogen activator inhibitor. J Exp Med 163:1595

Nawroth PP, Stern DM (1986) Modulation of endothelial cell hemostatic properties by tumor necrosis factor. J Exp Med 163:740

Nilsson IM, Ljunger H, Tengborn L (1985) Two different mechanisms in patients with venous thrombosis and defective fibrinolysis: low concentration of plasminogen activator or increased concentration of plasminogen activator inhibitor. Br Med J 290:1453

Nortamo P, Li R, Renkonen R et al. (1991) The expression of human leukocyte adhesion molecule intercellular adhesion molecule-2 is refractory to inflammatory cytokines. Eur J Immunol 21:2629–2632

Ogawa S, Gerlach H, Esposito C, Pasagian-Macaulay A, Brett J, Stern D (1990) Hypoxia modulates the barrier and coagu-

lant function of cultured bovine endothelium. J Clin Invest 85:1090–1097
Oldendorf WH, Cornford ME, Brown WJ (1977) The large apparent work capability of the blood–brain barrier: a study of the mitochondrial content of capillary endothelial cells in brain and other tissues of the rat. Ann Neurol 5:409–417
Palmer RM, Ferrige AG, Moncada S (1987) Nitric oxide release accounts for the biological activity of endothelium-derived relaxing factor. Nature 327:524–526
Palmer RM, Ashton DS, Moncada S (1988) Vascular endothelial cells synthesize nitric oxide from L-arginine. Nature 33:664–666
Patel KD, Zimmermann GA, Prescott SM, McIntyre TM (1991) Oxygen radicals induce human endothelial cells to express GMP-140 and bind neutrophils. J Cell Biol 112:749–759
Phillips ML, Nudelman E, Gaeta FC et al. (1990) ELAM-1 mediates cell adhesion by recognition of a carbohydrate ligand, sialyl-Lex. Science 250:1130–1132
Pober JS, Bevilacqua MP, Mendrick DL, Lapierre LA, Fiers W, Gimbrone MA jr (1986) Two distinct monokines interleukin 1 and tumor necrosis factor, each independdently induce biosynthesis and transient expression of the same antigen on the surface of cultured human vascular endothelial cells. J Immunol 136:1680–1687
Polley MJ, Phillips ML, Wayner E et al. (1991) CD62 and endothelial cell-leukocyte adhesion molecule 1 (ELAM-1) recognize the same carbohydrate ligand, sialyl-Lewis x. Proc Natl Acad Sci USA 88:6224–6228
Powell JS, Rouge M, Muller RK, Baumgarter HR (1991) Cilazapril suppresses myointimal proliferation after vascular injury: effects on growth factor induction in vascular smooth muscle cells. Basic Res Cardiol 86:65–74
Radomski MW, Palmer RM, Moncada S (1987a) The anti-aggregating properties of the vascular endothlium: interactions between prostacyclin and nitric oxide. Br J Pharmacol 92:632–646
Radomski MW, Palmer RMJ, Moncada S (1987b) Endogenous nitric oxide inhibits platelet adhesion to vascular endothelium. Lancet 2:1057
Radomski MW, Palmer RM, Moncada S (1990) Glucocorticoids inhibit the expression of an inducible, but not the constitutive, nitric oxide synthase in vascular endothelial cells. Proc Natl Acad Sci USA 87:10043–10047
Rapoport RM, Murad F (1983) Agonist induced endothelium-dependent relaxation in rat thoracic aorta may be mediated through cyclic GMP. Circ Res 52:352–357
Rees DD, Palmer RM, Moncada S (1989) Role of endothelium-derived nitric oxide in the regulation of blood pressure. Proc Natl Acad Sci USA 86:3375
Rodgers GM, Greenberg CS, Shuman MA (1983) Characterization of the effects of cultured vascular cells on the activation of blood coagulation. Blood 61:1155–1162
Rosenblum WI, Povlishock JT, Wei EP, Kontos HA, Nelson GH (1987) Ultrastructural studies of pial vascular endothelium following damage resulting in loss of endothelium-dependent relaxation. Stroke 18:195–199
Rubanyi GM, Vanhoutte PM (1985) Hypoxia releases a vasoconstrictor substance from the canine vascular endothelium. J Physiol 364:45–46
Ruoslahti E (1991) Integrins. J Clin Invest 87:1–5
Ryan US, Ryan JW, Whitaker C, Chiu A (1976) Localization of angiotensin-converting enzyme (kininase II). II. Immuno cytochemistry and immunoflurescence. Tissue Cell 8:125–146
Ryan US (1986) Pulmonary endothelium: a dynamic interface. Clin Invest Med 9:124–132
Sakariassen KS, Nievelstein PFEM, Coller BS, Sixma JJ (1986) The role of glycoprotein Ib and IIb-IIIa in platelet adherence to human artery subendothelium. Br J Haematol 63:681
Sakata Y, Curriden S, Lawrence D, Griffin JH, Loskutoff DJ (1985) Activated protein C stimulates the fibrinolytic activity of cultured endothelial cells and decreased antiactivator activity. Proc Natl Acad Sci USA 82:1121
Sandset PM, Abildgaard U, Larsen Ml (1988) Heparin induces release of extrinsic pathway inhibitor. Thromb Res 50:803
Shimada K, Ozawa T (1985) Evidence that surface heparin sulfate is involved in the high affinity thrombin binding to cultured porcine aortic endothelial cells. J Clin Invest 75:1308
Simmons D, Makgoba MW, Seed B (1988) ICAM, an adhesion ligand of LFA-1, is homologous to the neural cell adhesion molecule NCAM. Nature 331:624–627
Simonson MS, Wann S, Mene Pet al. (1989b) Endothelin stimulates phospholipase C, $Na^{+}H^{+}$ exchange, *c-fos* expression, and mitogen in rat mesangial cells. J Clin Invest 83:708–712
Simonson MS, Wann S, Mene P et al. (1989b) Endothelin stimulates phospholipase C, $Na^{+}H^{+}$ exchange, *c-fos* expression, and mitogenesis in rat mesangial cells. J Clin Invest 83:708–712
Smith D, Gilbert M, Owen WG (1985) Tissue plasminogen activator release in vivo in response to vasoactive agents. Blood 66:835
Spagnoli LG, Pietra GC, Villaschi S, Jones LW (1982) Morphometric analysis of gap junctions in regenerating endothelium. Lab Invest 46:139–148
Springer TA (1990) Adhesion receptors of the immune system. Nature 346:425–434
Springer TA, Lasky LA (1991) Sticky sugars for selectins. Nature 349:196–197
Steering Commitee of the Physician's Health Study Research Group (1989) Final report on the aspirin component of the ongoing Physician's Health Study. New Engl J Med 321:129–135
Stern DM, Nawroth PP, Harris K, Esmon CT (1986) Cultured bovine aortic endothelial cells promote activated protein C–protein S-mediated inactivation of factor Va. J Biol Chem 261:713
Takuwa N, Takuwa Y, Yanagisawa M, Yamashita K, Masaki T (1989) A novel vasoactive peptide endothelin stimulates mitogenesis through inositol lipid turnover in Swiss 3t3 fibroblasts. J Biol Chem 264:7856–7861
Tiemeyer M, Swiedler SJ, Ishihara M et al. (1991) Carbohydrate ligands for endothelial–leukocyte adhesion molecule 1. Proc Natl Acad Sci USA 88:1138–1142
Tomita K, Ujiie K, Nakanishi T et al. (1989) Plasma endothelin levels in patients with acute renal failure. New Engl J Med 321:1127
Trachtenberg JD, Sun S, Choi ET et al. (1993) Effect of endothelin-1 infusion on the development of intimal hyperplasia after balloon catheter injury. J Cardiovasc Pharm 22(Suppl 8) S355–S359
Valmu L, Autero M, Siljander P, Patarroyo M, Gahmberg CG (1991) Phosphorylation of the beta-subunit of CD11/CD18 integrins by protein kinase C correlates with leukocyte adhesion. Eur J Immunol 21:2857–2862
Varani J, Fligiel SEG, Til GO, Ryan US, Ward PA (1985) Pulmonary endothelial cell killing by human neutrophils: possible involvement of hydroxyl radical. Lab Invest 53:656–663
Vigne P, Marsault R, Breittmayer JP, Frelin C (1990) Endothelin stimulates phosphatidylinositol hydrolysis and DNA

synthesis in brain capillary endothelial cells. Biochem J 266:415–420

Wagner DD, Marder VJ (1984) Biosynthesis of von Willebrand protein by human endothelial cells: processing steps and their intracellular localization. J Cell Biol 99:2123

Walker FJ, Section PW, Esmon CT (1979) The inhibition of blood coagulation by activated protein C through the selective inactivation of activated factor V. Biochem Biophys Acta 571:333

Walz G, Aruffo A, Kolanus W, Bevilacqua M, Seed B (1990) Recognition by ELAM-1 of the sialyl-Lex determinant on myeloid and tumor cells. Science 250:1132–1135

Weiss HJ, Baumgarter HR, Tchopp TB, Turitto VT, Cohen D (1978) Correction by factor VIII of the impaired platelet adhesion to subendothelium in von Willebrand's disease. Blood 51:267

Weksler BB, Marcus AJ, Jaffe EA (1977) Synthesis of prostaglandin I2 (prostacyclin) by cultured human and bovine endothelial cells. Proc Natl Sci USA 126:365

Williams AF, Barclay AN (1988) The immunoglobulin superfamily-domains for cell surface recognition. Annu Rev Immunol 6:381–405

Yanagisgawa M, Kurihara H, Kimura S et al. (1988) A novel potent vasoconstrictor peptide produced by vascular endothelial cells. Nature 332:411–415

Yanagisawa M, Masaki T (1989) Molecule biology and biochemistry of the endothelins. Trends Pharmacol Sci 10:374–378

Zhou Q, Moore KL, Smith DF, Varki A, McEver RP, Cummings RD (1991) The selectin GMP-140 binds to sialylated, fucosylated lactosaminoglycans on both myeloid and non-myeloid cells. J Cell Biol 115:557–564

Ziv I, Fleminger G, Djaldetti R, Achiron A, Melamed E, Sokolosky M (1992) Increased plasma endothelin-1 in acute ischemic stroke. Stroke 23:1014–1016

13. Physical Factors in the Pathogenesis of Atheroma Formation

H. Schmid-Schönbein and K. Perktold

"The purpose of physics is
to make sense out of the
world around us"
DAVID RUELLE

This book is devoted to a review of interdisciplinary reinterpretations of scientific concepts explaining current understanding of the causes and consequences of atheroma formation. A chapter on physical factors must be more than just an addendum to an assembly of expert knowledge on the various distinct details of a complex process. Hemorheologists, in crossing the boundaries of established mathematical, biophysical, biochemical, cytological and hemodynamic disciplines, are forced to embrace the intricate details of vascular morphology, the effects of aging and of overt disease processes and, last but not least, the highly variable alterations in the physical properties of blood and its components. All of these factors can and must be put into common perspective in order to comprehend the physics that determines the natural history of atheromatosis, a pivotal pathogenetic process put into a fluid-dynamic perspective early in the 20th century by Aschoff (1912) who cooperated with Rehbock, a fluid dynamicist.

It is known that atherogenesis occurs when blood and its components are moving abnormally within arterial conduits, impinging against the walls and, at times, through walls. This sequence of events is highly localized, and can be visualised with color-coded Doppler scanners. These scanners often present the most complex time-dependent fluctuations of Doppler shifts, displayed as pulsatile sequences of transient patterns showing directional fluctuations (Fig.13.1). Direction-sensitive ultrasound Doppler scanning thus gives a new perspective of the disease state, which hopefully motivates physicians caring for potential victims of stroke and also intensifies the basic research efforts of fluid dynamicists.

In this joint effort a conceptual consensualization must be catalyzed: clinicians, trained by their experience as viewers of moving images, must familiarize themselves with the phenomenology of flow patterns obtained by fluid dynamicists in transparent models of pivotal vascular segments (branch-points, dilatations, stenoses and the peristenotic areas). Fluid dynamicists must be made aware of the fact that fluid mechanical factors might not just be important for localizing the sites of initial atheroma formation and the rate of their progression, but also that the flow disturbances observed there

Dedicated to the memory of H.P. Koepchen (1923–1993) whose keen interest in studying complex biological movements has immunized the senior author against the infection by contemporary reductionist research concepts.

become noxious to the adjacent wall both by mechanical impact and by the chemical consequences of the 'separated flow'. This chapter therefore puts the accumulated knowledge about the role of physical factors into a synergistic perspective.

Atherogenesis as the Result of Interdigitating Physical, Chemical and Cytological Cooperativity Coordinated by Vortex Flow as a Fluid-Dynamic Event

Both clinicians and theoreticians must use interactionist pathogenetic hypotheses to explain the preferential localization of atheroma formation as an intramural process, the analysis of which was initiated by Virchow (1856) in his macroscopic insudation theory, reinstituted by Hauss (1976) in his concept about atheroma formation as a global mesenchymal reaction, and culminating in the currently developed ramifications of Ross's theory viewing atheroma formation as a local response to injury (Ross 1986). A wide variety of interdependent interactions are revealed by uncovering previously hidden connections between distinctly different theories motivated by a growing awareness about the obvious limits of reductionist attempts to solve the riddles of a chronic multifactorial disease. Interpretation is helped by the emergence of interactionist biomedical theories derived from first physical principles, namely non-linear non-equilibrium thermodynamics (Prigogine 1979) and synergetics (Haken 1983).

These authors' thoughts about regularities in all 'driven systems' have unleashed the fruitful new scientific field of 'chaos research' (Gleick 1988; Coveney and Highfield 1992). In creating synergetics (the rules of 'joint efforts' in systems transferring energy and matter), Haken directs the attention of students of all scientific disciplines to some simple regularities essential for the transition into self-organized modes of operation. He has brought the attention of the scientific community to the fact that a very restricted set of boundary conditions has the potential to 'drive' physical, chemical, cellular and other ensembles on to a critical threshold, from where they fall into self-organized operational modes.

This chapter is based on the conviction that many complex systems, when subjected to multiple simultaneously operating 'energy-conversion processes', have the option of spontaneously finding a highly coordinated mode of operation, highlighted by temporospatial 'interdigitating synchronization' of energy conversion processes previously assumed to act separately. When 'driven' sufficiently (by continuously increasing transfer of energy and matter) such systems can be securely maintained, (i.e. in stable function); on the other hand they can reach the region where critical thresholds exist. As these are traversed, many systems will fall spontaneously into what was called 'synergetic cooperativity', where efficiency is optimized by minimizing unnecessary side movements. Interestingly, the initial state of turbulence, previously misconceived as a completely disordered mode of operation in moving fluid, falls into this important category. The formation of vortices is nowadays considered to be a framework for the spontaneous generation of highly ordered modes of operation. The emergence of vorticity in arterial blood flow is an event well known to be associated with systolic acceleration: synergetics teaches us that this is a situation where the heart is kicking a dynamic system into self-organization by rapidly driving it through many transients, which spontaneously recede during each diastole.

The essence of the contemporary fluid-dynamic interpretation of atheroma formation was developed in models of the arterial segments known to be predilection sites for atheroma formation, and the work of many pioneers in the field cannot be overestimated. However, we now know that the instruments used in this theoretical work (so-called hot film anenometers gauging fluid motion in regions very close to the wall) are as inappropriate for the analysis of separated flow as were the early versions of the ultrasound Doppler flowmeters used in clinical medicine. Both types of apparatus, in averaging-out the effects of complex, locally superimposed and divergent flows, literally blur the view into self-organized vortex movement and are therefore intrinsically useless for the study of multidirectional flows (Hutchison and Karpinski 1988).

Our knowledge about the true nature of self-organized vorticity is necessarily fragmentary, but rapid progress can be anticipated. In order to follow the development of the field detailed knowledge about the theory and practice of the basic research during the last two decades is less important than a correct comprehension of the ordered mode of operation found in vortices. Since mathematical modeling by computer, initially utilized to check the theoretical justification of model experiments, has now provided us with the most powerful tool for producing simplified yet very detailed computed moving images displaying the kinematic details of vortex flow, the physicians applying fluid-dynamic concepts in clinical settings are better advised to study a simple model that shows the peculiar properties of living vortices.

Vortex Shedding as a Daily Experience

In the following, terms like 'vortex flow', 'vorticity', 'self-organization' and 'pseudochaotic ordering' will frequently be used, none of which are part of conventional medical semantics, let alone conceptually familiar to neurologists. However, a fundamental phenomenon to which these terms apply is part of everyone's daily experience: the spontaneous ordering of moving water into a perfect whirl called a vortex, eddy, toroidal flow or sometimes 'dead water'. The highly volatile structure, more appropriately called 'dissipative structuring' (Schmid-Schönbein 1993a) is in fact spontaneously generated; it represents the sequelae of a 'consensualization process' in the mode of operation of a moving fluid. No external agency other than the interplay of three energy-conversion processes (pressure or potential energy, viscous energy dissipation and superimposed influences of kinetic energy) is acting, but the details of the dynamics driving the water, the geometry of the conduit taken by it and the previous history of the fluid motions play a dominant role. Water, when driven merely by a sufficiently high gradient of potential energy (in the present case by a hydrostatic pressure head), invariably falls into this ordered motion when running out of an ordinary bathtub or washbasin (Fig. 13.2).

Under these conditions the water starts to fall into 'curled' pathways: it is generally assumed that these are caused by the Coriolis effect, a consequence of terrestrial rotation. However, when the water is driven by some other current, the rotation can be anticoriolan; in each case the vortex movement becomes progressively stabilized in the initial direction owing to the kinetic energy of the rotating water. In either case, this is a mode of operation of a moving fluid basic to the kinematics of self-organized ordering emerging in the behavior of a driven fluid. The vortex movement is superimposed over a downward movement by gravitational forces acting on a system far displaced from the fluid-dynamic equilibrium.

The arterial vortices described here differ in many important aspects; physicians interested in the fundamental behavior of 'spontaneous vortex shedding' in flowing liquids may visualize for themselves the details by using ink injected through a narrow needle into an established vortex: the moving ink streaks show the essence of the phenomenon occurring in arterial segments with markedly non-cylindrical geometry (and sometimes in dilated left atria, see below). The kinematics and dynamics of highly ordered 'secondary movements' are thoroughly detailed in textbooks of advanced physics: they have recently been recognized as a universal process governing the physical behavior of fluid elements at all geometrical levels. Starting from the so-called 'quantum turbulence' involving individual molecules of superfluid helium (0.1 nm) to the movement of particles in electrical fields (1 mm), to the vortices behind moving vehicles (1 m) or in the earth's atmosphere (100 km) to the movements known to occur in spiraling galaxies, they are always the consequence of self-organized interactions between the effects of different forms of energy.

The following basic experiments are recommended to study vortex phenomenology.

1. A wash-basin should be filled and the water should be allowed to come to complete rest, as evidenced by the fact that a drop of ink remains stationary (Fig. 13.2A).
2. The outflow should then be very slightly opened by carefully pulling out the plug. If the basin empties only very slowly strictly laminar flow out of the sink occurs, a fact visible by the formation of one broad streak. (Fig. 13.2B).
3. After carefully pulling the plug completely open, a vortex will be seen to form, with circular streamlines (Fig. 13.2C). This is not due to any movement: the vortex turns as dictated by the Coriolis effect: clockwise in the southern, counterclockwise in the northern hemispheres of the earth. If the experiment is performed when the water is still in motion before the plug is lifted, the rotation of the

vortex may well be in the opposite direction i.e. clockwise in the northern hemisphere and vice versa (Fig. 13.2F).

4. Once established, there is prolonged residence of the dye in the toroidally moving water (Fig. 13.2D). If the plug is replaced after the vortex has formed, the toroidal movement in the water continues for a long time.
5. If, after the vortex has formed itself, the basin is carefully filled with running water so that a finite water level is maintained, the vortex can be kept alive virtually indefinitely.

The highly ordered yet extremely labile play of the fluid laminae can best be studied if a bathtub is used, which can better be kept in a dynamic flow equilibrium by careful replenishing of exactly the amount of water lost without disturbing the flow. This is a typical non-equilibrium system so called because it is operating far from the fluid-dynamic equilibrium prevailing just when the last drop of water has left the tub.

Playing with the possibility of preventing the Coriolis effect with small currents before pulling the plug, the duality of the directional options of the vortices can be experienced: theoretical physicists call this placing the system near to a 'bifurcation'. Once it has made up its mind, the preferred mode is not only maintained but amplifies itself so that a high degree of stability emerges irrespective of whether it is initiated by the Coriolis effect or by residual currents. Only very near the onset can the imminent toroidal movement be stopped: once established it will roll on, indicating the progressive preponderance of kinetic energy.

In the last analysis, the highly ordered and very stable state of a fully established vortex results from the reimportation of kinetic energy of whirling fluid. In the vortices that form in a pulsatile tube flow, the vorticity of the fluid has an entirely different reason: the rapidly accelerated but also rapidly decelerated primary import of kinetic energy; here, therefore, our bathtub model is no longer valid. (Kinematically very similar patterns are found in the examples of self-organized fluid-dynamic systems such as the Benard roles or Tayor vortices (Prigogine 1979; Haken 1983; Manneville 1990).

All vortices represent modes of movements where millions of fluid molecules are 'consensualized'. As a matter of fact, there is something like an inner circulation of fluid elements within the vortex: careful observation of immersed soap flakes allows the visualization within the toroidal flow of the relative movement of fluid trajectories with different velocities. The vorticity follows relatively simply trajectories, provided that there is sufficient time for the formation of a coherently moving macroscopic element, and that the momentum (or inertia) of the whirling fluid is conserved by incorporating more and more of the fluid elements into one large coherent pattern. This ordering of a movement requires essentially a minimum of downward flow and the initial import of some kind of local (current in the basin) or global (rotation of the earth) kinetic energy. Even the slightest alterations in the prevailing boundary conditions can have profound consequences, for example regarding the directionality of vortex rotation. It has become customary to refer to such systems as 'chaotic', a term consistent with the reasoning of traditional mathematics but not necessarily with the intuition of biologists and physicians. Hopefully, the latter will agree that the term 'pseudochaotic ordering' better suits the phenomenology of the observed forms of incipient turbulence never pushed to the random movements one observes when a river is rushing over a waterfall.

Transition and Turbulence: the 'Pseudochaotic' Order of Pulsatile Flows

It would be erroneous to assume that the stable ordering emerging spontaneously by way of 'self-organization' always produces beneficial effects (Schmid-Schönbein 1993a). This statement is the theme of this chapter, which is concerned with the spontaneous evolution and devolution of vortex flows at predilection sites for atheroma formation in arteries. Hitherto, their description has been entirely phenomenological. However, since the topic has deep roots in the new interdisciplinary scientific discipline of chaos research, our knowledge about the kinematic regularities can be embedded into a powerful scientific paradigm about processes previously considered to be totally random. We now know that many dynamic processes are determined but not predictable: this is a combination of features alien to the conventional wisdom of 'organized' systems. To do justice to both the traditional and the novel usage of the terms involved, one might call pulsatile vortex flows the result of pseudochaotic ordering in the mode of operation of a driven system. This terminology refers to the deterministic sequelae unavoidable in a fluctuating mode of operation, and not to any kind of arbitrariness in stationary states.

The fundamental work of Prigogine and Haken has emphasized that when amorphous materials (gases and fluids) are being progressively driven, i.e. when they are being exposed to continuously increasing gradients of pressure or temperature (in fluid dynamics), of voltage (in electrical systems such as a Laser), of chemical concentration (as they can be built up in chemical reactors) or of affinity (as follows from the emergence of catalytic agents by external addition or internal production), characteristic sequences of coherent movements can be induced. Movement in this context refers to the transfer of fluid elements (gaseous or liquid), charge (electrons or ions) and chemical species (micromolecules); in the latter the term movement refers either to the process of dislocation of individual mobile elements (molecule, ion, electron) or to the deformation of macromolecules. Collectively, provided it occurs in a highly ordered fashion, the sum of the individual microscopic movements represents the macroscopically visible mode of operation of a driven system. In order to see this mode of operation in fluids movement-sensitive detectors are required, such as those available in the so-called color-coded ultrasound Doppler anemometers, which are one of the few methods capable of displaying the emergence of complicated macroscopic order in blood as a driven fluid-dynamic system in medicine. This follows from the application of high-performance computers that can transfer not only information about the velocity and direction of moving particles in the blood, but also about the temporospatial distribution of the particle trajectories. It would be an error to average out these fluctuations, since the wavering aspect of the signals (or their mathematical representations on the computer screen) contains pivotal information about an important pathogenetic process. A bewildering liveliness springs to the eye, fugitive patterns coming and going with each heartbeat. (It is not inappropriate to emphasize that the terms chaotic (as recently used in non-linear dynamics) and pseudochaotic ordering describe modes of operation characterized by what might also be called intrinsic freedom (or synergetic liberty).) This can actually be derived from the causes acting and from procedures for macroscopic analysis. The possible trajectories (or their temporal alterations) taken by particles in the deterministic chaotic mode of operation determine the degree of freedom. The latter can be measured by appropriate methods such as the Fourier transformation. Liberty as a category can also be derived from the underlying causes of chaotic behavior in that the driven systems are fundamentally free to respond to very weak influences (so-called strange attractors) that would not even be felt by systems operating near equilibrium.)

In order to appreciate properly the information obtainable by such systems, it must be kept in mind that they represent an imperfect attenuated portrait somehow reflecting the temporal aspects of the oscillating mode of operation of the moving blood elements generating the Doppler-shift signals, which fluctuate in time and space (Steinke et al. 1990). What makes these instruments so interesting is that they provide a potential to diagnose in vivo whether or not rapid temporal and spatial fluctuations in blood movements actually take place in the individual patient, and they corroborate (however imperfectly) the emergence of one of the pivotal fluid-dynamic events identified by theoreticians within the atherogenetic process (vortex shedding at certain predilection sites of proximal arteries).

The future combination of these two assets promises scientific progress. By opening to the medical profession a routine method for hematokinematography, it will hopefully make users aware of the fact that we finally have ultrasound instruments that display indicators for the sequelae of a short-term deterministic yet long-term unpredictable (hence 'pseudochaotic' ordering) process with great pathogenetic significance. The instantaneous mode of operation of the driven blood from which vortex flows emerge depends on clearly definable cause and effect relations i.e. it is therefore deterministic. However, since the intensity of the accelerating forces (pressure gradients) is the subject of temporospatial fluctuations, and so are the responses of moving elements, in the form of complex three-dimensional flows, the latter are indeterminate: in this sense it must be accepted that they are not predictable and fall into the category of systems naively called 'chaotic'.

This follows from a characteristic combination of reasons: the intermittently perfused blood vessels are operating in an unstable dynamic steady state far displaced from fluid-dynamic equilibrium; such systems are known to be delicately sensitive to their previous history, the details of the presently acting forces and the trends in the kinematic responses; and they are unpredictable, because they can reach critical thresholds and undergo a spontaneous transition from one deterministic into the next deterministic (but different) mode of operation. This kind of repetitive pulsatile transition produces the vividly vacillating pattern seen on the screen of color-coded scanners, and it explains the ephemeral nature of the pattern, i.e. its potential to produce quite different patterns from beat to beat (Steinke et al. 1990).

What occurs when vorticity emerges is appropriately termed a 'fluid-dynamic instability' (Manneville 1990), a topic hotly debated among physicists studying the transition between the perfectly organized (or truly laminar) Poiseuille flow and the early manifestation of incipient turbulence (Ebeling and Klimonovic 1984; Ruelle 1991). Instabilities in accelerated fluids are known to exist only transiently; for this reason, earlier students of turbulence were often unable to detect them. Their regular emergence, their high degree of order (and, of course, their kinematic origin) were overlooked until the end of the 1970s, but are now placed at the center of interest. In theoretical physics it is this very topic that has proven false the time-honored idea that turbulence was always random. As one observes the sequelae of vortex emergence, one must appreciate that the energy transfer processes associated with the highly pulsatile flow of blood in geometrically peculiar regions contain the potential to disconnect for a limited period certain fluid elements from the rest of the blood. This occurs when pulsatile flow in vortex rings or vortex balls induces the coherent rolling movement of fluid elements dynamically and functionally separated from the remaining fluid by a boundary layer. Fig. 13.3A displays this fundamental mode of operation as seen in the eddies (in reality quite regular vortices with laminar flow within the eddying fluid) behind the pillars of bridges in a river. A different kind of eddy formation can be visualized in the clefts between cobblestones during rain: here, vortices are formed in the mouths of stagnant clefts by the current in the ones carrying water – or better, a little mud to allow flow visualization (Fig. 13.3B).

When flow is extremely pulsatile, as in arteries, and comes to a stop between two pulses, three-dimensional vortex rolls are born and automatically die out. In human arteries, individual vortices can be extremely short-lived but new ones can be regenerated with each systole. Interestingly, when described in terms of its energetics, a very similar scenario is found in biochemically self-organized systems which emerge wherever a non-linear biochemical process explodes into rapid reactions due to autocatalytic amplification loops (Schmid-Schönbein 1990a,b). Such self-organization is typically found intracellularly, where it occurs in compartments separated from their environment by stationary membranes (Hess and Markus 1987). However, as is known from non-stationary gas dynamics, a similar self-organization chemical can occur in vortex rolls separated by a fluid boundary layer from the remainder of a moving gas or fluid (Grassmann 1983).

For both types of self-organized order, very similar kinematic and energetic regularities are presently being discovered by theoretical physicists. Autocatalytically accelerated platelet-dependent procoagulant (and therefore atherogenic) processes can also occur in the compartments functionally separated (by a boundary layer rather than by a membrane) from the remainder of a moving fluid. As the catalytically accelerated processes in the separated fluid proceed, they also alter the cause- and-effect relationship for fluid movements, simply because – by platelet aggregation and/or fibrin formation – they alter the fluidity of the separated fluid elements. Here, another dynamic process with pseudochaotic ordering comes into play: a surprising inner congeniality of two functional principles cooperating in the atherogenetic process can be identified by utilizing contemporary concepts about 'joint efforts' in non-linear dynamic systems (Schmid-Schönbein 1990a).

Superimposed Physical and Chemical Self-Organization: Vortices as Short-Lived Flow Reactors

Self-organized coordination of highly distinct dynamic systems in cardiovascular physiology has only rarely been described (Schmid-Schönbein and Wurzinger 1986; Schmid-Schönbein 1990a). The scientific concepts of different disciplines can be combined into a new interactionist paradigm based on first physical principles (non-linear non-equilibrium thermodynamics). Its application to normal blood flow (Schmid-Schönbein 1993a), to the disturbed microcirculation (Schmid-Schönbein 1990b, 1993a) and to the chaotic reactions found in unstable angina (Schmid-Schönbein 1990b) proved to be fruitful in generating new experiments.

The mental process involved in the application of these principles to the present topic, however, requires not only a comprehensible explanation of ordering mechanisms but also the induction of an awareness and unrestricted acceptance of the fact that, with each systolic contraction of the heart, the blood moving at predilection sites for atherogenesis is not simply propelled, but rapidly driven through a sequence of transients positioned between laminar and turbulent flow. The pronounced difference between systolic flow pulse and diastolic standstill in the proximal arteries is the main cause for the quality of transiency, whereas a host of other factors is responsible for the highly variable extent and duration of vorticity. These factors include the exact geometric configuration of a conduit as a prime determinant, but also the distensibility of the conduit, the duration of the previous diastole, the flow behavior of the blood and the existence of separated matter (e.g. platelet aggregates). These considerations illustrate that an observed phenomenon is deterministic yet unpredictable in that it does not follow the trivial cause-and-effect relationship found in biological linear systems.

At present, the analysis of non-linear dynamic systems resides primarily in the hands of theoretical physicists, and the concepts elaborated are formulated in the abstract form of differential equations; modes of operation are being portrayed as phase diagrams and power spectra, allowing quantification of the action of attractors. It would be a hopeless enterprise to even start to introduce this world of important paradigms into a medical textbook. However, it is possible to display the important complex patterns of motions with the help of realistic moving pictures. Where it is essential to depict kinematics normally hidden to physicians as appliers of fluid-dynamic concepts, cinematography is the appropriate medium. Since a kinematography standard for inexpensive video norms (VHS using NTSC or PAL/SECAM) is now available, it is technically easy to distribute this important information. For this reason, the present authors have collected important kinematic recordings of the most prominent experimentalists (Goldsmith and Karino, Liepsch) and animated computer images (Perktold), kindly supplied for the present purpose into a short VHS-demonstration that can be borrowed for a nominal fee from the authors.

Readers interested in fluid dynamics are referred to the original literature (e.g. Nerem and Levesque 1983; Liepsch 1990; Caro 1981; Rodkievicz 1981; Perktold et al. 1987, 1988a, b; Naumann and Schmid-Schönbein 1983; Schmid-Schönbein and Naumann 1984).

Kinematics of Fluid Separation into Rolling Vortices

Localization of Vortices at Predilection Sites for Atherogenesis

For a long time (e.g. Aschoff 1912) fluid-dynamic considerations have been included in the pathogenetic theories about the initiation, progression and complications of thrombosis and atheromatosis. Such considerations can now be extended to a comprehensive functional explanation of why virtually all biochemical, enzymatic and cytological reactions comprising the atherogenic process preferentially take place at predilection sites. Various forms of long-term cooperativity of self-organized intravascular macroscopic abnormalities and transmural microscopic mass transfer (convective and diffusive processes) are assumed to occur, as well as intramural reactions subsequent to the exposure of cellular and non-cellular elements to the highly localized injurious effect of abnormal transmural transport. The generated atheroma then is considered as a fluid-dynamic obstacle which later determines both the site and extent of atheroma progression.

In producing flow separation, the cooperative effects of kinetic energy and frictional energy dissipation must determine the site of endovascular lesions, whereas the action of the arterial blood pressure is the factor that determines the phenomenon of transmural transport through walls denuded of endothelial cells. This simple hypothesis explains many established regularities, such as the localization of early and advanced lesions near branchings, as well as the role of aging and arterial hypertension in the progression of this disease. Fig. 13.4 shows a computer-generated false-color image of the directionality of flow in a carotid bifurcation at various intervals from late diastole through early and late systole and back to diastole. The picture should be viewed repetitively to grasp its dynamics.

The phenomenology of flow separation phenomena in models of arteries has been uncovered by the microrheological method (flow visualization of particle trajectories), almost exclusively by Goldsmith and Karino (Karino and Goldsmith 1989), as well as by Liepsch's work (Liepsch 1990) on macroscopic flow patterns in elastic models of arterial conduits, perfused in pulsatile fashion with transparent non-newtonian model fluids. Therefore, one can assume that arterial bifurcations are characterized by regions of accelerated systolic flow velocity, and hence potentially high fluid shear stresses, by flow separation and/or by formation of transiently generated recirculation zones, and lastly by the occurrence of stagnation point flow (see Fig. 13.6). Stagnation point flow can occur in two regions of arteries: as a physiological event at the flow divider, and under abnormal conditions at the reattachment point of separated streamlines. In the latter, particles carried downstream into the reattaching streamlines are moving forwards, whereas particles carried downstream into the closed streamlines of the circulation zone move backwards. Particles or molecules carried in the fluid lamellae between those impinged on to the wall. These very fluid lamellae originate in the recirculation zone, thus providing a rich source of material to be deposited on to the wall in the immediate neighborhood of the actual reattachment point.

For three reasons, these facts have pivotal significance in explaining the dynamics of a synergistic sequela of atherogenic events lasting many years:

1. As shown by Rodkievicz (1981), reattachment points in the natural arterial tree of man coincide with the predilection sites for enhanced permeability to albumin in early and, to a lesser extent advanced, atheromatous lesions.
2. The structure and biology of endothelial lining, i.e. the shape and orientation of endothelial cells and their rate of mitosis, differ significantly from that found elsewhere in the arterial trees, a fact generally held responsible for enhanced rates of filtration of macromolecules such as albumin and low-density lipoprotein (Sinzinger and Jellinek 1977).
3. The fluid-dynamic conditions in the areas of disturbed flow are capable of triggering or amplifying platelet activation. As shown by Wurzinger et al. (1983) high fluid stresses as they occur at the boundary between separated and laminar shear are capable of damaging platelets directly or indirectly via lysis of red blood cells, thereby triggering the release reaction and viscous metamorphosis, and demasking procoagulatory phospholipids. In the recirculation zone, ideal conditions exist for autocatalytic augmentation of platelet activation and release reactions, as well as the procoagulatory process, whereas the reattachment point provides appropriate conditions for platelet deposition (Goldsmith 1972; Schmid-Schönbein and Wurzinger 1986).

Fig. 13.1. Evolution and devolution of vortex flow in the pulsatile forward movement of blood in the carotid artery of a patient (kindly supplied and commented on by Dr Steinke and Professor Hennerici, Mannheim, Germany). Color-coded Doppler imaging of the velocity distribution in the bulb of the carotid artery at its transition from the bifurcation (at the right side) into the internal carotid artery (which extends beyond the left margin of each picture). Each photograph displays one instantaneous situation of a sequence of events accompanying one cardiac cycle, patterns being associated with the systolic acceleration (1–5) and diastolic deceleration (6–8). The dotted line (in photograph 1) (early systole) indicates the direction of ultrasound emission. All forward movements right to left in this figure) are displayed in the spectral colors red to yellow (vertical calibration bar),whereas the backward movements (left to right, as in the jugular vein (JV)) are represented in the spectral colors from dark blue to light blue and turquoise. During peak systole (photographs 2, 3, 4 and 5) a complicated array of evolving and devolving broad-spectrum patterns is displayed, including large regions of vortex formation represented by blue pixels, i.e. local backward flow despite the global forward movement of the arterial fluid element. These complex patterns disappear in diastole (photographs 6, 7, 8), indicating devolution of the previously evolved vortex balls (an event taking place between photographs 4 and 6) and concomitant restitution of uniform laminar flow during relatively slow diastolic forward movement (photographs 7 and 8).

The left enlargement of photograph 2 (representing peak systolic flow) more clearly shows the almost completely closed line of black pixels (arrows). The black pixel chains are the two-dimensional projection of a three-dimensional shell of fluid elements with net zero velocity, which proves the existence of a rotating vortex ball delineated by borderline shells separating regions with fluid movements in the opposite direction. In the right enlargement the sequence of white arrows marks the general outline of fluid trajectories in the outermost part of the closed vortex ball. The vortex ball crudely described in photograph 2 represents a much more complex distribution of velocities under these natural conditions: it is taken as in-vivo evidence of the existence of the vortex phenomenon described in this chapter. Due to the limitations of presently available US-Doppler anemometers, the exact nature of separated vortex balls cannot be detailed; their existence, however, and most importantly their prolonged survival, can be taken for granted whenever black pixel rings are seen in sequential pictures (a criterion also fulfilled in photographs 3, 4 and 5).

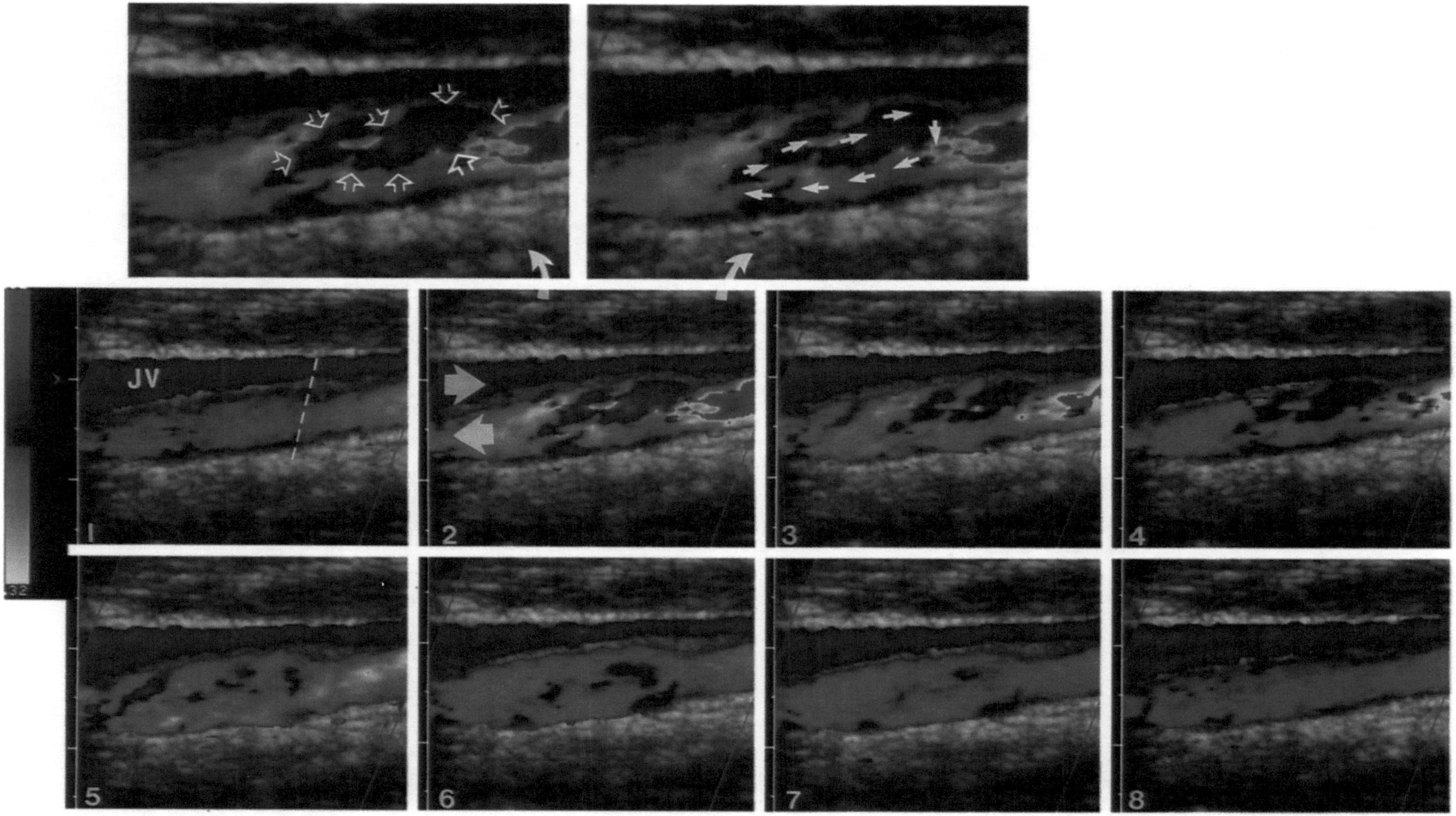

Fig. 13.1.

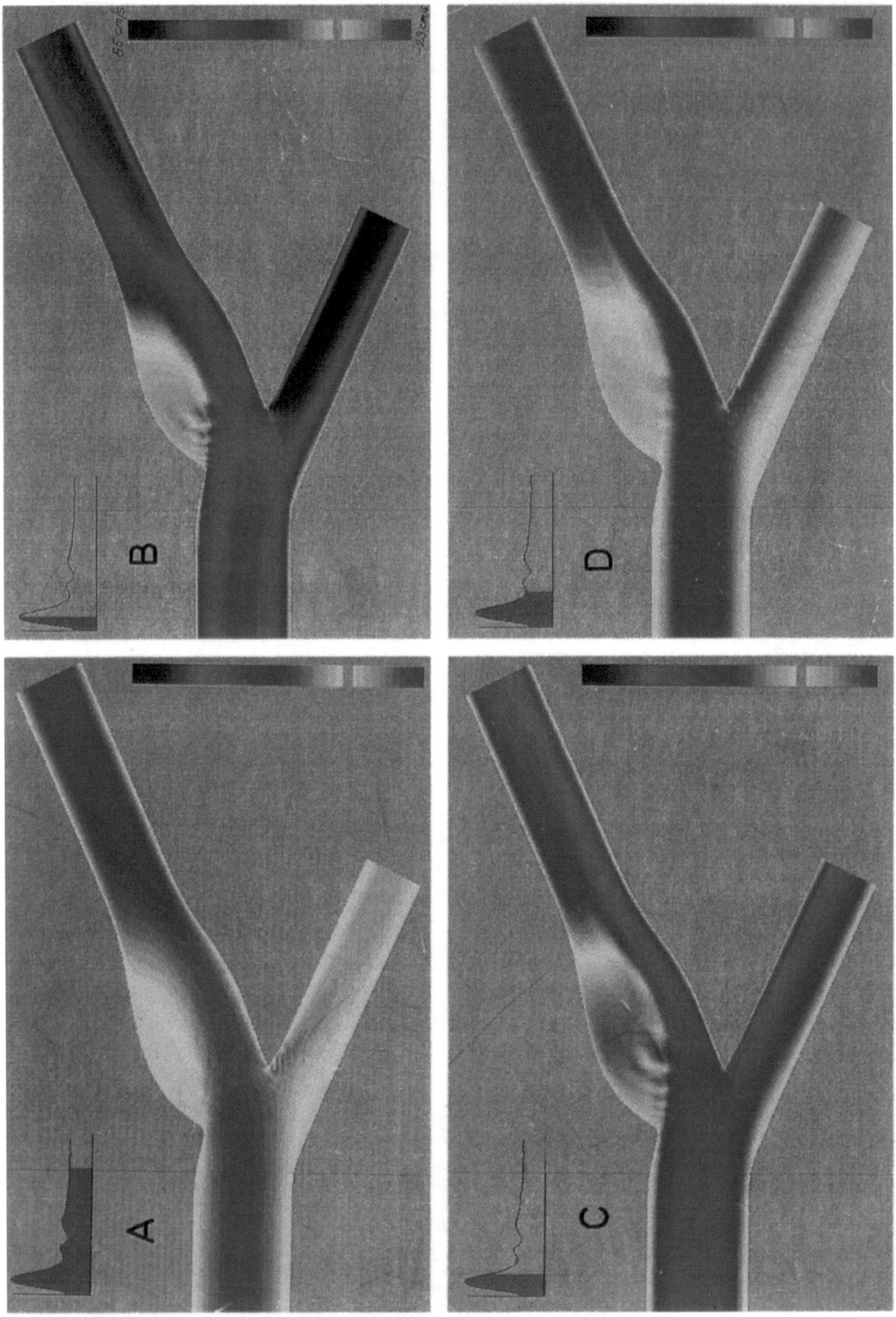

Fig. 13.4.

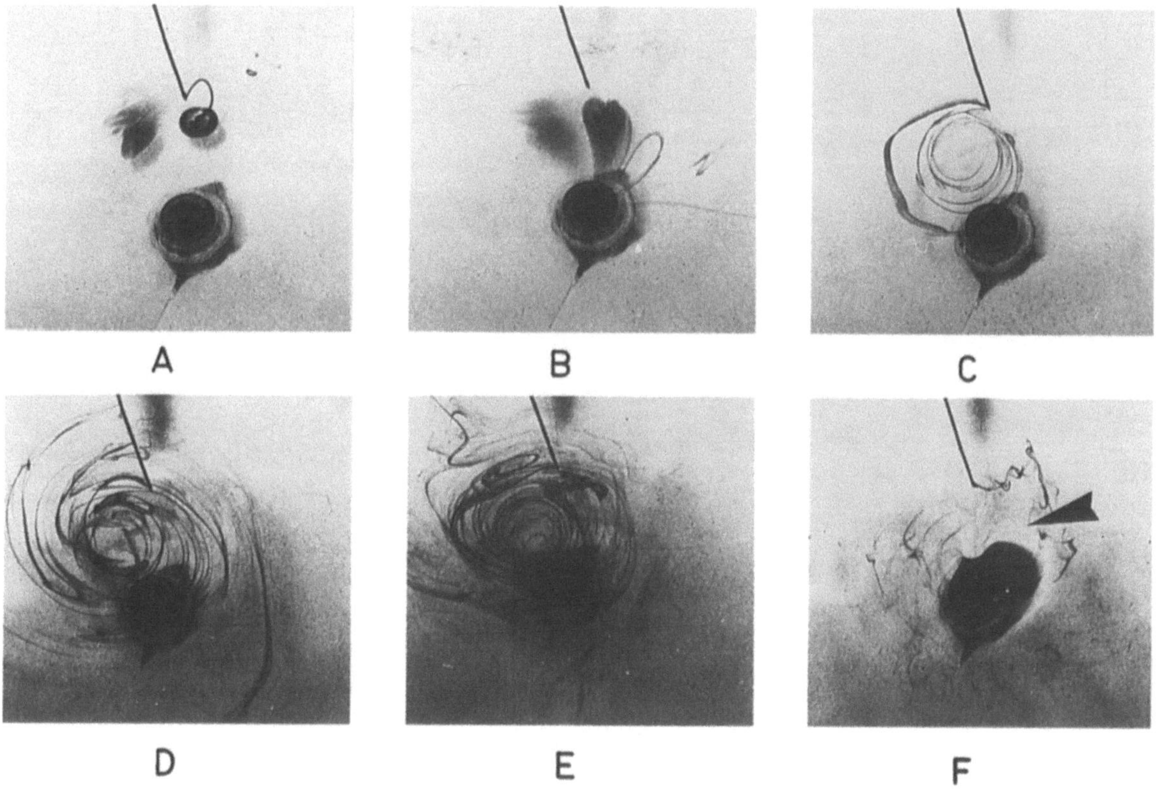

Fig. 13.2. Ink streamlines depicting the development of vorticity in a tub drained at varying flow rates through a gutter with controlled discharge. (Experiment suggested by Prof. D. Straub, Munich.) **A** The resting state: the dye remains stationary around the tip of the syringe needle. **B** Slow discharge: regular and linear movement of the dye directly towards the center of the gutter: strictly laminar flow (note: when performing this experiment, the stopper must be lifted very slowly to avoid secondary movements). **C** Discharge with high flow rate: spontaneous formation of a vortex, as indicated by the direction of the dyed streamlines passing by the tip of the needle (counterclockwise Coriolan movement, since experiment was performed in the northern half of the globe). **D** Fully developed vorticity: large-scale consensualization and long residence before colored water is discharged. **E** Development of secondary vortices (curled or curvilinear streamlines) at high rotatory movement. **F** Initiation of anticoriolan movement in fluid driven to regular movement (e.g. by asymetric inflow) before the plug is opened. Once the anticoriolan movement is established it is rapidly stabilized (due to kinetic energy influences). Careful play with input-derived movement (e.g. by appropriate direction of a shower) and spontaneous emergence of coriolan movements allow us to drive the system into frequent alterations of rotational direction of vortex: pseudochaotic ordering by variable external drive.

Fig. 13.4. Color-coded calculated velocity profile of carotid bifurcation with proximal stenosis and slight poststenotic dilatation. Sequential pictures showing late systole (**A**), early systole (**B**) late systole and early diastole (**C**), forward flow coded between +85 cm/s (dark blue) and 0 cm/s (yellow), backward flow depicted yellow (–0 cm), to red (–23 cm/s). Note coherent forward flow during diastole with large area of very low flow in the poststenotic area during early diastole, sharp demarcation of a separated region with pronounce backward flow and sharp boundary between rapid forward and vorticity with slow and backward flow. Late in systole the backward flow vanishes leaving separated stagnant area during the entire diastole reminiscent of **A**.

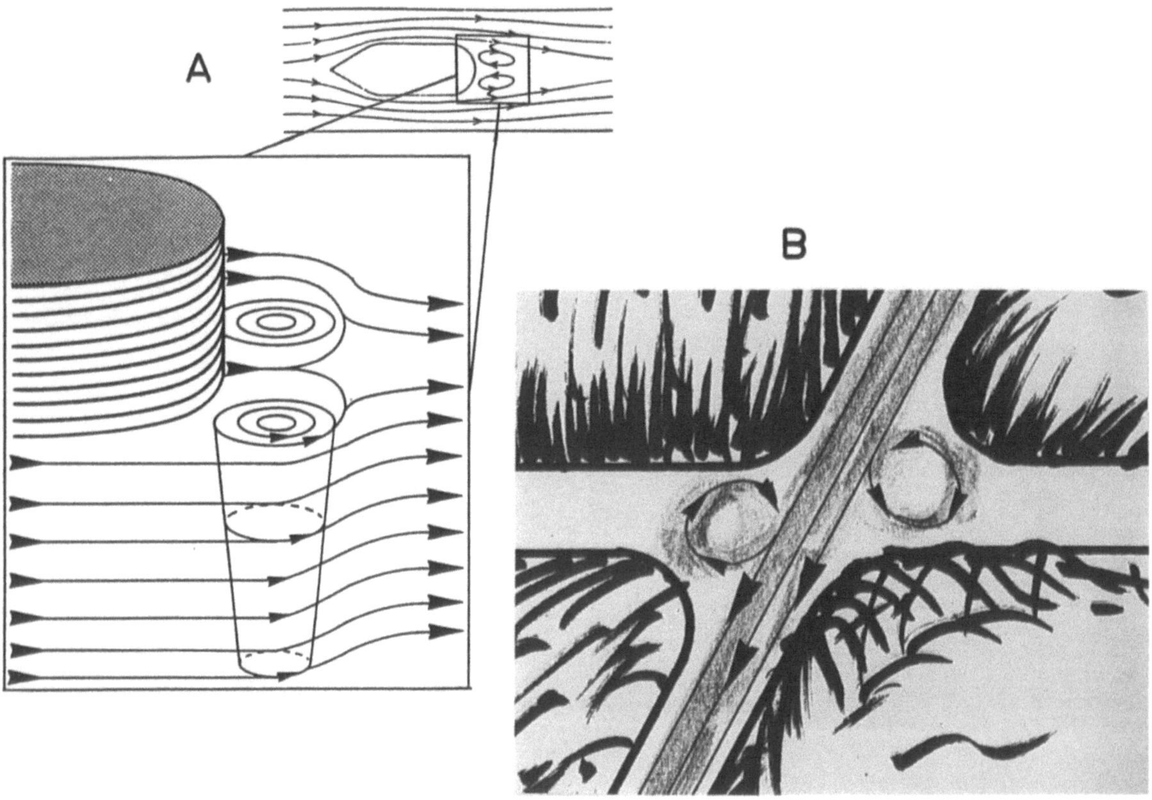

Fig. 13.3. Schematic representation of vortex formation in daily life. **A** Vortex pairs behind the pillar of a bridge: note that this example of self-organization is driven by a fluid layer separating a three-dimensional roll of fluid in which the individual water element is spiraling on trajectories closed upon themselves. The blown-up section depicts the viscous drag of a boundary layer of fluids which keeps the vorticity in perpetual motion. **B** Vortices in the mouths of clefts between cobblestones driven by the current of water in the thoroughfare.

Stagnation Point Flow as an Inevitable Epiphenomenon of Flow in Bifurcations and of Vortex Generation

In perfectly laminar flow in a cylindrical tube, no fluid elements are transported by convection towards the wall: all fluid elements travel on trajectories parallel to the inner surface of the conduit. Stagnation point flow is a fundamental and inevitable characteristic of fluid motion in branching (i.e. arterial) tube networks, whereas it does not occur in converging (i.e. venous) tube assemblies. The endothelial cells at arterial vertices (tips of the flow dividers) appear to be adapted to this: there is ample evidence (Stehbens 1983) that the flow near the apex of bifurcations is characterized by a combination of high shear stresses and high normal pressure due to the skewed velocity profile and the rapid pulse associated increase in the shear rates. It is therefore not surprising that arterial flow dividers show very early changes at these sites (which may even be present before birth), which have often been classified as pathological. However, while there is no doubt that there are degenerative alterations of the endothelium and in the subendothelial structures of so-called arterial cushions, we suggest that these cushions do not represent damage alone but rather biological adaptations to a uniquely occurring but normal dynamic load, which is experienced exclusively by these parts of the vasculature and to which they are exposed from the very beginning of vascular development (Fig. 13.5). The cuboidal shape of the endothelial cells found in the vicinity of flow dividers may well be of microrheological advantage, since the viscous drag force on each cell (shear stress multiplied by surface area exposed) is considerably smaller than in flat

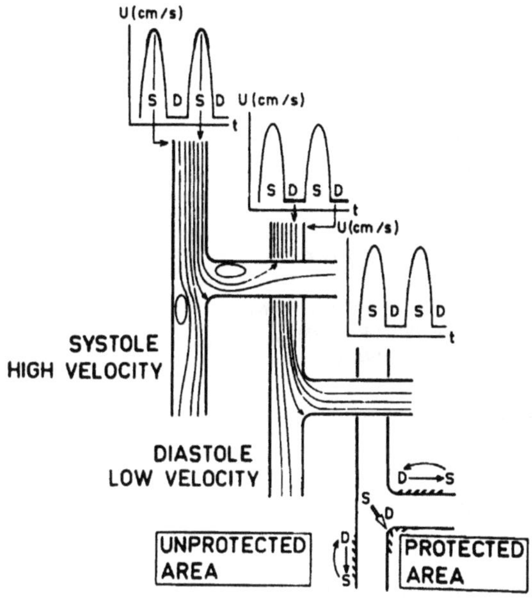

Fig. 13.5. Summary of the fluid dynamic and biological concepts in which flow towards the wall (which occurs in flow dividers and at reattachment points) is considered as the damaging factor for the endothelium. Near the flow divider, all fluid dynamic factors (flow rates, viscous drags, pressure and suction) vary considerably in time and space. It is assumed that the arterial cushions found at the tips of the flow dividers represent the results of lifelong adaptation to this situation, whereas the wall opposite the flow dividers (where flow separation and stagnation point flow occurs) are not properly adapted. The flow divider is protected from the progression of atherosclerosis because of adaptive mechanisms which develop early in life.

endothelial cells. Furthermore, the more cuboidal shape might allow elastic compliance (and/or viscous attenuation) or shear stresses fluctuating in direction, magnitude and duration during the systolic peak of arterial blood flow. As stressed by Zarins et al. (1983), the intimal pad in the arterial cushion has, in fact, a highly structured medial zone composed of oriented elastic and smooth muscle layers. Moreover, the adventitia underlying the flow divider contains many more vasa vasora than the neighboring straight vessel segments (Schmid-Schönbein and Naumann 1984). The work of Betz (1988) suggests that the growth of adventitial vasa vasorum may be secondary to local increase in endothelial permeability. Conceptually, this idea allows us to separate 'protected' from 'unprotected' arterial regions, the former being accustomed to fluid-dynamic stresses and hence structurally and functionally adapted to their load. Thus, they differ in the extent of their response to injury from those areas which are less drastically (or at a later time in life) affected by fluid-dynamic loads, but in which, without proper adaptation, a smaller stimulus leads to more pronounced reactions.

The situation is on the one hand similar, on the other hand very dissimilar, at the other possible site where reattachment flow can occur in arteries, namely in the region of the migrating reattachment points of dynamically separated vortices. These lie at the outer side of the branch arteries; they are exposed to stagnation point flow only later in life and they are obviously not properly adapted to the sequelae of stagnation point flow. Furthermore, the transmural transport of blood elements is related to a material previously activated in its biochemical potential. To understand this, the special kinematics of multidirectional flow near bifurcations must be understood.

Outline of the Fluid Dynamics of Flow Separation into Vortex Rolls

Any attempt to explain the localization of the atheromatous process near and in arterial bifurcations must consider the general fluid dynamic conditions created when a highly pulsatile pump (such as the left ventricle) phasically displaces a highly non-newtonian fluid into a tree of frequently branching distensible tubes. Owing to the fact that the heart generates a rapidly accelerated flow pulse, blood motion in the large central arteries is more or less restricted to the period of ventricular systole, followed by marked retardation or even standstill during diastole (Fig. 13.6). In the next systole, not only the stroke volume but also the blood volume held in the aorta and the first generation of arteries is suddenly accelerated (or loaded with kinetic energy). Note that, unlike the case of vortices in bathtubs, flow most likely comes to a complete halt in the subsequent diastole. Due to the geometrical features of the vascular tree, the volumetric flow rate and – even more so – the local instantaneous fluid velocities decrease rapidly as blood moves centrifugally from the proximal arterial bifurcations to the distal ones. As this occurs, the amount of kinetic energy of the individual blood element (which is proportional to the product of the moving mass and the square of the averaged velocity, $0.5m \times v^2$) rapidly vanishes in the more distal generations of arterial branches, which explains why only the

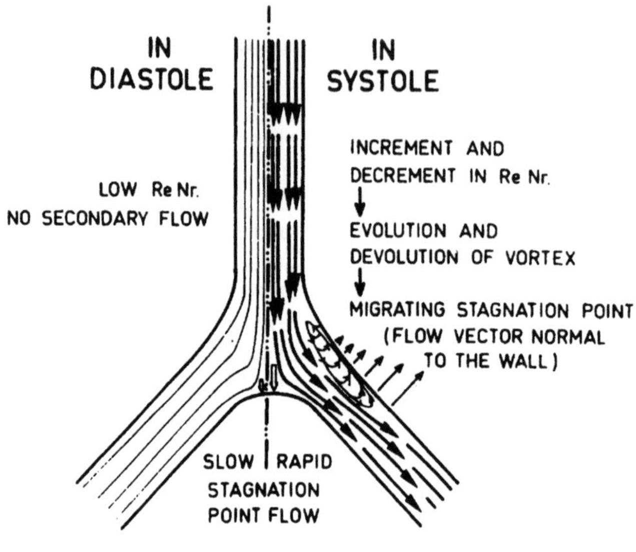

Fig. 13.6. Synoptic representation of the major flow components of pulsatile flow through an arterial bifurcation. The left side of the figure shows the situation of largely surface-parallel creeping flow during diastole. At the right side the pronounced separation of the mainstream from the outer wall in peak systole is shown with the development of backflow in the form of a vortex or recirculation zone. The latter comprises stagnation point flow at the reattachment point, which is moving back and forth with the systolic expansion and diastolic regression of the vortex.

proximal arteries and their bifurcations are expected to be sites of flow phenomena related to phasic shifts in kinetic energy.

Transfer rates of kinetic energy and rates of viscous energy dissipation are responsible for the stability of motions, the former destabilizing laminar flow, the latter attenuating fluctuations. This is illustrated (Fig. 13.7) by road traffic, an example taken from everyday life. Assume a straight highway blocked by a road construction site, and a short bypass leading over a parking loop. For good reasons a speed limit is strictly enforced, simply because orderly (stabilized) traffic in this tortuous path can only take place when the kinetic energy of the objects, forced into a sudden change in direction, is low. Whether or not the motion is orderly, that is to say follows the prescribed path of the conduit, however, also depends on the traction (or friction) of the tires, in other words on dissipative attenuation of an irregular motion. If the roads are icy, deviations from orderly motion (i.e. a skidding motion heading straight ahead) will take place even at very low velocities. Conversely, the broad tires of modern racing cars, which improve friction at the boundary with the road surface, allow them to speed through such a twisted or tortuous path, despite the action of very high kinetic energy associated with the sudden sequence of motions directed in different angles to the original path. Here, the inertia of moving masses begins to dominate the pathways: Fig. 13.7 shows that when fluid elements are forced to take tortuous paths past obstacles (i.e. are forced to deviate from the straight trajectory) a stabilized orderly motion will occur only when their kinetic energy is small in comparison to frictional energy provided that energy dissipation is associated with flow.

The mathematical relation between these energies is described by the time-honored Reynolds number:

$$Re = \frac{\text{average velocity x diameter}}{\text{kinematic viscosity } [(m/s \bullet m/m^2/s)]}$$

$$\left[\frac{\frac{m}{s} \bullet m}{\frac{m^2}{s}}\right]$$

a dimensionless ratio used conventionally in the applied engineering sciences to predict the occurrence of the transition from purely laminar to what appears to be fully turbulent flow in long cylindrical tubes. For the reasons outlined above, the traditional concept of a so-called critical Reynolds number for this kind of complete transition is irrelevant for the fluid-dynamic analysis

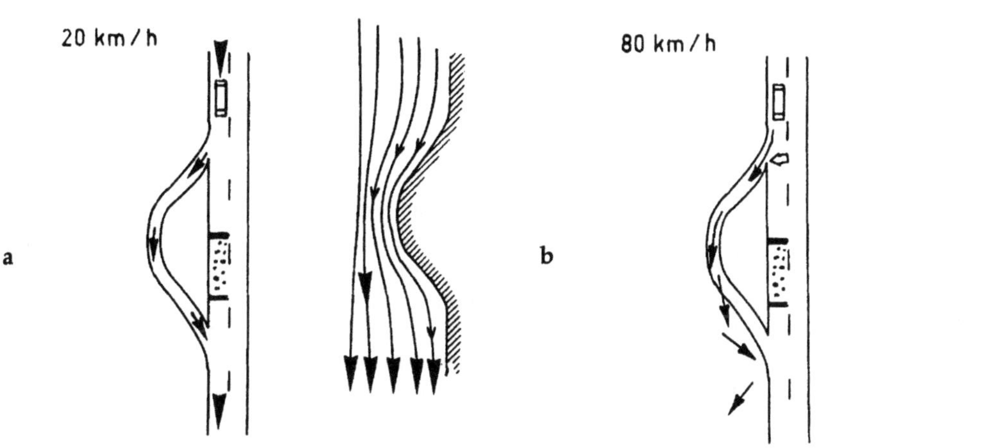

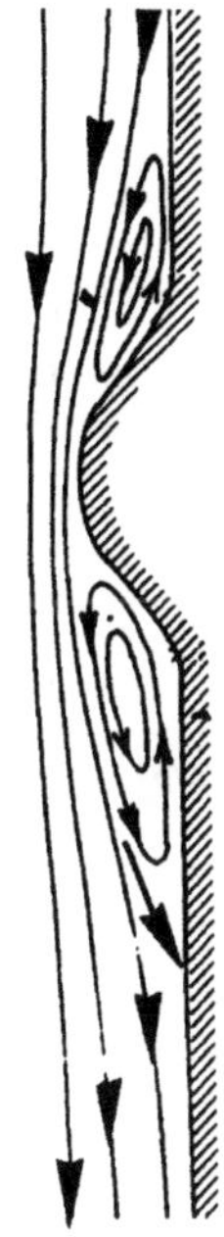

Fig. 13.7. The occurrence of secondary flow behind an obstacle as a function of flow velocity and viscosity is illustrated by the situation of a car bypassing a road obstruction through a winding bypass. The friction between tires and road corresponds to the viscosity of the fluid, which in turn must be considered the result of cohesive forces between the fluid molecules. **A** The movement of the car (and of fluid elements) is coherent and ordered if the kinetic energy is low and/or the frictional energy dissipation is high. **B** The movement is incoherent (skidding and tumbling) when transfer of kinetic energy is high and/or frictional energy dissipation is low.

of cardiovascular systems, simply because the development of full turbulence requires substantial time and higher Reynolds numbers (due to the early systolic acceleration) and occurs only for comparatively short instances (in the order of 100 ms).

Kinematically speaking, the logic of the Reynolds number (which in the context of cardiovascular physiology and pathophysiology should rather be called Reynolds ratio (Schmid-Schönbein 1993b) allows an estimate of the ratio of the kinetic energy of a moving fluid element and the frictional work done in conjunction with the movement. This logic can be easily made transparent by defining kinematic viscosity as the ratio of the dynamic viscosity:

$$\eta \; [Ns/m^2]$$

and the density:

$$\left[\sigma \frac{N \times s^2}{m^4}\right]$$

of a fluid and rearranging (using SI units) the determinants of the Re ratio thus read:

$$Re = \frac{\text{velocity x density x diameter}}{\text{dynamic viscosity}}$$

$$Re = \frac{\bar{u} \times \sigma \times d}{\eta}$$

Multiplying both the numerator and the denominator of the equation by the velocity (u [m/s]), the equation becomes:

$$Re = \frac{u^2 \times \sigma}{\eta \times \frac{u}{d}}$$

$$= \frac{\text{velocity squared x density}}{\text{dynamic viscosity x}\frac{\text{velocity}}{\text{diameter}}} \; \{\text{dimensionless}\}$$

The ratio u/d represents the averaged velocity gradients (or rates of shear $\dot{\gamma}$, ms/m or s^{-1}) induced in the moving fluid, a measure of the extent of local and global fluid deformations associated with forward displacement. If one now multiplies both numerator and denominator by a volume (m^3), the Re ratio reads

$$Re = \frac{u^2 \bullet \sigma \bullet V}{\eta \bullet \gamma \bullet V}$$

or

$$Re = \frac{\text{velocity squared x mass}}{\text{frictional work done per unit volume}}$$

$$\left[\frac{Nm}{Nm}\right]$$

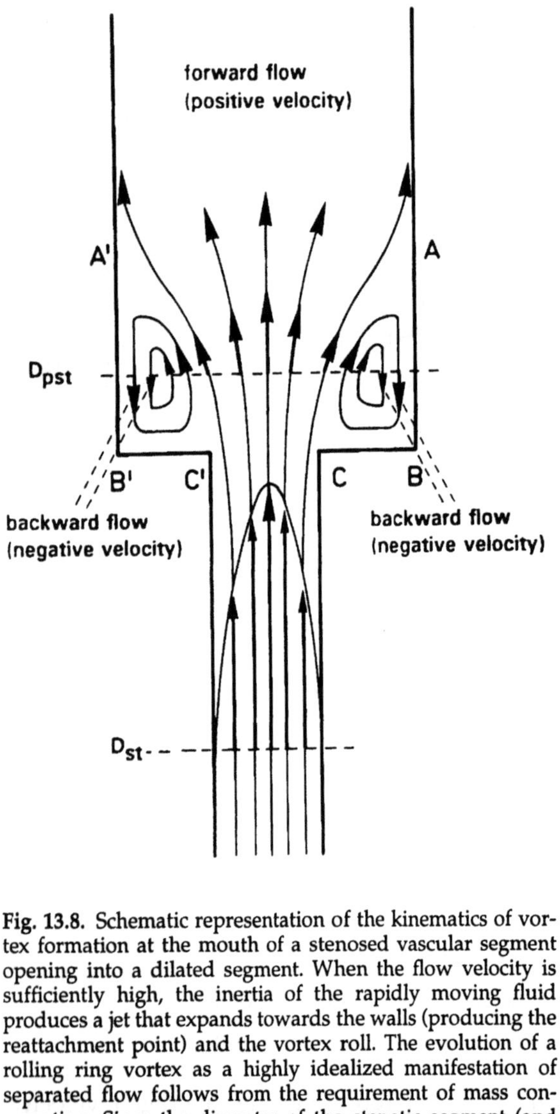

Fig. 13.8. Schematic representation of the kinematics of vortex formation at the mouth of a stenosed vascular segment opening into a dilated segment. When the flow velocity is sufficiently high, the inertia of the rapidly moving fluid produces a jet that expands towards the walls (producing the reattachment point) and the vortex roll. The evolution of a rolling ring vortex as a highly idealized manifestation of separated flow follows from the requirement of mass conservation. Since the diameter of the stenotic segment (and hence the cross-sectional area) is so much smaller than that of the poststenotic area, the average velocity is dramatically retarded as the fluid enters into the poststenotic area. Continuity conditions (same volumetric flow rate traverses the two cross-sectional areas) can only be fulfilled if, at the level of the poststenotic cross-sectional area, the very high forward velocity (inertial effect) is compensated for by backward flows, and, hence a negative velocity is established. Hence the effects of kinetic energy and the effects of continuity combine in producing a functionally separated region of a fluid, here shown as the cross-section through a vortex ring (see text).

which describes a ratio between kinetic energy transferred (*Nm*) and energy dissipated due to frictional work (*Nm*).

Even without penetrating into the very complicated details of its kinematics and dynamics, the logic of the Reynolds ratio aids in comprehending flow separation as a fluid is driven from a narrow to a wide tube, or when flowing on curved pathways, as in bifurcations.

This can be discussed in conjunction with the phenomenology of vortex formation (Fig. 13.8): as the rapidly flowing fluid (endowed with a high amount of kinetic energy when coming out of the stenosis) enters the widened segment with slow flow, the momentum imported from the upstream segment separates flow by much the same mechanism as operates if there is a sudden forced acceleration: local dominance of macroscopic momentum transfer. Obviously, an analogous situation prevails as the fluid is suddenly accelerated in systole from creeping flow in diastole. Local and temporal fluctuations in the locally operational Reynolds ratio destabilize the movement of the fluid (Fig. 13.9). This dynamic process is further augmented by a local variation in viscosity (as it occurs due to phase separation between erythrocytes and plasma. (see Fig 13.11). The latter process appears to be operational in organizing the preferential displacement of platelets into the separated blood elements so notorious in Goldsmith's films of poststenotic blood movement.

Separated flow is a deterministic yet highly organized mode of operation. Reynolds' original concept can be interpreted as the ratio of kinetic energy transferred to frictional energy dissipation. As the ratio crosses certain critical thresholds, a new type of highly ordered mode of fluid movements begins (Figs 13.9–13.12). It is deterministic but very unpredictable since many determinants of the locally operating Reynolds ratios vary (when velocity, diameter, density and viscosity fluctuate so vividly in time). Moreover, the individual parameters are interdependent: their alterations reflect non-linear hydrodynamic effects, hence it is understandable that to observers not familiar with the underlying causes, the pattern appears to be chaotic. (The simple fact that the pattern fluctuates in a fashion synchronized to the heartbeat (Figs 13.1 and 13.4) reveals its highly ordered nature: hence we refer to this fascinating mode of operation as pseudochaotic ordering.)

Most textbooks of circulatory physiology restrict their discussion of these problems to a rather irrelevant question, namely, the transition of truly laminar to fully turbulent, i.e. random, flow. Today we know that fully developed turbulence is a rare event in the circulation for two reasons: (1) because a so-called critical Reynolds ratio is not reached even in peak flow; and (2) because the short time of systolic peak velocity does not suffice to disperse the coarse grain of

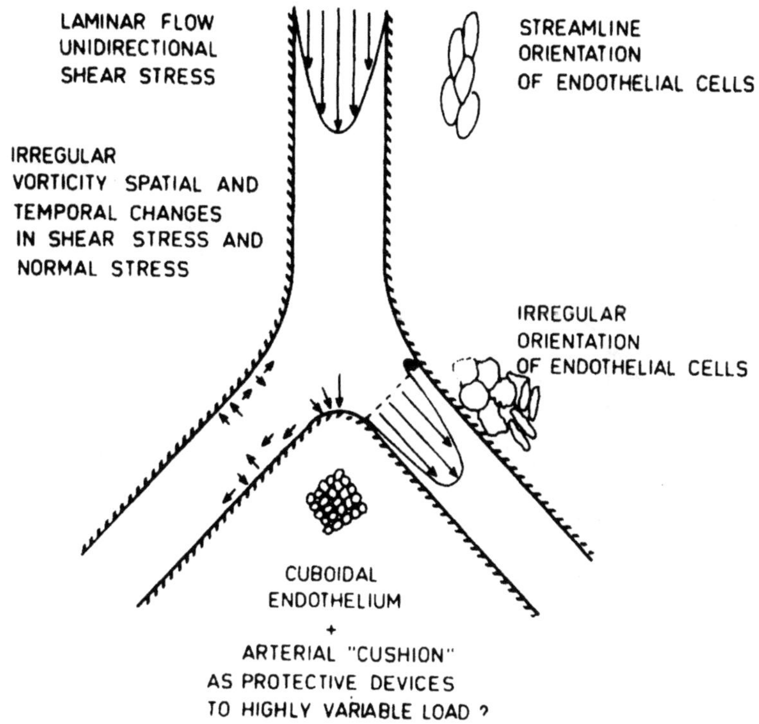

Fig. 13.9. Schematic representation of the fluid dynamics of pulsatile flow in an arterial bifurcation. Left half: during diastole, low flow velocity and hence low Reynolds ratios provide pure laminar flow (which can be oscillating in direction); there is a stagnation point flow directed towards the vortex of the flow divider with low normal stresses and shear stresses. In systole, the pulsating flow with an increment and subsequent decrement in Reynolds ratio leads to an evolution and subsequent devolution of a vortex or recirculation zone. At the reattachment point there is a flow vector directed against the wall, which migrates peripherally during the acceleration phase and centripetally during the deceleration phase of flow pulse. Depending on flow velocity, there is a more or less pronounced stagnation point flow directed against the divider.

macroscopic vortices to the fine grain of random movements (Naumann and Schmid-Schönbein 1983; Ebeling and Klimontovic 1984; Schmid-Schönbein and Naumann 1984; Manneville 1990; Ruelle 1991).

The local ratio of kinetic over dissipated energy undergoes rapid local fluctuations in any fluid travelling in non-cylindrical conduits, e.g. where the cross-sectional area rapidly increases and hence velocities (and especially their local distributions) can rapidly change both temporally and spatially; geometry is therefore an important risk factor first for flow separation and secondly for atheroma formation (Friedmann et al. 1983).

Why are bifurcations such a risk factor for flow separation? Because of their high kinetic energy in the central trajectories of a parabolic velocity profile, the fluid elements near the vessel axis tend to pursue the shortest (least tortuous) path. In a Y-shaped bifurcation they shoot as closely as possible along the two sides of the flow divider (Fig. 13.9). For this reason the velocity at the inner aspects of the flow divider is consistently higher than the average flow velocity across the vascular lumen (Liepsch 1990; Yamaguchi et al. 1990). However, not all fluid elements follow the fast path; that is to say, a minority become separated and take rather complicated tortuous paths, which depend on a variety of factors still awaiting detailed elucidation (but known to include angle of divergence, sharpness of edges, changes in total cross-sectional area, rate of acceleration and blood properties such as viscoelasticity and tendency to phase separation).

All of this is complicated flow direction and flow pulsatility: in short, there is a transient dissociation (high Re ratio) and subsequent reassociation (low Re ratio) of fluid elements, leading to the transient evolution (accompanying each systolic acceleration) and devolution of vortices (during diastolic deceleration). These appear as short-lived 'eddies' or whirlpools which spring up and vanish at characteristic sites near the outer or inner walls of the arteries. These phe-

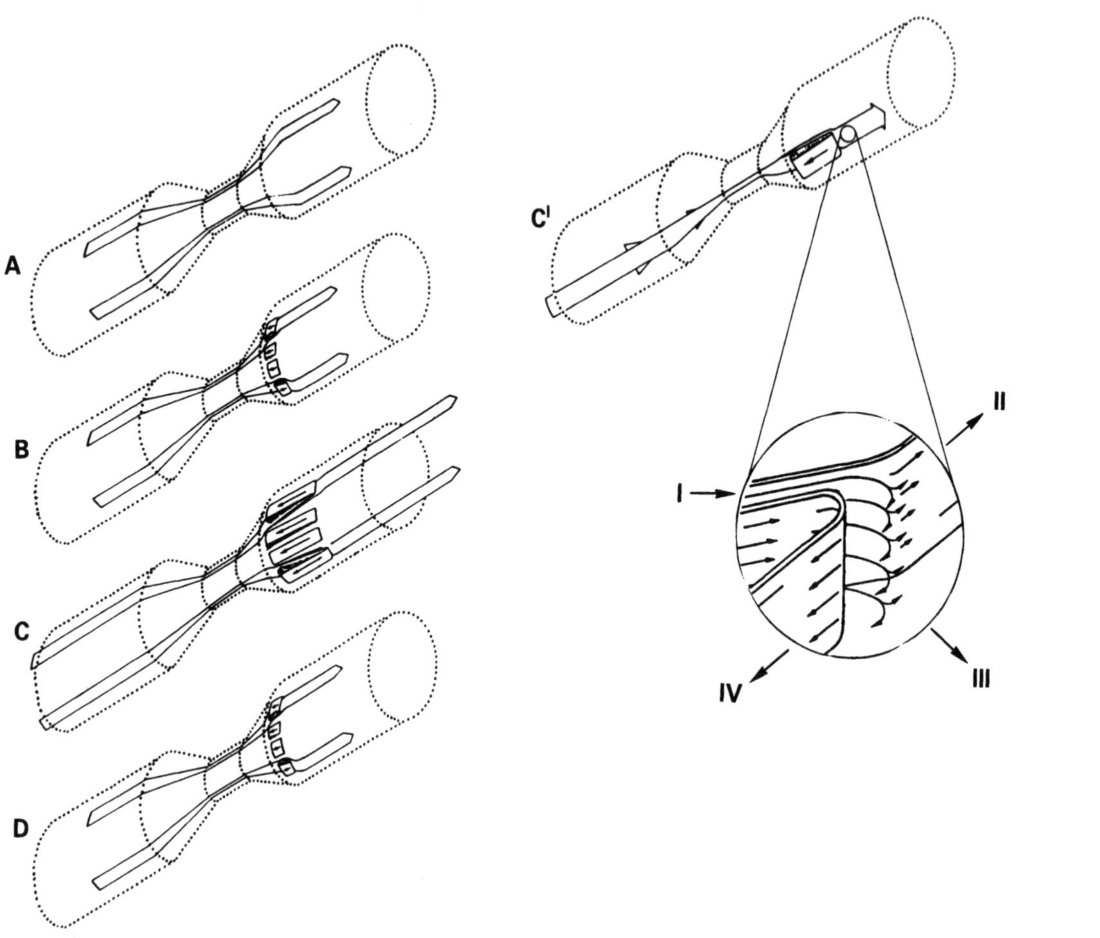

Fig. 13.10. Details of flow separation (Naumann vortex transport phenomenon) in poststenotic segment and in pulsatile flow. In diastole (A) the flow is fully laminar; in protodiastole (B) a ring vortex is developing, growing in size during peak systole (C) and followed by size reduction in late systole (D). C details one segment of the so-called relaminarization point. The streamlines are directed into separate pathways, part of them stemming from the boundary layer. The feeding streamlines (I) divided into those that proceed into laminar flow (II), those that flow back into the vortex (IV) and those that hit the wall. These streamlines in their totality generate the migrating stagnation points and in their totality operate like a jet that can be said to be washing the wall.

nomena representing pseudochaotic order have actually been consistently observed recently, not only in transparent models of arterial bifurcations in general, but specifically in models of the carotid sinus (Liepsch 1990).

Forward and Backward Velocities in Separated Flow: Effects Dictated by Continuity Requirements Near Stenoses

Although the details of the transient flow separation as a manifestation of a self-organized and deterministic fluid-dynamic response in all different kinds of non-cylindrical conduits can be extremely variable, they share common kinematic features. For this reason they can be described for the most simple case of a concentric stenosis feeding into a downstream expansion (Fig. 13.8). This situation is relatively well understood (Hutchinson and Karpinski 1988) and important hemorheological and hemostasiological sequences relevant to the pathogenesis and its sequelae have been studied under the conditions prevailing here. The temporospatial complexities of pulsatile flow patterns in the poststenotic velocity fields were recently covered from the point of view of ultrasound Doppler flow measurement by Hutchinson and Karpinski (1988) and by Phillips et al. (1989).

The situation is kinetically characterized by pronounced local flow acceleration in the sten-

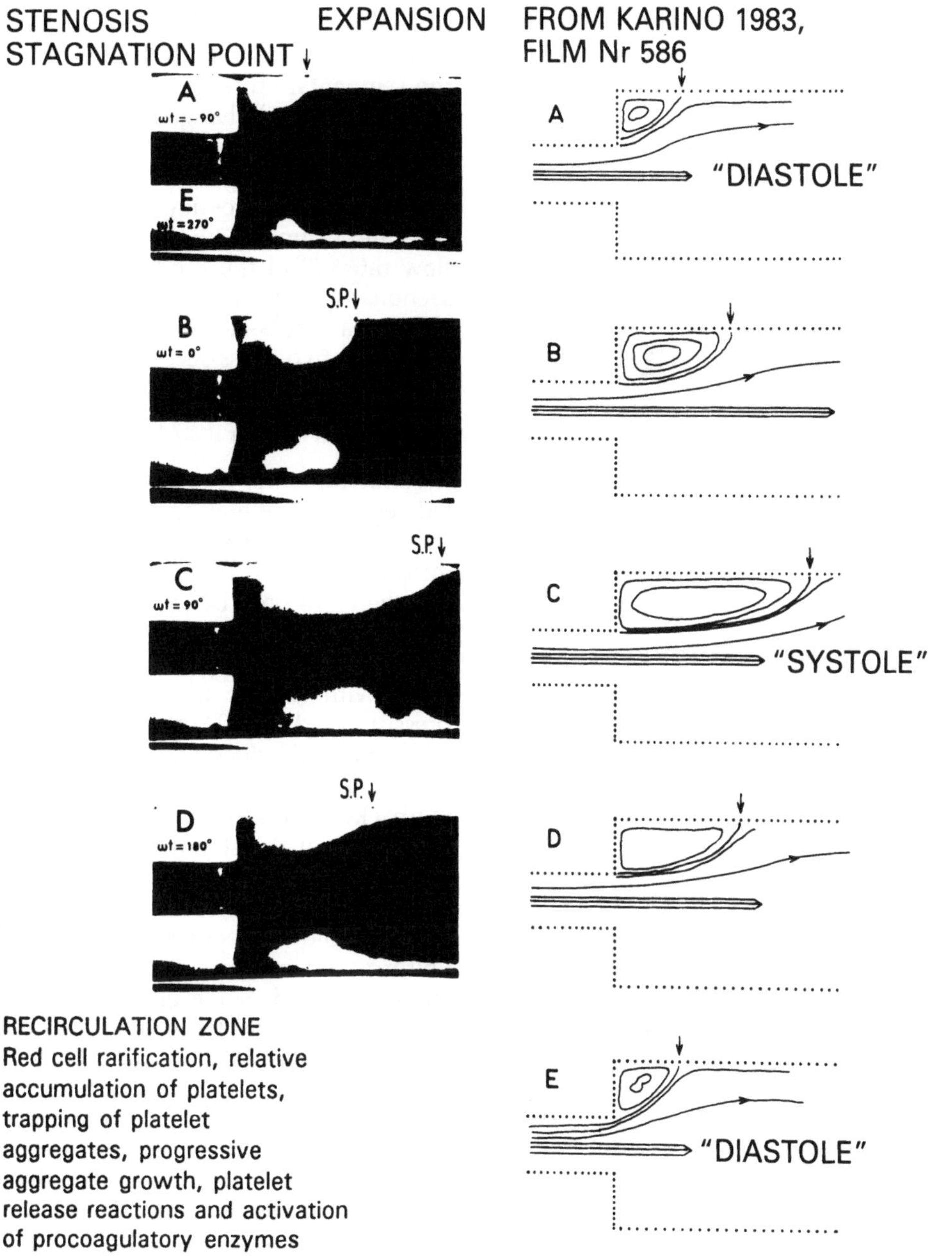

Fig. 13.11. Sequence of a motion picture produced by Karino (McGill University, Montreal, Canada) depicting pulsatile flow separation in the poststenotic segment of a model conduit perfused with human blood. The flow separation is made visible because the separated area contains fewer red cells (but more platelets) than the bulk of the blood. On the right, the evolution and devolution of the separated flow, and the migrating reattachment point (arrow) is schematically depicted.

osed segment and, since the poststenotic expansion is characterized by a very prominent and sudden increase in conduit diameter, a local flow deceleration ensues downstream. Speaking in terms of the locally variable Reynolds ratios, any fluid moving with high velocity possesses high kinetic energy. As fluid enters the dilated segment a sudden drop in average velocity occurs. This situation has often been interpreted from the point of view of pressure distribution (Naumann and Schmid-Schönbein 1983), but one can also relate it to the constraints dictated by the continuity equation: the spatially averaged velocity ($\bar{u}$) in the poststenotic segments must be reduced in proportion to the increase in the diameter ratio (d_{pSt}/d_{St}) squared. Conservation of

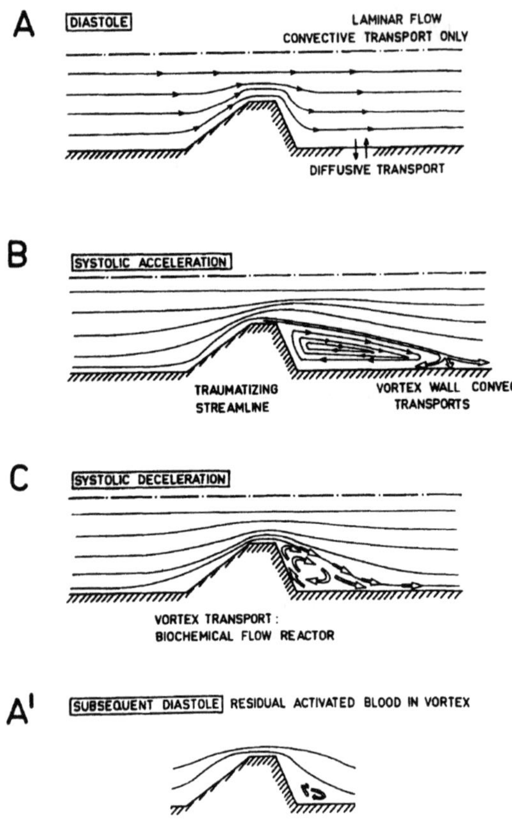

Fig. 13.12. Emergence of transient flow reactors in a vortex. **A** Diastole: laminar flow. **B** Systolic acceleration with hemostatic pivotal events taking place (progressive flow separation, generation of a traumatizing streamline, emergence of a transiently acting flow reactor, institution of stagnation point flow on the reattachment point). **C** during the late systolic deceleration there is washout of the previously activated content of the vortex via the receding boundary layer. $\mathbf{A}^1$ During the subsequent diastole diffusive exchange between blood and poststenotic arterial wall can occur. Provided that platelet aggregates were formed, they can remain in the poststenotic region and can be reintegrated in the vortex during the subsequent systole, thereby prolonging the activation of the newly entering blood. Whenever a vortex (or a small fraction of it) remains separated, the fluid becomes the site of multiple (autocatalytically amplified) activation processes so that aggregation, release reactions and thromboxane synthesis by the trapped platelets takes place, as well as thrombin formation and fibrin polymerization. All these reactions in the fluid phase can then lead to endothelial damage at the adjacent wall.

mass requires that in each sequential conduit there is the same flow rate ($\dot{Q}_{St-} = \dot{Q}_{pst}$), hence as the segmental cross-section is altered the averaged velocity ($\bar{u}$) changes in an inverse fashion. As the product of the cross-sectional area (πd^2, m^2) and the averaged velocities ($\bar{u}$, m/s) of the fluid trajectories crossing that area must remain constant (from which product the volumetric flow rate ($\dot{Q}$) of the fluid traversing that cross-sectional area is calculated) these products are equal at all cross-sections.

For the two cross-sectional areas depicted in Fig. 9 this is exemplified as follows:

$$\dot{Q}_{St} = dV/dt = \bar{u}_{St}\ \pi\ d^2_{St}/4 = \dot{Q}_{pSt} = \bar{u}_{pSt}\ \pi\ d^2_{pst}/4$$

Following entry into the poststenotically dilated segment, the high inertia (kinetic energy) of the rapidly moving fluid elements near the axis initially persists. The continuity requirements can therefore only be met in these poststenotic segments if there are also elements moving backwards (i.e. displaying negative velocity components with respect to the direction of acting pressure gradient). Therefore, in these segments self-organized velocity distribution results in the simultaneous institution of fluid lamellae moving backwards near the wall and forwards near the mouth of the stenoses. There are also fluid elements with intermediate velocities. Instead of the uniform forward movement of fluid elements found in regular laminar flow, a non-uniform and incoherent pattern of fluid movements is found. Some elements are carried on straight trajectories within the fluid moving forwards, others move on trajectories forming closed streamlines within a vortex ball (which in this case actually is a ring through which two sections are drawn) (Fig. 13.8).

Divergence follows from the fact that the accelerated central mainstream is no longer constrained and rapidly expands. It finds its way to the wall, but not immediately. Prelaminarization occurs at characteristic sites, referred to above as the reattachment points. Further downstream both the mainstream element velocity and its kinetic energy vanish, but here the continuity requirement can be met without background flow; therefore, as one proceeds downstream from the mouth of the stenosis, gradually less and less backward motion is needed at each cross-sectional area in order to meet continuity requirements. The pivotal consequence of all this is the automatic emergence of a vortex roll: rotating fluid as a spatially and kinematically separated system of strictly laminar flow. Those fluid elements forming the ring vortex are kinematically separated from the mainstream, but are driven by it into a rotating movement: fluid elements are travelling on closed streamlines. Considering the outermost shell, these start at or near the reattachment point (A in Fig. 13.8), move backwards (A****B) until they reach the corner of the poststenotic segment (B in Fig. 13.8), move up its face and reach the boundary layer of the mainstream exactly (C) at the site of highest velocity. In this boundary layer (C***A) very high velocity gradients or rates of shear are acting (Liepsch 1986), 1990). Mechanical forces, which exert significant effects on the blood elements, act in the boundary layer formed by an assembly of trajectories which contain the traumatizing streamlines (Schmid-Schönbein and Wurzinger 1986) as well as the separating streamlines that functionally disconnect the elements within the roll from those travelling in the central jet. The combined effects of all self-organized sequelae of separated flow can be considered as a functional unit called Naumann vortex transport (Schmid-Schönbein 1984) for the sake of semantic brevity and personal appreciation.

Phenomenology of Naumann Vortex Transport due to Flow Separation

Taking the poststenotic vortex roll as the example for all types of self-organized separated flow in the highly pulsatile arterial motion through stenoses, near bifurcations or behind unilateral protrusions, the fluid-dynamic and the derived microrheological peculiarities of the vortices, the boundary layers and the migrating reattachment point can claim special attention (Fig. 13.10).

The following considerations are based on a very clear film produced by Karino and Goldsmith (Fig. 13.11). The pulsating flow and the vortices growing in systole and receding in diastole are clearly visible, and in repeated pulsatile cycles the displacement of reattachment points and their impact on the wall are also easily seen. The dynamic process of relaminarization accompanying the evolution and devolution of vortices marks the site of the instantaneous transition of separated into parallel laminar flow, but forces fluid elements to travel towards (rather than parallel to) the wall (see Fig. 13.10).

Note that in order to appreciate the significance of the reattachment process, one must visualize it three-dimensionally: the combination of all reattaching streamlines can be viewed as sweeping transport jets that propel blood elements on to the wall, by the very effect of inertia that was shown to propel skidding cars on to the crash barriers of winding roads, as described in our example from traffic (see Fig. 7). This has a functional similarity with the preferential deposition of platelets near the stagnation point as demonstrated experimentally by Karino and Goldsmith (1979) in the case of perfused stenosis, thereby corroborating similar observations by Baldauf et al. (1978) about platelet deposition from diluted bovine platelet-rich plasma on to stagnation points as they occur in a T-shaped bifurcation. Similar findings explain early observations by Murphy et al. (1962) about platelet deposition opposite the flow dividers.

However, there is more than just collision with walls in these regions: fluid elements that are locally trapped are not only subject to an example of fluid-dynamic self-organization, but are also the subject of self-organization in a biochemical system, namely that conventionally described as the hemostatic reaction to vascular injury. This section is a combined biophysical–biochemical chain of events that triggers the endothelial damage that causes endothelial as well as subendothelial injury to the vascular wall. All partial functions of the hemostatic sequence (platelet adhesion, platelet release reaction, platelet aggregation and fibrin polymerization) are set in motion simply because transient flow reactors are formed in the vortices, which amplify autocatalytically accelerated processes and produce activated blood that washes the adjacent arterial wall.

The Carotid Tree Seen as a Trifurcation, Explaining the Lack of Atheroma in the Mouth of the External Carotid

The well understood phenomenology of secondary flow, in combination with our fragmentary knowledge of the kinetics that produce it, allows understanding of a surprising fact. At the carotid bifurcation, only the entrance into the internal branch is the site of atheroma formation, whereas that into the external carotid is spared, despite the fact that flow in the latter is more pulsatile. However, drainage into the thyroid artery strongly affects flow into the mouth of the external carotid artery. Flow separation, if it occurs at all, is much less pronounced, since it is disturbed in turn by strong mainstream currents into the thyroid branch immediately above the bifurcation. There must be a strong primary motion into the highly perfused thyroid artery driven by pressure gradients and inertial effects, which probably override all the secondary effects described above (Schmid-Schönbein 1988).

Pulsating Naumann Vortex Flow: Transient Flow Reactor Associated With Migrating Stagnation Points

From the above considerations it can be concluded that

1. The causes and consequences of self-organized flow separation in the form of vortex rings or vortex balls represent a form of self-referentiality typical for all synergetic processes operating far displaced from equilibrium.
2. The details of the movements can only be understood as the result of joint efforts exerted by inertial and frictional forces and dominated by the prevailing geometry.
3. The formation of a transient flow reactor and the preferential washing of wall segments with activated blood as two pivotal kinematic consequences emerge automatically.
4. In pulsatile flow, the reattachment point and hence the stagnation points perpetually migrate back and forth.

The combination of effects is depicted in Fig. 13.11 for the kinematically much simpler case of poststenotic vortex evolution. Here, as in the proximity of simple or complicated bifurcations, during each systole a dancing 'shower' of fluid elements impinges on to the wall. Each shower has a complicated 3-D configuration, sweeping in an oscillating fashion against the wall whenever both the velocity and the direction of the reattaching streamlines fluctuate in time (see Figs. 13.10 and 13.11). Each cardiac systole generates a kick that accelerates and straightens this kind of shower into the downstream direction, all the factors enhancing flow pulsatility extending the area reached by the sweeping jet (Fig. 13.11).

There are natural differences in flow pulsatility between different vascular regions, e.g. the coronary vascular bed, with its most pronounced pulsatility, characterized by peak flow in diastole. Ultrasound and Doppler velocimetry show that pulsatility in the cerebral bed changes with age (Hoeks et al. 1991). When blood flow becomes more or less restricted to systole, i.e. is strongly retarded in diastole, the strong protosystolic acceleration predisposes to flow separation, with all of its sequelae.

The material propelled by the jet originates from the separated vortex, i.e. differs fundamentally from normal blood (Fig. 13.12) in that it may contain activated platelets, activated coagulation enzymes and mediators released from platelets (Wurzinger and Schmid-Schönbein 1987). The possibility of true but partial coagulation of blood in or near the vortices due to platelet aggregation and fibrin polymerization should be considered.

Synergetic Interpretation of the Processes and the Products of Coagulation Events: First- and Second-Order Phase Transition in White Thrombus Formation

Thus far, we have discussed the self-organized movement of blood as a fluid without considering its rheological behavior. However, in order to understand the more advanced forms of cooperativity emerging between the energies shaping fluid motion and those acting in the moving fluid, we must include changes in the fluidity of the blood. The reaction of the latter to the highly variable flow conditions must be appreciated, reconciling the wide range of rheological potentials of the blood. Alterations following the activation of procoagulant and thromboplastic events must now be put into functional perspective (Schmid-Schönbein 1990a, b).

Local coagulation of blood (i.e. the polymerization of the soluble fibrinogen to strands of fibrin) and thrombus formation (dense deposition of blood-derived material such as platelets and fibrin, and occasionally other blood elements) on the blood vessel walls in themselves are flow-dependent processes. They are, in turn, associated with decisive changes in the physical and transport characteristics of the involved blood elements (Figs 13.11 and 13.12). the solidification of the fluid blood is due either to fibrin formation or to platelet aggregation, or to a combination of both called a first-order phase transition (Haken 1983), since it occurs at approximately constant temperature. Since it is associated with dissociation of the dispersed

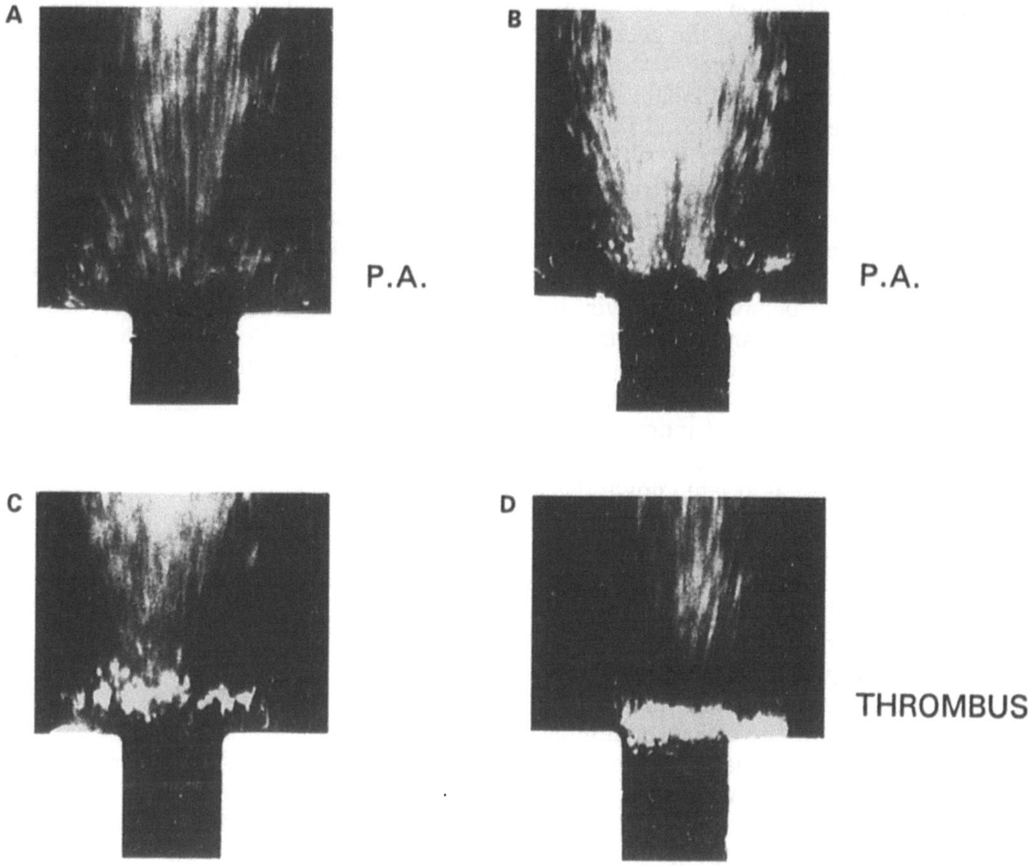

Fig. 13.13. Hydrodynamic trapping of a freely floating platelet aggregate (turning into a white thrombus) in a recirculation zone behind a stenosis (modified after a motion picture produced by Goldsmith. Despite the fact that the free platelets and the progressively growing platelet aggregates are initially not adherent to the wall, the aggregates grow in size in the rotationg vortex ring. In **D**, a ring-shaped white thrombus has emerged that rotates around its minor axis in a toroidal movement.

from the continuous phases, it also shares the characteristics of second-order phase transition. These phenomena and processes can produce substantial influences on the fluid-dynamic detail of flow separation. A vortex becomes the preferential site for platelet aggregation and accumulation, but aggregates also stabilize the vortex once it is formed (Fig. 13.13). These effects are in turn dependent upon the geometry and previous history of the separated region, and on the extent of cytological and enzymatic activation. There, we embark upon the join efforts of self-organized physical and chemical reactions (Kuramoto 1984).

There are multiple interdigitating modes of phase transitions related to an induced alteration of the microrheological, enzymatic and cytologic states, i.e. the adhesiveness, the aggregability and the catalytic activity of coagulation proenzymes and the equilibrium between prothrombotic and antithrombotic activities. The specific mode of blood movement is only felt by the separated blood elements: here, and only here, the comprehensive biological potentials of the blood elements (Fig. 13.13) can be triggered into activity in a self-organized fashion. Functional hemorheology, in transgressing the limits of simple material testing (i.e. measuring the apparent viscosity of cell suspensions, filtrability and aggregability of cells, or viscous coefficients and elastic moduli of cell membranes), forces biology to accept the fact that the transport characteristics of blood on its way into phase transitions that have finite time requirements cannot possibly be described by any type of fixed properties in the sense of continuum mechanics, but rather represent a behavior highly variable in time and highly susceptible to the details of the geometric hemorheological and fluid-dynamic conditions locally prevailing (Wurzinger et al. 1983).

We are also confronted with a second type of pseudochaotic order, this time in a self-organized enzymatic system: the events are deterministic as a matter of course, their occurrence is unpredictable as a matter of fact (Schmid-Schönbein 1990a). It is thus difficult if not impossible to infer from in-vitro tests a behaviour in vivo, in that the flow conditions are largely unpredictable. The rheological aspects of coagulation and thrombosis show a more or less perfectly measurable performance under a set of more or less fixed boundary conditions in any given test setting in vitro. We can at best associate known risk factors with pseudochaotic order and, in order to draw any useful conclusions about the rheological component of disturbed blood flow (or conversely, flow disturbance prevailing during coagulation and/or thrombosis), a strictly defined research strategy must be pursued in the future. The time-honoured assumption that blood standstill occurs prior to clot formation is probably the exception rather than the rule. This fact has yet to be incorporated into sophisticated biochemical concepts of the enzymatic processes governing fibrin formation and/or platelet–platelet interactions. Besides molecular and cellular interactions, transport processes (by diffusion and/or convection) that play a pivotal role in the regularly organized convective motion of blood in Naumann vortex transport, acting as a biological flow reactor will have to be more thoroughly investigated.

It is imperative to analyze as closely as possible the in-vivo fluid-dynamic conditions of the prothrombotic event studied. The arterial bed is characterized by high pressure and velocities, i.e. a state far displaced from fluid-dynamic equilibrium. Secondly, simplified models (with respect to geometry, shear forces, experimental times and other boundary conditions) must be constructed. In these, the interaction of defined rheological parameters with biochemical, metabolic and other biological events can be assessed, and also the inhibitory and excitatory influences on the biological system under study (Schmid-Schönbein 1990a).

Stagnation point flow produces a kind of preferential transport of high molecular weight species. Convective transport towards the wall is non-selective, but still prefers those with a larger mass, since spontaneous Brownian motion towards the wall is so much slower than that of low molecular weight species. Preferential motion of large molecules (e.g. VLDL, LDL, fibrinogen) and of particulate blood components (chylomicra, platelets and platelet debris, white and red cells) is therefore primarily due to convection (Naumann and Schmid-Schönbein 1983). Whether or not the propelled molecules or cells react permanently with the wall is another question, since this process depends primarily on the integrity of the species, which normally repel other negatively charged endothelial cell layers. Even when partially damaged at the site of stagnation point flow, highly effective transmural flow as well as filtration polarization is bound to occur, an ordering phenomenon driven by the very high transmural pressure acting in arteries.

Synergistic Influences of Platelet Activation and Wall Receptivity on Platelet Deposition, Platelet Aggregation and Aggregation and Release from Platelets

Karino and Goldsmith (1979) have not only demonstrated preferential platelet deposition in the immediate vicinity of stagnation points, but also shown that the process is augmented by red cells and wall adhesivity (modeled by attaching collagen). This finding is corroborated in extensive studies on native thrombotic deposition (Petschek et al. 1968). Many subsequent studies have shown that in stagnation point devices in vitro there are two preconditions for initial platelet deposition: a flow vector directed towards the wall, and the prior platelet activation by exposure to agents such as ADP (Baldauf et al. 1978).

Totally separate conditions thus act synergistically, in that the actual process requires both preconditions and in that the observed result of their cooperativity is higher than the sum of the two processes. Already the initiating platelet deposition is a multifunctional cooperative process governed by very specific interactions of fluid-dynamic, mural and cytologic determinants. This also holds true for progressive deposition: all arriving platelets use the initiating ones as 'landing places' (Richardson 1973; Begent and Born 1970). Whenever platelets are prevented from deposition and the initiating or primary adhesion is safely avoided, so are all subsequent steps. Once deposited, however, the maintained flow automatically leads to more and more thrombotic deposition.

Karino and Goldsmith (1979) in their model experiments shifted the site of platelet deposition in poststenotic areas by increasing or decreasing the flow rate (or Reynolds ratio) through the device. The almost exclusive deposition of platelets in the outside region of Y-shaped bifurcation models (Murphy et al. 1962) also shows that the microrheological behavior of blood elements is jointly determined by the details of prevailing flow conditions and by instantaneous platelet adhesivity when colliding with a wall.

By combining all presented concepts, new insights into the mechanics of white thrombus formation in arteries emerge; here too, recirculation zones in closed vortices can be assumed to act as flow reactors, an interpretation that has been amply confirmed by cytologic and enzymatic studies (Schmid-Schönbein and Wurzinger 1986), again showing striking symptoms of self-referentiality or circular causality:

1. The initiating fluid-phase enzymatic steps of the coagulation sequence can actually take place in the bulk of the fluid in motion, but only when flow is separated. The closed vortex in turn by providing rapid mixing, accelerates autocatalytic self-amplification.
2. The procoagulant and thrombotic deposition processes on to the wall can be both complications and the cause of platelet activation processes that occur elsewhere.

At this point one is confronted with the principle of circular causality so often seen in self-organized systems (Haken 1983). Moreover, one crosses the conceptual boundary between simple synergism and synergetic cooperativity (where they automatically emerge from a multitude of energy conversion processes). Here, the conversion of mechanical energy into heat, the conversion of chemical energies, and especially the generation of self-amplifying loops of autocatalytically accelerated chemical reactions are functionally linked, consensualized by only one hydrodynamic event, namely local vortex formation as described above.

Autocatalytically augmented enzymatic and mediator-transmitted steps of the hemostatic system cannot possibly take place in blood subjected to unidirectional laminar flow (Schmid-Schönbein 1990a, b) because convective dilution and the predominance of scavengers and antiproteases successfully suppress even the slightest initial steps of activation. Initial activation, however, when it emerges in the circumscript vortices as they emerge as flow reactors, becomes autocatalytically amplified because these combine highly efficient convective mixing with high levels of proenzyme substrate concentrations. Once started, catalytic activity rapidly accelerates in cytological alterations of the platelets and in the local fluidity of the moving blood. Especially when associated with rapid self-amplifying of platelet activation, this can easily lead to a situation highlighted by positive feedback (vicious circles or non-linear responses) with the potential to escalate into indeterminate chaos. Even when the fatal catastrophe of a full thrombosis does not occur, procoagulatory enzymatic events provide sufficient repetitive chemical noxious effects on the arterial wall neighboring the flow reactor to explain a highly localized vascular injury.

There are obviously many conceivable self-organized vicious circles at the predilection sites: in the presence of severe stenosis with pronounced pressure gradients, abnormally high shear stresses rupture cells, e.g. platelets and erythrocytes, and produce cell debris (Wurzinger and Schmid-Schönbein 1987). All these chemical agents are fed into the same streamlines that precede the recirculating and reattaching ones. Hence, activated blood components are flushed preferentially against those parts of the wall adjacent to short-lived vortices, and penetrate into discontinuities of the wall since they are driven by the high transmural arterial pressure. Note that those fluid elements originating from the immediate vicinity of the wall of the stenosis are carried in the boundary layer between coherent laminar flow and separated vortex, and thus carry damaged and thereby activated platelets towards the wall (Schmid-Schönbein and Wurzinger 1986).

These platelet-derived agents can therefore exert biological activity in the deeper regions of the vascular wall which, in turn, by accelerating atheroma growth, predispose to more pronounced flow separation. In this context, the action of released platelet proteases (e.g. platelet-derived elastase) might be important, as it would explain the notorious finding of poststenotic dilation. This highly consequential complication was classically attributed to the action of purely mechanical impact (Roach and Harvey 1964). More recently, however, the same phenomenon was interpreted by an interactionist theory (Gow et al. 1984; Greenwald et al. 1990).

Obviously, much more detailed information is necessary for a comprehensive understanding of the influences of hydromechanical, hemodynamic, geometric and hemorheological determinants of separated flow and their effects on the early and late phases of atherogenesis. However, as we are no longer in doubt about the existence and governing regularities of separated flow in the predilection sites, the term low shear is becoming equivocal; actually, it is as inappropriate as the term turbulence, and both should be replaced by vorticity, or separated fluid movements, which better describes the self-organized fluid-mechanical phenomena that seem to determine so many processes of atherogenesis and thrombogenesis.

White Thrombus Formation in Vortices

We now turn from the atheromatous to thrombotic events as late, often fatal, complications of atheromatous plaques. In uncomplicated pulsatile flow in normally configured bifurcations the vorticity is extremely transient, a fact explaining that escalation into full thrombosis is virtually impossible. However, as the atheromatous process alters the configuration and creates stenoses or poststenotic dilatations (or combinations of both), the volume as well as the longevity of the vortices increases, and it becomes more difficult to wash out the poststenotically dilated area, which becomes the site of a more and more persistent biological flow reactor. Whereas in laminar flow the reaction between antithrombin III and thrombin is dominant and prevents any form of autocatalytic amplification, separated flow favors all prothrombotic processes in a self-organized fashion, i.e. following the results of biological synergetics. The concepts of self-organized cooperativity can explain the initiation, the progression and the most important complications of atheroma formation. It provides a new theoretical basis for the well established risk factors for atherogenesis and thrombosis, in that all processes critically depend on the previous history and the extent of cooperativity emerging during the lifetime of a vortex (both of which are unknown in the individual case). This is where the presented theory assumes practical significance: as the stenosis increases or the arterial pressure falls, the afferent (traumatic) as well as the efferent (coagulatory) limb of the proposed scheme of white thrombus formation can be assumed to become more effective (Wurzinger et al. 1983). At this point, the present theory includes the possible interactions between platelet trauma and red cell trauma, such as is often assumed to occur in atheromatous plaques (Born 1985). Numerous biophysical and biochemical amplification loops emerge (Schmid-Schönbein et al. 1981; Wurzinger and Schmid-Schönbein 1987, 1990). One can predict that, as the stenosis narrows down, the pressure gradient and hence the local velocity gradient, as well as the shear stresses acting near the wall, are bound to increase and are likely to become more traumatic to the flowing blood elements (Schmid-Schönbein 1990a). Whether that event occurs in the freely flowing blood or in the plaque will remain probably forever undetermined. Blood trauma is sure to occur, but it is not clear when and where.

Geometric details of advanced lesions that have in the past been neglected are likely to assume pivotal significance in the future, e.g. the eccentricity and length of the stenosis on the one hand, and the extent of poststenotic dilatation on the other. Decisive functional consequences are likely to be related to such factors as the asymmetry of the expansion and the distribution of forward and backward velocities that can actually be measured in individual patients by color-coded Doppler anemometry. It is well known that very pronounced vortices are (in the extreme case) never washed out completely, neither in pulsatile nor in steady flow, a fact recognized by Fox and Hugh as early as 1966. As a first approximation, the combination of severe eccentric stenosis and a pronounced poststenotic dilation can be reduced to stenosis and subsequent aneurysm. Here, the effect of the primary trauma in the bulk fluid or in crevices of plaques is amplified in the vortex as flow reactor. If there is less and less washout of fluid elements from the latter, ample time is available for coagulation processes to occur. Such lesions carry a high risk of thromboembolic complications in the form of TIA and stroke. All these considerations require reconciliation of the hemodynamic determinants for the emergence of flow separation in the normal, the aging and the diseased arterial system in man: a topic that can and should be addressed with urgency (Naumann and Schmid-Schönbein 1983).

Physiology and Pathophysiology of Pulsatile Arterial Flow

We have presented an extension of Ross's response to injury theory of atherogenesis: our concept of self-organization requires the linking of conventional hemodynamics to fluid-mechanic forces. For example, changes in the pulsatility of pressure and flow in arteries, such as occur physiologically with aging, may represent important atherogenetic determinants requiring more attention in the future. Since secondary motions in pulsatile arterial flow depend on a great number of distinct determinants (O'Rourke 1982; Naumann and Schmid-Schönbein 1983), i.e. the function of the left ventricle as a pump, the composition and compliance of the arterial wall, the geometry of the arteries and bifurcations, the peripheral resistance and the viscoelasticity of the blood as a non-newtonian fluid, all of these can be said to exert decisive atherogenic influences when combined with other abnormalities favoring flow separation. All parameters listed undergo alterations during the normal process of aging of man (O'Rourke 1982).

The majority of animal experiments regarding atheroma formation must be regarded with caution. To the best of our knowledge, the in-vivo analysis of arterial fluid dynamics has usually been performed on the basis of measurements in young experimental animals, presumably with highly distensible arteries which physiologically dampen the pulsations in flow and pressure. Aging, in both animals and man, is associated with arterial hardening, a process that should be clearly differentiated from atheroma formation and will be called physiosclerosis, especially since the latter is usually associated with the enlargement of arteries.

As the *Windkessel-Funktion* (mechanical buffering by elastic walls) is lost during aging, not only is there well known augmentation of the pulse pressure but also changes occur in the propagation of both pressure and flow pulse into more peripheral parts of the arterial tree (Schmid-Schönbein and Naumann 1984). Suffice it to say that augmentation of flow and pressure pulsatility will not only damage the normal endothelium in the aging vasculature but will surely place an additional load upon the defective endothelium and the subendothelial space in advanced lesions (Naumann and Schmid-Schönbein 1983). Lastly, there are local differences in flow pulsatility that must be considered. Atherosclerosis increases with age, but the rate of progression seems to be higher in the coronary and carotid arteries than in other arteries. Assuming that flow separation is indeed a causative factor in the rate of progression, it is important to note that arterial fluid dynamics in the coronary and carotid arteries are significantly different from those in the arteries supplying other organs. They are both characterized by especially pronounced pulsatility in both pressure and flow, and hence in their susceptibility to secondary flows (Schmid-Schönbein 1990a). This also results in a more pronounced flow pulsatility and hence a higher systolic peak ReR and ReR-fluctuations. Since cerebrovascular resistance, i.e.. the input impedance to the internal carotid artery, increases with age, simply because cerebral blood flow is reduced with increasing age even in normal subjects, more and more complicated flow patterns are likely to occur at the level of the carotid bifurcation as subjects age.

Since the formation of secondary flow is greatly facilitated by macroscopic 'roughness' of arteries, i.e. by the consequences of existing atherosclerotic changes, the combination of local changes with microscopic general fluid-dynamic alterations can potentiate the effect of either abnormality alone.

References

Aschoff L (1912) Thrombose und Sandbankbildung. Ziegler's Betr Path Anat 52:205

Baldauf W, Wurzinger LJ, Kinder J (1978) The role of stagnation point flow in the formation of platelets on glass surfaces in tubes with various geometry. Pathol Res Pract 163:9

Begent N, Born GVR (1970) Growth rate in vivo of platelet thrombi, produced by iontophoresis of ADP, as a function of mean blood flow velocity. Nature 227:926–930

Betz E (1988) Experimental basis of the inhibition of atherogenesis in carotid arteries with polyanions and calcium antagonists. In: Hennerici M, Sitzer G, Weger HD (eds) Carotid artery plaques. Karger, Basel, p 171

Born GVR (1985) Adenosine diphosphate as a mediator of platelet aggregation in vivo: an editorial view. Circulation 72:741–746

Caro CG (1981) Arterial fluid-dynamics and atherogenesis. Rec Adv Cardiovasc Dis II:6–11

Coveney P, Highfield R (1992) Antichaos: the arrow of time. Allen, London 1990 German Edition: Leck (Dausen und Bosse 1992)

Ebeling W, Klimotnovic YL (eds) (1984) Self-organization and turbulence in liquids. Teubner, Leipzig

Fox JA, Hugh AE (1966) Localization of atheroma: theory based on boundary layer separation. Br Heart J 28:388

Friedmann MH, Deters OJ, Mark FF, Bargeron CB, Hutchins GM (1983) Geometric effects on the hemodynamic environment of the arterial wall: a basis for geometric risk factors? In: Schettler G, Nerem RM, Schmid-Schönbein H, Mörl H, Diehm C (eds) Fluid dynamics as a localizing factor for atherosclerosis. Springer, Berlin, p 71

Fry DL (1976) Hemodynamic forces in atherogenesis. In: Steinberg P (ed) Cerebrovascular diseases: Raven Press, pp 77–95

Gleick J (1988) Chaos: making of a new science. Droehmer-Knauer, München

Goldsmith HL (1972) The flow of model particles and blood cells and its relation to thrombogenesis. Prog Hemost Thromb 1:97–112

Gow BS, Devenish-Mears SE, Crosby DS, Legg MJ (1984) The role of vascular smooth muscle in post-stenotic dilatation. In: Hunyor S, Ludbrook J, Mcgrath M (eds) The peripheral circulation. Elsevier, Amsterdam, pp 248–255

Grassmann G (1983) Mehrphasige Strömungvorgänge. In Grassmann G (ed) Physikalische Grundlagen der Verfahrenstechnik. Salle-Verlag, Frankfurt, p 584

Greenwald SE, Kukongviriyapan U, Gow BS (1990) The influence of vascular smooth muscle on the development of post-stenotic dilation. In: Mosora F et al. (eds) Biomechanical transport processes. Plenum Press, New York, pp 347–356

Haken H (ed) (1983) Synergetics, an introduction. Springer, Berlin

Hauss WH (1976) Role of arterial wall cells in sclerogenesis, Ann NY Acad Sci 301:275

Hess B, Markus M (1987) Ordnung und Chaos in chemischen Uhren. In: Küppers BO (ed) Ordnung aus dem Chaos. Piper, München, pp 38–45

Hoeks APG, Dabekaussen A, Brands PJ, Reneman RS (1991) Technical limitations of the present color coded US-Doppler systems in analyzing flow anomalies in arteries. VASA (Suppl)32:72–84

Holen J, Nanna M, Lockhart J, Waag R (1990) Doppler color flow in echocardiography: analytical and in-vitro investigations of the quantitative relationship between orifice flow and color jet dimensions. Ultrasound Med Biol 16:543–551

Hutchison KJ, Karpinski E (1988) Stability of flow patterns in the in vivo poststenotic velocity field. Ultrasound Med Biol 14:269–275

Karino T, Goldsmith H (1979) Adhesion of human platelets to collagen on the walls distal to a tubular expansion. Microvasc Res 17:238

Karino T, Goldsmith H (1989) Microscopic structure of disturbed flow in the arterial and venous systems. Haemostasiology 9:53–65

Kuramoto Y (ed) (1984) Chemical oscillations, waves, and turbulence. Springer, Berlin.

Liepsch D (1986) Flow in tubes and arteries, a comparison. Biorheology 23:395–433

Liepsch D (1990) Blood flow in large arteries. Applications to atherogenesis and clinical medicine. Karger, Basel (Monographs on Atherosclerosis Vol. 15)

Manneville P (1990) Dissipative structures and weak turbulence. Academic Press, Boston

Murphy EA, Rowsell HC, Downie HG (1962) Encrustation and atherosclerosis: the analogy between early in vivo lesionsand deposits which occur in extracorporeal circulations. Can Med Ass J 87:259

Naumann A, Schmid-Schönbein H (1983) A fluid-dynamicist's and a physiologist's look at arterial flow and arteriosclerosis. In: Schettler G, Nerem RM, Schmid-Schönbein H, Mörl H, Diehm C (eds) Fluid dynamics as a localizing factor for atherosclerosis. Springer, Berlin, p 9

Nerem RM, Levesque MJ (1983) The case for fluid dynamics as a localizing factor in atherogenesis. In: Schettler G, Nerem RM, Schmid-Schönbein H, Mörl H, Diehm C (eds) Fluid dynamics as a localizing factor for atherosclerosis. Springer, Berlin, pp 26–37

O'Rourke MF (1982) Vascular impedance in studies of arterial and cardiac function. Physiol Rev 62:570

Perktold K, Resch M (1990) Numerical flow studies in human carotid bifurcations: basic discussion of the geometric factor in atherogenesis. J Biomed Eng 12:111

Perktold K, Florian H, Hilbert D (1987) Analysis of pulsatile blood flow: a carotid siphon model. J Biomed Eng 9:46:53

Perktold K, Kenner T, Hilbert D, Spork B, Florian H (1988a) Numerical blood flow analysis: arterial bifurcation with a saccular aneurysm. Basic Res Cardiol 83:24–31

Perktold K, Florian H, Hilbert D, Peter R (1988b) Wall shear stress distribution in the human carotid siphon during pulsatile flow. J Biomechanics 21:663–671

Petschek H, Adamis D, Kantrowitz AR (1968) Stagnation flow thrombus formation. Trans Am Soc Artif Intern Organs 14:356–366

Phillips DJ, Greene FM, Langlois YE, Roederer GO, Strandness DE (1983) Flow velocity patterns in the carotid bifurcation of young presumed normal subjects. Ultrasound Med Biol 9:39–49

Phillips DJ, Beach KW, Primozich J, Strandness DE (1989) Should results of ultrasound Doppler studies be reported in units of frequency or velocity? Ultrasound Med Biol 15:205–212

Prigogine I (1979) From being to becoming: time and complexity in physical sciences. Freeman, San Francisco

Richardson PD (1973) Effect of blood flow velocity on growth rate of platelet thrombi. Nature 245:103–104

Roach MR, Harvey K (1964) Experimental investigation of poststenotic dilation in isolated arteries. Can J Physiol Pharmacol 42:53

Rodkievicz CM (1981) Arteries and arterial blood flow. Biological and physiological aspects. Springer, Wien (CISM Courses and Lectures No. 270)

Ross R (1986) The pathogenesis of atherosclerosis – an update. New Engl J Med 314:488

Ruelle D (ed) (1991) Change and chaos. Princeton University Press

Schmid-Schönbein H, Born GVR, Richardson PD, Cusack N, Rieger H, Forst R et al. (1981) Rheology of thrombotic processes in flow: the interaction of erythrocytes and thrombocytes subjected to high flow forces. Biorheology 18:415–444

Schmid-Schönbein H (1984) Opening remarks to Alexander Naumann Memorial Symposium on rheology and atherogenesis. Biorheology 21:559–564

Schmid-Schönbein H, Wurzinger LJ (1988) Vortex transport phenomena of the carotid bifurcation: interaction between fluid-dynamic transport phenomena and hemostatic reactions. In: Hennerici M, Sitzer G, Weger HD (eds), Carotid artery plaques. Pathogenesis, development evaluation, treatment. Karger, Basel, pp. 64–91

Schmid-Schönbein H (1990a) Synergetic order and chaotic malfunctions of the circulatory systems in "multiorgan failure": breakdown of cooperativity of hemodynamic functions as cause acute microvascular pathologies. In: Vincent JL (ed) update in intensive care and emergency medicine, Vol 11. Springer, Berlin, pp 3–21

Schmid-Schönbein H (1990b) Synergetics of fluid-dynamic and biochemical catastrophe reactions in coronary artery thrombosis. In: Bleifeld W et al. (eds) Unstable angina: Springer, Berlin, pp 16–51

Schmid-Schönbein H (1993a) Synergetics of blood movement through microvascular networks: causes and consequences of nonlinear pressure–flow relationships. In: Haken S, Mikhailov AS (eds) Interdisciplinary approaches to nonlinear complex systems. Springer, Heidelberg (in press)

Schmid-Schönbein H (1993b) The rheology of blood and hemodynamics. In: International Textbook of Physiology. Springer, Heidelberg

Schmid-Schönbein H, Naumann A (1984) Fluiddynamische, zellphysiologische und biochemische Aspekte der atherogenese unter Strömungseinflüssen. Rhein.-Westf. Akade-

mie der Wissenschaften, Westdeutscher Verlag, p 6
Schmid-Schönbein H, Wurzinger LJ (1986) Transport phenomena in pulsating poststenotic vortex flow in arteries. An interactive concept of fluid-dynamic, hemorheological and biochemical processes in white thrombus formation. Nouv Rev Fr Hematol 28:257
Schwartz CJ, Sprague EA, Fowler SR, Kelley JL (1983) Cellular participation in atherogenesis: selected facets of endothelium, smooth muscle, and the peripheral blood monocyte. In: Schettler G, Nerem RM, Schmid-Schönbein H, Mörl H, Diehm C (eds) Fluid dynamics as a localizing factor for atherosclerosis. Springer, Berlin p 200
Sinzinger H, Jellinek H (eds) (1977) Atherogenesis:morphology, metabolism and function of the arterial wall. In: Schreer C (ed) Progress in Biochemical pharmacology Vol. 13, Karger, Basel
Stehbens WE (1983) Fluid dynamic approaches to atherosclerosis. In: Schettler G, Nerem RM, Schmid-Schönbein H, Mörl H, Diehm C (eds) Fluid dynamics as a localizing factor for atherosclerosis. Springer, Berlin, p 3
Steinke W, Kloetzsch Ch, Hennerici M (1990) Carotid artery disease assessed by color Doppler flow imaging: correlation with standard Doppler sonography and angiography. AJNR 11:259
Tamura T, Cobbold RSC, Johnston KW (1991) Quantitative study of steady flow using color Doppler ultrasound. Ultrasound Med Biol 17:595–605
Virchow R (1856) Über Thrombose und Embolie. In: Gesammelte Abh. zur Wiss. Medicin, Franfurt/M
Wurzinger LJ, Opitz R, Blasberg P, Eschweiler H, Schmid-Schönbein H (1983) The role of hydrodynamic factors in platelet activation and thrombotic events: the effects of shear stress of short duration. In: Schettler G, Nerem RM, Schmid-Schönbein H, Mörl H, Diehm C (eds) Fluid dynamics as a localizing factor for atherosclerosis. Springer, Berlin, p 91
Wurzinger LJ, Schmid-Schönbein H (1987) Surface abnormalities and conduit characteristics as cause of blood trauma in artificial internal organs: the interaction of fluid-dynamic, physico-chemical and cytological reactions in thrombus formation. Ann NY Acad Sci 516:316
Wurzinger LJ, Schmid-Schönbein H (1990) The role of fluid dynamics in triggering and amplifying hemostatic reactions in thrombogenesis. In: Liepsch DW (ed) Blood flow in large arteries: applications to atherogenesis and clinical medicine. Monogr Atheroscler 15:215–226
Yamaguchi T, Nakano A, Hanai S (1990) Three dimensional shear stress distribution around small atherosclerotic plaques with steady and unsteady flow. In: Mosor F et al. (eds) Biomechanical transport processes. Plenum Press, New York, pp 173–183
Zarins ChK, Taylor KE, Lundell MI, Glogov S (1983) Aortic ostial morphology and the localization of atherosclerotic lesions: preliminary observations. In: Nerem RM, Cornhill JF (eds) The role of fluid mechanics in atherogenesis, Springer, Berlin, p 24

14. Microvascular Pathology

M. Hommel and F. Gray

Pathology within the microvessels is poorly understood and is probably underestimated. Microvascular disease is very difficult to diagnose clinically, although some clues can be obtained from history, risk factors and neuroimaging. However, the diagnosis only can be established with certainty at autopsy. Moreover, autopsy, even with a thorough study of the vessels, often fails to demonstrate the mechanism of vascular occlusion. Various diseases may affect the microvasculature bed; in addition, aging may cause alterations in the small brain vessels and its effects are frequently associated with those of other vascular risk factors.

This chapter describes four main types of microvascular disease:

Small penetrating artery disease (intracranial branch atheromatous disease)

Microvascular effects of aging, lacunar infarcts and Binswanger's subcortical arteriosclerotic encephalopathy

Cerebral amyloid angiopathy

Familial subcortical strokes, leukoencephalopathy and dementia

Intracranial Branch Atheromatous Disease

This condition represents the occlusion of a penetrating artery at its origin due to atheroma of the parent artery. Fisher and Caplan (1971) first described two patients with basilar branch occlusions. The clinical features suggested a pontine stroke and at autopsy they had basilar artery atheromatous plaques obstructing or penetrating into the mouth of the perforating arteries. Fisher (1977) described a third patient with a 'locked-in syndrome' who at autopsy had bilateral occlusion of the pontine perforating arteries due to basilar artery atheroma. Two of these patients had acute intraplaque hemorrhages or microdissections probably promoting the arterial occlusion. Later, Caplan (1989) stressed the importance of the concept of branch disease and discussed its place in the field of small deep infarcts. However, this concept is supported by only sparse autopsy studies. Pathologically, the atheromatous disease has no specificity. Intracranial atheromatous disease has a racial preponderance in blacks, Japanese and Chinese, and is more frequent in women and diabetics. It lies at the origin of the perforating arteries: plaque of the parent artery extending into the penetrating artery, also called 'junctional plaque' or plaque of microatheroma at the origin of the penetrating artery (Fig. 14.1). Infarction is due to narrowing of the artery impeding flow or to occlusion, possibly promoted by microdissections, hemorrhage in plaques, platelets and platelet–fibrin plugs. The arteries affected by atheroma at the origin of branch disease are the main intracranial arteries at the level of the perforating arteries: the middle cerebral arteries at the level of the lenticulostriate arteries, the anterior cerebral arteries at the level of Heubner's arteries, the posterior

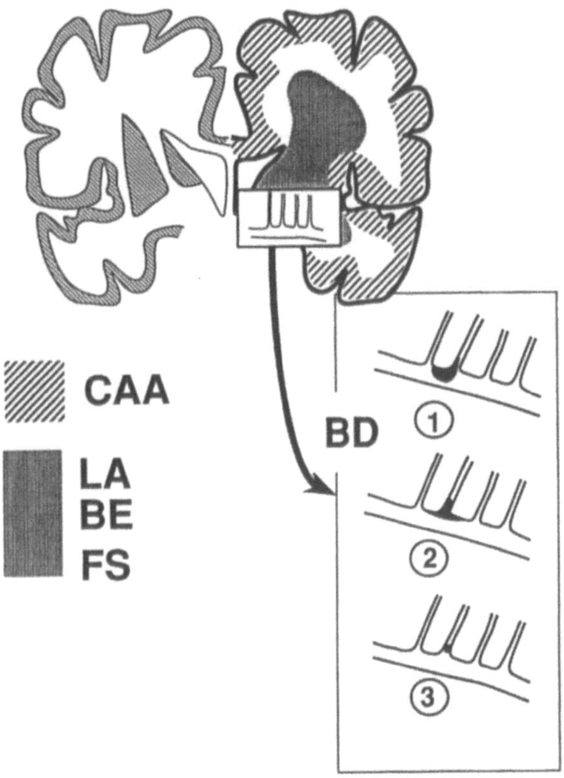

Fig. 14.1. Topography of arteriolar lesions. Coronal diagram of the brain. Branch disease affects the origin of the perforating arteries. Cerebral amyloid angiopathy affects the cortical and meningeal arterioles. Lacunar infarcts, Binswanger's encephalopathy and familial strokes affect the deep perforating arteries. BD: branch disease affecting the middle cerebral artery at its origin; 1, penetrating plaque; 2, atheromatous plaque occluding the penetrating artery; 3, microatheroma; CAA, cerebral amyloid angiopathy; LA, lacunar infarcts; BE, Binswanger's encephalopathy; FS, familial strokes, leukoencephalopathy and dementia.

cerebral arteries at the level of the thalamogeniculate arteries and the basilar artery (Fig. 14.1). The diagnosis should be considered in patients with small deep infarcts. Some clues may favor the presence of branch disease as the cause of the deep infarct: clinical features suggesting a gradual stenosis leading to occlusion (TIAs or gradual or stepwise progression); the absence of cardiac or arterial sources of embolism indicates a lower probability of embolism; and the absence of hypertension, indicating a low probability of distal perforating artery occlusion (lipohyalinosis). The clinical syndromes are related to the topography of small deep infarcts and specific syndromes have not so far been described. The topography of small deep infarcts due to branch artery occlusion can be different from lacunar infarcts because they extend to the basal surface, whereas lacunar infarcts usually lie within the parenchyma. Thus atheromatous branch disease can be suspected from neuroimaging (MRI or CT). Branch atheromatous disease has been neglected in the past in the field of small deep infarcts, and should certainly be the subject of more clinical and autopsy studies.

Microvascular Effects of Aging, Lacunar Infarcts and Binswanger's Subcortical Arteriosclerotic Encephalopathy

Aging, hypertension and other vascular risk factors, diabetes, hypercholesterolemia and smoking induce arteriolar changes. They can be schematically divided into three grades according to their severity.

Grade 1

Caliber Variations and Tortuosity

With aging, small arteries appear more tortuous and have an irregular diameter. The mechanisms for these changes are unclear. The arterioles could have an increase in their size or could appear elongated because of the shrinkage of the brain secondary to cerebral atrophy (Verny et al. 1991).

Arteriolar Wall Modifications

There is a sclerotic and sometimes hyalinotic thickening of the wall of the medullary arteries (endarteries from 100 to 200 μm in diameter), often with onion-skinning in the cerebral white matter arising from the cerebral surface branches and penetrating into the cerebral cortex perpendicularly, and of the perforating arteries arising from the large arteries at the base (Fig. 14.2). The sclerotic rate correlates well with the degree of ischemic white matter changes and blood pressure (Furuta et al. 1991). The walls of the arterioles, of a caliber between 100 and 300 μm are mineralized (calcium, phosphate, zinc, iron, aluminium), mostly in the basal ganglia (Verny et al. 1991) (Fig. 14.3). However, the clinical consequences of these changes are unknown and are probably mild, even if symptomatic.

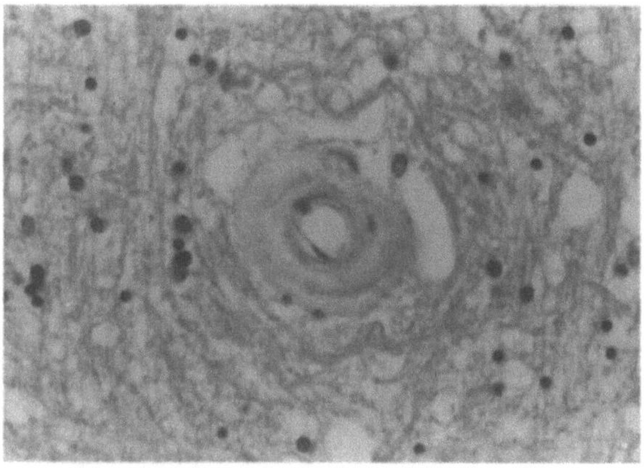

Fig. 14.2. Hyalinosis and arteriolar sclerosis reducing the lumen (grade 1).

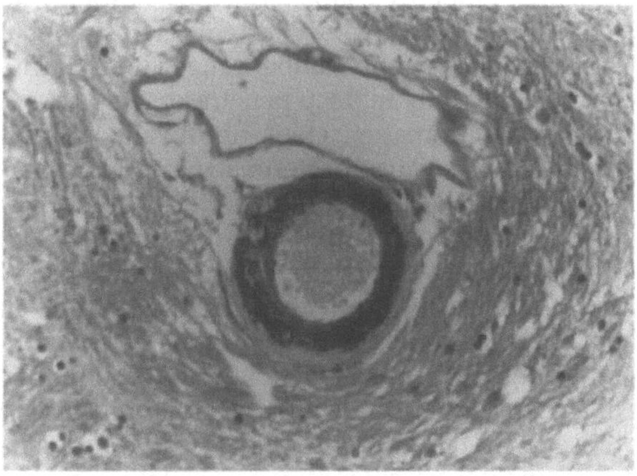

Fig. 14.3. Arteriolar mineralization in the elderly (grade 1).

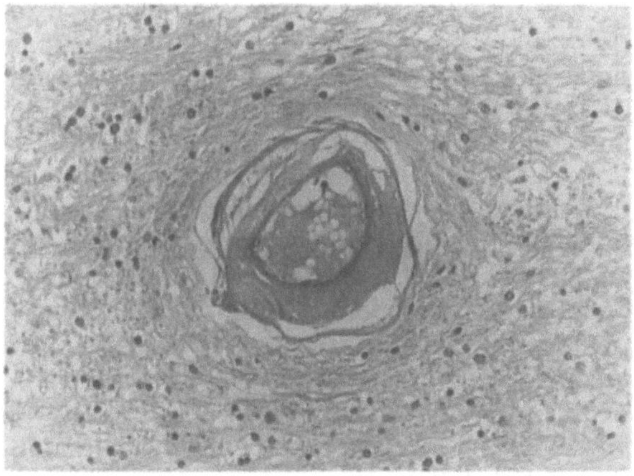

Fig. 14.4. Arteriolar sclerosis with hyalinosis and lipid-laden macrophages (lipohyalinosis) (grade 2).

It is difficult to differentiate the modifications related to aging from those due to associated risk factors. However, caliber changes and slight degree of arteriolar wall modification appear to be related mostly to aging.

Grade 2

Grade 2 represents a more severe microangiopathy, often due to the association of the effects of aging and vascular risk factors, mainly hypertension, diabetes and hypercholesterolemia. The microangiopathy is focal and affects the small perforating arteries between 50 and 400 μm in diameter. Sclerosis, hyalinosis and lipid deposits are also present in variable proportions. Grade 2 changes are characterized by disorganization of the vessel wall and disappearance of the smooth muscle. The lamina elastica is replaced by sclerosis and subintimal deposits of a pink-staining material (hyalinosis). Sometimes foam cells (lipids) are present within this material (Fisher 1969) (Fig. 14.4). Fisher coined the term lipohyalinosis, and many variants exist (hyalinosis, sclerohyalinosis etc.). This angiopathy is often associated with lacunes, small holes located in the deeper part of the brain in the territory of the perforating arteries. These have been classified into three types: type 1 are small old infarcts, type 2 are scars of small hematomas, and type 3 are dilatations of Virchow–Robin spaces (Poirier and Derouesné 1984). Only the type 1 lacunes are considered here. These are due to occlusion of the perforating arteries related to the grade 2 angiopathy. They are between 1 and 25 mm in size, with irregular edges. Microscopically they have all the characteristics of ischemic necrosis. Lacunes are sparsely trabeculated cavities that contain a few lipid-laden macrophages and, rarely, hemosiderin. In some a normal small blood vessel can be found. The cavities are surrounded by astrocytic gliosis (Fig. 14.5). They are located mostly in the basal ganglia, the corona radiata, the internal capsule, the thalamus and the pons (Fisher 1965) (Fig. 14.6). Fisher stressed the role of hypertension in the genesis of these arteriolar changes. Fisher and coworkers described the clinical autopsy correlations in lacunes, also called 'lacunar syndromes': pure motor hemiplegia, pure sensory stroke, ataxic hemiparesis and sensorimotor strokes. Small deep infarcts may correlate with a number of other syndromes. However, the mechanism of arterial occlusion has seldom been studied at autopsy. More necropsy studies are needed to study whether

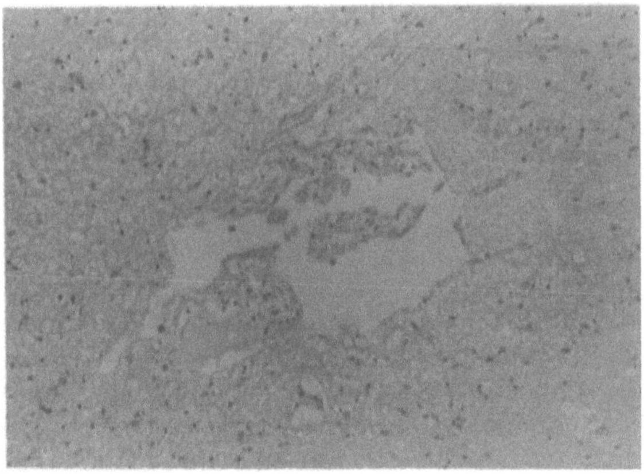

Fig. 14.5. Lacunar infarct (microscopy).

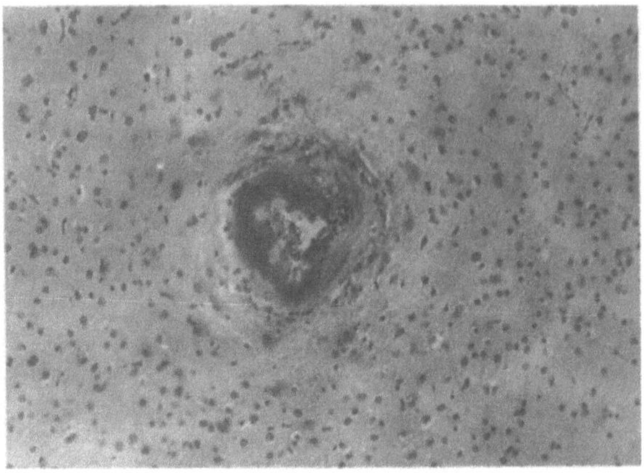

Fig. 14.7. Arteriolar fibrinoid necrosis (grade 3).

lacunes are usually due to the occlusion of a single arteriole (Fisher 1991).

Grade 3

Grade 3 represents the most severe degree of microangiopathy. It is especially related to severe chronic hypertension, and is characterized by fibrinoid necrosis, intrawall hemorrhages, aneurysms and occlusions. Often, patients with Binswanger's encephalopathy have a grade 3 microangiopathy. At microscopy, Brun et al. (1992) classified the small penetrating artery abnormalities in Binswanger's encephalopathy as type I microangiopathy characterized by fibrotic wall thickening (small-vessel arteriosclerosis) that occurs over the entire caliber range (10–700 μm), but severely only in deep penetrators to periventricular white matter. Here, fibrohyalinosis and lipohyalinosis caused arteriolar narrowing, particularly in the frontal and parietal lobes. This type I microangiopathy of Brun et al. (1992) corresponds to our grade 2 angiopathy. Type II microangiopathy (focal wall necrosis) occurs in penetrating arteries (400–100 μm diameter) within deep grey matter nuclei. Brightly eosinophilic fibrin strands, erythrocytes and siderophages are present in the necrotic vessel walls and surrounding perivascular regions. Dilatation, micro- and pseudoaneurysms are also present. Luminal stenosis is common, with thrombotic and fibrotic occlusions adjacent to small deep infarcts (Brun et al. 1992) (Fig. 14.7). Other abnormalities have been reported, such as intimal proliferation of the wall of small arteries which appears as onion-skinning, cystic changes in the media with disappearance of the internal lamina elastica, and calcifications (Caplan and Schoene 1978). De Reuck et al. (1980) stressed that segmental fibrinoid arterial degeneration and thrombosis predominate in the border zones. The type II angiopathy of Brun et al. (1992) corresponds to our grade 3 angiopathy. In Binswanger's encephalopathy, the large arteries may be affected by accompanying atherosclerosis (atheroma, sclerotic plaques) associated with complications of atheroma, i.e. occlusions causing deep infarcts (Brun et al. 1992).

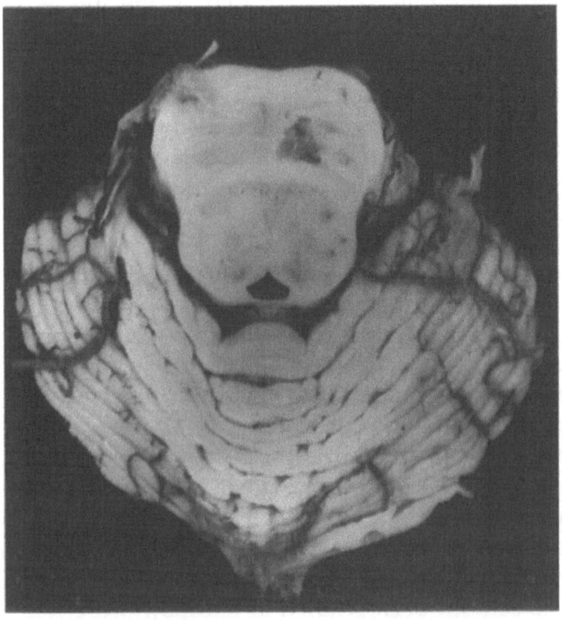

Fig. 14.6. Lacunar infarct in the pons (macroscopy).

At postmortem neuropathology, the macroscopic parenchymal abnormalities are ventricular dilation, multiple lacunar infarcts and large patch soft grayish discolored regions in the centrum semiovale. At microscopy, the brain parenchyma contains lacunes in the basal ganglia and frequently in the periventricular white matter (Fig. 14.8). Leukoencephalopathy specific for Binswanger's encephalopathy is an ill-defined myelin pallor of the deep hemispheric white matter, tending to spare the U fibers and the compact myelin bundles (corpus callosum, internal capsule, optic radiations). Oligodendroglial edema and astrocytic gliosis with occasional macrophages are associated with lacunar infarcts (Dubas et al. 1985). The perivascular spaces are dilated. Wallerian degeneration may be present in the corpus callosum and brain stem (Brun et al. 1992).

The clinical features, the severity of the microangiopathy and the presence of a leukoencephalopathy differentiates Binswanger's encephalopathy from the lacunar syndromes. Binswanger's encephalopathy was considered rare when autopsy was needed for the diagnosis. However, it is now possible, with the advent of imaging, to have clinical and radiological clues to the diagnosis during the life of the patient.

Clinically the first manifestations begin after the fifties, often in patients who have a history of hypertension. Evolution is characterized by acute discrete strokes and subacute progression of the deficits, sometimes with long plateau periods. Signs and symptoms are in part non-lateralized, i.e. lapses in memory, confusion, emotional lability, abulia, global cognitive impairment and, in the advanced state, dementia, seizures, urinary incontinence and cachexia. The other features are lateralized and consistent with small strokes, i.e. lacunar syndromes (mostly pure motor hemiplegia), limited upgaze, pseudobulbar palsy, small-stepped gait, dysarthria, dysphagia, spasmodic crying and laughing, disorders of language, spatial difficulties, hypokinesia, rigidity and cogwheel phenomenon (parkinsonism) (Caplan and Schoene 1978; Dubas et al. 1985; Babikian and Ropper 1987; Fredriksson et al. 1992).

The incidence of radiologic features of leukoencephalopathy is 1%–5% in the elderly, increasing with age, hypertension and severity of cerebrovascular disease (Babikian and Ropper 1987). CT shows periventricular low-density areas, widening of the lateral and third ventricles and cortical sulci, and lacunar infarcts (Lotz et al. 1986; Fredriksson et al. 1992). Magnetic resonance imaging is even more sensitive in showing the lacunar infarcts and the periventricular white matter changes (Kinkel et al. 1985).

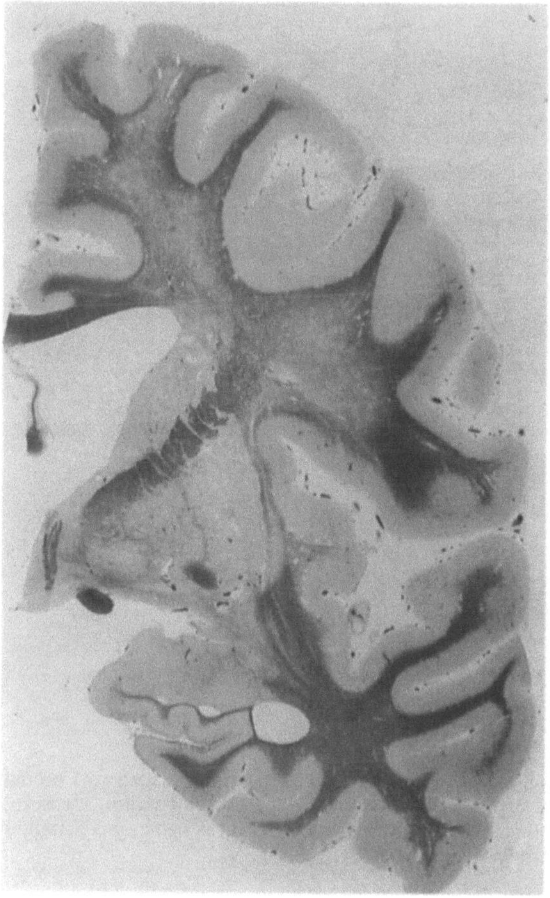

Fig. 14.8. Binswanger's encephalopathy. Coronal section in celloidin showing the myelin pallor sparing the compact myelin bundles associated with lacunar infarcts.

De Reuck et al. (1980) theorized that Binswanger's disease was the result of ischemia, due not to thrombosis as is usual with lacunar infarcts, but to a chronic watershed distal-field ischemia due to the combination of severe penetrating artery disease and hypoperfusion during hemodynamic crisis. This distal chronic ischemia could also be associated with an abnormality of the blood–brain barrier (Dubas et al. 1985).

Amyloid Angiopathy

Cerebral amyloid or congophilic angiopathy is not found in combination with systemic and visceral amyloidosis, and has gained increased interest because it is an important cause of intracerebral hematomas, and because there are links

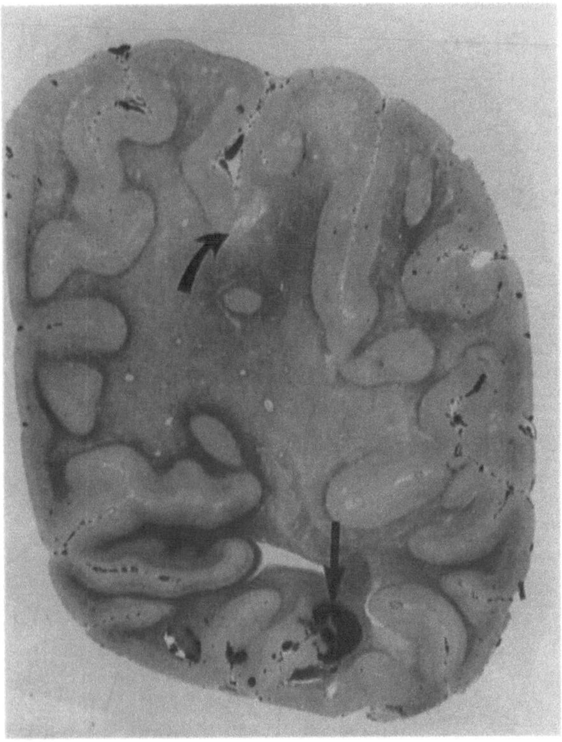

Fig. 14.9. Cerebral amyloid angiopathy macroscopy. Coronal section in celloidin showing the myelinic pallor, a cortical infarct (curved arrow) and a cortical hematoma (straight arrow).

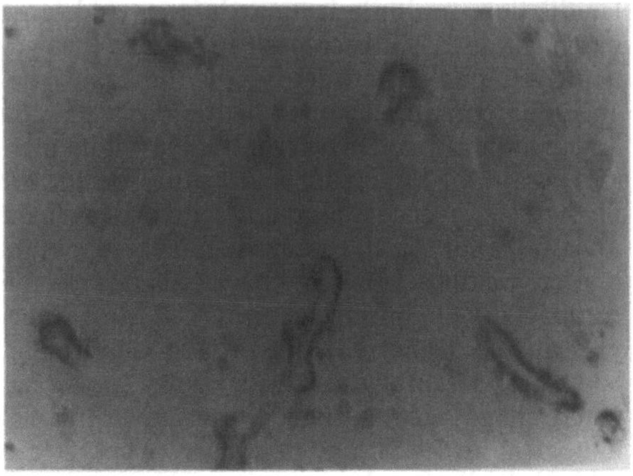

Fig. 14.10. Immunolabeling of the A4 protein in cortical capillaries in cerebral amyloid angiopathy.

with Alzheimer's disease (Vinters 1987). Amyloid angiopathy has a highly variable prevalence (15%–40%) according to different studies, increasing with age (Regli 1989): 80% of Alzheimer's demented patients are affected to some degree.

Clinically the signs and symptoms are related to acute stroke and progressive dementia. The strokes consist mostly of intracerebral hematomas, with lobar sites and a high recurrence rate in old patients without other cause for hemorrhage (Regli 1989) (Fig. 14.9). Sometimes the hemorrhage is preceded by transient ischemic attacks, treated with antithrombotic drugs, which prompt the hemorrhage. The dementia can be isolated, progressive or subacute, or punctuated by strokes suggesting a vascular or mixed dementia (Haan et al. 1990). The prognosis is poor, survival being usually no more than a few years.

At postmortem the leptomeningeal and cortical middle- and small-caliber arteries are affected by an acellular thickening of the walls, with amyloid deposits reacting to anti-β protein antibody (congophilic angiopathy) (Maruyama et al. 1990). The layers of the walls are often separated from one another and the internal elastic membrane is fragmented (Fig. 14.10). The deposits can also involve the capillaries and spread into the surrounding perivascular tissues adjacent to senile plaques (dyshauric angiopathy). The angiopathy is more severe in the posterior regions, sparing the white-matter vessels. The strokes are related to the angiopathy secondary to amyloidosis characterized by segmentary fibrinoid changes and microaneurysms (Dubas et al. 1985; Vinters 1987). With immunochemical technique, A4 protein and immunoglobulins can be detected. In the brain, multiple lobar cortical–subcortical hematomas or pale infarct scars of various ages are usually present. A leukoencephalopathy characterized by subacute or chronic lesions, i.e. spongiosis, swollen oligodendroglia, widening of perivascular spaces, incomplete demyelination and astrocytic gliosis, can be identified (Gray et al. 1985; Dubas et al. 1985). There probably is a link in the pathophysiology of leukoencephalopathy between Binswanger's encephalopathy and cerebral amyloid angiopathy.

Very rare forms of amyloid angiopathies that affect the young are genetically determined and characterized by similar types of deposits. In the Icelandic type the deposit consist of cystatin C (Grupp et al. 1984), which is decreased in the CSF (Shimode et al. 1991). The Dutch type is a protein A4 variant. The familial oculoleptomeningeal amylosis is characterized by transthyretin deposits.

The links with Alzheimer's disease are complex: both can occur independently. However, amyloid angiopathy is often associated with

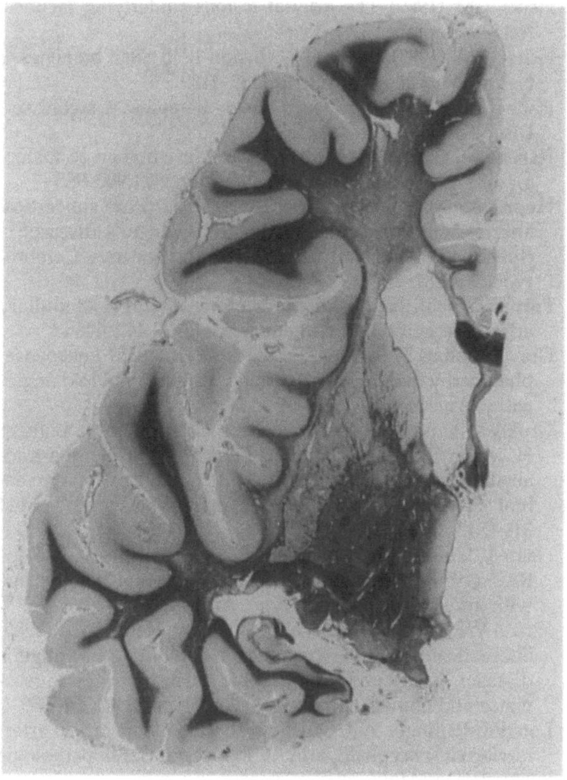

Fig. 14.11. Familial angiopathy. Coronal section in celloidin showing the myelin pallor of the deep white matter associated with lacunar infarcts.

senile plaques and neurofibrillary tangles, leading to the assumption of a common inducing mechanism.

Cerebral Autosomal Dominant Arteriopathy with Subcortical Infarcts and Leukoencephalopathy (CADASIL)

In recent years, different authors report the existence of a familial hereditary disorder characterized by a unique small-vessel disease (Wålinder 1981; Sonninen and Savontaus 1987; Tournier-Lasserve et al. 1991; Davous and Fallet-Bianco 1991; Mas et al. 1992). The clinical features show a dominant genetic transmission, the occurrence of recurrent subcortical strokes separated by more or less remission leading to pseudobulbar palsy and subcortical multi-infarct dementia of the subcortical type affecting young adults of both sexes. In the patients reported, neuroimaging showed cerebral atrophy and multiple lacunar infarcts associated with a leukoencephalopathy, a picture similar to Binswanger's encephalopathy. However, none of the affected patients was hypertensive. Some family members had imaging signs suggesting a leukoencephalopathy, despite the fact that they were free of signs and symptoms (Tournier-Lasserve et al. 1991; Mas et al. 1992). At autopsy there were multiple lacunar infarcts affecting the basal ganglia, thalamus, internal capsule, periventricular white matter and pons, accompanied by white-matter atrophy (Fig. 14.11) (Wålinder 1981; Davour and Fallet-Bianco 1991). On microscopy the large arteries were normal and the vascular changes affected small muscular arteries, predominating in the long perforating arteries. Small arteries showed hyaline degeneration of the fibrinoid type restricted to the inner layers, sometimes accompanied by mild inflammatory infiltration and granulomatous deposits. Lamellar elastosis and hyaline degeneration of the collagenous type were associated with fibroblastic proliferation of the intima with narrowing of the lumen leading to occlusion of the vessels and infarcts (Fig. 14.12). There was no amyloid, nor hemorrhages (Wålinder 1981; Davous and Fallet-Bianco 1991). The leukoencephalopathy was most severe in the periventricular white matter, and spared the U fibers. The white matter showed rarefaction of myelin, astrocytic glyosis and preservation of axons (Davous and Fallet-Bianco 1991). Some of the arteries stained with a foreign material that does not have the staining properties of amyloid. This

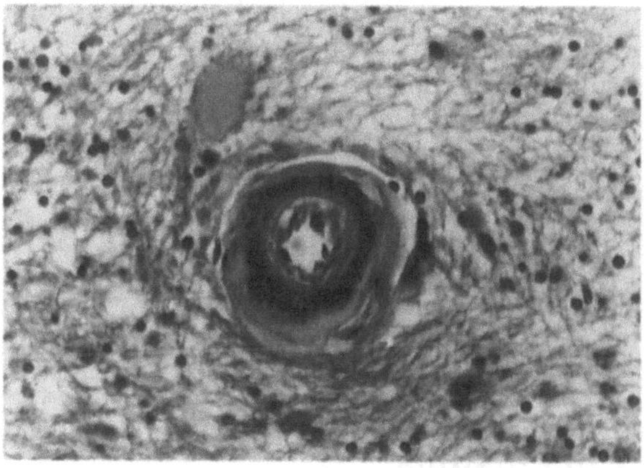

Fig. 14.12. Familial angiopathy: microscopy of a white-matter small artery showing the sclerosis and granular amphophylic arteriolar deposits.

entity is now called 'cerebral autosomal dominant arteriopathy with subcortical infarcts and leukoencephalopathy' (CADASIL) (Tournier-Lasserve et al. 1993). The pathophysiology of the disease is unknown. However, the affected gene has been located on chromosome 19 (Tournier-Lasserve et al. 1993). Nevertheless, even if all the reports suggest similarities, the existence of a common uniform disease is not established.

Conclusions

Many conditions can cause microvascular pathology. The effects of age and vascular risk factors can combine to make the diagnosis even more difficult. Microvascular pathology should be suspected when examining patients with clinical features and neuroimaging consistent with deep lesions: lacunes, hematomas, leukoaraiosis etc. Moreover, vascular risk factors such as hypertension and cardiac disease can give clues to the mechanism of the arteriolar disease. However, often only autopsy can establish the diagnosis. Nevertheless, our knowledge about microvascular pathology is still poor and further studies are essential for progress in this field.

References

Babikian V, Ropper AH (1987) Binswanger's disease: a review. Stroke 18:2–12

Brun A, Fredriksson K, Gustafson L (1992) Pure subcortical arteriosclerotic encephalopathy (Binswanger's disease): a clinicopathologic study. Part 2: Pathologic features. Cerebrovasc Dis 2:87–92

Caplan LR (1989) Intracranial branch atheromatous disease: a neglected, understudied, and underused concept. Neurology 39:1246–1250

Caplan LR, Schoene WC (1978) Clinical features of subcortical arteriosclerotic encephalopathy (Binswanger's disease). Neurology 28:1206–1215

Davous P, Fallet-Bianco C (1991) Démence sous-corticale familiale avec leucoencéphalopathie artériopathique observation clinico-pathologique. Rev Neurol 147:376–384

De Reuck J, Crevits L, De Coster W, Sieben G, van der Eecken H (1980) Pathogenesis of Binswanger chronic progressive subcortical encephalopathy. Neurology 30:920–928

Dubas F, Gray F, Roullet E, Escourolle R (1985) Leucoencéphalopathies artériopathiques. Rev Neurol 141:93–108

Fisher CM (1965) Lacunes: small, deep cerebral infarcts. Neurology 15:774–784

Fisher CM (1969) The arterial lesions underlying lacunes. Acta Neuopathol 12:1–15

Fisher CM (1977) Bilateral occlusion of basilar branches. J Neurol Neurosurg Psych 40:1182–1189

Fisher CM (1991) Lacunar infarcts – a review. Cerebrovasc Dis 1:311–320

Fisher CM, Caplan LR (1971) Basilar artery branch occlusion: a cause of pontine infarction. Neurology 21:900–905

Fredriksson K, Brun A, Gustafson L (1992) Pure subcortical arteriosclerotic encephalopathy (Binswanger's disease): a clinicopathologic study. Part 1: Clinical features. Cerebrovasc Dis 2:82–86

Furuta A, Ishii N, Nishihara Y, Horie A (1991) Medullary arteries in aging and dementia. Stroke 22:442–446

Gray F, Dubas F, Roullet E, Escourolle R (1985) Leucoencephalopathy in diffuse hemorrhagic cerebral amyloid angiopathy. Ann Neurol 18:54–59

Grubb A, Jensson O, Gudmundsson G, Arnason A, Löfberg H, Malm J (1984) Abnormal metabolism of gamma-trace alkaline microprotein. The basic defect in hereditary cerebral hemorrhage with amyloidosis. New Engl J Med 311:1547–1549

Haan J, Lanser JBK, Zijderveld I, van der Does IGF, Roos RAC (1990) Dementia in hereditary cerebral hemorrhage with amyloidosis – Dutch type. Arch Neurol 47:965–967

Kinkel WR, Jacobs L, Polachini I, Bates V, Heffner RR (1985) Subcortical arteriosclerotic encephalopathy (Binswanger's disease). Computed tomographic, nuclear magnetic resonance and clinical correlations. Arch Neurol 42:951–959

Lotz PR, Ballinger WE, Quisling RG (1986) Subcortical arteriosclerotic encephalopathy: CT spectrum and pathologic correlation. AJNR 7:817–822

Maruyama K, Ikeda SI, Ishihara T, Allsop D, Yanagisawa N (1990) Immunohistochemical characterization of cerebrovascular amyloid in 46 autopsied cases using antibodies to β protein and cystatin C. Stroke 21:397–403

Mas JL, Dilouya A, de Recondo J (1992) A familial disorder with subcortical ischemic strokes, dementia, and leucoencephalopathy. Neurology 42:1015–1019

Poirier J, Derouesné C (1984) Cerebral lacunae. A proposed new classification. Clin Neuropathol 3:266

Regli F (1989) Cerebral amyloid angiopathy. In: Toole JF (ed) Handbook of clinical neurology. Vascular disease Part II Vol 10. Elsevier, Amsterdam, pp 333–344

Shimode K, Fujihara S, Nakamura M, Kobayashi S, Tsunematsu T (1991) Diagnosis of cerebral amyloid angiopathy by the enzyme-linked immunosorbent assay of cystatin C in cerebro-spinal fluid. Stroke 22:860–866

Sonninen V, Savontaus ML (1987) Hereditary multi-infarct dementia. Eur Neurol 27:209–215

Tournier-Lasserve E, Iba-Zizen MT, Romero N, Bousser MG (1991) Autosomal dominant syndrome with strokelike episodes and leukoencephalopathy. Stroke 22:1297–1302

Tournier-Lasserve E, Joutel A, Melki J, Weissenbach J, Lathrop GM, Chabriat H, Mas JL, Cabanis EA, Baudrimont M, Maciazek J, Bach MA, Bousser MG (1993) Cerebral autosomal dominant arteriopathy with subcortical infarcts and leucoencephalopathy maps to chromosome 19q12. Nature Genetics 3:256–259

Verny M, Seilhean D, Duyckaerts C, Delaère P, Hauw JJ (1991) Vieillissement vasculaire cérébral (à l'exclusion de l'innervation). Ann Cardiol Angéiol 40:301–308

Vinters HV (1987) Cerebral amyloid angiopathy. A critical review. Stroke 18:311–324

Wålinder J (1981) Multi-infarct dementia. In: Vinken PJ, Bruyn GW (eds) Handbook of clinical neurology. Neurogenetic directory. Elsevier, Amsterdam, pp 283–285

Concluding Comments

The contributions in this Section should convince readers that the vascular lesions that cause brain ischemia are varied and very diverse. Moreover, many changes such as vasoconstriction, vasodilatation and endothelial-cell secretion are physiologic rather than morphological, and may vary from moment to moment. All of the authors have emphasized the intimate relationships and interactions between the blood vessels and the blood that courses through them.

Atherosclerosis and its Clinical Monitoring

Atherosclerosis is ubiquitous: nearly every adult has some degree of atherosclerosis by the time of death. Atherosclerosis is often severe and frequently threatens loss of limb, vision, brain function and life. Coronary artery, cerebrovascular, peripheral and renal vascular occlusive disease combined account for the majority of deaths in most developed countries, and an even larger share of the morbidity. I wish to emphasize here that atherosclerosis is a systemic disease. A wide variety of physician and non-physician health-care providers, including neurologists, vascular surgeons, cardiologists etc. serve as the entry points of patients into health-care systems. All have a duty, often ignored, to assess patients' risk factors for atherosclerosis, and to initiate, when possible, steps to reduce the risks found. Referral to a generalist is essential if the patient has no primary physician, as is often the case now in the United States.

But how will the presence of atherosclerosis be monitored? In many of the arteries involved by atherosclerosis there is no single non-invasive, reliable, relatively inexpensive method of documenting the presence of atherosclerotic lesions and monitoring them sequentially. Probably the most accessible arteries to monitor are the extracranial arteries to the brain, the carotid and vertebral arteries in the neck. Atherosclerotic lesions in these arteries closely parallel atherosclerosis in the coronary and peripheral arteries and in the aorta, although in any individual patient some vessels are involved much more extensively than others and some arteries may be selectively spared for some unknown reasons (Duncan et al. 1977; Mohr et al. 1978). Lesions in the extracranial arteries occur predominantly at two sites: in the internal carotid arteries along the posterior walls at the origin from the common carotid arteries and directly opposite the flow dividers between the internal and external carotid arteries, and at the origin of the vertebral arteries from the subclavian arteries. The morphology and physics of these arterial sites, mentioned in the prefatory comments to this Section, and described in detail by Schmid-Schönbein and Perktold, probably explain the predilection of these regions for the development of atheromatous plaques. These sites are excellent regions to monitor for the presence of plaques and sequential changes in them over time and after various therapeutic interventions.

The technology to monitor lesions clinically in these proximal extracranial arteries has improved greatly during the past 5–10 years. B-mode ultrasound yields two-dimensional images that accurately show the thickness of the arterial wall (Touboul et al. 1992) and encroachment by plaques on the arterial lumen. New technology makes it possible to reconstruct three-dimensional images of these arterial plaques (Hennerici and Steinke 1987). The addition of a multigated pulsed Doppler to high-resolution B-mode ultrasound scanners in a duplex system improves the accuracy of the evaluation and allows better identification of the artery being imaged in real time (Hennerici and Freund 1984). The images from duplex systems have proven very sensitive and specific for occlusive disease, compared to angiography and pathology (O'Donnell et al. 1985; Schenk et al. 1988). Duplex scans reliably divide patients into those with slight plaque disease, those with moderate stenosis (40%–75% luminal narrowing) and those with severe stenosis (>75% narrowing). Studies have shown a clear direct relationship between degree of stenosis and stroke risk at high degrees of stenosis (Norris and Zhu 1990).

Recently, it has become possible to predict the nature and composition of plaques with B-mode ultrasound (Reilly et al. 1983; Hennerici et al. 1985; O'Donnell et al. 1985; Bornstein and Norris 1989). Ultrasound characteristics and echogenicity can help separate five types of plaques: fibrous, soft, hard calcified, ulcerative and hemorrhage into plaques (Hennerici et al. 1985) (Table 1). Fibrous plaques made up mostly of collagen are very echogenic and have relatively homogeneous images. As the lipid content of a plaque increases the plaque becomes more echolucent, and localized regions of lipid accumulation image as discrete echolucent areas (O'Donnell et al. 1985). Similarly, in the aorta ultrasound in the form of transesophageal echocardiography can identify atheromas. Protruding, mobile and pedunculated aortic atheromas are more likely to be associated with intra-arterial embolism than flat plaques with broad attachments (Tunick et al. 1991a, b).

Table 1. Types of plaques and their ultrasound characteristics (From Hennerici et al. 1985; O'Donnell et al. 1985; Reilly et al. 1983)

Plaque type	Ultrasound (B-mode) features
Fibrous	Subintimal bright echo reflections Generally homogeneous
Soft	Heterogeneous structures protruding into lumen; lipid regions appear echolucent
Hard calcified	High-intensity echo reflections and shadows
Ulcerative	Discontinuous echo contour and a high reflecting echoniche representing the floor of the ulcer
Hemorrhage into plaques	Regions of reduced density surrounded by highly echoreflecting structures

When patients are followed sequentially most small plaques remain relatively unchanged or progress, but spontaneous regression occurs in about one-fifth of cases (Hennerici et al. 1985). The nature of the plaques shown by ultrasound does help to predict future changes. Fibrous and hard plaques usually remain stable or progress slowly; soft plaques can show a reduction in size, and ulcerative lesions do heal (Hennerici et al. 1985). Intraplaque hemorrhage is often associated with subsequent growth of the plaque, with luminal encroachment. As plaques enlarge and narrow the lumen, they reach a critical size when complications (hemorrhage, cracking, ulceration, mural thrombi etc.) are prone to occur (Norris and Zhu 1990). These complications promote the activation of platelets and the coagulation cascade, and both white platelet–fibrin aggregates and red clots can form on the surface of large complex protruding atheromas (Fisher and Ojemann 1986). When the artery occludes there is a higher risk of ischemic symptoms (Bornstein and Norris, 1989). Now color-flow Doppler helps show the flow characteristics and images functional changes in flow produced by the plaques (Steinke et al. 1990). Rapid or severe increase in the degree of stenosis on sequential ultrasound tests is predictive of a higher likelihood of the development of ischemic symptoms. Recent technology assessment concerning carotid evaluation from the American College of Physicians (Feussner and Matcher 1988a, b) made the assumption that carotid artery duplex examinations are warranted only if surgery is contemplated. I disagree strongly. Duplex scans and color-flow Doppler images are very helpful in prognosis, in studying and measuring the effects of various medical interventions, and for following patients after carotid surgery or angioplasty. Ultrasound provides clinicians from a variety of disciplines with a window into the presence and severity of atherosclerosis in the individual patient, and a way to follow the disease.

Reversible Vasoconstriction and Vasodilatation and Ischemia of Unknown Mechanism

Sloan, in his chapter, reviews in detail the situations and circumstances in which vasospasm, i.e. reversible vasoconstriction, has been shown to occur. The secretory capability of the endothelium as described by Trachtenberg and Ryan, and the important fibromuscular composition of the arterial media, account for the potential for vascular contractility. I believe that reversible vasoconstriction is more common than previously thought, and probably accounts for a significant number of instances of focal brain ischemia in which the etiology remains unsettled.

In any large series, infarcts of undetermined cause are quite common. In the NINCDS-sponsored Stroke Data Bank they accounted for 508 of 1273 infarcts (40%) and 28.1% of the total strokes (Foulkes et al. 1988; Sacco et al. 1989). In these patients no cardiac source was found and no arterial lesion was discovered that would explain the ischemia. There are a limited number of potential mechanisms that could account for the ischemia:

1. The brain dysfunction was not ischemic (seizure-related or other brain dysfunction)
2. Tests that would have shown the cardiac and arterial lesions, such as transesophageal echocardiography and contrast angiography, were not performed.
3. Full technological evaluation was performed but small cardiac or arterial lesions that could have given rise to emboli were too small to be detected by the existing technology, i.e. lesions were there but could not be seen.
4. The processes in the arteries supplying the ischemic zone were functional and reversible, i.e. affected vascular contractility.

The first three of these explanations are generally well accepted. It is likely that few patients with typical TIAs have seizures, hysteria or other non-ischemic etiologies after evaluation by neurologists and full investigations. Aspirin or other drugs that affect platelet functions might be useful in preventing embolization of fibrin–platelet clots in patients with small non-flow-reducing lesions. However, if the cause was vasospastic then other drugs, such as β-adrenergic blockers or calcium-channel blockers, might be more effective.

Table 2. Migrainous accompaniments

Timing. Attacks usually last 15–60 minutes (average 20 minutes); TIAs are usually shorter and strokes longer
Attacks usually start with 'positive' sensations affecting one modality (usually vision or somatic sensation). Brightness, colors, scintillations (vision); prickling, tingling, burning (somatosensory) are most common. Positive auditory, vestibular, olfactory, gustatory sensations occur, but less often
Positive sensations march or spread slowly within that modality; spots or stars move slowly across vision; paresthesias spread from digit to digit up limbs to other body parts. The spread is much slower than in Jacksonian seizures
Negative sensory phenomena follow in the wake of the positive. Loss of vision and numbness are noted after the train of scintillations and paresthesias have passed
After clearing of sensory abnormalities in the first modality, symptoms begin in a second modality. After vision has normalized paresthesias begin; when vision and sensation become normal, aphasia begins, etc. In TIAs and strokes, all usually occur concurrently
Attacks involve different anatomical vascular territories, left visual field in one attack, right visual field in another, and left paresthesias in a third attack
Spread is often not anatomical, e.g. from one side of the face to the other side
Vertebrobasilar territory is often involved, with dysarthria, ataxia, vertigo, diplopia and bilateral paresthesias

Fisher, in a series of articles, described transient episodes occurring in older patients as 'late life migraine accompaniments' (Fisher 1980, 1986). He called these attacks migrainous because they closely resembled episodes of dysfunction described by migraineurs prior to or accompanying their usual migraine headaches. Characteristic features relate to the timing, spread and character of the symptoms (Table 2). Most often the attacks are sensory as opposed to motor, and the sensory phenomena include so-called positive features (brightness and scintillation as opposed to blackness and loss of vision; paresthesias as opposed to numbness, etc.). Many of the attacks seem to involve posterior circulation-supplied structures. In my experience some attacks are precipitated by stimuli known to precede migraine. Many of the patients have had attacks compatible with migraine in their younger years. As arteries become more fibrotic and firm, they may be more susceptible to vasoconstriction.

I believe that transient global amnesia (TGA) is a disorder that is probably best explained by such temporary vasoconstriction. Table 3 lists reasons to link vasospasm with TGA, and I have elaborated elsewhere on this subject (Caplan 1985, 1990; Caplan et al. 1981). Particularly important, in my opinion, is the precipitation of attacks of TGA by stimuli known to precipitate migraine. Hot showers, a dip in cold water, sexual intercourse, arguments, sudden emotional stress, unusual exertion, can all precipitate TGA and migraine.

Table 3. Reasons to link TGA and vasoconstriction ('migraine')

TGA and migraine are both usually benign
Migraine patients have amnesic attacks
Migrainous auras or headache may precede, accompany or follow TGA attacks
TGA can be repeated (25%), occasionally on multiple occasions
Precipitants are similar: emotional stress, angiography, sexual intercourse, physical activity or effort, immersion in cold water, hot bath or shower, driving or riding in a motor vehicle
Each is a predominantly posterior circulation disorder
Headache often follows attacks, even in patients with no history of migraine
TGA patients seldom develop posterior-circulation strokes; patients with posterior circulation strokes seldom have a history of TGA
A significant number of TGA patients (35%) have a history of migraine
Focal temporal lobe EEG slowing can be found in both TGA and migraine
Decreased blood flow in the posterior cerebral artery (PCA) territories by SPECT can be found in both disorders
Repetitive queries may be the 'positive' phenomenon in TGA (comparable to positive visual and sensory phenomena in migraine); patients with amnesic strokes are more often unaware of the amnesia and are quiet
Other etiologies for TGA (seizures, strokes, psychiatric) are very unlikely

We now have available technology in the form of single photon emission computed tomography (SPECT) which can non-invasively show regions of altered blood flow. Preliminary studies in patients with TGA have shown variable decreased flow, usually in posterior circulation structures and usually on the left (Fayad et al. 1990). Transcranial Doppler (TCD) can now measure flow velocities in intracranial arteries and so study changes in flow during TGA or other transient attacks. I believe future investigations will provide evidence that a significant proportion of examples of TIAs and strokes of unknown mechanism relate to vasoconstriction.

Vascular Disease Affecting the Arterial Wall

Much attention, both clinically and in basic research, has been directed towards the inner juxtaluminal vascular structures, the endothelium, intima and subintima. Atherosclerosis affects these regions and is by far the commonest vascular disease recognized. The bulk of the contributions in this section have dealt with atherosclerosis and its development, the endothelium and its functions and the physical changes in arteries related to the development of plaques with luminal encroachment. Most of the investigations available give information about stenosis and luminal encroachment.

Neglected in general, and also in this section, are diseases that mainly affect the arterial media, the thickest of all the vascular coats and yet the layer least well studied and least well understood. The media is composed mostly of collagen, fibrous tissue and muscular elements and as individuals age is commonly affected by calcification and fibrosis. The media can be the site of deposition of biochemical substances such as amyloid. Arteritis and congenital and acquired diseases of connective tissue affect collagen within the vessel wall. We are now becoming aware of a host of relatively uncommon disorders affecting the vessel walls that have been difficult to investigate clinically.

Fibromuscular dysphasia (FMD) is a disorder of the walls of arteries that is poorly characterized and poorly understood. FMD was first recognized in the renal arteries as a cause of hypertension, but physicians soon recognized that it could affect a variety of different systemic arteries. The craniocerebral arteries are often involved, especially the carotid and vertebral arteries in the neck and, less often, the intracranial arteries. FMD is most common in middle-aged women (Corrin et al. 1981) and is found in about 0.6% of unselected carotid arteriograms (Corrin et al. 1981; So et al. 1981). The lesions most often involve the pharyngeal portions of the carotid arteries but spare the origins in the neck. Pathological changes are variable. The most commonly recognized form affects mostly the media. Bands of fibrous dysplastic tissue and proliferating smooth muscle cells constrict the lumen. These regions of luminal narrowing seem to alternate with regions of luminal dilatation in which medial thinning and disruption of the elastic membrane are found (Sandok 1983; Luscher et al. 1987). In these and other patients without medial involvement, hypertrophy of fibrous tissues occurs in the intima or adventitia, causing segmental regions of stenosis. The segmental changes explain the radiological feature of 'string of beads' type of sausage-like zones of constriction shown on angiography. FMD may in fact be a very heterogeneous disorder which can result from a variety of different etiologies. The etiology of FMD and the mechanism by which it causes brain ischemia are still obscure.

Arterial dissections have a predilection for the same regions of arteries involved by FMD. Most patients with dissections have single events, usually precipitated by trauma or stretching injuries even though the patient often cannot recall precipitating events. Stretch injuries cause tears in the vessel wall, usually between the media and intima or between the media and the adventitia. Tiny blood vessels are torn within the vessel wall and an intramural hematoma develops and dissects along the vessel. Some patients have recurrent dissections in different arteries (Caplan and Tettenborn 1992). In one patient with multiple dissections studied at necropsy, multiple regions of fibromuscular dysplasia were found (Bellot et al. 1985). These patients with chronic recurrent dissections must have some disorder of connective tissue or collagen. To date, ordinary investigations have not, in my experience, been able to predict which patients with dissections will go on to have recurrences. Some congenital disorders of connective tissue, such as Marfan's syndrome and pseudoxanthoma elasticum, clearly do predispose to dissections.

Calcium is often found within the arterial walls of large vessels as patients age. The original term atherosclerosis referred to 'hardening' of the arteries rather than atheromatous deposits. In the study by Baker and Iannone (1959a, b) covering the pathology of intracranial vascular disease, vascular calcification was given a prominent place in the morphological lesions studied. Do the calcification and fibrosis affect vascular function? Certainly in the aorta and the great vessels arising from the aortic arch calcification and rigidity lead to decreased distensibility and higher systolic and lower diastolic blood pressures.

We now know that the walls of some small penetrating and pial arteries can be infiltrated with foreign substances. In patients with amyloid angiopathy, the congophilic material causes thickening of the arterial walls and, at times, splitting or secondary reduplication of the wall, causing a distinctive 'double-barrel' lumen with amyloid found in either the inner or the outer media (Vinters 1987). The deposition of material leads to vascular fragility and recurrent brain and subarachnoid hemorrhages. Scattered small infarcts are also commonly found at necropsy. More recently, patients with amyloid angiopathy have been reported who have multiple infarcts and a striking leukoencephalopathy with subcortical, periventricular and corona radiata lesions shown on MRI (Gray et al. 1985; Loes et al. 1990; DeWitt and Louis 1991). The clinical findings resemble Binswanger's disease. Amyloid angiopathy can be familial. In the Icelandic variety, the abnormality is related to disordered metabolism of gamma-trace protein (Grubb et al. 1984). More recently there have been reports of a different familial disorder in which a Binswanger-like leukoencephalopathy is caused by a disease characterized by deposition of a still unidentified non-amyloid material in the wall of arteries. Morphologically, the arteries appear similar to those found in amyloid angiopathy and Binswanger's disease patients, except for the absence of congophilic staining. Curiously, in all of these conditions, although infarcts are very prominent, as are more diffuse ischemic white-matter abnormalities, narrowing of the arterial lumina as found at necropsy is insufficient to explain the ischemia (Fisher 1989). Might the infiltrative material and the fibromuscular and connective tissue changes in the media be producing functional flow alterations in these arteries not identifiable by strict morphological examination?

There is still very much to be learned about the walls of arteries, and unfortunately there are no available clinical investigations to identify the causes of these changes in the media.

Dilatative Arteriopathy

In some patients with brain ischemia vascular imaging shows elongated tortuous intracranial arteries. At times in these patients the extracranial arteries may also appear very tortuous and dilated. Although usually identified in adults, dilatative changes in multiple arteries have also been discovered in children (Read and Esiri 1979; Makos et al. 1987). When the dilatation is severe the elongated dilated arteries are called dolichoectatic or fusiform aneurysms. Most often the abnormality affects the intracranial vertebral and basilar arteries (Echiverri et al. 1989; Nishizaki et al. 1986; Pessin et al. 1989) but the MCA can also be involved and sometimes both anterior and posterior-circulation arteries are involved (Little et al. 1981).

At necropsy, the dolichoectatic arteries often contain calcified atherosclerotic plaques and thrombus formation. Clinical syndromes include stretching and pressure on the cranial nerves and brain structures, headache, recurrent brain ischemia and, rarely, rupture causing subarachnoid hemorrhage (Pessin et al. 1989). Many different mechanisms have been suggested for the brain ischemia, including distortion and functional blockage of branches because of vascular elongation and tortuosity; blockage of branches by plaque or thrombus in the lumen; and embolization of thrombus distally. TCD has recently proved helpful in diagnosis and may provide new insights into the mechanisms of ischemia. Mean flow velocities are usually reduced but peak flow velocities are relatively preserved (Hennerici et al. 1987). The physical properties of the dilated arteries may alter flow, creating turbulence and regions of reversed and stagnant flow. Distal ischemia could be due to reduced distal runoff because of functional flow alterations. The cause of dilatative arteriopathy is unknown. The frequent occurrence in diabetic hypertensive patients suggests that the changes are at least partially degenerative, while occurrence in childhood suggests that there may also be genetic factors.

Venous Occlusive Disease

Although this section is titled 'Blood Vessels', in fact it deals entirely with arteries and their diseases and physiology. Recently, there has been renewed interest in venous disease because the advent of CT and MRI has led to increased recognition of venous–dural sinus occlusions during life. Intracranial venous occlusive disease, for the greater part, is equated with occlusion of the dural sinuses. We know very little about the occlusion of deep and cortical veins in the absence of sinus occlusions. Sarcoidosis often causes a periphlebitis in the retina and brain, and infection can spread from the epidural space to the brain through infected veins. Non-inflammatory thrombosis limited to cortical or draining veins, and its sequelae, has not, to my knowledge, been reported.

Dural sinuses are very interesting and complex morphological structures (Browder et al. 1972). The vascular coats are quite different, venous sinuses having no well defined muscular medial layer. Valves, bridges, bands, cusps, flaps and divisions are visible when the dural sinuses are opened. Some of the divisions interconnect draining veins with the main sinuses. Flow is quite slow and under far lower pressure than in the arterial system. Intracranial pressure is transmitted to the venous system so that increased intracranial pressure is accompanied by high venous pressure.

Coagulopathy, with its increased tendency to form red fibrin-dependent thrombi, is usually thought of as a venous problem rather than arterial. Hypercoagulable states such as cancer, inflammatory bowel disease, pregnancy, the puerperium, oral contraceptive use, lupus erythematosis and antiphospholipid antibody syndromes can present with peripheral phlebothrombosis with pulmonary embolism. These disorders also often predispose to dural sinus occlusion (Bousser et al. 1985; Vidailhet et al. 1990). Why female sex hormones play such a key role in the epidemiology of dural sinus occlusion is unclear, but in India dural sinus occlusion during pregnancy or the puerperium is one of the most common causes of stroke in the young (Srinivasan 1984). Infections such as meningitis and epidural and subdural abscesses and dehydration also account for a significant number of cases of dural sinus occlusion in children and young adults (Bousser et al. 1985). Behçet's disease is an important cause of venous occlusive disease in the Middle East (Pamir et al. 1981; Bousser et al. 1985; Serdaroglu et al.).

The clinical implications vary. Some patients have a 'pseudotumor'-like syndrome with headache, increased intracranial pressure and papilledema. The spread of thrombus into the draining veins probably explains the development of venous infarction. Infarcts in patients with sagittal sinus occlusion are usually hemorrhagic, bilateral and most often posterior parietal in location. Hematomas may develop. Seizures, hemiplegia and increased intracranial pressure are the dominant clinical features (Bousser et al. 1985). CT usually reveals only non-specific abnormalities such as brain edema, with small ventricles or bilateral hemorrhagic infarcts. Occasionally the occluded sinus appears hyperdense on non-contrast CT, or may show a filling defect. MRI and MR angiography are much more definitive than CT in diagnosing and localizing dural sinus occlusion. Follow-up MRI studies show that the sinus may remain occluded (Mas et al. 1992). Heparin is now considered useful in treatment (Levine et al. 1988; Einhaupl et al. 1991). Some centers are exploring the use of thrombolytic agents infused locally through catheters placed in the venous system near the thrombi.

Intracranial Arterial Disease due to Changes in the Cerebrospinal Fluid

I close by mentioning one other interesting mechanism of arterial disease because of its theoretical interest. Meningeal infection, especially when chronic, and large amounts of blood in the CSF cause functional and morphological changes in the intracranial basal arteries which can lead to ischemia and infarction. Clinicians first became aware of this problem in patients with chronic tuberculous meningitis.

Oxyhemoglobin is now believed to be the most likely injurious substance in patients with SAH; it affects the function of platelet-derived growth factor released from platelets adherent to vessel walls and stimulates components of the coagulation cascade, especially thrombin, plasmin and fibrinogen (Kassell et al. 1985; White and Robertson 1985; MacDonald and Weir 1991). Arteries bathed by the exudate show changes characterized by thickening of the media and inflammation of the vascular wall, often with superimposed thrombosis. These changes were often seen in syphilis and were termed Heubner's arteritis. Later it became clear that patients with chronic fungal meningitis, especially cryptococcal, histoplasma and coccidioidal, also had identical changes in the intracranial arteries. Those most often involved are the basal arteries. The MCA and arteries penetrating into the brain from the anterior and posterior perforated spaces are most commonly thrombosed. Infarcts in the basal ganglia, thalamus and midbrain develop and may occur even after sterilization of the CSF. The arteries involved are often bathed in exudate in the basal cisterns. Similar secondary arterial pathology also probably occurs in patients with lymphomatous meningitis in which thick pockets of cells in the cisterns and Virchow–Robin spaces can surround penetrating arteries (Kieburtz et al. 1993).

It is now well known that aneurysmal subarachnoid hemorrhage is often complicated by vasospasm. Sloan discusses this phenomenon in detail in his chapter. Vasoconstriction seems to correlate roughly with the amount of blood, especially thick clot layers, in the basal and sylvian cisterns. Although the disorder is usually referred to as vasospasm, recently studies have shown that there are morphological changes that develop in the basal arteries. Patients with SAH who have vasospasm and die in less than 3 weeks show medial necrosis, whereas patients who survive more than 3 weeks show marked concentric thickening of the intima, subendothelial fibrosis and atrophy of the media of intracranial basal arteries (Hughes and Schianchi 1978; Conway and MacDonald 1972). Researchers believe that these changes are somehow induced by products released from erythrocytes that penetrate into the outer vascular wall. These biochemical changes may induce secretory changes within the arteries that cause vasoconstriction and also induce morphological changes in the arterial walls.

How exactly do infection, blood or tumor cells in the CSF induce changes in the arteries? Are there secretory or functional properties in the adventitial coat that have not been explored? Unlocking this story may help clarify some of the remaining mysteries about arterial physiology and pathology.

References

Baker AB, Iannone A (1959a) Cerebrovascular disease I. The large arteries of the circle of Willis. Neurology 9:321–332

Baker AB, Iannone A (1959b) Cerebrovascular disease II. The smaller intracerebral arteries. Neurology 9:391–396

Bellot J, Gheradi R, Porrier J et al. (1985) Fibromuscular dysplasia of cervicocephalic arteries with multiple dissections and a carotid cavernous fistula: a pathologic study. Stroke 16:255–261

Bornstein NM, Norris JW (1989) The unstable carotid plaque. Stroke 20:1104–1106

Bousser MG, Chiras J, Bories J, Castaigne P (1985) Cerebral venous thrombosis – a review of 38 cases. Stroke 16:199–213

Browder J, Browder A, Kaplan H (1972) The venous sinuses of the cerebral dura mater I. Anatomical structures within the superior sagittal sinus. Arch Neurol 26:175–180

Caplan LR (1985) Transient global amnesia. In: Vinken P,

Bruyn G (eds) Handbook of clinical neurology, Vol 1, Neurobehavioral disorders. Elsevier, Amsterdam, pp 205–218
Caplan LR (1990) Transient global amnesia: characteristic features and overview. In: Markowitsch HJ (ed) Transient global amnesia and related disorders. Hogrefe and Hubac, Toronto, pp 15–27
Caplan LR, Chedru F, Lhermitte F, Mayman C (1981) Transient global amnesia and migraines. Neurology 31:1167–1170
Caplan LR, Tettenborn B (1992) Vertebrobasilar occlusive disease: review of selected aspects. I. Spontaneous dissection of extracranial and intercranial posterior circulation arteries. Cerebrovasc Dis 2:256–265
Conway LM, MacDonald LW (1972) Structural changes in the intradural arteries following subarachnoid hemorrhage. J Neurosurg 37:515–723
Corrin LS, Sandok BA, Houser OW (1981) Cerebral ischemic events in patients with carotid artery fibromuscular dysplasia. Arch Neurol 38:619–622
DeWitt LD, Louis DN (1991) Case records of the Massachusetts General Hospital, Case 27–1991. New Engl J Med 325:42–54
Duncan GW, Leis RS, Ojemann RG, David SS (1977) Concomitants of atherosclerotic carotid artery stenosis. Stroke 8:665–669
Echiverri HC, Rubino FA, Gupta SR, Gujrati M (1989) Fusiform aneurysm of the vertebral basilar arterial system. Stroke 20:1741–1747
Einhaupl KM, Mehracin S, Haberl RL et al. (1991) Heparin treatment in sinus venous thrombosis. Lancet 338:597–600
Fayad P, Kain T, Hoffer P et al (1990) SPECT findings in transient global amnesia. Neurology 40 (Suppl 1):171
Feussner JR, Matcher DB (1988a) When and how to study the carotid arteries. Ann Intern Med 109:805–818
Feussner JR, Matcher DB (1988b) Diagnostic evaluation of the carotid arteries. Ann Intern Med 109:835–837
Fisher CM (1980) Late-life migraine accompaniments as a cause of unexplained transient ischemic attacks. Can J Neurol Sci 7:9–17
Fisher CM (1986) Late-life migraine accompaniments: further experience. Stroke 17:1033–1042
Fisher CM (1989) Binswanger's encephalopathy: a review. J Neurol 236:65–79
Fisher CM, Ojemann RG (1986) A clinico-pathologic study of carotid endarterectomy plaques. Neurology 142:573–589
Foulkes MA, Wolf PA, Price TR et al. (1988) The Stroke Data Bank: design, methods, and characteristics. Stroke 19:547–554
Gray F, Dubas F, Roullet E, Escourille R (1985) Leukoencephalopathy in diffuse hemorrhagic cerebral amyloid angiopathy. Ann Neurol 18:54–59
Grubb A, Jensson O, Gudmundsson G et al. (1984) Abnormal metabolism of γ-trace alkaline microprotein. The basic defect in cerebral hemorrhage with amyloidosis. New Engl J Med 311:1547–1549
Hennerici M, Freund HJ (1984) Efficacy of C-W Doppler and duplex system examination of the evaluation of extracranial carotid disease. J Clin Ultrasound 12:155–161
Hennerici M, Steinke W (1987) Durch Blutungsstorungen des Gehirns diagnostische moglichkeiten. Verlag Bertelsman Stiftung, Gutersloh
Hennerici M, Rautenberg W, Schwartz A (1987) Transcranial Doppler ultrasound for the assessment of intracranial arterial flow velocity. Part 2. Evaluation of intracranial arterial disease. Surg Neurol 27:523–532
Hennerici M, Trockel U, Rautenberg W, Kladetzky RG (1985) Spontaneous progression and repression of small carotid atheromas. Lancet 1:1415–1418
Hughes JT, Schianchi PM (1978) Cerebral artery spasm: a histological study at necropsy of the blood vessels in cases of subarachnoid hemorrhage. J Neurosurg 48:515–525
Kassell NF, Sasaki T, Colohan AR, Nazar G (1985) Cerebral vasospasm following aneurysmal subarachnoid hemorrhage. Stroke 16:562–572
Kieburtz K, Eskin TA, Ketonen L, Tuite MJ (1993) Opportunistic cerebral vasculopathy and stroke in patients with the acquired immunodeficiency syndrome. Arch Neurol 50:430–432
Levine SR, Twyman RE, Gilman S (1988) The role of anticoagulation in cavernous sinus thrombosis. Neurology 38:517–522
Little JR, St Louis P, Weinstein M et al. (1981) Giant fusiform aneurysms of the cerebral arteries. Stroke 12:183–188
Loes DJ, Biller J, Yuh WTC et al. (1990) Leukoencephalopathy in cerebral amyloid angiopathy: MR imaging in four cases. AJNR 11:485–488
Luscher TF, Lie JT, Stanson AW et al. (1987) Arterial fibromuscular dysplasia. Mayo Clin Proc 62:931–952
MacDonald RL, Weir B (1991) A review of hemoglobin and the pathogenesis of cerebral vasospasm. Stroke 22:971–982
Makos MM, McComb RD, Hart MN, Bennett DR (1987) Alpha-glucosidase deficiency and basilar artery aneurysm: report of a sibship. Ann Neurol 22:629–633
Mas JK, Meder JF, Meary E (1992) Dural sinus thrombosis: long term follow-up by magnetic resource imaging. Cardiovasc Dis 2:137–144
Mohr JP, Caplan LR, Melski JW (1978) The Harvard Cooperative Stroke Registry: prospective registry. Neurology 28:754–762
Nishizaki T, Tamaki N, Takeda N et al. (1986) Dolichoectatic basilar artery: a review of 23 cases. Stroke 17:1277–1281
Norris JW, Zhu CZ (1990) Stroke risk and critical carotid stenosis. J Neurol Neurosurg Psych 53:235–237
O'Donnell TF, Erdoes L, Mackey WC et al. (1985) Correlation of B-mode ultrasound imaging and arteriography with pathologic findings at carotid endarterectomy. Arch Surg 120:443–449
Pamir MN, Kansu T, Erbengi A, Zileli T (1981) Papilledema in Behçet's syndrome. Arch Neurol 38:643–645
Pessin MS, Chimowitz M, Levine SR et al. (1989) Stroke in patients with fusiform vertebrobasilar aneurysms. Neurology 39:16–21
Read D, Esiri MM (1979) Fusiform basilar artery aneurysm in a child. Neurology 29:1045–1049
Reilly LM, Lusby RJ, Hughes L et al. (1983) Carotid plaque histology using real-time ultrasonography: clinical and therapeutic applications. Am J Surg 146:188–193
Sacco RL, Ellenberg JH, Mohr JP et al. (1989) Infarction of undetermined cause: the NINCDS Stroke Data Bank. Ann Neurol 25:382–390
Sandok BA (1983) Fibromuscular dysplasia of the internal carotid artery. In: Barnett HJM (ed) Neurologic clinics, Vol 1. Cardiovascular disease. WB Saunders, Philadelphia, 17–26
Schenk EA, Bone G, Aretz T et al. (1988) Multicenter validation study of real-time ultrasonography, arteriography and pathology: pathologic evaluation of carotid endarterectomy specimens. Stroke 19:289–296
Seidarogler P, Yazici H, Ozdemir C et al. (1989) Neurologic involvement in Behçet's syndrome. A prospective study. Arch Neurol 46:265–269
So EL, Toule JF, Dalal P et al. (1981) Cephalic fibromuscular dysplasia in 32 patients. Arch Neurol 38:619–622
Srinivasan K (1984) Ischemic cardiovascular disease in the

young: two common cases in India. Stroke 15:733–735
Steinke W, Kloetzsch C, Hennerici M (1990) Carotid artery disease assessed by color Doppler flow imaging: correlation with standard Doppler sonography and angiography. AJNR 11:259–266
Touboul P-J, Prati P, Scarabin P-Y et al. (1992) Use of monitoring software to improve the measurement of carotid wall thickness by B-mode imaging. J Hypertension 10 (Suppl 5): 537–541
Tournier-Lasserve E, Iba-Zizon MT, Romero N, Bousser M-G (1991) Autosomal dominant syndrome with stroke-like episodes of leukoencephalopathy. Stroke 22:1297–1302
Tunick PA, Culliford AT, Lamparello PJ et al. (1991a) Atheromatosis of the aortic arch as an occult source of multiple systemic emboli. Ann Intern Med 114:392–392
Tunick PA, Perez JL, Kronzon I (1991b) Protruding atheromas in the thoracic aorta and systemic embolization. Ann Intern Med 115:423–427
Vidailhet M, Piette J, Wechsler B et al. (1990) Cerebral venous thrombosis in systemic lupus erythematosus. Stroke 21:1226–1231
Vinters HV (1987) Cerebral amyloid angiopathy. A critical review. Stroke 18:311–324
White RP, Robertson JT (1985) Role of plasmin, thrombin, and antithrombosis as etiological factors in delayed cerebral vasospasm. Neurosurgery 16:27–35

Section V

Blood Flow, Blood Pressure and Intracranial Pressure

Prefatory Comments

The job of the cerebral vascular system is to deliver sufficient fuel and metabolites to meet the metabolic needs of all parts of the brain. Clearly, lack of blood flow – ischemia – is the underlying mechanism of brain infarctions: infarction accounts for 80% of all strokes. The metabolic needs of different brain regions vary, and demand also depends on the level of metabolic activity of a given area. Roy and Sherrington (1890) (Friedland and Iadecola 1991) first studied the ability of the vascular system to change local regional blood flow in response to changes in neuronal activity. More recently, PET scan studies have confirmed that increased activity and glucose metabolism in a local region is usually accompanied by increased regional blood flow to that region.

Blood flow in the normal human brain is generally about 50 ml/100 g/min. When flow falls to about 20 ml/100 g/min, the electrical activity of some brain cells ceases (Jafar and Crowell 1987). Spontaneous EEG activity is affected when blood flow reaches 18 ml/100 g/min. Biochemical changes and oxygen and sugar metabolism are also seriously affected when regional cerebral blood flow drops below 20 ml/100 g/min. When flow is below 10 ml/100 g/min critical cell electrical and pump functions cease and cells are irreversibly damaged. Fig. 1 depicts these vulnerabilities graphically. Brain function can be simplistically thought of in three categories: normal, 'stunned', that is functioning, but not irreversibly damaged, and non-functioning and irreversibly injured (infarcted).

In this section the various contributors will discuss the factors and systems that normally maintain blood flow and how they respond to changes in activity, changes in blood pressure and various disease states. Blood pressure is clearly a very important factor. In a very simplistic delivery system, the flow in a tube will depend very heavily on the pressure of the fluid. Any increase in pressure will increase flow and a decrease in pressure will decrease flow. In the human body there is a complex protective system which attempts to maintain blood flow despite alterations in systemic blood pressure. This system, which keeps blood flow relatively constant over a wide range of arterial pressures, has customarily been referred to as autoregula-

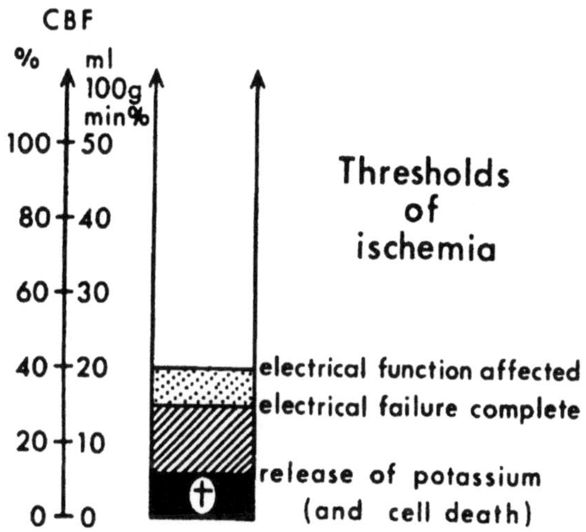

Fig.1. Thresholds of ischemia versus cerebral blood flow (CBF) in ml/g/min. From Jafar JJ, Crowell RM (1987) Focal ischemic thresholds. In: Wood JH (ed) Cerebral blood flow. Reproduced with permission of McGraw-Hill.

tion. Autoregulation is, however, ineffective if the blood pressure is too low or too high. Heistad, in his chapter, will discuss autoregulation and the relationships between blood pressure and blood flow and how the arterial system responds to chronically increased blood pressure.

The viscosity of the blood clearly affects its rheology, that is, the ability of a fluid to flow under various circumstances. The composition of the blood, including both its cellular and its serological components, vitally affects blood flow: the more viscous the blood, the more reduced the flow. Viscosity probably has a major effect on flow in the microvascular system, where the diameter of the vessels is quite small. Grotta, in his chapter, discusses the rheology of blood flow and the factors within the blood that are important in determining flow. A reduction in viscosity (hemodilution) is one strategy that has recently been employed in attempts to augment blood flow.

The brain is clearly very complex, and ideally there should be ways of altering blood flow to fit metabolic requirements. There should also be ways of diverting or directing blood flow towards regions in trouble due to ischemia or other diseases or injuries. Macfarlane and Moskowitz, in their chapter, discuss the innervation of the brain's blood vessels. Clearly, the transmission of messages through synaptic electrical activity is the traditional method of brain and nervous system activity. For a long time, the innervation of blood vessels was thought to be absent or only minimally important. Macfarlane and Moskowitz review the efferent sympathetic and parasympathetic control of blood vessel diameter and flow. They also review the sensory innervation of blood vessels, a mechanism which is probably important in letting the brain know what is happening in the blood vessels, and so initiating potential remedial efferent autonomic responses.

The brain can also achieve action via the secretion and release of various neurotransmitter substances. In their chapter, Reis and Golanov review some recent information on how the brain can alter blood flow by the action of various key sensor systems and brain nuclei which release neurotransmitters that play a role in the regulation of blood flow to various brain regions.

The brain and its vascular system reside in a relatively closed system. The pressure of blood and blood flow in the intracranial vessels is affected by the intracranial pressure since pressure inside the cranium is transmitted directly to the cerebral and dural veins. When intracranial pressure rises the venous pressure also rises. The arterial pressure must also rise in order to maintain an effective arteriovenous pressure gradient. Also, the volume and flow in the vascular compartment is a very important determinant of intracranial pressure, as is the cerebrospinal fluid volume. Ropper, in his chapter, discusses intracranial pressure and its relation to cerebral blood flow and blood pressure.

Modern technology, especially positron emission tomography (PET), single photon emission computed tomography (SPECT) and transcranial Doppler ultrasound (TCD) have given clinicians the means to measure blood flow and velocity during life. The physiology and pathophysiology of blood flow discussed in this section are very important in understanding the changes now measurable at the bedside and in the laboratory.

References

Friedland RP, Iadecola C (1991) Roy and Sherrington (1890) A centennial re-examination of, 'On the regulation of the blood supply of the brain'. Neurology 41:10–14

Jafar JJ, Crowell RM (1987) Focal ischemic thresholds. In: Wood JH (ed) Cerebral blood flow. McGraw-Hill Book Co, New York, pp 449–457

Roy CS, Sherrington C (1890) On the regulation of blood supply to the brain. J Physiol 11:85–108

15. Autoregulation, Hypertension and Regulation of the Cerebral Circulation

D. D. Heistad

Acute and chronic increases in blood pressure have important effects on cerebral blood vessels. In recent years, new concepts and hypotheses have been proposed concerning the mechanisms of autoregulation during acute increases in blood pressure, adaptive changes of cerebral blood vessels during chronic hypertension, and mechanisms that ultimately lead to cerebral vascular events that are typical of chronic hypertension. Some of these new concepts and hypotheses will be reviewed, as well as well accepted concepts about the effects of hypertension on cerebral blood vessels.

Autoregulation

Definition and Role

When perfusion pressure (i.e. arterial pressure minus venous pressure) increases or decreases, a proportional increase or decrease in cerebral blood flow would also be expected. Cerebral vascular 'tone', however, normally changes during such increases and decreases in blood pressure, so that blood flow remains relatively constant over a wide range of pressures (between approximately 60 and 150 mmHg). A reasonable definition of autoregulation, in relation to consequences for cerebral function, is that cerebral blood flow remains relatively constant during changes in blood pressure. A more precise definition is that cerebral vessels constrict and dilate during increases and decreases in perfusion pressure. The latter definition is more precise because changes in vascular diameter or resistance can be quantified more accurately than the absence of changes in blood flow. Both definitions are accurate, but it is important to use the more precise definition when mechanisms of autoregulation are studied (Heistad and Kontos 1983).

Substantial increases and decreases in arterial pressure occur in the daily life of normal humans (Mancia et al. 1986). Autoregulation plays a critical homeostatic role by protecting the brain against changes in perfusion. The dangers of hypoperfusion during hypotension are obvious, but the potential dangers of excessive perfusion are more subtle. It is possible that hyperperfusion, with increased delivery of oxygen, may generate oxygen radicals.

When autoregulation is impaired, or the autoregulatory capacity of cerebral blood vessels is exceeded by large changes in blood pressure, there are likely to be serious consequences. Large decreases in arterial pressure may reduce cerebral blood flow below a level that is sufficient to meet the metabolic needs of the brain, and thereby produce cerebral ischemia. Extreme increases in arterial pressure may produce cerebral vasodilation, which is mediated by oxygen radicals (Kontos 1985), with disruption of the blood–brain barrier; this occurs primarily in venules (Mayhan and Heistad 1986), not in capillaries or arterioles.

Regional Differences

There are large regional differences in autoregulatory capacity within the brain. During both increases and decreases in pressure, autoregulation appears to be more effective in white matter than in gray matter, and in the brain stem than in the cerebrum (Baumbach and Heistad 1985). During acute severe hypertension, the autoregulatory capacity is exceeded in the cerebrum before the brain stem, so that there may be selective increases in blood flow to the cerebrum with disruption of the blood–brain barrier (Fig. 15.1) (Baumbach and Heistad 1985). Fortunately sympathetic nerves attenuate increases in blood flow to the cerebrum during acute hypertension, and protect the blood–brain barrier (Bill and Linder 1976; Heistad and Marcus 1979).

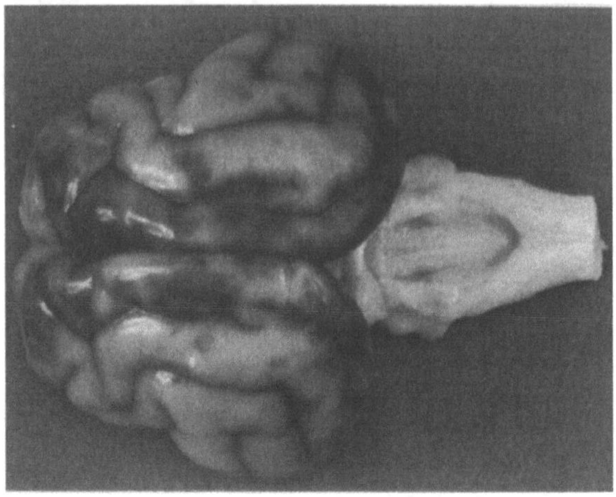

Fig. 15.1. Brain of a cat after sudden severe hypertension. Evans blue dye was given intravenously before hypertension. Cerebellum was removed to expose brain stem. Cerebral hemispheres were stained by Evans blue dye (dark areas), and brain stem was unstained. Because autoregulation is more effective in brain stem than in cerebrum, disruption of the blood–brain barrier in the brain stem is minimal during acute hypertension. From Baumbach and Heistad (1985).

Mechanisms

There has been a long-standing controversy concerning the mechanisms that account for autoregulation, with evidence that supports myogenic or metabolic mechanisms (Heistad and Kontos 1983; Johnson 1986). There now is convincing evidence that metabolic mechanisms play an important role in cerebral autoregulation, particularly during decreases in perfusion pressure (Kontos et al. 1978). The release of adenosine during tissue hypoxia may be one of the major mediators of metabolic autoregulation (Winn et al. 1980). Myogenic mechanisms (i.e. intrinsic responses of vascular muscle to changes in stretch) appear to be less important than metabolic mechanisms, but there is evidence that they play at least a modest role in autoregulation (Wei and Kontos 1984; Bohlen and Harper 1984).

A recent hypothesis is that the endothelium may play an important role in cerebral autoregulation. The contraction of cerebral blood vessels in response to stretch is impaired by the removal of endothelium in some (Harder 1987; Katusic et al. 1987), but not all (Faraci et al. 1989; McCarron et al. 1989), studies. Harder (1987) and Katusic et al. (1987) suggest that an endothelium-derived contracting factor (EDCF) may be released during stretching of the vessel, thereby producing contraction. There is also evidence that contraction of vessels during stretch may be produced by a reduction in release of endothelium-derived relaxing factor (EDRF) (Rubanyi 1988), and there is evidence against this mechanism (Harder 1987; Katusic et al. 1987).

Currently available evidence suggests that metabolic mechanisms play a predominant role in cerebral autoregulation, whereas myogenic mechanisms play a less important role and it is not clear whether endothelial vasoactive factors may also contribute.

Mechanisms that Protect Against Chronic Hypertension

A striking finding is that cerebral blood flow is usually not increased in humans or experimental animals with chronic hypertension (Kety et al. 1948; Strandgaard et al. 1974; Jones et al. 1976). The vessels often are able to withstand levels of arterial pressure which, if they occurred acutely, would produce increases in blood flow and disruption of the blood–brain barrier. The autoregulatory 'plateau' of the pressure–flow relation is shifted to the right in hypertensive patients and experimental animals (Kety et al. 1948; Strandgaard et al. 1974; Jones et al. 1976), so that cerebral blood flow may be normal despite very high levels of blood pressure.

Cerebral Vascular Hypertrophy

Hypertrophy of the cerebral blood vessels during chronic hypertension plays a critical role in shifting the autoregulatory curve, and in the protection of cerebral blood vessels. There is strong evidence for cerebral vascular hypertrophy during chronic hypertension (e.g. Baumbach et al. 1988).

Several mechanisms may contribute to vascular hypertrophy, including a 'trophic' effect of sympathetic nerves (Hart et al. 1980), angiotensin II (Hajdu et al. 1991) or other humoral mechanisms, and increases in pressure per se (Baumbach et al. 1991; Christensen 1991). Recent evidence suggests that increases in pulse pressure may be more important than increases in mean pressure as a stimulus for vascular hypertrophy (Baumbach et al. 1991; Christensen 1991). Some have proposed that increased 'pulsatility' during hypertension is not a risk factor for stroke (Darne et al. 1989): this observation is compatible with the hypothesis that increases in pulse pressure stimulate vascular hypertrophy (Baumbach et al. 1991; Christensen 1991), and that cerebral vascular hypertrophy protects against stroke (Sadoshima et al. 1981).

Cerebral Vascular Remodeling

A recent finding is that vascular 'remodeling' occurs in cerebral blood vessels during chronic hypertension (Baumbach and Heistad 1989). Remodeling is defined as a reduction in the external diameter of the vessels (Fig. 15.2). Support for the concept of remodeling during hypertension has been obtained in other vascular beds (Mulvany 1991). In the cerebral circulation, both hypertrophy and remodeling reduce the size of the vascular lumen, and both may have major functional consequences (Baumbach and Heistad 1991). Encroachment of the vessel wall on the lumen of cerebral arterioles, with an increase in cerebral vascular resistance, appears to be due primarily to vascular remodeling, with only a small contribution from hypertrophy (Baumbach and Heistad 1989, 1991).

The mechanisms that account for vascular remodeling are not entirely clear. Increases in pulse pressure, which contribute to vascular hypertrophy (Baumbach et al. 1991; Christensen 1991), apparently do not account for remodeling during hypertension, but humoral mechanisms, especially angiotensin II, may contribute (Hajdu et al. 1991).

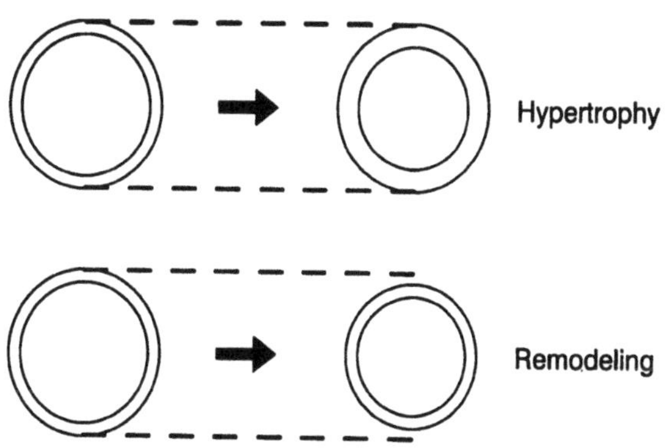

Fig. 15.2. Vascular hypertrophy vs. remodeling. Chronic hypertension produces hypertrophy of cerebral blood vessels, i.e. the cross-sectional area of the vascular wall increases. Vascular hypertrophy is accompanied by a reduction in the vascular lumen, with no reduction in the external diameter of the vessel. A recent finding is that cerebral blood vessels also undergo remodeling during chronic hypertension. The external diameter of the vessel is reduced, with a corresponding reduction in the vascular lumen. From Baumbach and Heistad 1989.

Protection by Hypertrophy and Remodeling

Vascular hypertrophy and remodeling, which are the two major structural changes in cerebral blood vessels during chronic hypertension, protect the cerebral vessels in several ways. First, increases in arterial pressure produce an increase in vascular wall stress. Wall stress is a tightly regulated variable, and an increase places the vessel at risk. Vascular hypertrophy and remodeling produce an increase in wall thickness and a reduction in vascular diameter, which return the wall stress to normal during hypertension (Heistad and Baumbach 1992). Secondly, the importance of regulation of cerebral microvascular pressure has recently been emphasized (Faraci and Heistad 1990). Pressure in cerebral arterioles is less than half of aortic pressure (Faraci and Heistad 1990) because the resistance of large cerebral arteries is surprisingly high. During hypertension, if large increases in aortic pressure were transmitted to the cerebral arterioles and capillaries, the microcirculation would be damaged. Hypertrophy and remodeling of large cerebral arteries attenuate increases in pressure in

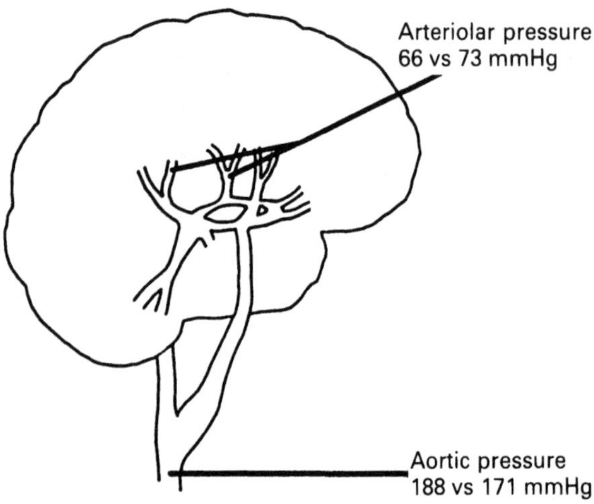

Fig. 15.3. Arteriolar pressure and aortic pressure measured in normotensive and hypertensive rats (Werber and Heistad 1984). The first value of each pair was measured in normotensive (WKY) rats, and the second value was measured in stroke-prone spontaneously hypertensive rats (SHRSP). Although aortic pressure is more than 50 mmHg higher in SHRSP, cerebral arteriolar pressure is only 7 mmHg higher. Thus, an increase in the resistance of large cerebral arteries in SHRSP protects against increases in cerebral microvascular pressure.

smaller vessels downstream (Fig. 15.3) (Weber and Heistad 1984), and thus protect the cerebral microcirculation during hypertension.

Thirdly, structural changes in the cerebral vessels play a critical role in shifting the autoregulatory curve during hypertension. Folkow (Folkow et al. 1973) has emphasized that vascular hypertrophy augments vasoconstrictor responses, and we have proposed that vascular remodeling augments cerebral vasoconstrictor responses (Baumbach and Heistad 1991). It is very probable that vascular hypertrophy and remodeling account for augmented cerebral vasoconstriction during increases in arterial pressure, and thus produce the shift to the right of the autoregulatory curve.

'Costs' of Hypertrophy and Remodeling

There are costs of structural changes in cerebral blood vessels during hypertension. Vascular hypertrophy leads to encroachment on the vascular lumen and thereby impairs maximal vasodilator responses (Folkow et al. 1973; Weber and Heistad 1984). Remodeling, with a reduction in external vascular diameter, also reduces the vascular lumen (Baumbach and Heistad 1989). We have suggested that encroachement on the lumen of cerebral arterioles during hypertension is produced primary by remodeling, and that vascular hypertrophy plays a smaller role (Baumbach and Heistad 1989; Baumbach and Heistad 1991).

The impairment of maximal vasodilator responses has the potential to reduce responses to metabolic stimuli, hypercapnia, hypoxia, hypotension and flow-mediated vasodilation (Fujii et al. 1991), as well as compensatory vasodilator responses after vascular obstruction or occlusion (Coyle and Heistad 1987). Impairment of vasodilator responses would presumably produce tissue ischemia, because blood flow may be inadequate to meet tissue needs. Although structural changes in blood vessels interfere with maximal vasodilator responses, it is not clear that responses to a submaximal stimulus are significantly impaired. This unanswered question is quite important, because submaximal stimuli occur far more frequently than maximal stimuli.

The magnitude of the shift to the right of the cerebral autoregulatory curve in hypertension is usually assumed to be symmetrical, i.e. the shift is similar at both high and low levels of pressure. Data of Harper and Bohlen (1984), however, suggest that the shift to the right during hypertension is greater at high levels of pressure than at low levels (Fig. 15.4). Data obtained in patients, as well as experimental animals, suggest that the shift to the right at low levels of pressure is relatively modest. Thus, the protective effect of the shift of the autoregulatory curve at high levels of pressure may be much greater than the small but deleterious shift to the right at low pressure. A paradoxical increase in the compliance of hypertrophied vessels (Baumbach et al. 1988) may occur when the vessels are dilated, at low levels of pressure, and thus explain the finding of a modest shift at low levels of pressure.

Fibrinoid necrosis of the cerebral vessels is a common finding during chronic hypertension. It has been suggested that vascular necrosis is one of the costs of vascular hypertrophy, as the increased workload of vessels cannot be met because the thickness of the vessel wall exceeds the diffusion path of oxygen and substrate. This mechanism is not an adequate explanation for vascular necrosis. Oxygen and substrate can supply arteries that are far thicker than cerebral arteries (Werber and Heistad 1985), and it is extremely unlikely that vessel thickness would ever increase in cerebral arteries beyond a distance that could be supplied easily by diffusion.

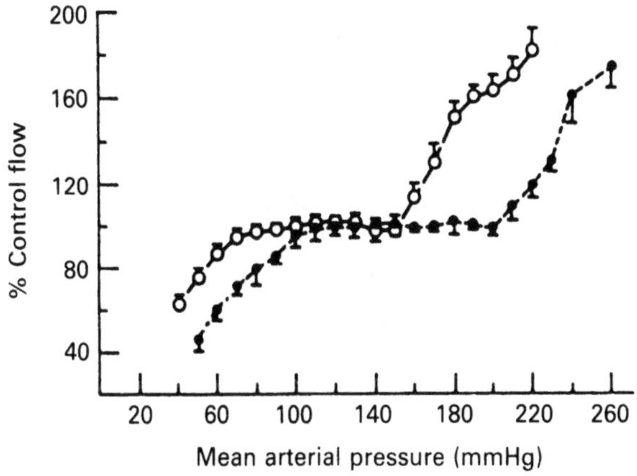

Fig. 15.4. Shift in the cerebral pressure–flow relation during hypertension. Values were obtained in normotensive rats (WKY: ○) and hypertensive rats (SHR: ●). There is a large shift to the right in SHR at high levels of blood pressure, but only a small shift to the right at low levels of pressure. From Harper and Bohlen (1984). Microvascular adaptation in the cerebral cortex of adult spontaneously hypertensive rats. Reproduced with permission of Hypertension 6:408–419.

Failure of Protective Mechanisms During Chronic Hypertension

Although there are several complex and relatively effective protective mechanisms, these often fail during chronic hypertension, resulting in typical cerebral vascular events, including hypertensive encephalopathy, cerebral hemorrhage and ischemic stroke.

Endothelial dysfunction may be of primary importance in the major cerebral vascular complications of hypertension. Evidence will be summarized below that disruption or dysfunction of the endothelial blood–brain barrier mediates hypertensive encephalopathy. Endothelial dysfunction also appears to produce abnormal modulation of vascular tone in hypertension, with release of EDCF and impaired release of EDRF.

Hypertensive Encephalopathy

In the past, it was thought that hypertensive encephalopathy was caused by cerebral vasospasm (Meyer et al. 1960). A 'sausage-string' appearance of retinal and cerebral arterioles may be seen in both acute hypertension and in malignant hypertension (Byrom 1954). This appearance led to the assumption that constricted segments of cerebral vessels were in spasm. There is substantial evidence, however, that the constricted segments of arterioles are manifesting autoregulatory constriction and that blood flow is not reduced below normal levels. In contrast, the dilated segments of the vessels are abnormal, as the vessels dilate and the blood–brain barrier in these segments is disrupted (Baumbach and Heistad 1988).

Studies of a variety of experimental models, including hypertensive rats with signs of encephalopathy (Tamaki et al. 1984), support the concept that vasodilatation and disruption of the blood–brain barrier are primary events in hypertensive encephalopathy. It is not clear whether signs of encephalopathy after disruption of the blood–brain barrier are produced by focal edema, with a reduction in blood flow and tissue necrosis, or by changes in local concentrations of neurotransmitters and ions (including K^+), which may interfere with neuronal transmission.

Cerebral Hemorrhage

It is predictable that hypertension predisposes to cerebral hemorrhage. Rupture of microaneurysms accounts for many or most cerebral hemorrhages in hypertensive patients, but there is no consensus about local mechanisms that lead to the formation and rupture of such microaneurysms. Hemodynamic mechanisms may include elevations in pulse pressure, which would probably be more important than mean pressure in producing expansion and rupture of aneurysms. The finding that there is no correlation between pulsatility and risk of stroke (Darne et al. 1989), however, provides evidence against a predominant role of pulse pressure in leading to rupture of cerebral aneurysms.

Ischemic Infarction

It is surprising that hypertension predisposes to cerebral ischemia and infarction: how does an *increase* in blood pressure lead to a *decrease* in blood flow? A likely answer is that endothelial dysfunction in chronic hypertension contributes to cerebral ischemia and infarction. Cerebral vasodilatation in response to endothelium-dependent stimuli is impaired by hypertension (Mayhan et al. 1987). Altered responses are due in part to

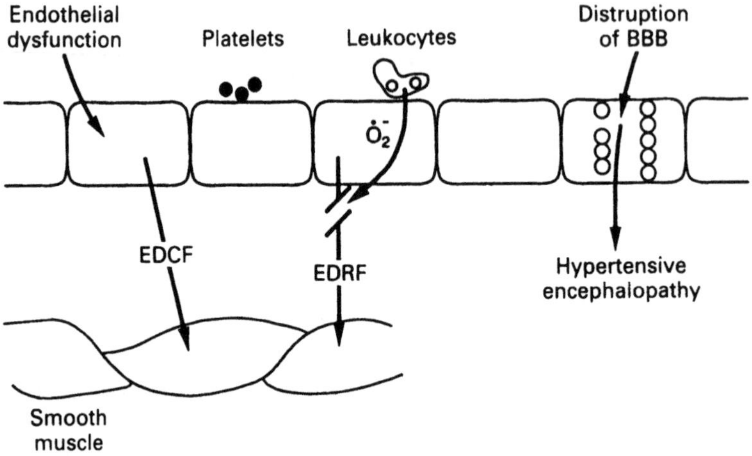

Fig. 15.5. Endothelial dysfunction plays a critical role in cerebral vascular complications of hypertension. Release of an abnormal endothelium-derived contracting factor (EDCF), combined with impaired synthesis/release of endothelium-derived relaxing factor (EDRF), results in abnormal vascular responses to a variety of stimuli, including the potent vasoactive agonists released by platelets. The abnormality in endothelium-dependent relaxation may be due in part to the abnormal adherence of leukocytes to endothelium in hypertension, with the release of superoxide anion ($\dot{O}_2^-$), which inactivates EDRF. Endothelial dysfunction also appears to contribute to hypertensive encephalopathy: disruption of the endothelial blood–brain barrier may account for hypertensive encephalopathy.

impaired synthesis/release of EDRF in hypertension (Yang et al. 1991) and, perhaps more importantly, to release of EDCF (Fig. 15.5) (Mayhan et al. 1988), which acts primarily through thromboxane/PGH_2 receptors. An EDCF has been demonstrated in cerebral arterioles (Mayhan et al. 1988), but not in the basilar artery (Mayhan 1990) during chronic hypertension.

Leukocytes may contribute to endothelial dysfunction during hypertension. Leukocytes are abnormal in hypertension, with increased release of oxygen radicals (Schmid-Schönbein et al. 1991). Superoxide anion, an oxygen radical which is normally released by leukocytes, inactivates EDRF (Ohlstein and Nichols 1989). Thus, increased release of oxygen radicals by leukocytes may inactivate EDRF and produce endothelial dysfunction.

One possible implication of endothelial dysfunction in hypertension relates to cerebral vascular responses to the activation of platelets. Human platelets release large quantities of adenosine diphosphate (ADP), smaller amounts of serotonin and tiny amounts of thromboxane (Meyers et al. 1982). Platelet aggregation produces vasodilatation, as platelets release ADP which in turn releases EDRF (Houston et al. 1986). Endothelial dysfunction during hypertension reduces vasodilator responses to ADP (Mayhan et al. 1988). Thus, when platelets are activated, vasoconstrictor responses to serotonin and thromboxane are not masked by the normal vasodilator response to ADP, and the usual vasodilator response to platelets may be reversed to constriction in hypertensive vessels.

EDRF is a potent inhibitor of activation of platelets and leukocytes (Moncada et al. 1991). Prostacyclin is also a potent inhibitor of platelet aggregation. Endothelial dysfunction is associated with adherence of platelets to cerebral vessels (Rosenblum et al. 1987). Thus, endothelial dysfunction in hypertension, in addition to its effect on vascular tone, may lead to increased adherence and activation of platelets and leukocytes.

Recent studies have attempted to determine whether EDRF affects infarct size. Inhibition of synthesis of EDRF has been reported to reduce infarct size after MCA occlusion in the mouse (Nowicki et al. 1991) and increase infarct size in hypertensive rats (Yamamoto et al. 1992). This important question obviously has not yet been resolved.

Finally, susceptibility to ischemic infarction may be related to impairment of collateral vessels in hypertension. Occlusion of the middle cerebral artery (MCA) produces cerebral infarction in stroke-prone spontaneously hypertensive rats, but not in normotensive rats (Coyle and Heistad 1987). Occlusion of the MCA produces a larger decrease in blood flow in hypertensive rats than in normotensive rats, because collateral flow is less (Coyle and Heistad 1986). Impaired collateral blood flow in chronic hypertension would almost certainly increase propensity to ischemic infarction after vascular occlusion, by

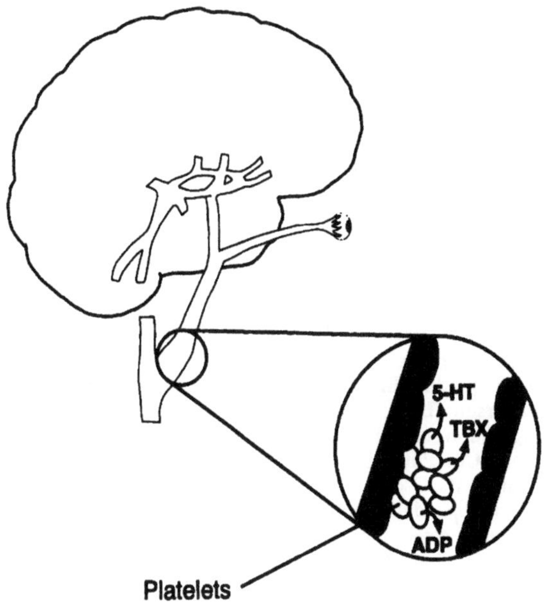

Fig. 15.6. Endothelial dysfunction in chronic hypertension may contribute to transient ischemic attacks (TIAs). When platelets aggregate in a normal artery the response is vasodilatation, as adenosine diphosphate (ADP) and serotonin (5HT) released by the platelets stimulate the release of EDRF from the underlying endothelium and mask the vasoconstrictor response to thromboxane (TBX). In chronic hypertension, cerebral vasodilatation in response to ADP is impaired and serotonin produces vasoconstrictor responses. Thus, when platelets adhere to a plaque in the carotid artery, they may initiate vasoconstriction or perhaps vasospasm in hypertensive vessels. Endothelial dysfunction may thereby contribute to TIAs and perhaps stroke in chronic hypertension. From Heistad and Baumbach 1992.

allowing blood flow to decrease below the amount required to prevent infarction.

Transient Ischemia

There may be important clinical implications of endothelial dysfunction in chronic hypertension, in relation to vasoconstrictor responses to the activation of platelets. Most transient ischemic attacks (TIAs) are attributed to retinal and cerebral emboli from platelets that adhere to plaques in the carotid arteries. We have proposed that in such a situation, endothelial dysfunction due to atherosclerosis (Williams et al. 1989) may lead to vasospasm produced by adherent platelets, as well as emboli. Activation of leukocytes may also produce vasoconstriction and perhaps TIAs (Faraci et al. 1991).

The hypothesis that vasospasm may contribute to TIAs was favored a century ago, lost favor, was revived by studies of atherosclerotic monkeys (Williams et al. 1989) and received support recently by findings in patients (Burger et al. 1991). A similar mechanism is possible in hypertension. Thus, endothelial dysfunction in hypertensive carotid arteries may lead to vasospasm when platelets or leukocytes are activated, and contribute to TIAs and perhaps ischemic stroke. (Fig. 15.6).

Conclusion

The impairment of endothelium-dependent relaxation and the responses of collateral vessels may both contribute to ischemic infarction in chronic hypertension. For example, based on studies summarized in this chapter, endothelial dysfunction may predispose to the adherence of activated platelets to the carotid artery. Activated platelets release ADP, but endothelial dysfunction in hypertension leads to impaired vasodilation in response to ADP. Thus, serotonin and thromboxane, which are released by platelets, may produce vasoconstriction and a reduction in cerebral blood flow. Impairment of collateral circulation by hypertension may then compromise cerebral blood flow. In summary, the combination of endothelial dysfunction and impaired collateral circulation in hypertension may convert an innocuous insult into an ischemic infarct.

References

Baumbach GL, Heistad DD (1985) Heterogeneity of brain blood flow and permeability during acute hypertension. Am J Physiol: Heart Circ Physiol 18:H629–H637

Baumbach GL, Heistad DD (1988) Cerebral circulation in chronic arterial hypertension. Hypertension 12:89–95

Baumbach GL, Heistad DD (1989) Remodeling of cerebral arterioles in chronic hypertension. Hypertension 13:968–972

Baumbach GL, Heistad DD (1991) Adaptive changes in cerebral blood vessels during chronic hypertension. J Hypertension 9:987–991

Baumbach GL, Dobrin PB, Hart MN, Heistad DD (1988) Mechanics of cerebral arterioles in hypertensive rats. Circ Res 62:56–64

Baumbach GL, Siems JE, Heistad DD (1991) Effects of local reduction in pressure on distensibility and composition of cerebral arterioles. Circ Res 68:338–351

Bill A, Linder J (1976) Sympathetic control of cerebral blood flow in acute arterial hypertension. Acta Physiol Scand

96:114–121
Bohlen HG, Harper SL (1984) Evidence of myogenic vascular control in the rat cerebral cortex. Circ Res 55:554–559
Burger SK, Saul RF, Selhorst JB, Thurston SE (1991) Transient monocular blindness caused by vasospasm. New Engl J Med 325:870–873
Byrom FB (1954) The pathogenesis of hypertensive encephalopathy and its relation to the malignant phase of hypertension. Experimental evidence from the hypertensive rat. Lancet 2:201–211
Christensen K (1991) Reducing pulse pressure in hypertension may normalize small artery structure. Hypertension 18:722–727
Coyle P, Heistad DD (1986) Blood flow through cerebral collateral vessels in hypertensive and normotensive rats. Hypertension 8:II67–II71
Coyle P, Heistad DD (1987) Blood flow through cerebral collateral vessels one month after middle cerebral artery occlusion. Stroke 18:407–411
Darne B, Girerd X, Safar M, Cambien F, Guize L (1989) Pulsatile versus steady component of blood pressure: a cross-sectional anlaysis and a prospective analysis on cardiovascular mortality. Hypertension 13:392–400
Faraci FM, Heistad DD (1990) Regulation of large cerebral arteries and cerebral microvascular pressure. Circ Res 66:8–17
Faraci FM, Baumbach GL, Heistad DD (1989) Myogenic mechanisms in the cerebral circulation. J Hypertension 7:S61–S65
Faraci FM, Lopez JAG, Breese K, Armstrong ML, Heistad DD (1991) Effect of atherosclerosis on cerebral vascular responses to activation of leukocytes and platelets in monkeys. Stroke 22:790–796
Folkow B, Hallback M, Lundgren Y, Sivertsson R, Weiss L (1973) Importance of adaptive changes in vascular design for establishment of primary hypertension, studied in man and in spontaneously hypertensives. Circ Res 32(Suppl I):I2–I-16
Fujii K, Heistad DD, Faraci FM (1991) Flow-mediated dilatation of the basilar artery in vivo. Circ Res 69:697–705
Hajdu MA, Heistad DD, Baumbach GL (1991) Effects of antihypertensive therapy on mechanics of cerebral arterioles in rats. Hypertension 17:308–316
Harder DR (1987) Pressure-induced myogenic activation of cat cerebral arteries is dependent on intact endothelium. Circ Res 60:102–107
Harper SL, Bohlen HG (1984) Microvascular adaptation in the cerebral cortex of adult spontaneously hypertensive rats. Hypertension 6:408–419
Hart MN, Heistad DD, Brody MJ (1980) Effect of chronic hypertension and sympathetic denervation on wall/lumen ratio of cerebral vessels. Hypertension 2:419–423
Heistad DD, Baumbach GL (1992) Cerebral vascular changes during chronic hypertension: good guys and bad guys. J Hypertens 10:571–575
Heistad DD, Kontos HA (1983) Cerebral circulation. In: Abboud FM, Shepherd JT (eds) Handbook of physiology: the cardiovascular system, Vol III. American Physiological Society, Bethesda, pp 137–182
Heistad DD, Marcus ML (1979) Effect of sympathetic stimulation on permeability of the blood-brain barrier to albumin during acute hypertension in cats. Circ Res 45:331–338
Houston DS, Shepherd JT, Vanhoutte PM (1986) Aggregating human platelets cause direct contraction and endothelium-dependent relaxation of isolated canine coronary arteries: role of serotonin, thromboxane A_2, and adenine nucleotides. J Clin Invest 78:539–544
Johnson PC (1986) Autoregulation of blood flow. Circ Res 59:483–495
Jones JV, Fitch W, MacKenzie ET, Strandgaard S, Harper AM (1976) Lower limit of cerebral blood flow autoregulation in experimental renovascular hypertension in the baboon. Circ Res 39:555–557
Katusic ZS, Shepherd JT, Vanhoutte PM (1987) Endothelium-dependent contraction to stretch in canine basilar arteries. Am J Physiol: Heart Circ Physiol 21:H671–H673
Kety SS, Hafkenschiel JH, Jeffers WA, Leopold IH, Shenkin HA (1948) The blood flow, vascular resistance, and oxygen consumption of the brain in essential hypertension. J Clin Invest 27:511–514
Kontos HA (1985) Oxygen radicals in cerebral vascular injury. Circ Res 57:508–516
Kontos HA, Wei EP, Raper AJ, Rosenblum WI, Navari RM, Patterson JL Jr (1978) Role of tissue hypoxia in local regulation of cerebral microcirculation. Am J Physiol: Heart Circ Physiol 3:H582–H591
McCarron JG, Osol G, Halpern W (1989) Myogenic responses are independent of the endothelium in rat pressurized posterior cerebral arteries. Blood Vessels 26:315–319
Mancia G, Parati G, Pomidossi G, Casadei R, DiRenzo M, Zanchetti A (1986) Arterial baroreflexes and blood pressure and heart rate variabilities in humans. Hypertension 8:147–153
Mayhan WG (1990) Impairment of endothelium-dependent dilatation of basilar artery during chronic hypertension. Am J Physiol: Heart Circ Physiol 28:H1455–1462
Mayhan WG, Heistad DD (1986) Role of veins and cerebral venous pressure in disruption of the blood–brain barrier. Circ Res 59:216–220
Mayhan WG, Faraci FM, Heistad DD (1987) Impairment of endothelium-dependent responses of cerebral arterioles in chronic hypertension. Am J Physiol: Heart Circ Physiol 22:H1435–H1440
Mayhan WG, Faraci FM, Heistad DD (1988) Responses of cerebral arterioles to adenosine 5'-diphosphate, serotonin, and the thromboxane analogue U-46619 during chronic hypertension. Hypertension 12:556–561
Meyer JS, Waltz AG, Gotoh F (1960) Pathogenesis of cerebral vasospasm in hypertensive encephalopathy: II. The nature of increased irritability of smooth muscle of pial arterioles in renal hypertension. Neurology 10:859–867
Meyers KM, Holmsen H, Seachord CL (1982) Comparative study of platelet dense granule contents. Am J Physiol: Reg Int Comp Physiol 12:R454–R461
Moncada S, Palmer RMJ, Higgs EA (1991) Nitric oxide: physiology, pathophysiology, and pharmacology. Pharmacol Rev 43:109–142
Mulvany MJ (1991) Abnormalities of resistance vessel structure in essential hypertension: are these important? Clin Exp Pharmacol Physiol 18:13–20
Nowicki JP, Duval D, Poignet H, Scotton B (1991) Nitric oxide mediates neuronal death after focal cerebral ischemia in the mouse. Eur J Pharmacol 204:339–340
Ohlstein EH, Nichols AJ (1989) Rabbit polymorphonuclear neutrophils elicit endothelium-dependent contraction in vascular smooth muscle. Circ Res 65:917–924
Rosenblum WI, Nelson GH, Povlishock JT (1987) Laser-induced endothelial damage inhibits endothelium-dependent relaxation in the cerebral microcirculation of the mouse. Circ Res 60:169–176
Rubanyi GM (1988) Endothelium-dependent pressure-induced contraction of isolated canine carotid arteries. Am J Physiol: Heart Circ Physiol 24:H783–H788
Sadoshima S, Busija D, Brody M, Heistad DD (1981) Sympathetic nerves protect against stroke in stroke-prone

hypertensive rats. Hypertension 3:I-124–I-127
Schmid-Schönbein GW, Seiffge D, DeLano FA, Shen K, Zweifach BW (1991) Leukocyte counts and activation in spontaneously hypertensive and normotensive rats. Hypertension 17:323–330
Strandgaard S, MacKenzie ET, Sengupta D, Rowam JO, Lassen NA, Harper AM (1974) Upper limit of autoregulation in cerebral blood flow in the baboon. Circ Res 34:435–550
Tamaki K, Sadoshima S, Baumbach GL, Iadecola C, Reis DJ, Heistad DD (1984) Evidence that disruption of the blood–brain barrier precedes reduction in cerebral blood flow in hypertensive encephalopathy. Hypertension 6:I-75–I-81
Wei EP, Kontos HA (1984) Increased venous pressure causes myogenic constriction of cerebral arterioles during local hyperoxia. Circ Res 55:249–252
Werber AH, Heistad DD (1984) Effects of chronic hypertension and sympathetic nerves on the cerebral microvasculature of stroke-prone spontaneously hypertensive rats. Circ Res 55:286–294
Werber AH, Heistad DD (1985) Diffusional support of arteries. Am J Physiol: Heart Circ Physiol 17:H901–H906
Williams JK, Baumbach GL, Armstrong ML, Heistad DD (1989) Hypothesis: vasoconstriction contributes to amaurosis fugax. J Cerebr Blood Flow Metab 9:111–116
Winn HR, Welsh JE, Rubio R, Berne RM (1980) Brain adenosine production in rat during sustained alteration in systemic blood pressure. Am J Physiol: Heart Circ Physiol 8:H636–H641
Yamamoto S, Golanov EV, Berger SB, Reis DJ (1992) Inhibition of nitric oxide synthesis increases focal ischemic infarction in rat. J Cerebr Blood Flow Metab 12:717–726
Yang ST, Mayhan WG, Faraci FM, Heistad DD (1991) Mechanisms of impaired endothelium-dependent cerebral vasodilatation in response to bradykinin in hypertensive rats. Stroke 22:1177–1182

16. The Innervation of Pial Blood Vessels and their Role in Cerebrovascular Regulation

R. Macfarlane and M. A. Moskowitz

Thomas Willis noted in his *Cerebri Anatome* (1664) that brain vessels were accompanied by fibers. Hovelaque (1927) and Sheehan (Northfield 1938) identified neural connections between the trigeminal ganglion and the internal carotid artery, but were unable to determine whether they represented sympathetic innervation of the trigeminal ganglion or a sensory innervation to the cerebral vessels. Levine and Wolff (1932) demonstrated that electrical stimulation of cat pial arteries increased galvanic skin responses, and that this effect was blocked by the application of procaine to the vessel. Motor end-plates were observed microscopically in the adventitia of pial arteries, and were thought to represent afferent nerve endings (McNaughton 1938). In primates, Wall and Pribram (1950) found that the hypertensive response to electrical stimulation of pial arteries was blocked by trigeminal neurotomy. In man, Fay (1932), Penfield and McNaughton (1940) and Ray and Wolff (1940) elicited forehead pain by electrical stimulation of the pial and dural arteries. However, it was not until the axonal tracing studies of Mayberg and colleagues (1981) that direct connections between the trigeminal nerve and the circle of Willis were documented.

We now know that, in addition to sensory fibers (Mayberg et al. 1981, 1984; Arbab et al. 1986; Suzuki et al. 1989a, b), parasympathetic (Chorobski and Penfield 1932; Vasquez and Purves 1979; Hara and Weir 1986; Suzuki et al. 1988) and sympathetic (Nielsen and Owman 1967; Edvinsson et al. 1972) axons innervate large cranial vessels and their branches. (Projections from intrinsic brain neurons have also been proposed based on experimental evidence, but this important topic will not be addressed as part of this chapter.) Naked undifferentiated axons belonging to each of these fiber types form a plexus on the vessel adventitia within which vasoactive neurotransmitters reside, usually within vesicles. Upon their release into the perivascular space, these neurotransmitter molecules mediate important adaptive responses of cranial blood vessels during health and disease.

Despite this knowledge, the precise role of perivascular axons in the neurogenic control of the cerebral circulation is poorly understood. Based on their constituent transmitters, sympathetic fibers promote vasoconstriction, whereas sensory and parasympathetic fiber mediate vasodilation. So, for example, electrical stimulation of sensory (Suzuki et al. 1990b) or parasympathetic nerve fiber (Chorobski and Penfield 1932; Linder 1981; Seylaz et al. 1988; Suzuki et al. 1990a) causes cerebral vasodilation and an increase in cerebral blood flow (CBF). Neurogenically mediated vasodilation is important in the blood flow responses accompanying focal and global ischemia and possibly migraine and subarachnoid hemorrhage. The evidence for each fiber type will be reviewed briefly in this chapter.

Autonomic Innervation

Sympathetic

The sympathetic innervation to the cerebral vasculature has been studied extensively. A dense innervation derives from the superior cervical ganglion, with small additional contributions from the middle cervical sympathetic and stellate ganglia (Edvinsson and Mackenzie 1977; Heistad and Marcus 1982).

As well as norepinephrine, sympathetic nerve terminals secrete a second potent vasoconstrictor, neuropeptide Y. Feline pial arterioles respond to topical norepinephrine with vasoconstriction, and this response is abolished by phenoxybenzamine, a non-competitive α-adrenergic blocking agent (Nielsen and Owman 1971; Wahl et al. 1972). However, application of the latter alone does not influence arteriolar diameter when applied in the same concentration. Thus, under physiological conditions, there is no evidence to suggest tonic sympathetic control of CBF. Other studies have shown that the sympathetic nervous system does not contribute to CBF regulation during arterial hypotension (Mueller et al. 1977), hypoxia (Edvinsson 1987) or hypercapnia (Wei et al. 1980).

Most studies have concluded that the response to electrical stimulation of sympathetic nerves is much smaller in the cerebral circulation than in other vascular beds. This seems somewhat paradoxical in view of the relatively dense plexus of sympathetic fibers contained within pial blood vessels. Only modest reductions in blood flow (of the order of 5%–10%) follow electrical stimulation of postganglionic cervical sympathetic fibers, primarily by constricting the large proximal arteries (Kobayashi et al. 1971; Harper et al. 1972; Wei et al. 1975). Harper et al. (1972) observed that, whereas cervical sympathetic stimulation or norepinephrine exert minimal effects upon CBF at rest, a pronounced fall in CBF follows stimulation if the cerebral vessels have been dilated by hypercapnia. A 'dual control' hypothesis was proposed whereby the cerebral circulation is composed of two resistances in series. Extraparenchymal vessels are influenced by the autonomic nervous system, but intraparenchymal vessels are regulated primarily by intrinsic metabolic or myogenic mechanisms. Under physiological conditions pial arteries contribute less to vascular resistance than do intracortical arterioles (Shima et al. 1983). Vasomotor responses to topical norepinephrine are more pronounced in pial veins than arteries, suggesting that norepinephrine-containing nerve fibers may have considerable influence upon cerebral blood volume (Edvinsson et al. 1982).

Sympathetic nerves may have an important protective effect during severe hypertension (Bill and Linder 1976). When blood pressure is increased above the limits of autoregulation, sympathetic stimulation blunts the anticipated rise in CBF and reduces the extravasation of albumin into the brain which results from breakdown of the blood–brain barrier (Bill and Linder 1976). Increases in blood flow during acute hypertension are more pronounced in gray than in white matter, and the effects of sympathetic stimulation are similarly greater in gray matter (Heistad and Marcus 1979). Activation of sympathetic nerves resets the autoregulatory curve such that both the upper and the lower pressure thresholds are raised. For this reason, drug-induced hypotension by sympathetic blockade provides a somewhat higher CBF at the same pressure than does hemorrhagic hypotension, in which circulating levels of catecholamines are high (Fitch et al. 1975b). It has been suggested that the most important role of the sympathetic innervation is to protect vessels during surges in arterial blood pressure within the physiological range (Heistad et al. 1982). Although cerebral vessels 'escape' the vasoconstrictor response to prolonged sympathetic stimulation under conditions of normotension, this does not occur during acute hypertension (Marcus et al. 1982).

Sympathetic nerves may also exert trophic influences upon the vessels which they innervate. Chronic sympathectomy in young rats results in a significant reduction in the wall–lumen ratio of intraparenchymal vessels (Mueller and Rusterholtz 1982). Cerebral vessels undergo hypertrophy in response to chronic hypertension, and this protects the blood–brain barrier in stroke-prone spontaneously hypertensive rats (SPSHR) (Mueller and Heistad 1980). Sympathetic denervation attenuates the development of cerebral vascular hypertrophy (Hart et al. 1980), and the strokes that occur in unilateral sympathectomized SPSHR develop primarily in the denervated hemisphere.

Parasympathetic

The parasympathetic innervation to the cerebral vasculature derives from multiple sources. In addition to contributions from the sphenopalatine, otic and associated miniganglia (Walters et al. 1986; Uemura et al. 1988), small aggregates of ganglion cells have been isolated within the cavernous plexus, vidian and lingual nerves (for review see Gibbins et al. 1984). The density of innervation is greatest in the proximal vessels, but innervation is present in vessels as small as 20 μm (Gibbins et al. 1984).

The multiplicity of parasympathetic ganglia innervating the cerebral vasculature, and their anatomical inaccessibility, has hampered investigations using surgical sectioning to examine the influence of parasympathetic fibers. Pharmacological studies are also difficult, for two reasons. First, numerous effector systems are sensitive to the effect of cholinergic drugs, whereas neurogenic vasodilation is rarely abolished by atropine (Duckles 1983). An early transient component that is potentiated by physostigmine and reduced by atropine is accompanied by a slower, more prolonged phase that is unaffected by cholinergic agents (Lee et al. 1978; Bevan et al. 1984). Secondly, specific antagonists for vasoactive intestinal polypeptide (VIP) are only now being developed. VIP, a 28-amino acid polypeptide, is a potent dilator of cerebral arteries which has been localized immunohistochemically within parasympathetic nerve endings (Larsson et al. 1976). It exerts its action through a specific receptor (Suzuki et al. 1984) via non-endothelium-dependent mechanisms in pial vessels (Edvinsson et al. 1985). Nitric oxide synthase (NOS), the enzyme which forms nitric oxide from L-arginine, has been identified recently in perivascular nerve fibers of parasympathetic origin from the rat sphenopalatine ganglia (Nozaki et al. 1993). In these nerve fibers, NOS immunoreactivity colocalizes with immunoreactive VIP. More than 70% of cell bodies within sphenopalatine ganglia contain immunoreactive NOS, thereby suggesting an important role for neuronally derived nitric oxide in target tissues, including blood vessels. Nitric oxide is a potent vasodilator and it and/or a related molecule (e.g. nitrosothiol) may well be an endothelium-dependent relaxing factor.

As with sympathetic and sensory innervation, there is little evidence to suggest that cholinergic mechanisms contribute significantly to CBF regulation under either physiological conditions or during hypercapnia (Hardebo et al. 1982). Studies by Lambert et al. (1984) and Goadsby and Duckworth (1987) indicate that the carotid vasodilator response to stimulation of the trigeminal ganglion is mediated in part via parasympathetic connections derived from the greater superficial petrosal and vidian nerves, and that activation of central trigeminal connections is involved. The central trigeminal effect in the cat comprises a bilateral flow increase in the frontal and parietal cortex, which is abolished by bilateral section of the facial nerve. Stimulation of the parasympathetic nerves increases CBF (Pinard et al. 1979; Suzuki et al. 1990a) by a mechanism independent of brain glucose utilization (Goadsby et al. 1988).

Selective stimulation of parasympathetic nerves in the rat elicits a rise in cortical CBF (Suzuki et al. 1990a), and stimulation of the peripheral cut end of the petrosal nerve in the cat results in a rise in CBF which is blocked by atropine (Purves 1978). Recent work by Kano et al. (1991) has shown that chronic parasympathetic denervation in the rat increases infarct volume after permanent MCA occlusion, primarily by a reduction in CBF under conditions in which perfusion pressure is decreased (Koketsu et al. 1992).

Sensory Innervation

Axonal tracing techniques using horseradish immunoperoxidase, wheatgerm agglutinin and nerve transsection has established that pial, dural and extracranial cephalic blood vessels are surrounded by an adventitial plexus of sensory axons arising from the trigeminal and upper dorsal root ganglia. The density of sensory axons is greatest along proximal arteries of the circle of Willis, and diminishes over the convexity of the hemispheres. The overall pattern is well conserved across species (Mayberg et al. 1981; 1984; Liu-Chen et al. 1984; Arbab et al. 1986; Ruskell and Simons 1987). Studies, primarily in cats and more recently in rats, have established that the circle of Willis and the rostral basilar artery, together with their major tributaries, receive a sensory innervation primarily from the trigeminal nerve ('the trigeminovascular system'). Most cell bodies reside within the ophthalmic division of the ganglion (Mayberg et al. 1981, 1984), except in primates, where the maxillary division provides a small supplementary contribution

(Ruskell and Simons 1987). A further additional sensory innervation to pial arteries arises, in rodents at least, from a miniganglion located on the intrapetrous carotid artery just proximal to the cavernous sinus (Suzuki et al. 1988). Individual sensory neurons project widely within the cranial cavity, but no single nerve cell innervates both intracranial and extracranial vessels (O'Connor and van der Kooy 1986).

The caudal basilar and vertebral arteries, and their major tributaries, are innervated primarily by the upper three cervical dorsal root ganglia, although there is considerable overlap between the trigeminal and the spinal sensory innervation (Norregaard and Moskowitz 1985; Keller et al. 1985; Arbab et al. 1986; Saito et al. 1988).

The distribution of sensory nerves is predominantly ipsilateral, with the exception of midline vessels such as the anterior cerebral arteries and major venous sinuses, which receive a significant contralateral innervation. Each ganglion cell projects divergent axon collaterals to innervate multiple large vessels supplying both brain parenchyma and the overlying dura mater (Arbab et al. 1986; O'Connor and van der Kooy 1986). However, the trigeminal fibers innervating intracranial and extracranial cephalic vessels appear to project from discrete trigeminal ganglion cells (McMahon et al. 1985).

Very little is known about the central projections of the trigeminovascular system. Central projections do terminate within the trigeminal nuclear complex, which extends throughout the greater part of the brain stem (Afshar and Dykes 1984). Specific relays from this nuclear complex include the nucleus of the tractus solitarius (a relay for visceral afferent information), the dorsal motor nucleus of the vagus and the C2 dorsal horn (Arbab et al. 1988). Afshar and Dykes (1984) have used computer-assisted graphics to create superb three-dimensional visualizations of the brain stem trigeminal nuclear complex in humans. A full account of the central connections can be found in Young (1990). Some inferential information was recently published about the central connectivity of trigeminovascular afferents using antisera directed against the early immediate response gene, c-*fos* (Nozaki et al. 1992a, b). The data indicate that cells within rexed lamina I and II_o express the protein antigen following meningeal stimulation and c-*fos* expression in these postsynaptic cells can be blocked by denervation or treatment with antimigraine drugs (Nozaki et al. 1992a, b). Because lamina I and II_o are known to contain the synaptic terminations of nociceptive fibers, there is the strong possibility that trigeminovascular afferents terminate within this region as well.

The described anatomical projections to pial vessels account for the strictly unilateral distribution of many vascular headaches, such as migraine, and certain types of stroke (Mayberg et al. 1981), and for their diffuse quality. It also explains the phenomenon of referred head pain, whereby lesions (e.g. tumors) of the posterior fossa, for example, may be accompanied by forehead pain (Moskowitz et al. 1989b). Following experimental subarachnoid hemorrhage (a painful stimulus), peptide and messenger RNA content increase within the trigeminal ganglion, indicating increased metabolic and possibly neuronal activity. This response is consistent with a role for sensory neurons in transmitting nociceptive information (Linnik et al. 1989).

Trigeminovascular nerves are predominantly unmyelinated C fibers which form a network on the adventitial surface of cerebral vessels. Nerve endings are present in the adventitia and at the border between adventitia and media, but not within the media or endothelium (Liu-Chen et al. 1986). All major cerebral arteries and cortical arterioles in the cat are invested in fine varicose nerve fibers (McCulloch et al. 1986). Several polypeptides are contained within vesicles in the 'naked' nerve endings. These include the tachykinins (a family of peptides sharing a similar C-terminal amino sequence) substance P (SP) (Liu-Chen et al. 1983; Yamamoto et al. 1983; Liu-Chen et al. 1984) and neurokinin A (NKA) (Saito et al. 1987). Both are produced by proteolytic cleavage from a common precursor, β-preprotachykinin. A third substance, calcitonin gene-related peptide (CGRP) (Hanko et al. 1985; Uddman et al. 1985), is a 37-amino acid-containing neuropeptide whose existence was predicted from analysis of the DNA sequence for the calcium regulatory peptide calcitonin (Amara et al. 1982). Other putative neurotransmitters include galanin and perhaps adenosine triphosphate (ATP) (see Burnstock 1989 and Moskowitz et al. 1989b for review). Brain natriuretic peptide, another vasodilator, has also been found recently within perivascular nerves in the rat, although the origin of the fibers and their significance has yet to be characterized (Saper et al. 1990).

CGRP, SP and NKA are all vasodilators. SP and NKA relax vessels by endothelium-dependent mechanisms (Furchgott 1983; Zawadzki et al. 1983), presumably by releasing nitric oxide or a related substance from the endothelium; CGRP relaxes cerebral vessels predominantly via receptors on vascular smooth muscle. In addition, SP promotes plasma protein extravasation. CGRP and SP are colocalized within nerve terminals (McCulloch et al. 1986; Suzuki et al. 1989b). CGRP is a more potent vasodilator than either SP or NKA (Edvinsson 1989). Unlike SP, CGRP inhibits neurogenic vasoconstriction mediated by the activation of adrenergic nerves, and shortens the duration of the vasoconstrictor response to norepinephrine (McCulloch et al. 1986).

Release of SP from pial vessels has been demonstrated in vitro after capsaicin-induced depolarization of sensory fibers (Moskowitz et al. 1983), and this occurs presumably for NKA and CGRP as well. SP is released from vesicles by calcium-dependent mechanisms (Moskowitz et al. 1983) and binds to the endothelium (Edvinsson et al. 1985), promoting vasodilation and an increase in vascular permeability (Markowitz et al. 1987). Both antidromic (Couture and Cuello 1984; Markowitz et al. 1987) and orthodromic mechanisms (Lang and Zimmer 1974; Gonzalez et al. 1975; Lambert et al. 1984) can dilate cephalic blood vessels.

Neurogenic vasodilation develops in facial skin during thermocoagulation of the trigeminal ganglion in humans (Sweet and Wepsic 1974), and is associated with an increase in SP levels in the jugular vein (Goadsby et al. 1988). Following acute or chronic lower limb ischemia, venous levels of SP also increase (Henriksen et al. 1986). Moreover, CGRP levels rise in the superior sagittal sinus during electrical stimulation of the trigeminal ganglion (Buzzi et al. 1991). High levels of venous CGRP have been found in patients with subarachnoid hemorrhage, and may represent activation of the trigeminovascular system (Juul et al. 1990). Vasodilation and plasma protein leakage can be elicited in rat dura mater by either electrical stimulation of the trigeminal ganglion or by the infusion of SP or NKA (Markowitz et al. 1987). The development of neurogenically mediated plasma extravasation is not dependent on mast cell degranulation (Markowitz et al. 1989).

The cutaneous vasodilation that follows release of arterial occlusion is attenuated by 60% in rat hindpaw after either chronic denervation or pretreatment with capsaicin, an agent with toxicity specific for primary afferent sensory nerves (Lembeck and Donnerer 1981), and is mimicked by intra-arterial SP infusion (Lembeck and Holzer 1979). Selective electrical activation of trigeminovascular nerves in the rat induces a transient 17% increase in CBF (Suzuki et al. 1990b).

Role of Perivascular Sensory Nerves in the Regulation of Cerebral Blood Flow

The functional significance of the trigeminovascular system is unclear. However, it is likely to be involved in the transmission of pain and in the cerebrovascular disturbances that accompany migraine (Moskowitz 1984, 1991; Moskowitz et al. 1989b). The trigeminovascular system is also likely to be involved in complex autonomic reflexes. In diving mammals, for example, immersion of the face in cold water induces changes in regional blood flow that include cerebral vasodilation (Miles 1979; Zapol et al. 1979).

Several studies have shown that the trigeminovascular system is not involved in the moment-to-moment regulation of CBF (McCulloch et al. 1986; Moskowitz et al. 1989a; Sakas et al. 1989; Macfarlane et al. 1991a). Neither basal CBF nor the hyperemic response to hypercapnia are attenuated by sensory denervation. This is in keeping with the nature of CGRP, which is a large polypeptide (37 amino acid residues) with long-lasting actions on the cerebral vasculature (McCulloch et al. 1986).

It has been shown recently that perivascular sensory nerves participate in the regulation of CBF under certain pathological conditions. Trigeminal ganglionectomy attenuates the increases in blood flow which accompany acute severe hypertension (beyond the limits of autoregulation) and generalized seizures by around 30% in cats (Sakas et al. 1989), and diminishes the extravasation of radiolabeled albumin which results from disruption of the blood–brain barrier (Moskowitz et al. 1988). Feline trigeminal sensory fibers limit the constrictor response of pial cortical vessels to topical norepinephrine at 50 μM concentration (Moskowitz et al. 1988), but not at 100 μM (McCulloch et al. 1986), but restore constricted vessels more quickly to their resting caliber (McCulloch et al. 1986; Moskowitz et al. 1988).

During the early reperfusion period following 10 minutes of global cerebral ischemia in cats, postocclusive hyperemia is attenuated by around 50% in cortical gray matter ipsilateral to the side of chronic trigeminal ganglionectomy (Moskowitz et al. 1989a; Macfarlane et al. 1991a). Insignificant differences in CBF develop in cortical white matter or deep gray matter nuclei which, although supplied by the circle of Willis, receive a negligible trigeminal innervation. This demonstrates that reactive hyperemia is not merely a passive phenomenon resulting from changes in perfusion pressure exclusively, but that the decrease in cerebrovascular resistance is mediated in part by the release of neuropeptides. No significant change in CBF was observed between the intact and the denervated hemispheres in the ensuing phase of postischemic hypoperfusion (Macfarlane et al. 1991a).

The changes in postischemic CBF which follow chronic trigeminal ganglionectomy do not occur following rhizotomy (Sakas et al. 1989). Although both denervation procedures block central transmission, division of the trigeminal root does not result in destruction of the cell body, or induce Wallerian degeneration in peripheral axons, and local reflex responses are therefore preserved by root section. The increase in postischemic CBF mediated by sensory neurons must therefore occur by antidromic stimulation via 'axon reflex-like' mechanisms (Langley 1923).

The way in which sensory nerves promote increases in CBF is presumably via the axonal release of vasoactive neuropeptides. Release may be triggered by agents released from brain tissue or the vessel wall. Nitric oxide (NO) released from brain cells remains an exciting candidate because it diffuses readily, and relaxes cerebrovascular smooth muscle predominantly by releasing CGRP from sensory fibers. (NO-mediated CGRP release from sensory fibers mediates the vasodilating actions of nitroglycerin and nitroprusside in cerebral vessels (Wei et al. 1992). In fact, the mechanism by which nitric oxide relaxes pial vessels is primarily via the release of CGRP from trigeminovascular fibers (Kontos et al. unpublished observations). Bradykinin is synthesized by vessel walls during severe hypertension (Ellis et al. 1987), and is capable of depolarizing unmyelinated C fibers at a concentration of 100 nM (Armstrong 1970). During seizures, high levels of extracellular potassium (Leniger-Folert 1984) may provide the coupling mechanism by which trigeminovascular fibers become activated (Moskowitz et al. 1983). In addition to potassium and hydrogen ions, molecules such as adenosine, ATP, leukotrienes, prostaglandins and arachidonate metabolites accumulate in brain during ischemia, and are either stimulators or potentiators of C-fiber activation (Kontos et al. 1981; Moskowitz 1984; Martin et al. 1987; and see Pulsinelli et al. 1982 for a review). The likely stimulus for the release of vasoactive peptides is 'real or threatened tissue injury' (Sherrington 1900). However, depolarization of sensory axons is not an obligatory mechanism because neuropeptide release (in response to acute intravenous capsaicin administration) can occur in the presence of tetrodotoxin.

Studies of reperfusion after global cerebral ischemia in cats have shown that EDRF-dependent responses are lost during the early reperfusion period (Macfarlane 1991b), suggesting that CGRP rather than SP or NKA is likely to mediate the hyperemic response.

The Role of Perivascular Nerves in Cerebrovascular Disease

Sympathetic, parasympathetic and sensory perivascular nerves probably exert only a limited influence upon cerebral blood vessels under physiological conditions. However, the sensory innervation in particular may be involved in mediating pain and in influencing cerebral blood flow under a variety of pathological conditions.

Cerebral Hyperperfusion Syndromes

Restoration of blood flow after a period of ischemia results in vasodilation and an increase in blood flow in many organs, including the brain (Kagstrom et al. 1983). Complete cessation of blood flow is not a prerequisite for the development of reactive hyperemia. Hyperperfusion has been demonstrated during recovery from systemic hypotension and from hypocapnia (Raichle et al. 1971; Symon et al. 1972; Fitch et al. 1975a). Ischemia of as brief a duration as 30 seconds is sufficient to elicit a three- to fourfold increase in CBF (Gourley and Heistad 1984). Carotid endarterectomy for high-grade stenosis (Sundt et al. 1981), migration of an embolus some time after impaction (Olsen and Lassen 1984; Olsen 1986), and the obliteration of large arteriovenous malformations (Spetzler et al. 1978) may all result in fulminant hyperemia and the subsequent development of intraparenchymal hemorrhage, cerebral edema and/or seizures. Other conditions associated with hyperperfusion (defined as CBF in excess of metabolic requirements) include seizures (Plum et al. 1968), acute severe hypertension (Dinsdale 1983), a rapid rise in intracranial pressure (Wolff and Forbes 1928) and the removal of large mass lesions such as tumors or intracranial hematomas (Broderson and Gjerris 1975). Hyperperfusion has also been

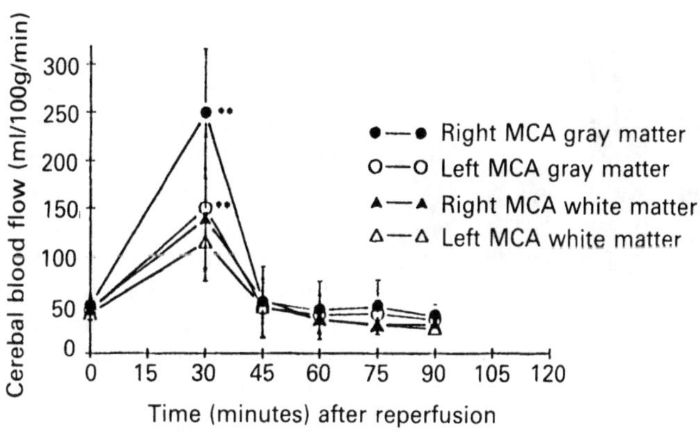

Fig. 16.1. CBF in middle cerebral artery (MCA) gray and white matter after 10 minutes of global cerebral ischemia in animals subjected to chronic left trigeminal ganglionectomy (n=8). Values are expressed as mean ±SD. Hyperemia in gray matter from the denervated hemisphere is attenuated by 48% (** $P < 0.01$), but white matter CBF is affected slightly. Trigeminalectomy neither influences the duration of hyperemia nor alters CBF significantly during the posthyperemic phase. From Macfarlane et al. (1991) Reproduced with permission of Journal of Cerebral Blood Flow and Metabolism 11:261–271.

linked with the development of edema and the likelihood of a poor outcome after severe head injury (Enevoldsen et al. 1976).

The trigeminovascular system was shown to be an important mediator of some hyperperfusion syndromes (Fig. 16.1). Chronic trigeminalectomy attenuates postischemic hyperperfusion in feline cortical gray matter by around 50% (Moskowitz et al. 1989b; Macfarlane et al. 1991a), and attenuates rises in CBF induced by acute severe hypertension and bicuculline-induced seizures by about 30% (Sakas et al. 1989). The pial arteriolar constrictor response to topical norepinephrine is also prolonged by trigeminal ganglionectomy (Moskowitz et al. 1988). However, these changes do not occur after trigeminal rhizotomy (Moskowitz et al. 1989a; Sakas et al. 1989), indicating that axon reflex-like mechanisms are involved.

However, the trigeminal nerve is not involved in all hyperperfusion states. Neither ganglionectomy (Moskowitz et al. 1989a) nor topical capsaicin application (Macfarlane et al. 1991a) attenuates the cerebrovascular response to hypercapnia.

Cerebral Ischemia

The role of sensory nerves in mediating reperfusion injury after embolic stroke has been mentioned in the preceding section, as has the protective influence of sympathetic nerves in promoting vascular hypertrophy in chronic hypertension and thereby protecting against disruption of the blood–brain barrier and the development of hemorrhagic infarction. Recently, Kano et al. (1991) established that vasodilator nerves may exert a protective influence in focal ischemia. Infarct volume was measured after unilateral branch occlusion of the rat middle cerebral artery. Whereas chronic section of sensory nerve fibers had no influence upon infarct volume, ablation of perivascular parasympathetic nerve fibers increased infarct size by a mean of 37% (Fig. 16.2). This suggests a unique role for parasympathetic fibers in the pathophysiology of focal ischemia that is not shared by sensory nerves, and may explain the propensity of patients with autonomic neuropathy to develop stroke. Although the precise mechanism remains unclear, preliminary studies following parasympathetic fiber sectioning suggest that parasympathetic fibers may increase blood flow to the ischemic penumbra (Koketsu et al. 1992) (Fig. 16.3). For example, studies using laser Doppler flowmetry found that regional cerebral blood flow was significantly lower on the ipsilateral side during reduced perfusion pressure (controlled hemorrhage) following chronic parasympathetic sectioning, an observation not made after selective sensory sectioning. Greater than expected reductions in blood flow in the penumbra or perinfarct zone may account for the significantly larger infarct volumes during focal ischemia in these animals.

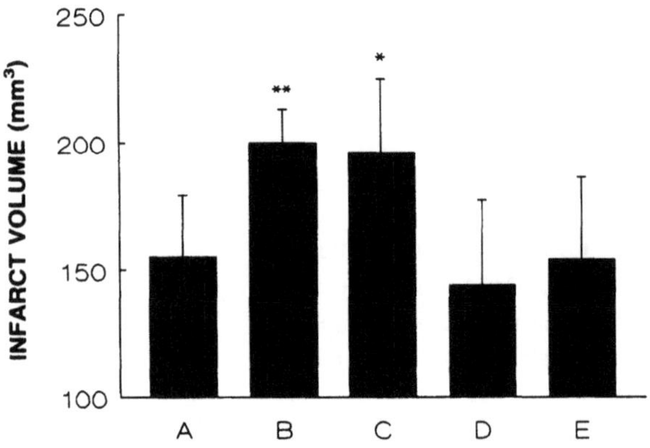

Fig. 16.2. Larger infarction volumes were recorded in SHR after chronic parasympathetic denervation of the circle of Willis following MCA and ipsilateral CCA occlusion. Infarct volume was increased in Group B (chronic parasympathetic and sensory sectioning; n=10) and Group C (selective parasympathetic sectioning; n=10) compared to the sham-operated group (Group A; n=30), but did not increase after selective sensory denervation (Group D; n=9). Animals in Groups B and C demonstrated reductions in the density of VIP-containing perivascular fibers in the ipsilateral MCA of ≥40%, compared to the non-operated side. Group E animals showed ≤30% reduction in the density of VIP-containing fibers within the ipsilateral MCA despite surgical sectioning at the ethmoidal foramen (n=6). Infarction volume in this group was the same as the sham group. *P<0.005, ** < 0.0005 compared to the sham-operated animals. From Koketsu N et al. (1992) Reproduced with permission of Journal of Cerebral Blood Flow and Metabolism 12:613–620.

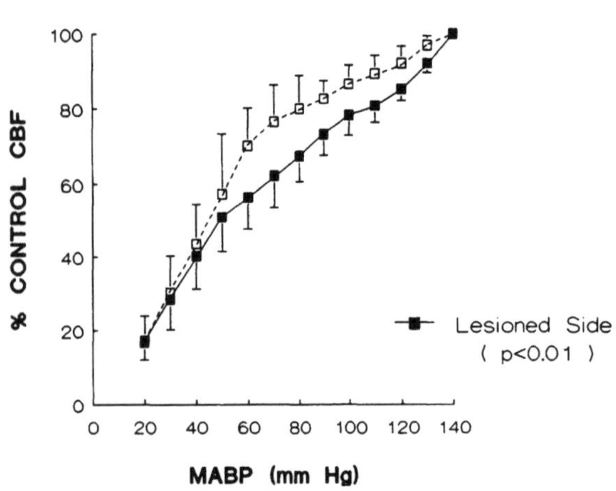

Fig. 16.3. Reductions in rCBF within the cortical barrel fields were larger on the lesioned side (right) during controlled exsanguination as blood pressure decreased from 140 mmHg after selective unilateral sectioning of parasympathetic efferents and removal of ipsilateral sphenopalatine ganglion (n=9). The two curves were statistically different at $P < 0.01$. From Koketsu N et al. (1992) Reproduced with permission of Journal of Cerebral Blood Flow and Metabolism 12:613–620.

The Trigeminal Nerve and Subarachnoid Hemorrhage

The trigeminovascular system may provide a protective mechanism to enhance cerebral blood flow early after subarachnoid hemorrhage. Trigeminal tachykinin gene expression is altered in this condition, and is consistent with increased neuronal activity and enhanced neuropeptide release (Linnik et al. 1989). Perivascular nerves show evidence of axonal degeneration about 7 days after subarachnoid hemorrhage, and this coincides with the development of delayed vasospasm in some (Duff et al. 1986), but not all (Hara et al. 1986), studies. McCulloch et al. (1986) proposed that the trigeminovascular system provides an immediate response to prevent excessive vasoconstriction of large arteries. Of interest is a recent clinical trial which concluded that Glasgow coma scores after subarachnoid hemorrhage improve transiently by the systemic administration of CGRP 1.15 μg/min (Johnston et al. 1990). However, this finding has not been confirmed by the preliminary results of a European multicenter study, in which 99 patients with delayed ischemic deficits after subarachnoid hemorrhage were randomized to either CGRP (0.6 μg/min) or standard best management. Three months after treatment, there was no benefit in outcome in the CGRP group (European CGRP in Subarachnoid Haemorrhage Study Group (1992).

The Trigeminovascular System and Headache

Vascular headaches may develop in association with many conditions, including hypertension, seizures, hyperperfusion syndrome and stroke. On other occasions there is no recognizable pathophysiology, as for example in migraine or cluster headache.

The pial and dural arteries are innervated by sensory nerve fibers which contain vasoactive neuropeptides mediating vasodilation, neurogenic inflammation and pain. Vascular headache shares several features in common with pain originating from other viscera, such as liver, heart and bowel (Moskowitz 1991). The pain is poorly localized and is referred to superficial structures such as the scalp or face, and to muscles (e.g. temporalis and neck). Like other viscera, the brain is sparsely innervated by thin unmyelinated or poorly myelinated fibers with extensive branching and relatively large receptive fields (Cervero 1987). Headache, too, is often accompanied by intense autonomic responses such as sweating, hypertension, tachycardia and tachypnea. These similarities can be accounted for by the central connections of the trigeminal nerve. Referred pain may represent the convergence of information from visceral and somatic afferents within the trigeminal nucleus caudalis, whereas a trigeminal projection to the nucleus of the tractus solitarius (mediating autonomic functions) has been defined (Kerr 1970). Potential activators of the trigeminovascular system are many, and include biochemical (e.g. bradykinin, serotonin and histamine), mechanical (stretch), immunological (allergy), ionic (potassium) and neural (sympathetic or opiate-containing fibers mechanism (for review see Moskowitz et al. 1989b).

Migraine

Considerable evidence is growing to link the trigeminovascular system with the pathogenesis of migraine. For many years it was thought that migraine headache was caused by cerebral vasodilation. Vasodilators such as nitroglycerin are associated with headaches and ergot alkaloids, known to be effective in the treatment of migraine but which have no analgesic properties, are vasoconstrictors (Graham and Wolff 1938). However, vasodilation is not a consistent feature of migraine. Indeed, blood flow is often decreased during classic migraine, and normal in cluster headache (Andersen et al. 1988). Furthermore, we now know that nitroglycerin and sodium nitroprusside directly stimulate trigeminovascular fibers, and may cause headache by this mechanism rather than in response to vasodilation. In fact, nitrate-induced vasodilation is mediated by neuropeptides released from perivascular sensory fibers (Wei et al. 1992).

The term 'sterile inflammation' has been used to describe the appearance of the superficial temporal arteries in migraine (Ostfeld et al. 1957). Neurogenic inflammation (local vasodilation and plasma protein extravasation that occurs within the peripheral distribution of C fibers following stimulation) is related to the phenomenon of tissue sensitization and the prolongation of pain. Cephalic blood vessels are potent generators of nociceptive molecules, including bradykinin, prostaglandins, histamine and serotonin (for review, see Moskowitz 1991). Specific receptor populations exclusive to cephalic blood vessels may explain the development of headache in response to certain provocative stimuli. $5\text{-HT}_{IB/D}$-like receptors reside on peptide-containing trigeminovascular axons within the dura mater, but not on extracranial vascular tissues innervated by the same nerve (Saito et al. 1988; Buzzi and Moskowitz 1990). The antimigraine drugs sumatriptan and ergot alkaloids, both $5\text{-HT}_{IB/D}$ agonists, block the development of neurogenic inflammation following trigeminal antidromic stimulation (Saito et al. 1988) and attenuate the concomitant increase in CGRP within the superior sagittal sinus (Buzzi and Moskowitz 1990). However, neurogenic inflammation elicited by the administration of substance P or neurokinin A remains unaffected. Nor do vasoconstrictors such as angiotensin or phenylephrine block the neurogenic inflammation induced by trigeminal stimulation, confirming that vasoconstriction cannot be the mechanism of action of these drugs. Together, these data suggest that the pain-relieving actions of sumatriptan and ergot alkaloids in migraine are mediated through neurogenic effects on the trigeminovascular system (see Moskowitz 1992 for review).

The mechanisms which activate the trigeminovascular system in migraine remain unclear. However, others have implicated cortical spreading depression (Leao 1944a, b) as an initiating event in migraine. It has recently been postulated that substances released from brain following depolarization (e.g. hydrogen ions, extracellular potassium, arachidonate metabolites (Hansen et al. 1980) are of sufficient magnitude to depolarize trigeminovascular nerves, and thus initiate vascular headache (Moskowitz 1984; Moskowitz et al. 1993). Of note is that recurrent spreading depression activates trigeminovascular fibers, as evidenced by an increase in the phosphoprotein *c-fos* (immunoreactive) within postsynaptic cells of lamina I and II_o in trigeminal nucleus caudalis. Sumatriptan, a new antimigraine drug and 5-HT_1 agonist, blocks the number of cells expressing this protein, as does trigeminal denervation of meningeal afferents (Moskowitz et al. 1993). However, the wave of spreading depression itself was not inhibited under these conditions.

Conclusion

Cerebral perivascular nerves appear not to play a significant role in CBF autoregulation. However, recent evidence suggests that sensory, parasympathetic and sympathetic nerves contribute to cerebral blood flow and blood vessel integrity in a variety of pathological conditions. Strategies designed to modulate the activity of these nerves, or their constituent neurotransmitters, may be of benefit in reducing the morbidity and mortality associated with a variety of disorders affecting the cerebral vasculature.

Acknowledgement This work was supported by grants NS 26361 and NS 21558 from the United States National Institute of Neurological and Communicative Disorders and Stroke.

References

Afshar F, Dykes E (1984) Computer-generated three-dimensional visualization of the trigeminal nuclear complex. Surg Neurol 22:189–196

Amara SG, Jonas V, Rosenfeld MG, Ong ES, Evans RM (1982) Alternative RNA processing in calcitonin gene expression generates mRNAs encoding different polypeptide products. Nature (London) 298:240–244

Andersen AR, Frieberg L, Skyhoj Olesen TJ (1988) Delayed hyperaemia following hypoperfusion in classic migraine. Arch Neurol 45:154–160

Arbab MA-R, Wiklund L, Svendgaard NA (1986) Origin and distribution of cerebral vascular innervation from superior cervical, trigeminal and spinal ganglia investigated with retrograde and anterograde WGA-HRP tracing in the rat. Neuroscience 19:695–708

Arbab MA-R, Delgado T, Wiklund L, Svendgaard NA (1988) Brain stem terminations of the trigeminal and upper spinal ganglia innervation of the cerebrovascular system: WGA-HRP transganglionic study. J Cerebr Blood Flow Metab 8:54–63

Armstrong D (1970) Pain. Hbk Exp Pharmacol 25:434–481

Bevan JA, Moskowitz MA, Said SI, Buga G (1984) Evidence that vasoactive intestinal polypeptide is a dilator transmitter to some cerebral and extracerebral cranial arteries. Peptides 5:385–388

Bill A, Linder J (1976) Sympathetic control of cerebral blood flow in acute arterial hypertension. Acta Physiol Scand 96:114–121

Broderson P, Gjerris F (1975) Regional cerebral blood flow patients with chronic subdural haematomas. Acta Neurol Scand 51:233–239

Burnstock G (1989) Vascular control by purines with emphasis on the coronary system. Eur Heart J 10 (Suppl F):15–21

Buzzi MG, Moskowitz MA (1990) The antimigraine drug sumatriptan (GR 43175) specifically blocks neurogenic plasma extravasation from blood vessels in dura mater. Br J Pharmacol 99:202–206

Buzzi MG, Carter B, Moskowitz MA, Shimizu T, Heath HH III (1991) Dihydroergotamine and sumatriptan attentuate the increase in plasma CGRP levels within rat superior sagittal sinus during electrical trigeminal ganglion stimulation. Neuropharmacology 30:1193–1200

Cervero F (1987) Visceral pain. In: Dubner R, Gebhart GF, Bond MR (eds) Proceedings of the Vth World Congress on Pain. Elsevier, Amsterdam, pp 216–226

Chorobski J, Penfield W (1932) Cerebral vasodilator nerves and their pathways from the medulla oblongata. Arch Neurol Psych (Chicago) 28:1257–1289

Couture R, Cuello C (1984) Studies on the trigeminal antidromic vasodilation and plasma extravasation in the rat. J Physiol (London) 346:273–285

Dinsdale HB (1983) Hypertensive encephalopathy. Neurol Clin 1.1:3–16

Duckles SP (1983) Innervation of the cerebral vasculature. Ann Biomed Eng 11:599–605

Duff TA, Scott G, Feilbach JA (1986) Ultrastructural evidence of arterial denervation following experimental subarachnoid hemorrhage. J Neurosurg 64:292–297

Edvinsson L (1987) Innervation of the cerebral circulation. Ann NY Acad Sci 519:334–348

Edvinsson L (1989) Dilatory neuropeptides in the cerebral circulation. In: Seylaz J, Mackenzie ET (eds) Neurotransmission and cerebrovascular function. Elsevier, Amsterdam, pp 229–242

Edvinsson L, Mackenzie ET (1977) Amine mechanisms in cerebral circulation. Pharmacol Rev 28:275–348

Edvinsson L, McCulloch J, Uddman R (1982) Noradrenaline-, substance P-, and vasoactive intestinal polypeptide-containing nerve fibers and vasomotor responses of cerebral arteries and veins. In: Heistad DD, Marcus M (eds) Cerebral blood flow: effects of nerves and neurotransmitters. Elsevier, New York, pp 219–222

Edvinsson LC, Owman E, Rosengren E, West KA (1972) Concentration of noradrenaline in pial vessels, choroid plexus, and iris during two weeks after sympathetic ganglionectomy or decentralization. Acta Physiol Scand 85:201–206

Edvinsson L, Fredholm BB, Hamel E, Jansen I, Verrechia C (1985) Perivascular peptides relax cerebral arteries concomitant with stimulation of cyclic adenosine monophosphate accumulation or release of an endothelium-derived relaxing factor in the cat. Neurosci Lett 58:213–217

Ellis E, Wei EP, Holt SA, Kontos HA (1987) Evidence that bradykinin stimulates the cyclooxygenase-dependent cerebral arteriolar abnormalities following concussive brain injury. (Abstract) Fed Proc 46:800

Enevoldsen EM, Cold G, Jensen FT, Malmros R (1976) Dynamic changes in regional CBF, intraventricular pressure, CSF, pH and lactate levels during the acute phase of head injury. J Neurosurg 44:191–214

European CGRP in Subarachnoid Haemorrhage Study Group (1992) Effect of calcitonin-gene-related peptide in patients with delayed postoperative cerebral ischaemia after aneurysmal subarachnoid haemorrhage. Lancet 339:831–834

Fay T (1932) Atypical facial neuralgia, a syndrome of vascular pain. Ann Otol, Rhinol, Laryngol 41:1030–1062

Fitch W, Ferguson GG, Sengupta D, Garibi J (1975a) Autoregulation of cerebral blood flow during controlled hypotension. In: Langitt TW, McHenry LC, Reivich M, Wollman H (eds) Cerebral circulation and metabolism. Springer-Verlag, New York, pp 18–20

Fitch W, McKenzie ET, Harper AM (1975b) Effects of decreasing arterial blood pressure on cerebral blood flow in the baboon; influence of the sympathetic nervous system. Circ Res 37:550–557

Furchgott RF (1983) Role of the endothelium in response of vascular smooth muscle. Circ Res 53:557–573

Gibbins IL, Brayden JE, Bevan JA (1984) Perivascular nerves

with immunoreactivity to vasoactive intestinal polypeptide in cephalic arteries of the cat: distribution, possible origins and functional implications. Neuroscience 13:1327–1346

Goadsby PJ, Duckworth JW (1987) Effect of stimulation of trigeminal ganglion on regional cerebral blood flows in cats. Am J Physiol 253:R270–R274

Goadsby PJ, Edvinsson L, Ekman R (1988) Release of vasoactive peptides in the extracerebral circulation of humans and the cat during activation of the trigeminovascular system. Ann Neurol 23:193–196

Gonzalez G, Onofrio BM, Kerr FWL (1975) Vasodilator system for the face. J Neurosurg 42:696–703

Gourley JK, Heistad DD (1984) Characteristics of reactive hyperemia in the cerebral circulation. Am J Physiol 246:H52–H58

Graham JR, Wolff HG (1938) Mechanism of migraine headache and action of ergotamine tartrate. Arch Neurol Psych 39:737–763

Hanko J, Hardebo JE, Kahrstrom J, Owman C, Sundler F (1985) Calcitonin gene-related peptide is present in mammalian cerebrovascular nerve fibres and dilates pial and peripheral arteries. Neurosci Lett 57:91–95

Hansen AJ, Quistorff B, Fjedde A (1980) Relationship between local changes in cortical blood flow and extracellular K^+ during spreading depression. Acta Physiol Scand 109:1–6

Hara H, Weir B (1986) Pathway of acetylcholinesterase-containing nerves to the major cerebral arteries in rats. J Comp Neurol 250:245–252

Hara H, Nosko M, Weir B (1986) Cerebral perivascular nerves in subarachnoid hemorrhage. A histochemical and immunohistochemical study. J Neurosurg 65:531–539

Hardebo JE, Lindvall O, Nilsson B (1982) On the possible influence of adrenergic and cholinergic mechanisms in normo- and hypercapnia. In: Heistad DD, Marcus M (eds) Cerebral blood flow: effect of nerves and neurotransmitters. Elsevier, New York, pp 377–383

Harper AM, Deshmukh VD, Rowan JO, Jennett WB (1972) Influence of sympathetic nervous activity in cerebral blood flow. Arch Neurol 27:1–6

Hart M, Heistad DD, Brody MJ (1980) Effect of chronic hypertension and sympathetic denervation on wall/lumen ratio of cerebral arteries. Hypertension 2:419–423

Heistad DD, Marcus M (1979) Effects of sympathetic stimulation on permeability of the blood–brain barrier to albumin during acute hypertension in cats. Circ Res 45:331–338

Heistad DD, Marcus M (1982) Cerebral blood flow: effects of nerves and neurotransmitters. Elsevier, New York

Heistad DD, Marcus M, Busija D, Sadoshima S (1982) Protective effects of sympathetic nerves in the cerebral circulation. In: Heistad DD, Marcus ML (eds) Cerebral blood flow: effect of nerves and neurotransmitters. Elsevier, New York, pp 267–273

Henriksen JH, Bulow JB, Schaffalitzky de Muckadell O, Fahrenkrug J (1986) Do substance P and vasoactive intestinal polypeptide (VIP) play a role in the acute occlusive or chronic ischaemic vasodilatation in man? Clin Physiol 6:163–170

Hovelaque A (1927) Anatomie des nerfs craniens et rachidiens et du système grand sympathétique chez l'homme. Paris: Gaston Doin. Cited in Ruskell GL, Simons T (1987) Trigeminal nerve pathways to the cerebral arteries in monkeys. J Anat 155:23–37

Johnston FB, Bell BA, Robertson IAJ et al. (1990) Effect of calcitonin gene-related peptide on postoperative neurological deficits after subarachnoid haemorrhage. Lancet 335:869–872

Juul R, Edvinsson L, Gisvold SE et al. (1990) Calcitonin gene-related peptide-LI in subarachnoid haemorrhage in man. Signs of activation of the trigemino-cerebrovascular system? Br J Neurosurg 4:171–180

Kagstrom E, Smith M-L, Siesjo BK (1983) Local cerebral blood flow in the recovery period following complete cerebral ischemia in the rat. J Cerebr Blood Flow Metab 3:170–182

Kano M, Moskowitz MA, Yokota M (1991) Parasympathetic denervation of rat pial vessels significantly increases infarction volume following middle cerebral artery occlusion. J Cerebr Blood Flow Metab 11:628–637

Keller JT, Beduk A, Saunders MC (1985) Origins of fibers innervating the basilar artery of the cat. Neurosci Lett 58:263–268

Kerr FWL (1970) The organization of primary afferents in the sub-nucleus caudalis of the trigeminal: a light and electron microscopic study of degeneration. Brain Res 23:127–165

Kobayashi S, Waltz AG, Rhoton AL (1971) Effects of stimulation of cervical sympathetic nerves on cortical blood flow and vascular reactivity. Neurology 21:297–320

Koketsu N, Moskowitz MA, Kontos HA, Yokota Y, Shimizu T (1992) Chronic parasympathetic sectioning decreases regional cerebral blood flow during hemorrhagic hypotension and increases infarct size after middle cerebral artery occlusion in spontaneously hypertensive rats. J Cerebr Blood Flow Metab 12:613–620

Kontos HA, Wei EP, Ellis EF, Dietrich WD, Povlishock JT (1981) Prostaglandins in physiological and in certain pathological responses of the cerebral circulation. Fed Proc 40:2326–2330

Lambert GA, Bogduk N, Goadsby PJ, Duckworth JW, Lance JW (1984) Decreased carotid arterial resistance in cats in response to trigeminal stimulation. J Neurosurg 61:307–315

Lang R, Zimmer R (1974) Neurogenic control of cerebral blood flow. Exp Neurol 43:143–161

Langley JN (1923) Antidromic action. J Physiol (London) 57:428–446

Larsson L-I, Edvinsson L, Fahrenkrug J et al. (1976) Immunohistochemical localization of a vasodilatory polypeptide (VIP) in cerebrovascular nerves. Brain Res 113:400–404

Leao AAP (1944a) Spreading depression of activity in the cerebral cortex. Br J Neurophysiol 7:359–390

Leao AAP (1944b) Pial circulation and spreading depression of activity in the cerebral cortex. J Neurophysiol 7:391–396

Lee TJ-F, Hume WR, Su C, Bevan JA (1978) Neurogenic vasodilation of cat cerebral arteries. Circ Res 42:535–542

Lembeck F, Donnerer J (1981) Postocclusive cutaneous vasodilatation mediated by substance P. Naunyn-Schmiedenberg's Arch Pharmacol 316:165–171

Lembeck F, Holzer P (1979) Substance P as neurogenic mediator of antidromic vasodilation and neurogenic plasma extravasation. Naunyn-Schmiedenbergs Arch Pharmacol 310:175–183

Leniger-Folert E (1984) Mechanisms of regulation of cerebral microflow during bicuculline-induced seizures in anesthetized cats. J Cerebr Blood Flow Metab 4:150–165

Levine M, Wolff HG (1932) Cerebral circulation: afferent impulses from the blood vessels of the pia. Arch Neurol Psych Chicago 28:140–150

Linder N (1981) Effects of facial nerve section and stimulation on cerebral and ocular blood flow in haemorrhagic hypotension. Acta Physiol Scand 112:185–193

Linnik MD, Sakas DE, Uhl GR, Moskowitz MA (1989) Subarachnoid blood and headache: altered trigeminal tachykinin gene expression. Ann Neurol 25:179–184

Liu-Chen L-Y, Han DH, Moskowitz MA (1983) Pia arachnoid

contains substance P originating from trigeminal neurons. Neuroscience 9:803–808
Liu-Chen L-Y, Gillespie SA, Norregaard TV, Moskowitz MA (1984) Co-localization of retrogradely transported wheat germ agglutinin and the putative neurotransmitter substance P within trigeminal ganglion cells projecting to cat middle cerebral artery. J Comp Neurol 225:187–192
Liu-Chen L-Y, Liszczak TM, King JC, Moskowitz MA (1986) Immunoelectron microscopic study of substance P-containing fibers in feline cerebral arteries. Brain Res 369:12–20
McCulloch J, Udmann R, Kingman TA, Edvinsson L (1986) Calcitonin-gene related peptide: functional role in cerebral regulation. Proc Natl Acad Sci USA 83:5731–5735
Macfarlane R, Tasdemiroglu E, Moskowitz MA et al. (1991a) Chronic trigeminal ganglionectomy or topical capsaicin application to pial vessels attenuates postischemic hyperemia but does not influence postischemic hypoperfusion. J Cerebr Blood Flow Metab 11:261–271
Macfarlane R, Moskowitz MA, Tasdemiroglu E, Wei EP, Kontos HA (1991b) Postischemic cerebral blood flow and neuroeffector mechanisms. Blood Vessels 28:46–51
McMahon MD, Norregaard TV, Beyerl BD, Borges LF, Moskowitz MA (1985) Trigeminal afferents to cerebral arteries and forehead are not divergent axon collaterals in cat. Neurosci Lett 60:63–68
McNaughton F (1938) The innervation of the intracranial blood vessels and dural sinuses. Assoc Res Nerv Dis 18:178–200
Marcus ML, Busija DW, Gross PM, Brooks LA, Heistad DD (1982) Sympathetic escape in the cerebral circulation during normotension and acute severe hypertension. In: Heistad DD, Marcus ML (eds) Cerebral blood flow: effect of nerves and neurotransmitters. Elsevier, New York, pp 281–289
Markowitz S, Saito K, Moskowitz MA (1987) Neurogenically mediated leakage of plasma protein occurs from blood vessels in dura mater but not brain. J Neurosci 7:4129–4136
Markowitz S, Saito K, Buzzi MG, Moskowitz MA (1989) The development of neurogenic plasma extravasation in the rat dura mater does not depend upon the degranulation of mast cells. Brain Res 477:157–165
Martin HA, Basbaum AI, Kwiat GC, Goetzel EJ, Levine JD (1987) Leukotriene and prostaglandin sensitization of cutaneous high threshold C- and A-delta mechanonociceptors in the hairy skin of rat hindlimbs. Neurosci 22:651–659
Mayberg MR, Zervas NT, Moskowitz MA (1984) Trigeminal projections to supratentorial pial and dural blood vessels in cats demonstrated by horseradish peroxidase histochemistry. J Comp Neurol 223:46–56
Mayberg MR, Langer RS, Zervas NT, Moskowitz MA (1981) Perivascular meningeal projections from cat trigeminal ganglia: possible pathway for vascular headache in man. Science 213:228–230
Miles TS (1979) Features peculiar to the trigeminal innervation. Can J Neurol Sci 6:95–103
Moskowitz MA (1984) The neurobiology of vascular head pain. Ann Neurol 16:157–168
Moskowitz MA (1991) The visceral organ brain: implications for the pathophysiology of vascular head pain. Neurology 41:182–186
Moskowitz MA (1992) Neurogenic versus vascular mechanisms of sumatriptan and ergot alkaloids in migraine. Trends Pharmacol Sci 13:307–311
Moskowitz MA, Brody M, Liu-Chen L-Y (1983) In vitro release of immunoreactive substance P from putative afferent nerve endings in bovine pia arachnoid. Neuroscience 9:809–814
Moskowitz MA, Sakas DE, Wei EP et al. (1989a) Postocclusive hyperemia in feline cortical gray matter is mediated by trigeminal sensory axons. Am J Physiol 257:H1736–H1739
Moskowitz MA, Buzzi MG, Sakas DE, Linnik MD (1989b) Pain mechanisms underlying vascular headaches. Rev Neurol (Paris) 145:181–193
Moskowitz MA, Wei EP, Saito K, Kontos HA (1988) Trigeminalectomy modifies pial arteriolar responses to hypertension or norepinephrine. Am J Physiol 255:H1–H6
Moskowitz MA, Nozaki K, Kraig RP (1993) Neocortical spreading depression provokes the expression of c-*fos* protein-like immunoreactivity within trigeminal nucleus caudalis via trigeminovascular mechanisms. J Neurosci 13:1167–1177
Mueller SM, Heistad DD (1980) Effect of chronic hypertension on the blood–brain barrier. Hypertension 2:809–812
Mueller SM, Rusterholtz DB (1982) 'Trophic' influence of sympathetic nerves on the peripheral and cerebral vasculature. In: Heistad DD, Marcus ML (eds) Cerebral blood flow: effects of nerves and neurotransmitters. Elsevier, New York, pp 317–325
Mueller SM, Heistad DD, Marcus ML (1977) Total and regional cerebral blood flow during hypotension, hypertension and hypocapnia: effect of sympathetic denervation in dogs. Circ Res 41:350–359
Nielsen KC, Owman C (1967) Adrenergic innervation of pial arteries related to the circle of Willis of the cat. Brain Res 6:773–776
Nielsen KC, Owman C (1971) Contractile response and amine receptor mechanisms in isolated middle cerebral artery of the cat. Brain Res 27:33–42
Norregaard TV, Moskowitz MA (1985) Substance P and the sensory innervation of intracranial and extracranial feline cephalic arteries. Implications for vascular pain mechanisms in man. Brain 108:517–533
Northfield DWC (1938) Some observations on headache. Brain 61:133–162
Nozaki K, Boccalini P, Moskowitz MA (1992a) Expression of c-*fos*-like immunoreactivity in brainstem after meningeal irritation by blood in the subarachnoid space. Neuroscience 49:669–680
Nozaki K, Moskowitz MA, Boccalini P (1992b) CP-93,129, sumatriptan, dihydroergotamine block c-*fos* expression within rat trigeminal nucleus caudalis caused by chemical stimulation of the meninges. Br J Pharmacol 106:409–415
Nozaki K, Moskowitz MA, Maynard KI et al. (1993) Possible origins and distribution of immunoreactive nitric oxide synthase-containing nerve fibres in rat and human cerebral arteries. J Cerebr Blood Flow Metab 13:70–79
O'Connor TP, van der Kooy D (1986) Pattern of intracranial and extracranial projections of trigeminal ganglion cells. J Neurosci 6:2200–2207
Olsen TS (1986) Regional cerebral blood flow after occlusion of the middle cerebral artery. Acta Neurol Scand 73:321–337
Olsen TS, Lassen NA (1984) A dynamic concept of middle cerebral artery occlusion and cerebral infarction in the acute state based on interpreting severe hyperemia as a sign of embolic migration. Stroke 15:458–468
Ostfeld AM, Reis DJ, Goodell H, Wolff HG (1957) Headache and hydration. Arch Intern Med 96:142–152
Penfield W, McNaughton F (1940) Dural headache and innervation of the dura mater. Arch Neurol Psych Chicago 4:43–75
Pinard E, Purves MJ, Seylaz J, Vasquez JV (1979) The cholinergic pathway to cerebral blood vessels. II. Physiological studies. Pflugers Arch 379:165–172
Plum F, Posner JB, Troy B (1968) Cerebral metabolic and

circulatory responses to induced convulsions in animals. Arch Neurol 18:1–13
Pulsinelli WA, Levy DE, Duffy TE (1982) Regional cerebral blood flow and glucose metabolism following transient forebrain ischemia. Ann Neurol 11:499–509
Purves MJ (1978) Do vasomotor nerves significantly regulate cerebral blood flow? Circ Res 43:485–493
Raichle ME, Posner JB, Plum F (1971) Cerebral blood flow during and after hyperventilation. In: Russell RWR (ed) Brain and blood flow. Proceedings of the 4th International Symposium on regulation of cerebral blood flow. Pitman, London, pp 223–228
Ray BS, Wolff HG (1940) Experimental studies on headache: pain-sensitive structures of the head and their significance in headache. Arch Surg 41:813–856
Ruskell GL, Simons T (1987) Trigeminal nerve pathways to the cerebral arteries in monkeys. J Anat 155:23–37
Saito K, Greenberg S, Moskowitz MA (1987) Trigeminal origin of beta-preprotachykinin products in feline pial blood vessels. Neurosci Lett 76:69–73
Saito K, Markowitz S, Moskowitz MA (1988) Ergot alkaloids block neurogenic extravasation in dura mater: proposed action in vascular headaches. Ann Neurol 24:732–737
Sakas DE, Moskowitz MA, Wei EP et al. (1989) Trigeminovascular fibers increase blood flow in cortical gray matter by axon reflex-like mechanisms during acute severe hypertension and seizures. Proc Natl Acad Sci USA 86:1401–1405
Saper CB, Kibbe MR, Murley KM, Spencer S (1990) Brain natriuretic peptide-like immunoreactive innervation of the cerebrovascular system in the rat. Stroke 21 (Suppl III):III-166–167
Seylaz J, Hara H, Pinard E et al. (1988) Effect of stimulation of the sphenopalatine ganglion on cortical blood flow in the rat. J Cerebr Blood Flow Metab 8:875–878
Sherrington CS (1900) In: Schafer EA (ed) Textbook of physiology. Portland, London, pp 920–1001
Shima T, Hossmann K-A, Date H (1983) Pial arterial pressure in cats following middle cerebral artery occlusion. I. Relationship of blood flow, regulation of blood flow and electrophysiological function. Stroke 14:713–719
Spetzler RF, Wilson CB, Weinstein P (1978) Normal perfusion pressure breakthrough theory. Clin Neurosurg 25:651–672
Sundt TM Jr, Sharbrough FW, Piepgras DG et al. (1981) Correlation of cerebral blood flow and electroencephalographic changes during carotid endarterectomy. With results of surgery and hemodynamics of cerebral ischemia. Mayo Clin Proc 56:533–543
Suzuki N, Hardebo JE, Owman C (1988) Origins and pathways of cerebrovascular vasoactive intestinal polypeptide-positive nerves in rat. J Cerebr Blood Flow Metab 8:697–712
Suzuki N, Hardebo JE, Owman C (1989a) Origins and pathways of cerebrovascular nerves storing substance P and calcitonin gene-related peptide in rat. Neuroscience 31:427–438
Suzuki N, Hardebo JE, Owman C (1989b) Trigeminal fiber collaterals storing substance P and calcitonin gene-related peptide associate with ganglion cells containing choline acetyltransferase and vasoactive intestinal polypeptide in the sphenopalatine ganglion of the rat. An axon reflex modulating parasympathetic ganglionic activity? Neuroscience 30:595–604
Suzuki Y, McMaster D, Lederis K, Rostad OP (1984) Characterization of the relaxant effects of vasoactive intestinal polypeptide (VIP) and PHI on isolated brain arteries. Brain Res 322:9–16
Suzuki N, Hardebo JE, Kahrstrom J, Owman C (1990a) Selective electrical stimulation of postganglionic cerebrovascular nerve fibers originating from the sphenopalatine ganglion enhances cortical blood flow in the rat. J Cerebr Blood Flow Metab 10:383–391
Suzuki N, Hardebo JE, Kahrstrom J, Owman C (1990b) Effect on cortical blood flow of electrical stimulation of trigeminal cerebrovascular nerve fibres in the rat. Acta Physiol Scand 138:307–316
Sweet WH, Wepsic JG (1974) Controlled thermocoagulation of trigeminal ganglion and rootlets for differential destruction of fibers. Part I: Trigeminal neuralgia. J Neurosurg 40:143–156
Symon L, Ganz JC, Dorsch NWC (1972) Experimental studies of hyperaemic phenomena in the cerebral circulation of primates. Brain 95:265–278
Uddman R, Edvinsson L, Ekman R, Kingman T, McCulloch J (1985) Innervation of the feline cerebral vasculature by nerve fibers containing calcitonin gene-related peptide: trigeminal origin and co-existence with substance P. Neurosci Lett 62:131–136
Uemura Y, Sugimoto T, Kikuchi H, Mizuno N (1988) Possible origins of cerebrovascular nerve fibers showing vasoactive intestinal polypeptide-like immunoreactivity: an immunohistochemical study in the dog. Brain Res 448:98–105
Vasquez J, Purves MJ (1979) The cholinergic pathway to cerebral blood vessels. 1. Morphological studies. Pfluegers Arch 379:157–163
Wahl M, Kuschinsky W, Bosse O et al. (1972) Effect of L-epinephrine on the diameter of pial arterioles and arteries in the cat. Circ Res 31:248–256
Wall PD, Pribram KH (1950) Trigeminal neurotomy and blood pressure responses from stimulation of lateral cerebral cortex of *Macaca mulatta*. J Neurophysiol 13:409–412
Walters BB, Gillespie SA, Moskowitz MA (1986) Cerebrovascular projections from the sphenopalatine and otic ganglia to the middle cerebral artery of the cat. Stroke 17:488–494
Wei EP, Kontos HA, Patterson JL (1980) Dependence of pial arteriolar response to hypercapnia on vessel size. Am J Physiol 238:H697–H705
Wei EP, Moskowitz MA, Boccalini P, Kontos HA (1992) Calcitonin gene-related peptide mediates nitroglycerin and sodium nitroprusside-induced vasodilation in feline cerebral arterioles. Circ Res 70:1313–1319
Wei EP, Raper AJ, Kontos HA, Patterson JL (1975) Determinants of response of pial arteries to norepinephrine and sympathetic nerve stimulation. Stroke 6:654–659
Wolff HC, Forbes HS (1928) The cerebral circulation. V. Observations of the pial circulation during changes in intracranial pressure. Arch Neurol Psych Chicago 20:1035–1047
Yamamoto K, Matsuyama T, Shiosaka S et al. (1983) Overall distribution of substance P-containing nerves in the wall of the cerebral arteries of the guinea pig and its origins. J Comp Neurol 215:421–426
Young RF (1990) The trigeminal nerve and its central pathways. Physiology of facial sensation and pain. In: Rovit RL, Murali R, Jannetta PJ (eds) Trigeminal neuralgia. Williams and Wilkins, Baltimore, pp 27–51
Zapol WM, Liggins GC, Schneider RC, Qvist J, Sider MT, Creasy RK et al. (1979) Regional blood flow during simulated diving in the conscious Weddell seal. J Appl Physiol: Resp Env Exercise Physiol 47:968–973
Zawadzki JV, Furchgott RF, Cherry PD (1983) Endothelium-dependent relaxation of arteries by octa-substance P, kassinin and octa-cholecystokinin. (Abstract) Fed Proc 42:619

17. Rheology of Flow and its Effects

J. C Grotta

The characteristics of blood which determine its ability to flow under various conditions are termed its 'rheology'. Obviously, since rheology determines the ability of blood to perfuse the brain, it is an important variable in understanding cerebral blood flow under normal conditions and in pathological states, particularly brain ischemia. This chapter will define the normal rheologic characteristics of blood, in particular its viscosity; will describe the relationship of viscosity to cerebral blood flow in normal and pathological states; will explain how viscosity can be manipulated as part of a therapeutic approach to cerebral ischemia; will describe the relationship of blood viscosity to increased stroke risk; and finally, will describe some specific disorders of blood viscosity which are associated with stroke.

Blood Rheology

Fluids can be either newtonian, such as water, which has a constant viscosity under all flow conditions, or non-newtonian, such as paint which, because it is a particulate suspension increases in viscosity in inverse relation to flow velocity (Merrill 1969a). Consequently, under conditions of relative stasis, blood will become more resistant to flow and a vicious cycle can arise leading to greater stasis and perhaps thrombosis. Such low-flow situations in the cerebral circulation can occur distal to an occluded artery, proximal to an occluded vein, in the distal supply of end-arterioles or stenotic arteries in the setting of hypotension, or in eddy currents in regions of turbulence distal to stenotic vessels (Dormandy 1970).

The viscosity of blood is mainly dependent on the red blood cell mass or hematocrit, which represents the overwhelming majority of the particulate component of blood (Begg and Hearns 1966), and fibrinogen, which represents the largest quantity of high molecular weight proteins in the plasma (Merrill et al. 1963). Under pathological conditions other blood elements such as white blood cells, platelet aggregates, immunoglobulins or lipids can contribute to blood viscosity. The non-newtonian nature of blood is caused by greater aggregation of red blood cells into rouleaux as flow decreases. With higher concentrations of fibrinogen there is greater formation of rouleaux and, since fibrinogen is an acute-phase reactant, this accounts for the elevated sedimentation rate (and whole-blood viscosity) found in most acute disease states and inflammatory disorders.

Whole-blood viscosity is usually estimated from the hematocrit and fibrinogen concentration and can be estimated as a function of $[\mathrm{FIB} \times (\mathrm{HCT})^3]$. However, as previously descri-

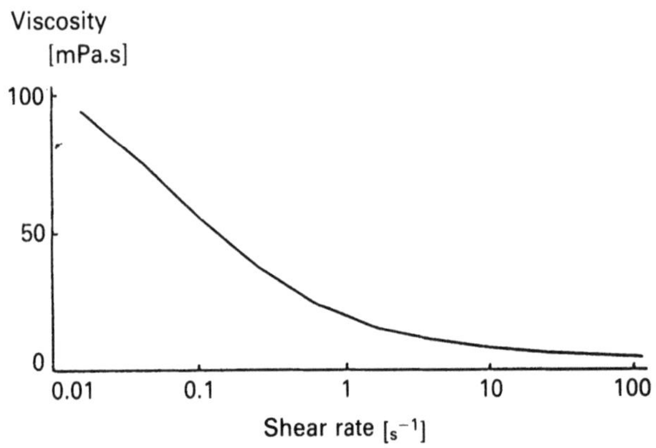

Fig. 17.1. Normal viscosity curve measured at a hematocrit of 42%. The increase in viscosity at lower shear rates is obvious. From Goslinga H (1990) Blood viscosity and circulatory filling pressure in clinical practice. Reproduced with permission of Dynamo Pers, Amsterdam.

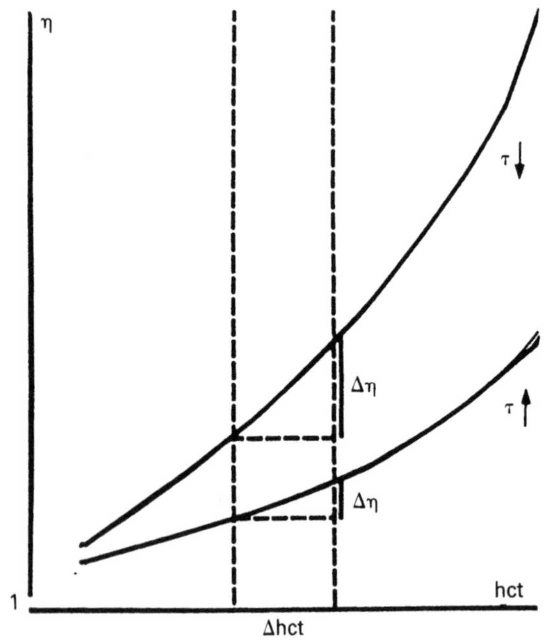

Fig. 17.2 Relationship between hematocrit (hct) and blood viscosity (η) under high (τ↓) and low (τ↑) shear stress. The viscosity increase (Δ↑) due to a 5% rise in hematocrit is substantially greater when shear stress is low. Similarly, a reduction in hematocrit (i.e. hemodilution) of 5% will cause a markedly stronger fall in blood viscosity when shear stress is low than when it is high. It follows that hemodilution may be more effective where the flow conditions are such that blood flows at low shear. From Ernst E et al. (1987) Blood cells and the control of circulation. In: Hartmann A, Kuschinsky W (eds) Cerebral ischemia and hemorheology. Reproduced with the permission of Springer-Verlag, Berlin, pp 51–56.

bed, a viscosity measurement only has meaning in relationship to flow. The precise way to measure whole-blood viscosity is to determine the resistance to flow at various shear rates (Fig. 17.1) (Goslinga 1990). The shear rate is the rate of movement of one plane of liquid in relation to an adjacent plane, and can be adjusted in a viscometer by adjusting its speed of rotation. In vessels with laminar flow, the highest shear rates are found near the vessel wall. Another term which is used in describing the rheology of a non-newtonian liquid is the 'yield shear stress' (YSS), which is the force required to begin the movement of a liquid in a state of zero flow, i.e. the viscosity of the fluid in a condition of stasis. The YSS can be calculated as $\text{YSS} = 13.5\ (10^{-6})\ (\text{FIB})^2\ (\text{HCT} - 6)^3$ (Merrill 1969b). The relationship between viscosity and hematocrit is critically dependent on the shear rate, with a more profound effect occurring with low shear when red blood cell aggregation is more prominent (Fig. 17.2) (Messmer 1975).

Viscosity and Cerebral Blood Flow (CBF)

Viscosity both contributes to flow and is affected by flow as described above, and this is certainly true in the cerebral circulation. In large vessels, the flow of newtonian fluids is proportional to perfusion pressure and vessel radius and inversely proportional to vessel length and blood viscosity, according to the Hagen–Poiseuille equation

$$\text{Flow} = \frac{\Delta P\ \pi\ r^4}{8LV}$$

where ΔP is the pressure gradient, r is the radius, L the length and V the viscosity. In the microcirculation other factors play a larger role and make this relationship more complex (Wood and Kee 1985). For instance, whereas red cell aggregation and plasma viscosity contribute, as in large vessels, shear-induced changes in red cell deformability and platelet aggregation, and the plugging of capillaries by white blood cells, begin to play more critical roles (Chen 1987; Hellums et al. 1987). Furthermore red cells are 'screened' out of the microcirculation so that the hematocrit within the cerebral microcirculation is lower than in conduction vessels. This may account for the Fahreus effect, where viscosity

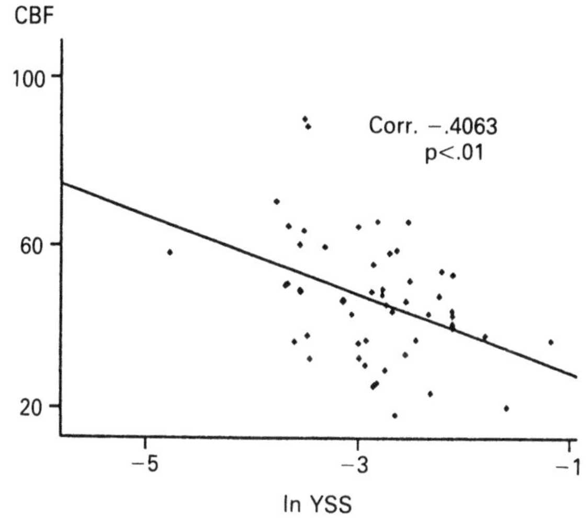

Fig. 17.3. Regression of lnYSS vs CBF in 53 patients. From Grotta JC et al. (1982) Whole blood viscosity parameters and cerebral blood flow. Reproduced with permission of Stroke 13:296–301.

falls as blood flows through tubes of progressively smaller diameters until it reaches the capillary level, where resistance again increases.

Several studies of CBF in normal individuals have demonstrated an inverse correlation with blood viscosity. Grotta et al., using xenon inhalation to measure CBF, found a close relationship to the YSS as well as hematocrit (HCT) and fibrinogen (FIB) levels (Fig. 17.3). They found that CBF=103−40 (FIB−HCT (Grotta et al. 1982). Consequently, it can be assumed that viscosity contributes to CBF and that in carrying out measurements of CBF the viscosity of the blood should be taken into consideration. However, to some extent the brain appears to be protected against fluctuations in viscosity by an ability to autoregulate its flow, much as it is autoregulated in response to fluctuations in perfusion pressure (Muizelaar et al. 1986).

Although it is unlikely that heightened viscosity leads directly to stroke, except in rare situations, elevated viscosity has been found in stroke patients in several studies. More recently, Fisher and Meiselman investigated 100 consecutive patients with transient ischemic attacks or ischemic stroke within 72 hours of the onset of symptoms, and in 66 cases again 2 months later (Fisher and Meiselman 1991). Measurements included viscosity, HCT, FIB, white blood cell (WBC) count, red blood cell (RBC) aggregation, erythrocyte deformability (which related to the ability of red blood cells to squeeze through capillaries smaller than their own diameter), and two measures of platelet function (B-thromboglobulin and platelet factor 4). Both whole-blood and plasma viscosity were measured. Patients were compared to non-vascular neurological disease controls and normal controls. Whole-blood viscosity, fibrinogen and RBC aggregation were all elevated in stroke patients. Abnormalities were greatest in patients with severe strokes, and tended to return to normal during the 2 months following the stroke. Elevated fibrinogen was also found in non-stroke neurological patients. The authors concluded that their data indicated that hemorheological abnormalities are common but non-specific in patients with neurological diseases, and are probably the result of acute increases in fibrinogen. Other investigators have described similar rheological abnormalities in stroke patients (Ott and Lechner 1982). More recently decreased leukocyte filtrability and consequent microcirculatory 'plugging' have been described in experimental models (Schmid-Schönbein 1987; del Zoppo et al. 1991) and may be responsible for impaired microcirculatory flow during reperfusion ('no-reflow').

Whether it is the cause or the effect of stroke (and based on the data described above it appears that a non-specific effect is most likely), elevated viscosity could contribute to impaired reperfusion following stroke. This had led to interest in lowering blood viscosity as a therapeutic measure in an attempt to raise CBF after acute stroke (Wood and Kee 1985).

Therapeutic Reduction of Viscosity

The easiest, quickest and most effective way to lower whole-blood viscosity is by reducing the hematocrit. Fibrinogen and other blood proteins can be lowered by plasmapheresis, and white blood cells by leukophoresis, but except in experimental stroke models these techniques have not been used clinically for stroke therapy. The biology of leukocyte adhesion, in particular, has been intensively studied in recent years (Kochanek and Hallenbeck 1992), although the results of leukocyte depletion or binding inhibitors on outcome after experimental stroke have been variable and model-specific (Clark et al. 1991). Fibrinogen has been lowered therapeutically after acute stroke using snake venom (Ancrod), with a suggestion of benefit in patients achieving the most profound reduction in fibrinogen levels (Sherman 1993).

Another promising technique for lowering viscosity, but one which again has not yet been extensively tested in human stroke patients, is blood substitutes (Peerless et al. 1981). Artificial or modified hemoglobin can be either added to or substituted for red blood cells, and will both reduce viscosity and increase the oxygen-carrying capacity of blood (Cole et al. 1992).

The reduction of hematocrit and consequently viscosity is termed hemodilution. There have been many experimental and clinical studies attempting to improve CBF and clinical outcome after stroke using various hemodilution regimens. All regimens have in common some lowering of hematocrit, and Wood has shown that an approximate 15% lowering of hematocrit is necessary to achieve an increase in CBF (Wood and Kee 1985). Of course, when the hematocrit is lowered there will be a concomitant drop in oxygen-carrying capacity, and there is still some debate as to whether the increase in CBF attendant upon lowering hematocrit is due to the reduction in viscosity or to vasodilation in response to lowered tissue oxygenation (Messmer et al. 1973; Wade 1983; Brown and Marshall 1985; Korosue and Heros 1991). It is most likely that both mechanisms are important, the former when baseline hematocrit is high, i.e. above 45%. However, when hematocrit is reduced below the mid-30% range, it is likely that tissue oxygenation falls. Surprisingly, this critical issue has not been thoroughly studied in ischemic brain. Messmer has shown that oxygen delivery to normal brain tissue is optimal at a hematocrit around 33% (Messmer et al. 1973), but this may not apply to ischemic tissue, where oxygen extraction is greater. Further study using positron emission tomography before and after hemodilution at various hematocrit levels could answer this question, and is urgently needed before further hemodilution trials are carried out on humans. Ideally, one might guide the hemodilution regimen in an individual patient by the continuous monitoring of oxygen extraction, determined by either PET or jugular bulb monitoring of arteriovenous oxygen difference.

There have now been numerous clinical trials of hemodilution for acute ischemic stroke, and they have differed greatly on a number of critical design issues. Consequently, the results have not been uniform. After some initial enthusiasm from preliminary positive results (Strand et al. 1984) a large Scandinavian trial employed an isovolemic technique where, after a small loading dose of dextran, patients were given a phlebotomy of the whole blood alternating with the administration of a comparable volume of dextran in order to keep the circulating blood volume the same (Scandinavian Stroke Study Group 1987). In many cases this technique did not result in a 15% reduction of hematocrit within the critical first 24 hours, and finally patients were entered into the therapeutic trial up to 48 hours after the onset of symptoms. Experimental data have now shown that, at least in animal models, reperfusion must be established within 3–6 hours to reduce infarct size. The Scandinavian trial and an Italian trial (Italian Acute Stroke Study Group 1988) using this study design were decidedly negative, even if subgroups started on therapy within 12 hours and receiving substantial hematocrit reduction were identified (Scandinavian Stroke Study Group 1988).

Critics of this technique, in addition to decrying the delayed entry of patients and the modest reduction in hematocrit, pointed out that many stroke patients are relatively dehydrated, which might be aggravated by an isovolemic regimen. In 9 consecutive stroke patients undergoing Swan–Ganz catheterization, low pulmonary wedge pressures were found in most cases (mean 6.3, range 1–11) (Grotta et al. 1985). Furthermore, early relatively hypervolemic trials had been encouraging (Gottstein 1987; Vorstrup 1989). Consequently, a hypervolemic hemodilution trial was carried out in several US centers, administering an analog of the synthetic starch hetastarch while monitoring pulmonary wedge pressure (PWP) and cardiac output (Hemodilution in Stroke Study Group 1989). Pentastarch was administered without phlebotomy until either a PWP of 15 mmHg was reached or 1500 ml were given. If the hematocrit was not lowered to 33%–35%, then phlebotomy and another dose of starch was given. However, because the emphasis was on achieving hypervolemia, phlebotomy and substantial hematocrit reduction was delayed and sometimes did not occur. Therefore, this was more a trial of volume expansion than of viscosity reduction. Treatment was started within 24 hours of symptom onset. Although improvement was noted in subgroups of patients having increased cardiac output, hematocrit reduction of at least 15% of baseline and treatment within 12 hours, the protocol was cumbersome to carry out, and most importantly there was a large number of deaths from cerebral edema in the treatment group. The study was terminated prematurely, however, and it appeared that randomization resulted in an excessive number of patients with severe stroke being allocated to the treatment group, which possibly explains the higher incidence of cerebral edema. Nevertheless, it became clear that such aggressive hypervolemic therapy, although possibly beneficial in some subgroups, was dangerous in others. More recently, Koller et al. found that a hypervolemic regimen using dextran, which accomplished rapid hematocrit reduction by simultaneous phlebotomy, was associated with improved outcome (Koller et al. 1990).

Most recently, Goslinga and colleagues from Amsterdam, The Netherlands and Aachen, Germany (Goslinga et al. 1992) have studied 'normovolemic' hemodilution where, under Swan–Ganz monitoring, patients received a combination of phlebotomy and an infusion of 20% albumin and crystalloid, thereby rapidly achieving a hematocrit of 32% and PWP of 12, i.e. therapy was tailored to the volume status of the patient, aiming for rapid normovolemic hemodilution. With this 'custom-tailored' protocol, results showed improved outcomes in the patients who at baseline were not volume depleted and/or had a high hematocrit. Hemodilution did not help patients admitted with dehydration, since even the control group received sufficient crystalloid to correct their volume depletion.

At present, universal hemodilution cannot be recommended. However, it is clear that volume depletion in stroke patients is common and should be carefully corrected by crystalloids. In other patients normovolemic hemodilution may be effective, and deserves further study. Caution should be exercised in patients with large strokes at risk for cerebral edema, and in those with cardiac compromise. Volume status should be carefully monitored and the amount of volume expander and phlebotomy adjusted to maintain normovolemia. Hematocrit should be lowered as fast as possible by simultaneous phlebotomy, but probably should not be lowered below 32%, which means that only those patients with hematocrit above 40% should be considered for treatment. The development of blood substitutes and hemoglobins which replace blood without sacrifice (and possibly with augmentation) of oxygen delivery while at the same time lowering viscosity, may result in renewed interest in the broad area of therapeutic viscosity reduction.

There is one clinical condition of cerebral ischemia where hemodilution is widely accepted and commonly used. Vasospasm occurring 3–10 days after aneurysmal subarachnoid hemorrhage results in a relatively chronic state of ischemia which, if untreated, often leads to infarction. This can be ameliorated somewhat by vasodilators such as calcium antagonists, and newer techniques such as free radical scavengers and thrombolytic dissolution of the subarachnoid clot may also help. However, hypervolemic hemodilution is now considered standard therapy in patients suspected of vasospasm, based on the experimental data of Kindt and colleagues, and the clinical observations of Spetzler and colleagues (Pritz et al. 1978; Awad et al. 1987). Because clinical improvement can be seen as a direct result of volume expansion, neurosurgeons have been reluctant to submit this therapy to a randomized trial. The intention of this treatment is to volume expand patients by the administration of a colloid, crystalloid, synthetic starch or whole blood to try to increase cardiac output and the perfusion of regions with borderline CBF. Because the clinical condition is more one of ischemia than infarction, cerebral edema, although a concern, is not as much of a problem as in acute stroke trials. Lowering hematocrit and viscosity occurs as a result of volume expansion but is not the primary objective. If the aneurysm has been clipped, induced hypertension is often added to the regimen.

Pentoxifylline improves the deformability of red cells, thereby improving their ability to transit the capillary bed whose smallest diameter (4–6 μm) is smaller than that of the average RBC (8 μm) (Ott and Lechner 1983). A clinical trial of this drug in acute stroke was negative, but patients were begun on therapy late and no rheological studies were used to guide therapy. More importantly, while contributing to whole-blood viscosity it is not clear how much RBC deformability contributes to impaired microcirculatory perfusion after stroke. Recent experimental evidence implicates white blood cells as a more important determinant of 'no-reflow' through previously unperfused capillaries (del Zoppo et al. 1991). Pentoxifylline may be more useful as a preventive therapy, particularly in those patients with red blood cell disorders such as sickle-cell disease, which reduce deformability, or in patients with vasculopathies such as Moya Moya disease or chronic hypertension, in which there is increased resistance to flow in the microcirculatory bed.

Viscosity as a Risk Factor for Cerebrovascular Disease

Several epidemiological studies have identified increased viscosity as a risk factor for stroke. In the Framingham study, for instance, elevated fibrinogen was a risk factor for cardiovascular disease (Kannel et al. 1987), and was also found in patients at risk for stroke compared to controls (Coull et al. 1991). In patients with carotid atherosclerosis, elevated fibrinogen is an independent risk factor for worsening of the disease. In a cohort of 38 patients with carotid stenosis identified by ultrasound, fibrinogen, low-density lipoprotein and the presence of coexisting cardiac disease were the three variables associated with progression of atherosclerosis on follow-up ultrasound (Grotta et al. 1989). Other investigators have confirmed the association of elevated fibrinogen and carotid atherosclerosis. One theory is that increased fibrinogen, either by a direct effect on the platelet membrane or by increasing serum viscosity, increases the residence time of platelets and perhaps other atherogenic blood factors at the vessel wall.

Pathological Conditions Associated With Increased Viscosity

Sickle-Cell Disease

This condition is the prototype of abnormal red blood cell deformability because of the polymerization of the abnormal hemoglobin S molecules, resulting in the characteristic rigid sickle shape of the cell. This of course makes it impossible for the cells to transit the microcirculation, resulting in vaso-occlusive crises. Strokes are common, and can be prevented by exchange transfusion to reduce the quantity of hemoglobin S (Grotta et al. 1986).

Polycythemia

Polycythemia vera is associated with increased whole-blood viscosity in direct proportion to the elevation in hematocrit. Neurological symptoms are usually non-specific, such as headache, lethargy, difficulty in concentrating and tinnitus. Strokes are actually rare, and are more likely to be the result of platelet abnormalities associated with the primary pathologic process in the bone marrow (Grotta et al. 1986). When strokes occur they are more frequently in the venous circulation than usual, indicating that heightened viscosity contributes to thrombosis in regions of relative stasis. Phlebotomy is a useful therapeutic measure in patients who are symptomatic, and should be carried out prophylactically with hematocrit ›52%.

Secondary polycythemia rarely leads to cerebrovascular symptoms unless associated with other hematological disturbances.

Hyperlipidemia and Dysproteinemias

Marked elevation in triglycerides and immunoglobulins can lead to increased blood viscosity (Somer 1987), but rarely if ever leads to cerebrovascular (arterial or venous) occlusion. Stroke in the setting of dysproteinemia such as Waldenstrom's macroglobulinemia or multiple myeloma would be more likely to be caused by an associated coagulopathy.

In summary, the brain is normally well autoregulated to changes in blood viscosity resulting from most pathological conditions. However, if brain injury has occurred, heightened viscosity and disturbed blood rheology could certainly aggravate the damage by reducing flow to borderline perfused regions. This argues for the correction of pathological viscosity states in the setting of acute cerebrovascular disease.

References

Awad IA et al. (1987) Clinical vasospasm after subarachnoid hemorrhage: response to hypervolemic hemodilution and arterial hypertension. Stroke 18:365–372

Begg TB, Hearns JB (1966) Components in blood viscosity – the relative contribution of haematocrit, plasma fibrinogen and other proteins. Clin Sci 13:87–93

Brown MM, Marshall J (1985) Regulation of cerebral blood flow in response to changes in blood viscosity. Lancet 2:604–609

Chen S (1987) The role of white blood cells in the control of blood rheology. In; Hartmann A, Kuschinsky W (eds) Cerebral ischemia and hemorheology. Springer-Verlag, Berlin, pp 57–68

Clark WM et al. (1991) Reduction of central nervous system ischemic injury in rabbits using leukocyte adhesion antibody treatment. Stroke 22:877–883

Cole DJ et al. (1992) Hemodilution during cerebral ischemia in rats: effects of stroma-free hemoglobin on brain injury. Anesth Analg 74:368

Coull BM et al. (1991) Chronic blood hyperviscosity in subjects with acute stroke, transient ischemic attack, and risk factors for stroke. Stroke 22:162–168

del Zoppo GJ et al. (1991) Polymorphonuclear leukocytes occlude capillaries following middle cerebral artery occlu-

sion and reperfusion in baboons. Stroke 22:1276–1283
Dormandy JA (1970) Clinical significance of blood viscosity. Ann Roy Coll Surg 47:211–228
Fisher M, Meiselman J (1991) Hemorheological factors in cerebral ischemia. Stroke 22:1165–1169
Goslinga H (1990) Blood viscosity and circulatory filling pressure in clinical practice. Dynamo Pers, Amsterdam, p 7
Goslinga H et al. (1992) Custom-tailored hemodilution with albumin and crystalloids in acute ischemic stroke. Stroke 23:181–188.
Gottstein U (1987) Hemodilution in cerebral infarction. In: Hartmann A, Kuschinsky, W (eds) Cerebral ischemia and hemorheology. Springer-Verlag, Berlin, pp 416–422
Grotta JC et al. (1982) Whole blood viscosity parameters and cerebral blood flow. Stroke 13:296–301
Grotta JC et al. (1985) Baseline hemodynamic state and response to hemodilution in patients with acute cerebral ischemia. Stroke 16:790–795
Grotta JC et al. (1986) Red blood cell disorders and stroke. Stroke 17:811–817
Grotta JC et al. (1989) Prediction of carotid stenosis progression by lipid and hematologic measurements. Neurology 39:1325–1331
Hellums JD et al. (1987) Studies on the mechanisms of shear-induced platelet activation. In: Hartmann A, Kuschinsky W (eds) Cerebral ischemia and hemorheology. Springer-Verlag, Berlin, pp 80–89
Hemodilution in Stroke Study Group (1989) Hypervolemic hemodilution treatment of acute stroke. Stroke 20:317–323
Italian Acute Stroke Study Group (1988) Haemodilution in acute stroke: results of the Italian haemodilution trial. Lancet 1:318–321
Kannel WB et al. (1987) The Framingham Study: fibrinogen and risk of cardiovascular disease. JAMA 258:1183–1186
Kochanek PM, Hallenbeck JM (1992) Polymorphonuclear leukocytes and monocytes/macrophages in the pathogenesis of cerebral ischemia and stroke. Stroke 23:1367–1379
Koller M et al. (1990) Adjusted hypervolemic hemodilution in acute ischemic stroke. Stroke 21:1429–1434
Korosue K, Heros RC (1991) Mechanism of increase in cerebral blood flow with hemodilution. 17th International Joint Conference on Stroke and Cerebral Circulation, p 137
Merrill EW et al. (1963) Non-newtonian rheology of human blood – effect of fibrinogen deduced by 'subtraction'. Circ Res 13:48–55
Merrill EW (1969a) Rheology of blood. Phys Rev 49:863–887
Merrill EW (1969b) Yield stress of normal human blood as a function of endogenous fibrinogen. J Appl Physiol 26:1–3
Messmer K et al. (1973) Oxygen transport and tissue oxygenation during hemodilution with dextran. Adv Exp Med Biol 37:669–680
Messmer K (1975) Hemodilution. Surg Clin North Am 55:659–678
Muizelaar JP et al. (1986) Cerebral blood flow is regulated by changes in blood pressure and in blood viscosity alike. Stroke 17:44–48
Ott E, Lechner H (1982) Hemorheologic and hemodynamic aspects of cerebrovascular disease. Pathol Biol 30:611–614
Ott E, Lechner H (1983) Changes of flow properties of the blood in cerebrovascular disease and their medical treatment with pentoxifylline. J Cerebr Blood Flow Metab 3:S530–S531
Peerless SJ et al. (1981) Protective effect of fluosol-DA in acute cerebral ischemia. Stroke 12:558–563
Pritz MB, Giannotta SL, Kindt GW et al. (1978) Treatment of patients with neurological deficits associated with cerebral vasospasm by intravascular volume. Neurosurgery 3:364–368
Scandinavian Stroke Study Group (1987) Multicenter trial of hemodilution in acute ischemic stroke – results in the total patient population. Stroke 18:691–699
Scandinavian Stroke Study Group (1988) Multicenter trial of hemodilution in acute ischemic stroke – results of subgroup analyses. Stroke 19:464–471
Schmid-Schönbein GW (1987) Capillary plugging by granulocytes and the no-reflow phenomenon in the microcirculation. Fed Proc 46:2396–2401
Sherman D (1993) Ancrod for the treatment of acute ischemic stroke. Ann Neurol 34:251
Somer T (1987) Rheology of paraproteinaemias and the plasma hyperviscosity syndrome. Baillière's Clin Haematol 1:695–723
Strand T et al. (1984) A randomized controlled trial of hemodilution therapy in acute ischemic stroke. Stroke 15:980–988
Vorstrup S et al. (1989) Hemodilution increases cerebral blood flow in acute ischemic stroke. Stroke 20:884–889
Wade JPH (1983) Transport of oxygen to the brain in patients with elevated haematocrit values before and after venesection. Brain 106:513–523
Wood JH, Kee DB (1985) Hemorheology of the cerebral circulation in stroke. Stroke 16:765–772

18. Pathophysiology of Raised Intracranial Pressure

A. H. Ropper

Normal intracranial pressure (ICP) reflects the integration of pressures from the cerebral veins and cerebrospinal fluid. Once a mass such as a cerebral hemorrhage is added to the intracranial compartment several processes occur simultaneously, but at different rates, to accommodate the increased volume. Pressure first rises in the region around the mass because of the restrictions of the semiclosed skull and dural folds. Subsequently, the raised pressure is distributed throughout the cranium, including the posterior fossa, and to the spinal axis. Generalized raised intracranial pressure itself causes few clinical changes except for headache, vomiting and papilledema, but tissue shifts at a distance from the mass produce the dramatic signs traditionally associated with raised ICP. These signs and ICP are therefore parallel barometers of the way in which the intracranial contents compensate for a mass. This is amply demonstrated by pseudotumor cerebri, in which there is no compartmentalization of pressures, no secondary compression of the upper midbrain, and therefore none of the signs associated with a mass in the cranium, and similarly by the absence of symptoms when intracranial pressure is experimentally elevated to 50 mmHg by infusion of saline into the spinal subarachnoid space (Schumacher and Wolff 1941).

In the presence of a mass, ICP reflects the accommodations that the brain through tissue shifts, the vasculature through reduced blood flow, and the spinal fluid system through the shifting of fluid, all make to compensate for the addition of volume to the restricted cranium. Each of these can also participate in a complex decompensation of intracranial dynamics which, if unchecked, ultimately results in brain death. The system is complex because features of systemic physiology, particularly blood pressure, have a major effect on cerebral blood flow in damaged regions of brain, and because of the exponential compliance relationship between volume and pressure in the intracranial compartment (Fig. 18.1). There is a paucity of material from which to understand the pathophysiology of ICP in patients, and most clinical assumptions have therefore been inferred from animal work.

Intracranial pressure can be measured relatively safely (Barnett 1993). The most recent innovation has been a fiberoptic device that is inserted into the brain parenchyma and gives continuous waveforms and mean pressures. Although ICP measurements greatly assist management, there has been no clear demonstration that guiding therapy by these measurements improves outcome (Ropper 1992, 1993).

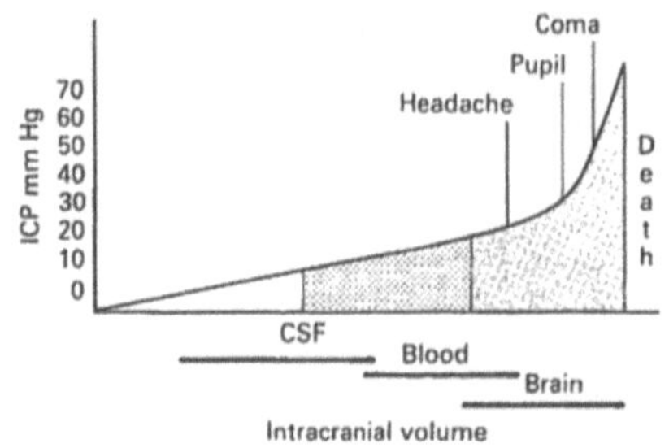

Fig. 18.1. The relationship between volume and pressure in the intracranial compartment.

The Incompressible Brain and Tissue Shifts

Acutely raised ICP causes headache, vomiting and visual obscurations in awake patients. The headache is presumably due to the distension of cerebral vessels, vomiting to the distortion of the dorsal medullary tegmentum through transmitted fourth ventricular pressure, and visual obscurations possibly to ischemia in the posterior circulation. Hydrocephalus, a special case of increased intracranial pressure, occurs initially without tissue shifts but eventually acts as a supratentorial mass and results in the displacement of tissue towards the infratentorial compartment, with a syndrome of decreased consciousness, increased leg tone, miosis and upgaze paresis.

The clinical signs of herniation, coma and pupillary enlargement are 'false localizing signs' due to the distortion of brain or vascular structures at a distance from a mass. The clinical herniation syndromes, as codified in McNealy and Plum's (1962) classic paper, suggest that the pathologic phenomenon of herniation of tissue through the tentorial opening causes these clinical changes, but many signs, including diminished level of consciousness, appear before herniation, and others, such as pupillary enlargement, may have several causes, including horizontal displacement of the upper midbrain (Ropper et al. 1991).

Cerebrospinal Fluid

The 100–150 ml of CSF (slightly less in children) in the ventricles, cisterns and spinal subarachnoid space are in a dynamic equilibrium because of matched production and absorption. The choroid plexus produces the bulk of CSF at a rate of 0.3 ml/min or 20 ml/h. Raised ICP causes a variable – but usually slight – reduction in CSF production. Absorption of CSF is less well understood, but most seems to occur via arachnoid granulations and then through pressure-sensitive valves to the large cerebral veins. There is also transependymal movement of CSF from the ventricles to the convexity surface, and possibly through direct venous absorption. Reabsorption increases slightly with raised ICP, and many diseases that primarily expand the CSF space, particularly subarachnoid hemorrhage and meningitis, have the potential to impede reabsorption and increase the resistance to outflow.

Secondary compression of the CSF spaces by a mass in one cerebral hemisphere produces several radiologic signs that appear serially: compression of the convexity, subarachnoid space and ventricle adjacent to the mass; enlargement of the perimesencephalic cistern on the side of the mass, accompanied by obliteration of the cistern on the other side; obliteration of both cisterns; and finally, enlargement of the ventricle opposite the mass. Compression of the upper brain-stem cisterns is a predictor of both pupillary dilation and poor clinical outcome in head injury (Toutant et al. 1984). As CSF is displaced from these regions to the spinal CSF space, a potential source of compensation for the mass is lost. Several early workers with an interest in herniation suggested that obstruction of CSF pathways at the base of the brain, or occlusion of the cerebral aqueduct, are important elements in the further decompensation of ICP because they impede the further egress of CSF from the cranium.

Subarachnoid hemorrhage, meningitis and choroid plexus papillomas producing excess CSF are the only truly primary diseases of the CSF pathways resulting in raised ICP, and all typically produce hydrocephalus. In most acute circumstances the ventricles communicate with each other and with the spinal space, and drainage with a shunt or lumbar puncture is rapidly effective in reversing the signs of stupor, small pupils and posturing. Pharmacologic means of reducing CSF pressure by slowing production (acetazolamide, furosemide, corticosteroids) are only temporarily useful.

The Vascular Compartment and Cerebral Blood Flow (CBF)

The vascular compartment is the most important in clinical ICP management because it is ultimately compromised as pressure rises, and because it causes rapid changes in ICP. The vasculature drives CSF pressure under normal conditions and is probably the main determinant of the pressure–volume relationship (compliance) of the intracranial compartment (Fig. 18.1). Under most circumstances, an increase in CBF to above normal levels is deleterious to ICP because there is a corresponding increase in cerebral blood volume (CBV). In pathological circumstances such as head trauma, the changes in the cerebral vasculature account for two-thirds of the elevation in ICP, and the CSF space one-third (Marmarou et al. 1987). One of the main effects of mannitol in the treatment of raised ICP is mild cerebral vasoconstriction, which improves compliance. This effect occurs before the diuresis and net water reduction in brain that reduces ICP (Miller et al. 1971).

CBF is normally 75 ml/100 g/min in gray matter and half that in white matter, together representing 15% of the total cardiac output for the brain. In children an even greater proportion of cardiac output is committed to the brain. Blood flow below approximately 18 ml/100 g/min causes neuronal damage, and below 12 ml/100 g/min neuronal death, depending also on the duration of ischemia. These thresholds are reflected in progressive slowing of EEG rhythms and loss of cortical somatosensory evoked potential amplitudes (Astrup et al. 1981). CBF is normally regulated to match the metabolic needs of adjacent cortical regions, probably through local hydrogen ion concentration (Lassen and Christensen 1976). There is also a pressure-dependent autoregulatory mechanism which maintains a constant flow through a wide range of mean blood pressures between approximately 50 and 150 mmHg, in part mediated by vascular myogenic tone. Chronic hyper- or hypotension, and sympathetic tone in particular, alter these limits. Autoregulation is not instantaneous and lags behind changes in blood pressure by up to several minutes. Perfusion is maintained more effectively at elevated ICP than at equivalently lower blood pressure (Miller et al. 1971), and prolonged hypotension has been shown to impair autoregulation long after blood pressure has been restored (Bleyart et al. 1980).

CBF is very responsive to ambient CO_2, rising 4% for each mmHg drop in P_aCO_2, therefore almost doubling CBF when arterial P_aCO_2 is raised from 40 mmHg to 60 mmHg (Harper and Glass 1965). CBF and CBV are raised to a lesser extent by hypoxia and hyperthermia, but in tenuous circumstances minor alterations may cause large rises in ICP. Cerebral vasoconstriction to hypocapnia is probably mediated by extracellular and smooth muscle hydrogen ion concentration. Hyperventilation therefore provides a rapidly effective way to reduce ICP, but its effect lasts only as long as the CSF alkalosis it causes – approximately several hours. Excessive hyperventilation is not thought to produce ischemia (Harp and Wolman 1973). One study has found that CBF did not reach ischemic levels during hyperventilation (Muizelaar et al. 1991), although another found slightly increased CSF lactate with extreme hyperventilation (Plum and Posner 1967). In a randomized study, a small subgroup of head-injured patients had poorer short-term outcomes if they were hyperventilated (Muizelaar et al. 1991). One reason proposed for this difference was the higher and more variable ICP attained in hyperventilated patients. Most inhalation anesthetics uncouple regional CBF from local metabolism, leaving CBF and CBV at higher levels than necessary to support local metabolism, and usually raising ICP.

Plateau waves are spontaneous or induced elevations in ICP that are sustained for more than several minutes and may result in clinical decompensation. They are the result of spontaneous cerebral vasodilation and increased CBV. Many waves seem to follow patient manipulation, and it has been thought that hypertension induces some of these decompensations. Experimental observations, however, suggest that plateau waves may be incited by mild, often unnoticed, hypotension that causes an autoregulatory vasodilation, followed by recovery of blood pressure before the vessels constrict again (Rosner and Becker 1984). Abrupt elevation of blood pressure can abort some plateau waves in animals.

The relationship between hypertension, ICP and edema is therefore complex. Excessive blood pressure is known to exaggerate edema in damaged regions of brain. Hypotension, however, risks cerebral ischemia when ICP is elevated. The ideal is to maintain perfusion pressure at approximately normal levels in damaged regions of brain, recognizing that CBF fluctuates passively with blood pressure in those areas. By measuring ICP, it may be possible to estimate the worst perfusion pressure in the cranium and adjust the blood pressure. In clinical practice, when ICP is elevated the risk of hypotension is probably greater than the risk of hypertension, and it is best to treat all hypotension but only extreme hypertension. This is particularly true because most acute hypertension is due to a sympathoadrenal discharge that subsides over days after a severe brain injury (Robertson et al. 1983).

The Cushing Response

The Cushing response of hypertension, bradycardia and slow irregular breathing (Cushing 1901) is a traditional sign of rapidly elevated ICP and has an interesting physiology. Fragments of the syndrome are more common than the complete triad. The hypertension caused by the Cushing response has the same deleterious secondary effects on CBF and brain edema as does hypertension from other sources. The teleologic explanation usually offered for the hypertension is that the response maintains CBF when ICP is very high, but there is no evidence for a protective role. Acute distortion of the lower brain stem appears to be necessary and sufficient to elicit the response (Hoff and Reis 1970). Direct pressure applied to a restricted area of the paramedian caudal medulla in animals produces all the features. In practice, supratentorial masses, typically intracerebral hemorrhage, cause the response as a secondary feature of brain-stem compression and raised ICP in the posterior fossa. Pontomesencephalic section greatly exaggerates the response and vasomotor areas in the upper spinal cord explain the hypertension elicited by raising intraspinal pressure (Pasztor and Pastzor 1980). In contrast, compression of a nearby ventral medullary vasodepressor region causes profound hypotension. Many patients with subacutely evolving posterior fossa masses, such as cerebellar infarction, develop bradycardia without much hypertension. Paroxysmal hypertension as an isolated feature of the Cushing response is sometimes difficult to distinguish from the more frequent sympathoadrenal discharge of acute head injury (Robertson et al. 1983).

References

Astrup J, Siejso BK, Symon L (1981) Thresholds in cerebral ischemia – the ischemic penumbra. Stroke 12:723–725

Barnett GH (1993) Insertion and care of intracranial pressure monitoring devices. In: Ropper AH (ed) Neurological and neurosurgical intensive care, 3rd edn. Aspen, Rockville MD, pp 53–68

Bleyart AL, Sands PA, Safar P et al. (1980) Augmentation of postischemic brain damage by intermittent hypotension. Crit Care Med 8:41–47

Cushing H (1901) Concerning a definite regulatory mechanism of the vaso-motor centre which controls blood pressure during cerebral compression. Johns Hopkins Hosp Bull 12:290–292

Harp JR, Wolman H (1973) Cerebral metabolic effects of hyperventilation and deliberate hypotension. Br J Anaes 45:256–261

Harper AM, Glass HI (1965) Effect of alterations in arterial carbon dioxide tension on the blood flow through the cerebral cortex at normal and low arterial blood pressure. J Neurol Neurosurg Psych 28:449–462

Hoff JT, Reis DJ (1970) Localization of regions mediating the Cushing response in CNS of cat. Arch Neurol 23:228–240

Lassen NA, Christensen MS (1976) Physiology of cerebral blood flow. Br J Anaes 48:719–734

McNealy DE, Plum F (1962) Brainstem dysfunction with supratentorial mass lesions. Arch Neurol 7:10–32

Marmarou A, Maset AL, Ward JL et al. (1987) Contributions of CSF and vascular factors to elevation of ICP in severely head injured patients. J Neurosurg 66:883–890

Miller JD, Stanek A, Langfitt TW (1971) Concepts of cerebral perfusion pressure and vascular compression during intracranial hypertension. Prog Brain Res 35:411–419

Muizelaar JP, Marmarou A, Ward JD et al. (1991) Adverse effects of prolonged hyperventilation in patients with severe head injury. J Neurosurg 75:731–739

Pasztor A, Pasztor E (1980) Spinal vasomotor reflex and Cushing response. Acta Neurochir 52:85–97

Plum F, Posner JB (1967) Blood flow and cerebrospinal fluid lactate during hyperventilation. Am J Physiol 212:869–871

Robertson CS, Clifton GL, Taylor AA, Grossman GG (1983) Treatment of hypertension associated with head injury. J Neurosurg 59:455–460

Ropper AH, Cole D, Louis D (1991) Clinicopathologic correlation in a case of pupillary dilatation from cerebral hemorrhage. Arch Neurol 48:1166–1169

Ropper AH (1992) Acute increased intracranial pressure. In: Asbury AK, McKhann GM, McDonald WI (eds) Disease of the nervous system. WB Saunders, Philadelphia, 2nd edn, pp 978–988

Ropper AH (1993) Treatment of intracranial hypertension. In: Ropper AH (ed) Neurological and neurosurgical intensive care, 3rd edn. Aspen, Rockville MD, pp 29–52

Rosner MJ, Becker DP (1984) Origin and evolution of plateau waves. J Neurosurg 60:312–324

Schumacher GA, Wolff HG (1941) Experimental studies on headache. Arch Neurol Psych 45:199–214

Toutant SM, Klauber MR, Marshall LF et al. (1984) Absent or compressed cisterns on first CT scan: ominous predictors of outcome in severe head injury. J Neurosurg 61:691–694

19. Central Neurogenic Regulation of Regional Cerebral Blood Flow (rCBF) and Relationship to Neuroprotection

D. J. Reis and E. V. Golanov

Traditionally, rCBF has been seen as being controlled by an interplay between cerebrovascular autoregulation, metabolic signals generated as products of neuronal activity but also by inspired CO_2 or O_2, and by peripheral nerves with cell bodies outside the CNS but with processes innervating cerebral arteries (Heistad and Kontos 1983). However, there is increasing evidence that neurons contained entirely within the CNS may modify rCBF. This central neurogenic regulation (Reis and Iadecola 1989) may be focal or diffuse. It may, as a primary central neurogenic regulation (Reis and Iadecola 1989), occur independently from changes in metabolic activity as measured by the usual indices of cerebral glucose utilization (rCGU) or O_2 consumption. Also, as metabolically associated or secondary central neurogenic regulation, it may be associated with appropriate proportional and temporally and topographically related changes in cerebral metabolism, although the association need not necessarily result from diffusion of metabolites but could represent the parallel coactivation of blood flow by transmitter-mediated release and local metabolism.

The possibility that rCBF might be modified in response to activity in the CNS was recognized by early workers in the field (for review see Purves 1972; Reis 1979; Seylaz et al. 1986; Raichle 1987) who noted that rCBF could be increased by sensory stimuli or (e.g. Cobb and Talbott 1927) electrical stimulation of the brain. Many of the older studies, however, are in retrospect difficult to evaluate, since other experimental variables influencing rCBF, in particular arterial pressure (AP) and blood gases, were not always controlled. Such studies, nevertheless, raised the possibility (not always accepted, e.g. Schmidt (1945)) that the CNS might, through intrinsic networks, regulate its own blood flow.

The more recent era of inquiry began with the studies of Stavraky (1936), who demonstrated in cats that electrical stimulation of anatomically undefined areas of rostral brain stem would modify pial vascular diameter independently of changes in AP and sympathetic nerves. Later, Ingvar (1958) and Söderberg (Ingvar and Söderberg 1958), demonstrated that electrical stimulation of the mesencephalic reticular formation in cats, lightly anesthetized or awake with spinal cord transection at the level of the cervicomedullary junction (encéphale isolé), resulted in substantial increases in CBF in the cerebral cortex independent of AP and ventilation. The changes were usually, but not invariably associated with desynchronization (i.e. activation) of the EEG, leading to the proposal that the centrogenic vasodilation was coupled to arousal (Ingvar and Söderberg 1958).

Studies by Molnar and Szanto (1964) and others in cats and rabbits demonstrated that stimulation of sites within the bulbar reticular formation could both increase and decrease CBF in cortical and subcortical regions without changes in EEG or systemic AP, an effect not mediated by sympathetic nerves. Others showed that rCBF was elevated by brain-stem stimuli independent of EEG, the sympathetic innervation and not mediated by adrenergic or cholinergic mechanisms (Langfitt and Kassell 1968; Meyer et al. 1969, 1971; Lang and Zimmer 1974).

A novel concept has also recently emerged from studies on the central neural regulation of rCBF, namely, that the brain has the capacity, by excitation of the same pathways, to reduce the size of focal cerebral infarctions (Reis et al. 1989; Berger et al. 1990), a phenomenon of central neurogenic neuroprotection. The evidence, moreover, suggests that the neural substrates for flow regulation and protection from ischemic infarctions may, in part, be distinct.

Contemporary studies in which physiological variables have been carefully controlled and in which the anatomical substrates have been documented have clearly established the fact that the brain can potently, and with substantial finesse, regulate its own blood flow and/or metabolism. The distribution of the control is anatomically discrete and is represented at all levels of the neuraxis. The era is relatively new and understanding only partial, yet the case for central neurogenic regulation is strong. Here we consider it with reference to its regional representation.

Levels of Organization of Intrinsic Control of CBF

Spinal Cord

It is evident that the spinal cord must participate in the regulation of rCBF and mediate changes detected in the cerebral cortex and elsewhere in response to somatosensory stimuli (see below and Raichle 1987). Hosobuchi (1985, 1991) observed in man that low-intensity electrical stimulation of the cervical, but not thoracic, segments of the spinal cord increased CBF in the contralateral cerebral cortex. It is not known whether the increase is associated with a parallel increase in metabolism, but is reminiscent of older animal studies demonstrating that stimulation of peripheral nerves increased flow in spinal cord (Field et al. 1951).

Cerebellum

Fastigial Pressor Response

Miura and Reis (1969) demonstrated that electrical stimulation of the rostral ventromedial pole of the cerebellar fastigial nucleus (FN) elicited a potent elevation of AP and heart rate (HR): the fastigial pressor response (FPR) (Doba and Reis 1972). The FPR is associated with widespread changes in regional organ blood flow, simulating the cardiovascular adjustments to upright posture, i.e. the orthostatic reflex (Doba and Reis 1972). Associated are hormonal changes including the release of pituitary vasopressin (Del Bo et al. 1983b), adrenomedullary catecholamines (Del Bo et al. 1983a) and, in unanesthetized animals, a repertoire of motivated behaviors (Reis et al. 1973).

Primary central neurogenic vasodilation from the FN Electrical stimulation of the FN also elevates rCBF. The elevation has been observed in anesthetized rats, rabbits, and cats and anesthetized and unanesthetized monkeys, and measured by a range of methodologies (Doba and Reis 1972; McKee et al. 1976; Reis et al. 1982; Mraovitch et al. 1986b; Goadsby and Lambert 1989). The increase in rCBF elicited in anesthetized rats from the FN by a brief stimulus train and measured by laser Doppler flowmetry (Fig. 19.1) appears after a few seconds and remains elevated for minutes after the stimulus is terminated, suggesting persistence of vasodilators and/or continuing neuronal activity (Iadecola and Reis 1990). The increases in rCBF are widespread, including all segments of the brain and spinal cord with the greatest elevations – up to 250% of control – within the frontal cortex (Nakai et al. 1983).

The fact that the elevations in rCBF elicited by FN stimulation persist following bilateral cervical sympathectomy, spinal cord transection, adrenalectomy or transection of the greater superficial petrosal nerve (Nakai et al. 1982; Reis et al. 1982; Iadecola et al. 1986) but are abolished in the cerebral cortex rostral to a unilateral electrolytic lesion of the basal forebrain (Iadecola et al. 1983a) indicates that the response is mediated by activation of the neural pathways and not by excitation of the extrinsic nerves innervating the cerebral vessels, nor by substances released into the CSF.

The vasodilation elicited from the FN is independent from changes in cerebral glucose utilization (rCGU) in most areas when rCBF or rCGU

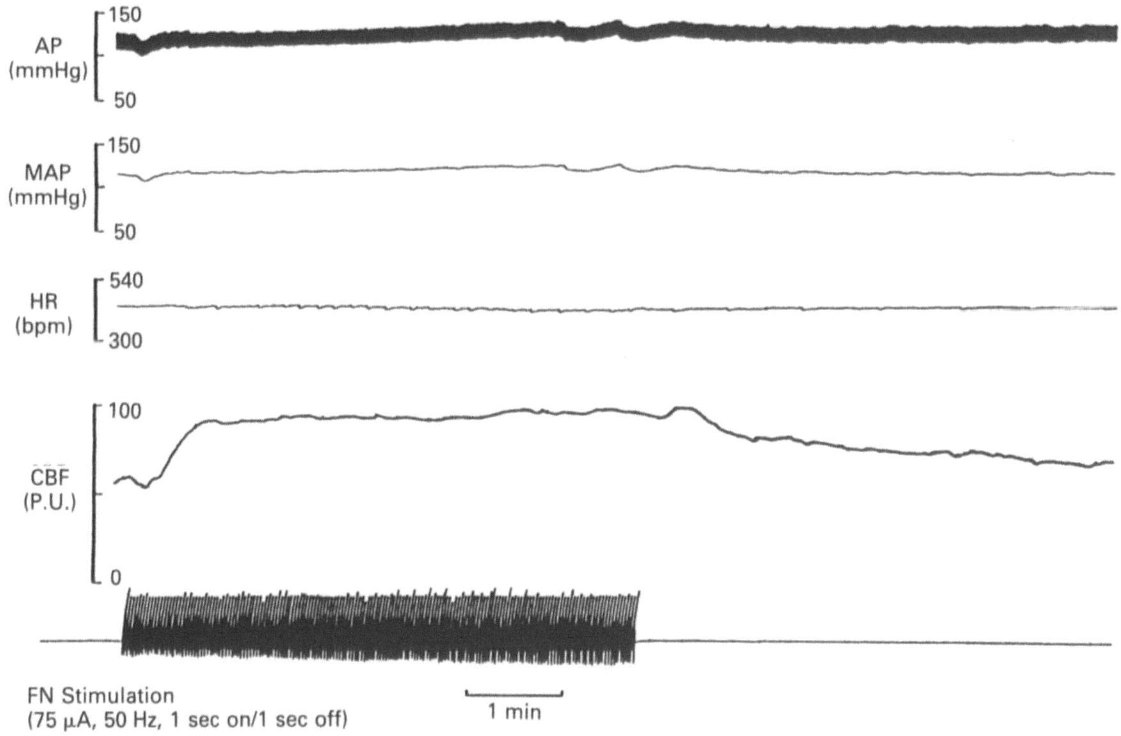

Fig. 19.1. Effect of stimulation of the cerebellar fastigial nucleus (FN) on cortical cerebral blood flow (CBF) as measured by laser Doppler flowmetry in spinalized rat with arterial pressure maintained by i.v. infusion of phenylephrine. CBF, expressed in arbitrary units (perfusion units, PU), was continuously monitored over the parietal cortex by laser Doppler flowmetry. FN stimulation produced a large and sustained increase in CBF that was independent of AP. Note that the increase reached plateau after about 1 min of stimulation (75 µA, 50 Hz, 1s on/1 s off) and that it persisted after termination of the stimulus. (From Iadecola and Reis 1990, with permission of Journal of Cerebral Blood Flow and Metabolism)

are measured autoradiographically (Nakai et al. 1983) (Fig. 19.2). In unanesthetized monkeys, the evoked elevation in rCBF is not associated with increased oxygen utilization (McKee et al. 1976). Indeed, the elevations of rCBF from the FN were the first evidence for experimentally induced primary central neurogenic vasodilation.

The pharmacology of the response is, however, poorly understood. Local release of acetylcholine in the target appears to play some role, since the cerebrovascular vasodilation elicited from the FN is abolished by systemic and attenuated by topical atropine (Arneric et al. 1986, 1987; Iadecola et al. 1986). Whether the cortical acetylcholine is synthesized in afferent fibers from the basal forebrain, local neurons or capillary endothelium, all of which are sources of acetylcholine biosynthesis, is not clear. The neurotransmitters in the ascending pathways are not known.

Mechanism in target The mechanism within the target by which a neural signal is transduced into a vascular event is not known, but must result from the relaxation of vascular smooth muscle. Hemodynamically, the elevation in rCBF could reflect capillary recruitment, an increase in flow velocity or a combination of both, but appropriate studies to distinguish between these possibilities have yet to be undertaken. The fact that central neurons do not project into the subarachnoid space (Virchow–Robin space), whereas arterioles lose their muscular sheath before penetrating the brain, raises the intriguing question of how vascular smooth muscle is signaled to dilate. Conceivably, diffusible agents might be activated by neural signals to activate extraparenchymal arterioles. Alternatively, pericytes might act to open up capillary beds to increase flow. The contribution of nitric oxide (NO) generated in response to the signal has been proposed as one mediator (Iadecola 1992), although the fact that NO is extremely short-lived suggests that the prolongation of the response results from the production of stable nitrosylated products and/or continuous neuronal activity.

Cerebrovascular vasodilation within the cerebral cortex, however, appears to depend upon

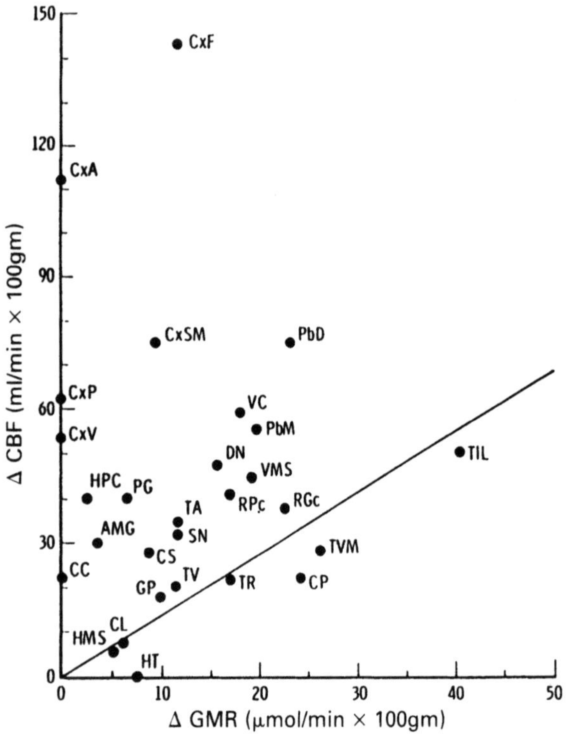

Fig. 19.2. Relation between changes in rCBF (ΔCBF) and changes in regional cerebral glucose utilization (ΔGMR) in different brain areas of the rat. Straight line represents the relation between rCBF and rΔGMR observed in unstimulated rats. Note that, in most brain regions, changes in CBF are disproportional to changes in GMR. In the forebrain elevated flow is unaccompanied by changes in GMR, representing a primary vasodilation so that most of the points lie away from the proportionality line. AMG, amygdala; CC, corpus callosum; CL, inferior colliculus; CP, caudate–putamen; CS, superior colliculus; CxA, auditory cortex; CxF, frontal cortex; CxP, parietal cortex; CxSM, sensory motor cortex; CxV, visual cortex; DN, cerebellar dentate nucleus; GP, globus pallidus; HMS, cerebellar hemisphere; HPC, hippocampus; HT, hypothalamus; PbD, dorsal parabrachial nucleus; PbM, medial parabrachial nucleus; PG, pontine gray matter; RGc, gigantocellular reticular nucleus; RPc, parvocellular reticular nucleus; SN, substantia nigra; TA, thalamus anterior nucleus; TIL, thalamic intralaminar nuclei (centromedial and paracentral); TR, thalamus reticular nucleus; TV, thalamus ventral nucleus; TVM, thalamus ventromedial nucleus; VC, vestibular complex; VMS, cerebellar vermis. From Reis et al. 1985, with permission of Raven Press.

the activity of local neurons, since their destruction by excitotoxins abolishes the vasodilation from the FN (Iadecola et al. 1987a). Moreover, the excitation of neurons whose activity generates spontaneous cerebrovascular waves (see below) also seems necessary for the appearance of the vasodilation from the FN. That K^+ is the active signal has been excluded (Iadecola and Kraig 1991). However, the role of non-neuronal elements, transmitters, receptors and signal transduction pathways subserving the vascular responses is as yet unexplored.

Pathways The pathway by which stimulation of the FN elicits a primary vasodilation is only partly understood. The fact that the elevation of rCBF persists after destruction of fastigial neurons by excitotoxins and/or surgical ablation of the vermis, but is abolished by lesions of the rostral ventrolateral reticular nucleus (RVL), has suggested that the cerebrovascular response is the result of antidromic stimulation in the FN of collateralized brain-stem neurons with branches projecting to the RVL and an RVL projection which mediates the vasodilation (Chida et al. 1986, 1989a, b, 1990a, b). Since the RVL does not innervate the cerebral cortex directly (Ruggiero and Reis 1988), the pathway for excitation must engage pathways which innervate the cerebral cortex.

Fastigial Depressor Response

In contrast to the effects of electrical stimulation, chemical stimulation of the FN with excitatory amino acids elicits a fall in AP and HR (Chida et al. 1986, 1990a, b): the fastigial depressor response (FDR). Associated is a widespread reduction in rCBF coupled to a proportional reduction in metabolism (Chida et al. 1989b) (Fig. 19.3). The sympathetic and cerebrovascular responses are both abolished by lesions of the RVL (Chida et al. 1990a). Since the FDR is a consequence of excitation of intrinsic FN neurons, none of which project to the C1 area (Ruggiero and Reis 1988), the response must be dependent upon a polysynaptic pathway projecting to brain-stem neurons themselves projecting into the RVL. The importance of this finding is that it demonstrates that rCBF (and rCGU) are normally not at basal levels but, like AP, maintained at midrange; also that the brain has networks capable of widely and actively suppressing its own metabolism. The relationship of such systems to behavior is unknown.

Medulla Oblongata

Three areas within the medulla oblongata have been demonstrated to have an action upon CBF in experimental animals: the nucleus of the tractus solitarii (NTS); an area of the RVL containing sympathoexcitatory reticulospinal neurons of the C1 group; and the dorsal medullary reticular formation (DMRF).

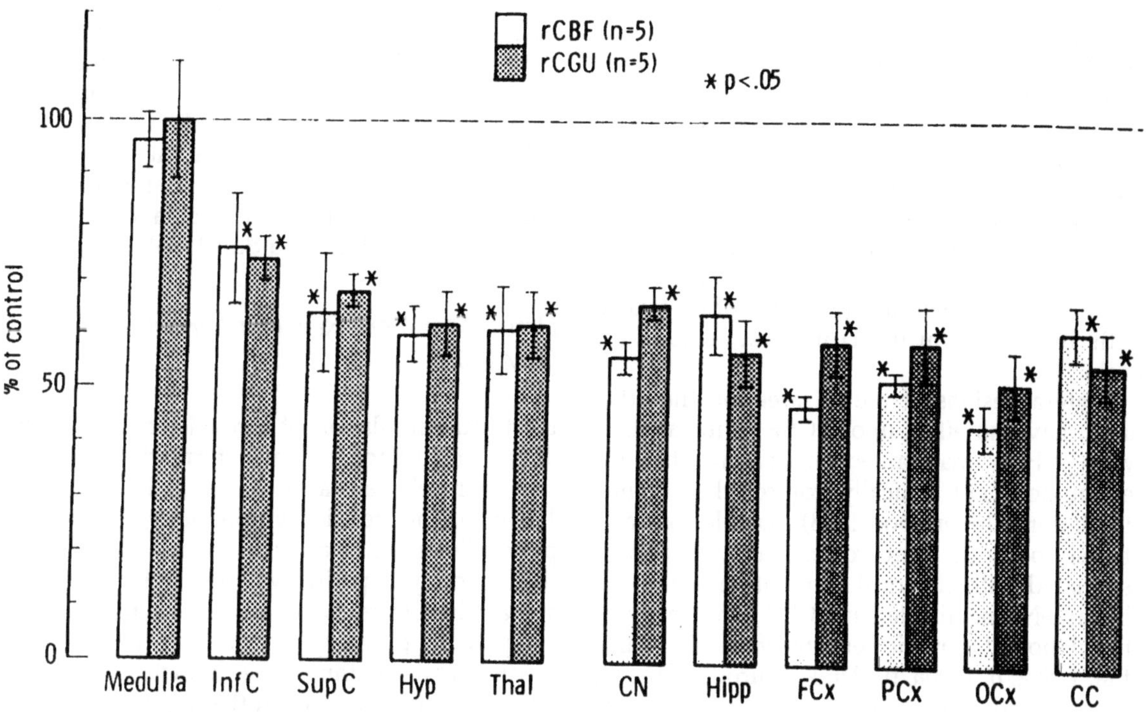

Fig. 19.3. Excitation of neurons of FN by kainic acid reduces rCBF and rCGU in all brain regions except the medulla (as a percentage of vehicle-injected control). Reductions in the cerebral cortex are greatest. Note that reductions in rCBF are proportional to rCGU. InfC, inferior colliculus; SupC, superior colliculus; Hyp, hypothalamus; Thal, thalamus; CN, caudate–putamen; Hipp, hippocampus; FCx, PCx, OCx, frontal, parietal and occipital cortices; CC, corpus callosum. From Chida et al. 1989a, with permission of Brain Research.

NTS

The NTS is the principal site of termination of visceral afferents of the IXth and Xth cranial nerves, including those arising from arterial baro- and chemoreceptors and other cardiopulmonary afferents. As such, it is the principal site of integration of all visceral cardiovascular and pulmonary reflexes (Reis et al. 1988). Although it is well established that rCBF is not regulated by arterial baro- and chemoreceptors (Heistad and Kontos 1983; Ishitsuka et al. 1986), electrical stimulation of the NTS will increase rCBF throughout the rat brain, maximally to 75% of control in the cerebral cortex (Nakai 1985). Since the increase in rCBF occurs after transection of the spinal cord or vagosympathetic trunks, or pharmacological blockade of associated release of vasopressin from the pituitary, and is not elicited by stimulation of associated structures, the effect is anatomically selective and mediated by central pathways, conceivably through the RVL. Whether the increase in rCBF is associated with increases in rCGU is unknown.

That the NTS may also regulate cerebrovascular autoregulation has been demonstrated in animal experiments in which bilateral lesions have been placed in the nucleus. Thus, although such lesions do not affect resting rCBF or rCGU, they do abolish cerebrovascular autoregulation (Ishitsuka et al. 1986), an effect not replicated by combined baro- and chemoreceptor denervation. Thus, cerebrovascular autoregulation may be an independently regulated parameter, as suggested by recent pharmacologic studies (Sandor et al. 1986, 1987, 1988).

C1 Area of the RVL

Systemic circulation The RVL and, in particular, a subdivision containing reticulospinal sympathoexcitatory neurons, the C1 area, plays a major role in controlling the systemic circulation (Reis et al. 1988). These neurons provide a monosynaptic excitatory input to preganglionic sympathetic neurons of the intermediolateral column, thereby maintaining resting (tonic) levels of AP and HR. They are critical in the reflex regulation of sympathetic neurons as part of arterial baro- and chemoreceptor, somatosympathetic and exercise reflexes. They are also directly stimulated by hypoxia (Sun et al. 1992).

rCBF The C1 area of the RVL can also regulate rCBF (Underwood et al. 1992b, 1994; Saeki et al. 1989). Electrical or chemical stimulation of the RVL elevates rCBF two- to three-fold throughout most of the brain when measured regionally by autoradiography. The increase in rCBF occurs within seconds of the stimulus and, as with stimulation of FN, recovers slowly over minutes after termination of the stimulus train. The increase in rCBF persists after spinal cord transection and adrenalectomy, and is not associated with an increase in rCGU in most areas of the brain.

Bilateral lesions of the C1 area, on the other hand, have no effect upon rCBF or metabolism when AP is maintained, indicating that the region does not tonically control the cerebral circulation (Underwood 1988). Such lesions also abolish both the cerebrovascular vasodilation associated with electrical stimulation of the cerebellar FN and the fall in rCBF accompanying stimulation of intrinsic neurons of the FN (Chida et al. 1989a, b, 1990a, b). Thus, neurons of the C1 area also appear to be essential for the expression of the widespread cerebrovascular effects of excitation of the FN, either as a primary relay nucleus or, possibly, by playing a permissive rather than an obligatory role.

Pathways The pathways through which RVL stimulation elicits global cortical vasodilation, particularly within the cerebral cortex, is not known. Since the RVL does not directly project to the cortex (Ruggiero and Reis 1988), it must engage a cortical afferent projection. Promising candidate projections would include diffusely projecting nuclei of the midline and intralaminar–parafascicular complex (for reviews see Jones 1985; Saper 1985) and, possibly, the ventromedial thalamic nucleus innervating lamina I of the cortex (Herkenham 1980). In support of this are the findings that stimulation of the parafascicular thalamic nucleus increases rCBF, but not rCGU, in the cortex in rats (Mraovitch and Seylaz 1987) and that midline thalamic nuclei are innervated by the RVL (Ruggiero et al. 1989). Less likely areas are the lateral hypothalamus (Saper 1985) and the nuclei of the dorsal pontine tegmentum, including the peripedeuncular/lateral dorsal tegmental complex (Ruggiero et al. 1989).

Hypoxic vasodilation Lesions of the RVL globally reduce, by 50%, primary cerebrovascular vasodilation elicited by hypoxia but not hypercarbia (Underwood et al. 1987; Underwood 1988). The finding suggests that the cerebrovascular response to hypoxia is, in part, reflex and initiated by the stimulation of receptors in the brain stem sensitive to hypoxia. This contention is supported by the recent demonstration that sympathoexcitatory neurons of the RVL are excited both in vivo and in vitro by cyanide or hypoxia, and are responsible for the sympathetic excitation elicited by cerebral ischemia or hypoxia (Sun et al. 1992; Reis and Sun 1993).

Dorsal Medullary Reticular Formation (DMRF)

Electrical stimulation of a restricted area of the dorsal medullary reticular formation (DMRF) also diffusely elevates rCBF (Iadecola et al. 1983b, c). The area lies within the confines of the parvicellular reticular nucleus, and also within the trajectory of pathways traversed by reticulospinal fibers from the C1 area projecting both to sympathetic preganglionic neurons in the intermediolateral column of the thoracic spinal cord (IML) and to structures in the upper brain stem (Ruggiero et al. 1985). It also resides within the central tegmental tract of the brain stem, a major thoroughfare for aminergic fibers. The increase in rCBF elicited from the DMRF is tightly coupled to changes in rCGU, in turn resulting from the release of epinephrine from the adrenal medulla (Iadecola et al. 1987b; Lacombe et al. 1990) since bilateral adrenalectomy, adrenal demedullation or systemic administration of β-adrenergic antagonists will significantly reduce the increase in rCBF in some brain areas and abolish it in others, thereby revealing adrenal-dependent and adrenal-independent components of the response, the former mediated via β-adrenergic receptors.

The mechanism by which adrenal catecholamines increase cerebral metabolism and blood flow is unclear. Normally, adrenal catecholamines do not penetrate the blood–brain barrier and, when elevated in blood, have little effect upon rCBF or rCGU (Iadecola et al. 1987b). When introduced directly into brain by the intraventricular route, however, they will markedly increase rCBF (MacKenzie et al. 1976). Yet, in adrenalectomized rats (Iadecola et al. 1987b; Lacombe et al. 1990) elevating circulating epinephrine to levels obtained in intact animals with stimulation of the DMRF has no effect upon rCBF; when epinephrine is comparably elevated and the DMRF stimulated, rCBF is increased to the same extent as seen in intact rats with DMRF stimulation (Fig. 19.4); and DMRF stimulation does not increase the permeability of the blood–brain barrier, at least to the non-selective indicator aminoisobutyric acid.

The findings suggest that, in some manner, the stimulation of neurons in or passing through the DMRF may 'sensitize' the brain to circulating catecholamines. Concordant evidence for a 'state-dependent' effect of catecholamines on cerebral metabolism and/or rCBF is seen in the adrenal-dependent increases in rCBF elicited in rabbits by immobilization stress (Lang and Zimmer 1974; Pinard et al. 1985). These represent a secondary cerebral vasodilation (Iadecola et al. 1983b).

Pons

Two nuclei within the pons have been identified which may influence cerebral flow through intrinsic neural pathways: the parabrachial nucleus and the locus ceruleus.

Parabrachial Nucleus

The parabrachial nucleus (PBN) is a major autonomic nucleus of the pons, receiving afferent inputs from the NTS and projecting broadly to a variety of cortical and subcortical areas, including the basal forebrain, many involved in visceral and emotional control (Fulwiler and Saper 1984; Cechetto et al. 1985; Vertes 1988). Electrical stimulation of the medial PBN, as with chemical stimulation of the FN, globally decreases rCBF up to 35% of control throughout brain and maximally in the cerebral cortex (Mraovitch et al. 1985). The pathways and mechanisms for this are unknown.

Locus Ceruleus

The nucleus locus ceruleus (LC) provides the principal noradrenergic innervation of the neuraxis through highly collateralized branches of relatively few neurons in the dorsal pons (Foote et al. 1983; Grant et al. 1988). It may function as a complex integrating system in brain, particularly with encoding of the the emotional content of stimuli.

Studies on the effects of LC upon CBF have not always been in agreement. In part, discrepancies reflect the fact that the LC lies in a region of the dorsal pons close to other autonomically active centers (e.g. the parabrachial nucleus) and near

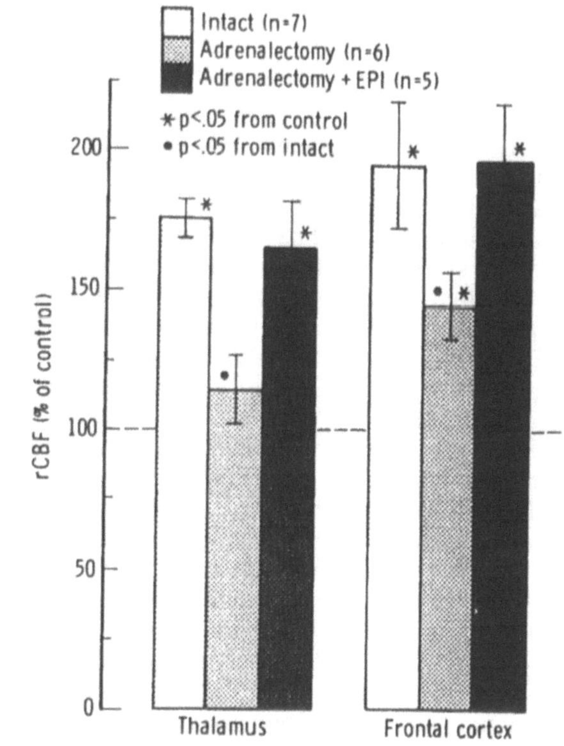

Fig. 19.4. Effect of total adrenalectomy with and without epinephrine (EPI) infusion on increases in rCBF in the thalamus and frontal cortex elicited by DMRF stimulation in the rat. Note that DMRF stimulation increases rCBF, that the increase is abolished by adrenalectomy in the cortex and significantly attenuated in the thalamus, and that with i.v. EPI concurrently infused, the response recovers. See text for details. From Reis and Iadecola, 1991, with permission of Raven Press.

major ascending visceral autonomic bundles in the periventricular gray matter. Such proximity raises issues with respect to the anatomical effects of lesions and/or stimulation of the nucleus. Nevertheless, most studies have suggested that electrical stimulation of the LC in monkeys, rats or cats (Raichle et al. 1975; Bates et al. 1977; de la Torre et al. 1977; Goadsby et al. 1982, 1985a, b; Goadsby and Lance 1988; Goadsby and Duckworth 1989) increases cerebral vascular resistance and reduces rCBF. Lesions of the LC do not affect resting rCBF (Bates et al. 1977; Dahlgren et al. 1981) and fail to influence the response of CBF to changes in blood gases (Dahlgren et al. 1981). Thus, LC does not seem to exert a tonic influence on rCBF nor modulate the cerebrovasodilation elicited by hypercapnia.

Mesencephalon

Dorsal Raphe Nucleus

Within the mesencephalon, the principal area which has been identified as having cerebrovascular control is the raphe complex, a collection of nuclei within the midline which contains neurons that synthesize, store and release serotonin (5-HT), often in association with colocalized substance P. 5-HT fibers innervate both parenchymal and extraparenchymal vessels (Chan-Palay 1977; Reinhard et al. 1979; Edvinsson et al. 1983; Itakura et al. 1985).

The effects of electrical stimulation of the most rostral nucleus in the group, the dorsal raphe nucleus (DRN), which provides the principal serotonergic innervation of the telencephalon, has been studied by several investigators (Goadsby et al. 1985a,b; Bonvento et al. 1989; Cudennec et al. 1989; Underwood et al. 1992a). Goadsby et al. (1985a,b) observed that, in monkeys and cats, stimulation of the DRN increased blood flow in the internal carotid arteries, as measured with electromagnetic flowmeters. The response was not abolished by transection of the sympathetic nerves or the spinal cord, but was attenuated by the depletion of central serotonin stores with parachlorophenylalanine. In unanesthetized freely moving rats, electrical stimulation of the DRN has also been reported to elicit increases in rCBF, measured autoradiographically with ^{14}C-iodoantipyrine (IAP) in the telencephalon; these were greatest in the cortex (Cudennec et al. 1989). Changes in rCGU were also seen but did not invariably correlate with rCBF changes. On the other hand, a recent study in anesthetized rats (Bonvento et al. 1989) claimed that electrical stimulation elicited a decrease in rCBF, largely in limbic forebrain nuclei. These differences may reflect experimental variables.

Thalamus

Centromedian–Parafascicular Complex

Little is known about the effects of thalamic stimulation or lesions on rCBF. The only region of the thalamus which has been systematically studied with contemporary techniques is the parafascicular–centromedian complex (Iadecola et al. 1982; Mraovitch and Seylaz 1987). However, it is likely that primary sensory relay thalamic nuclei must also influence cortical blood flow, in view of the evidence that, in man, somatosensory stimulation elicits both primary and secondary vasodilation in the cerebral cortex (Fox and Raichle 1986; Fox et al. 1988).

Electrical stimulation of the centromedian–parafascicular complex of the midline thalamus in anesthetized paralyzed rats increases rCBF throughout most of the brain (Mraovitch et al. 1986a; Mraovitch and Seylaz 1987). The major increases are within the cortex, where elevations up to 180% of control were seen. In most brain regions the increases in rCBF are not paralleled by alterations in rCGU, indicating that the effect is largely a primary vasodilation. Interestingly, as for the DMRF, the increases in rCBF evoked by stimulation of the midline thalamus are substantially attenuated or abolished by bilateral acute adrenalectomy (Mraovitch et al. 1986a). On the assumption that the adrenal contribution to the elevations in flow is mediated by catecholamines, the finding is of interest as it indicates that the adrenal influence on rCBF is not exerted solely via an action upon metabolism.

Forebrain

Basal Forebrain

The basal forebrain contains a network of cholinergic neurons in a horizontally distributed network which innervates the cerebral cortices (Saper 1984). These provide a major source of the cholinergic innervation of the cortex. It also contains the major extension of the medial forebrain bundles, the conduit for the major extrathalamic projections to the cerebral cortex (Vertes 1988). In the cortex, acetylcholine (ACh) is synthesized not only in axons from the basal forebrain but also in local neurons and, possibly, the capillary endothelium (Eckenstein and Baughman 1984; Arneric et al. 1988). In the cortex, vasoactive intestinal polypeptide (VIP) is also contained in cholinergic ACh fibers in close apposition to blood vessels (Eckenstein and Baughman 1984). In the cortex and amygdala, the contact of cholinergic fibers may be on capillaries without apposition of basal lamina (Armstrong 1986; Milner et al. 1989).

That the basal forebrain may affect cerebral cortical blood flow was first observed by Iadecola et al. (1983a), who noted that a unilateral lesion of the basal forebrain, although not altering cerebrovascular autoregulation nor the reactivity of the vessels to hypercarbia, abolished the primary vasodilation in the ipsilateral cerebral cortex elicited by electrical stimulation of the FN. Interestingly, resting rCBF was decreased by about 20% in the ipsilateral cerebral cortex (Iadecola et al. 1983a), indicating that this region of the basal forebrain may participate in maintaining resting cerebrocortical vascular tone.

Electrical and/or chemical (L-glutamate) stimulation of the basal forebrain will increase cortical rCBF ipsilaterally (Arneric 1989; Biesold et al. 1989; Lacombe et al. 1989; Adachi et al. 1990). The increase in rCBF, up to 250% of control, is not associated with substantial changes in AP and is attenuated by nicotinic (Arneric 1989; Biesold et al. 1989) and also possibly muscarinic (Biesold et al. 1989) cholinergic antagonists. It has been established that increases in blood flow elicited from the basal forebrain are not coupled to metabolism (Kimura et al. 1990). Nor is it certain that the ACh released from the axons of basal forebrain neurons mediates the change in flow. In the cerebral cortex ACh is also synthesized in local neurons, many of which also contain VIP (Eckenstein and Baughman 1984), in endothelial cells of local capillaries (Arneric et al. 1988) and in parasympathetic fibers arising largely from the sphenopalatine ganglion (Seylaz et al. 1988). Whether intrinsic neurons of the basal forebrain or axons projecting through the region are responsible for the vasodilation elicited from the FN is not certain, since the cerebellar response appears to be entirely muscarinic (Iadecola et al. 1986). Since the basal forebrain receives afferents from several regions from which neural modulation of CBF can be elicited, including raphe nuclei, parabrachial nuclei, locus ceruleus and, to a lesser extent, NTS and ventral medulla (Semba et al. 1988; Vertes 1988), it may serve as a common output for the cerebrovascular responses in the cortex from autonomic brain-stem centers.

Cerebral Cortex

Spontaneous Cerebrovascular Waves and EEG

Intrinsic neurons of the cerebral cortex also appear to participate in generating spontaneous waves of CBF in deeply anesthetized rats. These spontaneous cerebrovascular waves (SCW) consist of irregular elevations of rCBF, as measured by laser Doppler flowmetry (LDF), with an irregular frequency of approximately 6/min (Fig. 19.5). The appearance of the waves is invariably preceded, approximately 1.5 s, by a burst of EEG activity with a pattern identical to the burst-suppression or barbiturate-spindle EEG. The burst–SCW complex appears synchronously across the cerebral cortex, is directly dependent upon the depth of anesthesia, and can be triggered by single shocks to the FN or RVL, thereby indicating that it is probably not coupled to metabolism. In view of the fact that the EEG bursts are generated in the thalamus to trigger intrinsic cortical neurons to discharge, we have proposed that the neural elements in the cortex responding to the thalamic waves are also coupled to the cerebral circulation, and may represent a neuronal element in the cortex which is the transducer between ascending inputs and the cerebral vasculature. The observation raises the intriguing question of whether the spindle bursts represent the electrical fingerprint of neuronal activity whose function is primarily to increase blood flow to the brain.

Sensory Inputs and Special Senses

It is evident from studies in man using contemporary imaging methods, and from studies in animals, that stimulation of the somatic sensory pathways (Fox and Raichle 1986; Ngai et al. 1988) as well as visual (Fox et al. 1986) and auditory pathways (LeDoux et al. 1983; Lauter et al. 1985) can elicit changes in rCBF (Raichle 1991). Such stimuli are also assumed to elicit changes in metabolism.

In the somatosensory system innocuous tactile stimulation elicits topographically selective increases in CBF in man (Fox and Raichle 1986) and animals (Ngai et al. 1988). In rats, stimulation of single vibrissae can increase rCBF within the territory of a single cortical barrel field (Greenberg et al. 1979), an effect which is paralleled by a comparably restricted change in rCGU (McCasland and Woolsey 1988). Thus, it has long been assumed that the increase in CBF within the somatotopically corresponding cortical region reflected metabolic activation from increased local neuronal activity. However, there is now evidence in man to suggest that the initial event may be a primary vasodilation uncoupled from metabolism (Fox and Raichle 1986), with the metabolic changes in response to sensory stimulation being a later contribution. Conceivably, the coupling of flow and rCGU recorded autoradiographically within barrel fields in experimental animals may be artefactual, reflecting differences in timing between measurements of rCBF (usually 30–60 seconds) and glucose utilization (30–45 minutes).

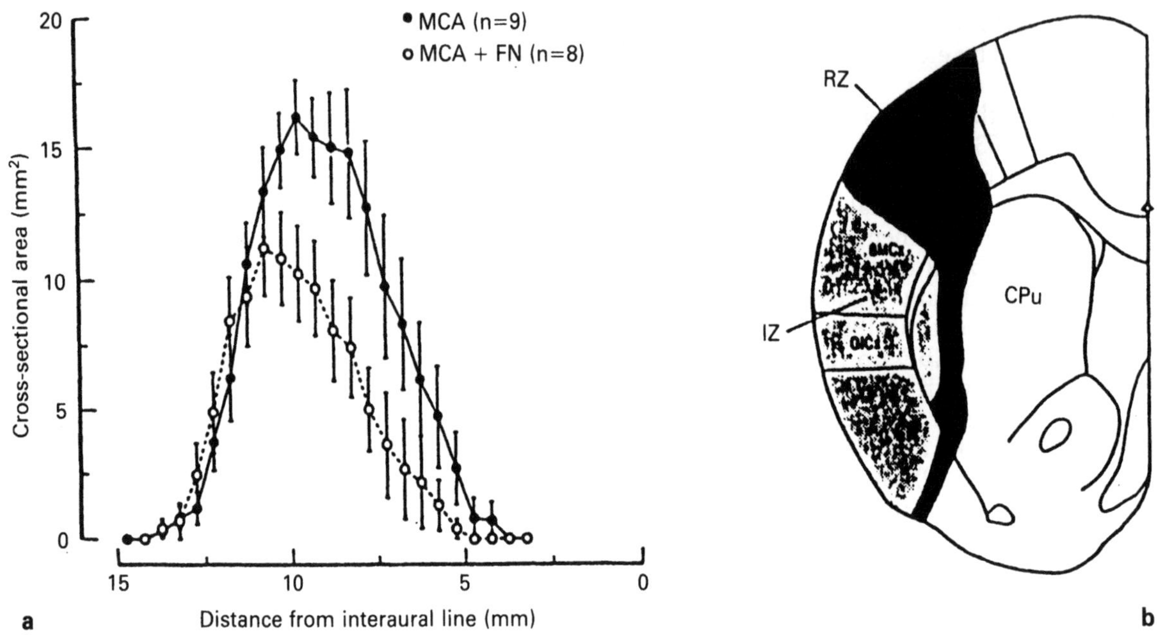

Fig. 19.5. Effect of electrical stimulation of the cerebellar fastigial nucleus (FN) on the distribution of lesions elicited by occlusion of the middle cerebral artery (MCA) in the rat (Sprague–Dawley). **a** area in mm^2 of lesions at various levels of brain rostral from the intra-aural line. Closed circles are controls with volume of 193.7 ± 34.8 mm^3 (SD). Open circles represent animals with FN stimulation. The volume is reduced by almost 64% to a mean of 64.5 mm^3 ($P<0.05$). **b** Cross-section of brain at the area of greatest salvage. The black and shaded areas together represent control lesions (summed from nine animals). The black area represents the zone which was salvaged by FN stimulation, the shaded area that which was not. From Reis et al. 1991, with permission of Journal of Cerebral Blood Flow and Metabolism.

The increase in tissue perfusion in the somatosensory cortex elicited by electrical stimulation of the sciatic nerve is associated with an increase in the diameter of pial vessels on the cortical surface (Ngai et al. 1988). The increase in vascular diameter is localized only to those arterial branches supplying the cortical representation of afferents from the sciatic nerve, and is unaffected by sympathectomy. These observations are particularly intriguing since they raise the question of how restricted neural activity within brain can elicit compensatory and retrograde dilation in the afferent artery to the cortex. Less clearly established is the pattern of activation of the brain in response to painful stimuli. According to Ngai et al. (1988), electrical stimulation of the sciatic nerve with stronger stimuli activating pain fibers increases the magnitude of vasodilation in cortical arterioles over that obtained with weaker stimuli. Studies by Tsubokawa et al. (1981) with H_2 washout suggest that pain may elicit widespread activation of flow, although in that study the possibility that the associated elevations in AP exceeded the autoregulated range was not evaluated.

Relationship to Neuroprotection

The fact that electrical stimulation of the FN could elevate rCBF in the cerebral cortex without associated changes in rCGU raises the prospect that activation of the network might reduce the volume of focal ischemic infarctions (Underwood et al. 1989; Berger et al. 1990; Reis et al. 1991). The rationale was based on observations that in experimental infarctions elicited by occlusion of the middle cerebral artery (MCA) in rats, the area of brain surrounding a maximally hypoperfused core in which rCGU is depressed is surrounded by an area of diminished rCBF, in which metabolism is increased consequent to stimulation of anerobic metabolism (Nedergaard et al. 1986). This zone, the ischemic penumbra (Nedergaard et al. 1986), remains viable for hours after vascular occlusion and can be salvaged by a number of treatments aimed at reducing Ca^{2+} entry, for example by blocking *N*-methyl-D-aspartate (NMDA) receptors or Ca^{2+} channels (Germano et al. 1987; Maiese et al. 1992; Wahlestedt et al. 1993). It was argued that

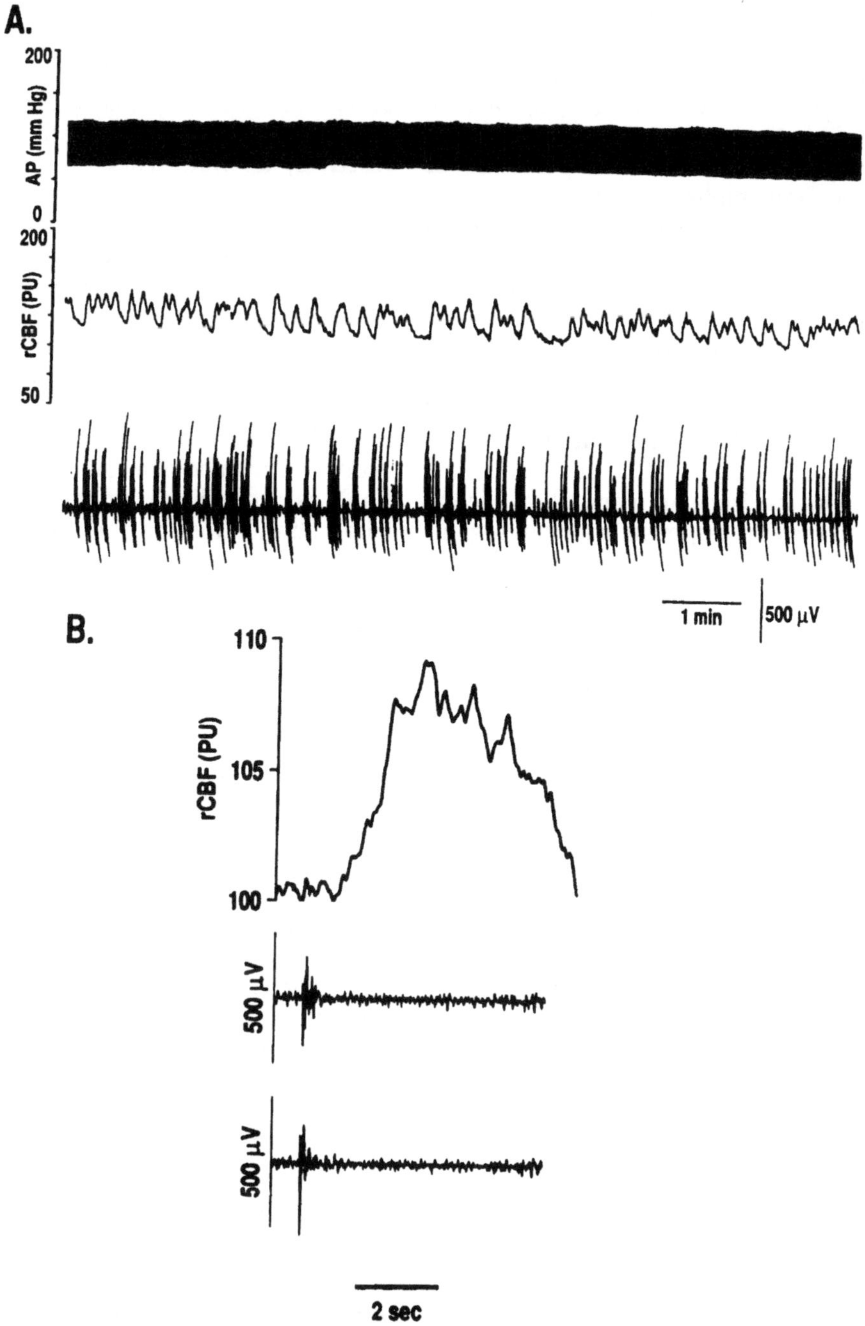

Fig. 19.6. Spontaneous cerebrovascular waves (SCWs) in association with vascularly associated bursts of electrocorticographic (ECoG) activity recorded by laser Doppler flowmetry from the frontal cortex of anesthetized spinalized rat. Anesthesia was with 1.5% isoflurane. **A** AP (upper trace), rCBF (middle trace) and ECoG (lower trace). Note that SCWs are independent of changes in AP. **B** Averaged (n=20) bursts of ECoG activity recorded from right and left hemispheres and associated SCWs. Note stable initial component of burst of ECoG activity and following stable wave of rCBF.

increasing rCBF but not metabolism in the penumbra might serve to reduce lesion size.

In a test of the hypothesis, Reis et al. (1989) observed that electrical stimulation of the FN for 1 hour, coincident with MCA occlusion, resulted 24 hours later in reduction in infarction volume by over 50% on average in different strains of rats (Reis et al. 1989; Underwood et al. 1989; Berger et al. 1990) (Fig. 19.6). The area of salvage (the retrievable zone) corresponded to the ischemic penumbra. However, subsequent studies have now demonstrated that electrical stimulation of the RVL, a region from which stimulation comparably increased rCBF not rCGU, and through which the elevations of rCBF from the FN are relayed (Chida et al. 1990a), fails to reduce infarct volume, as do treatments normally increasing rCBF, including hypercarbia or hypertension (Yamamoto et al. 1993). Comparably, rCBF was not elevated from its depressed level in the ischemic cortex during the period of FN stimulation, nor was metabolism reduced in the penumbral zone.

The results therefore indicate that the brain has the capacity to reduce the size of ischemic infarctions by the activation of specific neuronal pathways, and that the mechanism of salvage is not the result of elevating rCBF nor decreasing metabolism in the ischemic penumbra. Hence, the neuronal systems responsible for centrogenic primary neurogenic vasodilation and centrogenic neural protection are, in part, distinct.

The biological relevance of a central neuroprotective network is not known, but it may relate to neuronal systems that initiate oxygen-conserving reflexes, which are highly developed in diving vertebrates. It is of interest that diving mammals may sustain levels of hypoxia sufficient to lower Pa_{O_2} into the near-pathological range without producing tissue damage.

Conclusions and Some Proposed General Principles of Intrinsic Regulation of the Cerebral Circulation

1. The CBF may be regulated by intrinsic neural networks in the brain acting independently of the extrinsic innervation of cerebral vessels.

2. Intrinsic neuronal systems tonically maintain rCBF in the midrange, since stimulation of the parabrachial or fastigial cerebellar neurons can reduce rCBF, in parallel with rCGU, by 50%. The source of tonic drive is unknown, but differs from the generation of sympathetic tone in the ventral medulla.

3. The effects of intrinsic neuronal networks upon the cerebral circulation are many, and probably represent the action of distinct systems:

(a) They can increase rCBF globally and independently from metabolism, i.e. by a primary vasodilation. Such vasodilation may involve pathways represented in the FN, parafascicular thalamus and basal forebrain and may, in part, be cholinergic. They may function to produce increases in rCBF in anticipation of, rather than reactively to, brain work.

(b) They can globally increase metabolism and, secondarily, flow (secondary vasodilation) by networks represented in DMRF.

(c) They can increase flow in topographically highly restricted distributions which may be linked to, or partly independent from, metabolism, as exemplified in somatosensory projections to the cortex.

(d) They can reduce flow globally by an action on metabolism (secondary vasoconstriction), as exemplified by the responses to electrical stimulation of the medial parabrachial nucleus.

4. The representation of networks governing rCBF encompasses all segments of the neuraxis and is closely allied with the brain nuclei involved in central autonomic regulation (the central autonomic core) (for review see Reis and Iadecola 1989).

5. Central neural mechanisms may activate cerebral metabolism and, indirectly, CBF by sensitizing brain through β-adrenergic receptors to concentrations of circulating epinephrine which themselves would have no cerebral effects when excluded from brain.

6. The central neurogenic elevations in flow unassociated with metabolism depend upon the integrity of neurons within the target areas. These neurons may increase flow either by a direct action of neurotransmitter and/or mediators upon vascular smooth muscle, or through the intervening elements of glia and/or vascular endothelium.

7. Cerebrovascular vasodilation in response to hypoxia may, in part, be regulated by intrinsic neural systems represented in the rostral ventrolateral medulla oblongata. Thus, these variables in flow regulation are not, as traditionally believed, entirely dependent upon responses confined to the blood vessel wall and/or endothelium.

8. Central neurogenic mechanisms, by modifying autoregulation, may dynamically vary the dependence of the cerebral circulation upon systemic AP.

9. The brain, through its intrinsic neural networks, may also protect itself from ischemic damage. Intrinsic neuroprotective mechanisms may relate to central networks which elicit integrated responses to hypoxia.

References

Adachi T, Biesold D, Inanami O, Sato A (1990) Stimulation of the nucleus basalis of Meynert and substantia innominata produces widespread increases in cerebral blood flow in the frontal, parietal and occipital cortices. Brain Res 514:163–166

Armstrong DM (1986) Ultrastructural characterization of choline acetyltransferase-containing neurons in the basal forebrain of rat: evidence for a cholinergic innervation of intracerebral blood vessels. J Comp Neurol 250:81–92

Arneric SP (1989) Basal forebrain neurons modulate cortical blood flow: increases by nicotinic cholinergic mechanisms. J Cerebr Blood Flow Metab 9 (Suppl 1):S502

Arneric SP, Iadecola C, Underwood MD, Reis DJ (1987) Local cholinergic mechanisms participate in the increase in cortical cerebral blood flow elicited by electrical stimulation of the fastigial nucleus in the rat. Brain Res 411:212–225

Arneric SP, Iadecola C, Honig MD, Underwood MD, Reis DJ (1986) Local cholinergic mechanisms mediate the cortical vasodilation elicited by electrical stimulation of the fastigial nucleus. Acta Physiol Scand 127 (Suppl 552):70–73

Arneric SP, Honig MA, Milner TA, Greco S, Iadecola C, Reis DJ (1988) Neuronal and endothelial sites of acetylcholine synthesis and release associated with microvessels in rat cerebral cortex: ultrastructural and neurochemical studies. Brain Res 454:11–30

Bates D, Weinshilboum RM, Campbell RJ, Sundt TM Jr (1977) The effect of lesions in the locus coeruleus on the physiological responses of the cerebral blood vessels in cats. Brain Res 136:431–443

Berger SB, Ballon D, Graham M et al. (1990) Magnetic resonance imaging demonstrates that electric stimulation of cerebellar fastigial nucleus reduces cerebral infarction in rats. Stroke 21 (Suppl III):III-172–III-176

Biesold D, Inanami O, Sato A, Sato Y (1989) Stimulation of the nucleus basalis of Meynert increases cerebral cortical blood flow in rats. Neurosci Lett 98:39–44

Bonvento G, Lacombe PM, Seylaz J (1989) Effects of electrical stimulation of the dorsal raphe nucleus on local cerebral blood flow in the rat. J Cerebr Blood Flow Metab 9:251–255

Cechetto DF, Standaert DG, Saper CB (1985) Spinal and trigeminal dorsal horn projections to the parabrachial nucleus in the rat. J Comp Neurol 240:153–160

Chan-Palay V (1977) Innervation of cerebral blood vessels by norepinephrine, indolamine, substance P, and neurotensin fibers and leptomeningeal indolamine axons: their role in vasomotor activity and local alterations of brain-blood composition. In: Owman C, Edvinsson L (eds) Neurogenic control of brain circulation. Pergamon Press, Oxford, pp 39–53

Chida K, Iadecola C, Reis DJ (1989a) Global reduction in cerebral blood flow and metabolism elicited from intrinsic neurons of fastigial nucleus. Brain Res 500:177–192

Chida K, Iadecola C, Reis DJ (1989b) Differences in selective cardiovascular characteristics of vasopressor responses elicited from the cerebellar fastigial nucleus and the rostral ventrolateral medulla in rats. Ther Res 10:37–55

Chida K, Iadecola C, Reis DJ (1990a) Lesions of rostral ventrolateral medulla abolish some cardio- and cerebrovascular components of the cerebellar fastigial pressor and depressor responses. Brain Res 508:93–104

Chida K, Underwood MD, Miyagawa M et al. (1990b) Participation of the rostral ventrolateral medulla in the cerebral blood flow of rats: effects of stimulation and lesions on systemic and cerebral circulations. Ther Res 11:77–85

Chida N, Iadecola C, Underwood MD, Reis DJ (1986) A novel vasodepressor response elicited from the rat cerebellar fastigial nucleus: the fastigial depressor response. Brain Res 370:378–382

Cobb S, Talbott JH (1927) Studies in cerebral capillaries. II. A quantitative study of cerebral capillaries. Trans Assoc Am Physicians 45:255–262

Cudennec A, Duverger D, MacKenzie ET (1989) Nature of the regional increases in cerebral blood flow induced by raphe stimulation in the conscious rat. J Cerebr Blood Flow Metab 9 (Suppl 1):S375

Dahlgren N, Lindvall O, Sakabe T, Stenevi U, Siesjö BK (1981) Cerebral blood flow and oxygen consumption in the rat brain after lesions of the noradrenergic locus coeruleus system. Brain Res 209:11–23

de la Torre JC, Surgeon JW, Walker RH (1977) Effects of locus coeruleus stimulation on cerebral blood flow in selected brain regions. Acta Neurol Scand Suppl 64:104–105

Del Bo A, Ross CA, Pardal JF, Saavedra JM, Reis DJ (1983a) Fastigial stimulation in rats releases adrenomedullary catecholamines. Am J Physiol 244:R801–R809

Del Bo A, Sved AF, Reis DJ (1983b) Fastigial stimulation releases vasopressin in amounts that elevate arterial pressure. Am J Physiol 244:H687–H694

Doba N, Reis DJ (1972) Changes in regional blood flow and cardiodynamics evoked by electrical stimulation of the fastigial nucleus in the cat and their similarity to orthostatic reflexes. J Physiol (London) 227:729–747

Eckenstein F, Baughman RW (1984) Two types of cholinergic innervation in cortex, one co-localized with vasoactive intestinal polypeptide. Nature 309:153–155

Edvinsson L, Degueurce A, Duverger D, MacKenzie ET, Scatton B (1983) Central serotonergic nerves project to the resistance (pial) vessels of the brain. Nature 306:55–57

Field EJ, Grayson J, Rogers AF (1951) Observations on blood flow in spinal cord of rabbit. J Physiol (London) 114:56–62

Foote SL, Bloom FE, Aston-Jones G (1983) Nucleus locus coeruleus: new evidence of anatomical and physiological specificity. Physiol Rev 63:844–914

Fox PT, Raichle ME (1986) Focal physiological uncoupling of cerebral blood flow and oxidative metabolism during somatosensory stimulation in human subjects. Proc Natl Acad Sci USA 83:1140–1144

Fox PT, Raichle ME, Mintun MA, Dence C (1988) Nonoxidative glucose consumption during focal physiologic neural activity. Science 241:462–464

Fox PT, Mintun FM, Raichle ME, Miezin FM, Allman JM, Van Essen DC (1986) Mapping human visual cortex with positron emission tomography. Nature 323:806–809

Fulwiler CE, Saper CB (1984) Subnuclear organization of the efferent connections of the parabrachial nucleus in the rat. Brain Res 319:229–259

Germano IM, Bartkowski HM, Cassel ME, Pitts LH (1987) The therapeutic value of nimidipine in experimental focal cerebral ischemia. Neurological outcome and histopathological findings. J Neurosurg 67:81–87

Goadsby PJ, Duckworth JW (1989) Low frequency stimulation of the locus coeruleus reduces regional cerebral blood flow in the spinalized cat. Brain Res 476:71–77

Goadsby PJ, Lambert GA (1989) Electrical stimulation of the fastigial nucleus increases total cerebral blood flow in the monkey. Neurosci Lett 107:141–144

Goadsby PJ, Lance JW (1988) Brain stem effects on intra- and extracerebral circulations. Relation to migraine and cluster headache. In: Olesen J, Edvinsson L (eds) Basic mechanisms of headache. Elsevier Science Publishers, Amsterdam, pp 413–427

Goadsby PJ, Lambert GA, Lance JW (1982) Differential effect on the internal and external carotid circulation of the monkey evoked by locus coeruleus stimulation. Brain Res 249:247–254

Goadsby PJ, Lambert GA, Lance JW (1985a) The mechanism of cerebrovascular vasoconstriction in response to locus coeruleus stimulation. Brain Res 326:213–217

Goadsby PJ, Piper RD, Lambert GA, Lance JW (1985b) Effect of stimulation of nucleus raphe dorsalis on carotid blood flow. Am J Physiol 248:R257–R269

Grant SJ, Aston-Jones G, Redmond DE Jr (1988) Responses of primate locus coeruleus neurons to simple and complex sensory stimuli. Brain Res Bull 21:401–410

Greenberg J, Hand P, Sylvestro A, Reivich M (1979) Localized metabolic-flow couple during functional activity. Acta Neurol Scand 60 (Suppl 72):12–13

Heistad DD, Kontos HA (1983) Cerebral circulation. In: Shepherd JT, Abboud FM (eds) Handbook of physiology. Circulation. Vol. III. Peripheral circulation and organ blood flow. American Physiological Society, Bethesda, pp 137–182

Herkenham M (1980) Laminar organization of thalamic projections to the rat neocortex. Science 207:532–535

Hosobuchi Y (1985) Electrical stimulation of the cervical spinal cord increases cerebral blood flow in humans. Appl Neurophysiol 48:372–376

Hosobuchi Y (1991) Treatment of cerebral ischemia with electrical stimulation of the cervical spinal cord. PACE, Pacing Clin Electrophysiol 14:122–126

Iadecola C (1992) Nitric oxide participates in the cerebrovasodilation elicited from cerebellar fastigial nucleus. Am J Physiol 263:R1156–R1161

Iadecola C, Kraig RP (1991) Focal elevations in neocortical interstitial K^+ produced by stimulation of the fastigial nucleus in rat. Brain Res 563:273–277

Iadecola C, Reis DJ (1990) Continuous monitoring of cerebrocortical blood flow during stimulation of the cerebellar fastigial nucleus: a study by laser-Doppler flowmetry. J Cerebr Blood Flow Metab 10:608–617

Iadecola C, Underwood MD, Reis DJ (1986) Muscarinic cholinergic receptors mediate the cerebrovasodilation elicited by stimulation of the cerebellar fastigial nucleus in rat. Brain Res 368:375–379

Iadecola C, Arbit E, Nakai M, Mraovitch S, Tucker LW, Reis DJ (1982) Increased regional cerebral blood flow and metabolism elicted by stimulation of the dorsal medullary reticular formation in the rat: evidence for an intrinsic neural system in brain regulating cerebral metabolism. In: Heistad DD, Marcus ML (eds) Cerebral blood flow: effects of nerves and neurotransmitters. Elsevier, Amsterdam, pp 485–492

Iadecola C, Mraovitch S, Meeley MP, Reis DJ (1983a) Lesions of the basal forebrain in rat selectively impair the cortical vasodilation elicited from cerebellar fastigial nucleus. Brain Res 279:41–52

Iadecola C, Nakai M, Arbit E, Reis DJ (1983b) Global cerebral vasodilatation elicited by focal electrical stimulation within the dorsal medullary reticular formation in anesthetized rat. J Cerebr Blood Flow Metab 3:270–279

Iadecola C, Nakai M, Mraovitch S, Ruggiero DA, Tucker LW, Reis DJ (1983c) Global cerebral increase in cerebral metabolism and blood flow produced by focal electrical stimulation of dorsal medullary reticular formation in rat. Brain Res 272:101–114

Iadecola C, Arneric SP, Baker HD, Tucker LW, Reis DJ (1987a) Role of local neurons in cerebrocortical vasodilation elicited from cerebellum. Am J Physiol 252:R1082–R1091

Iadecola C, Lacombe PM, Underwood MD, Ishitsuka T, Reis DJ (1987b) Role of adrenal catecholamines in cerebrovasodilation evoked from brain stem. Am J Physiol 252:H1183–H1191

Ingvar DH (1958) Cortical state of excitability and cortical circulation. In: Jasper HH, Proctor LD, Knighton RS, Noshay WC, Costello RT (eds) Reticular formation of the brain. Little, Brown, Boston, pp 381–408

Ingvar DH, Söderberg U (1958) Cortical blood flow related to EEG patterns evoked by stimulation of the brain stem. Acta Physiol Scand 42:130–143

Ishitsuka T, Iadecola C, Underwood MD, Reis DJ (1986) Lesions of nucleus tractus solitarii globally impair cerebrovascular autoregulation. Am J Physiol 251:H269–H281

Itakura R, Yokote H, Kimura H et al. (1985) 5-hydroxytryptamine innervation of vessels in the rat cerebral cortex. J Neurosurg 62:42–47

Jones EG (1985) The thalamus. Plenum, New York

Kimura A, Sato A, Takano Y (1990) Stimulation of the nucleus basalis of Meynert does not influence glucose utilization of the cerebral cortex in anesthetized rats. Neurosci Lett 119:101–104

Lacombe PM, Iadecola C, Underwood MD, Sved AF, Reis DJ (1990) Plasma epinephrine modulates the cerebrovasodilation evoked by electrical stimulation of dorsal medulla. Brain Res 506:93–100

Lacombe PM, Sercombe R, Verrecchia, C, Philipson V, MacKenzie ET, Seylaz J (1989) Cortical blood flow increase induced by stimulation of the substantia innominata in unanaesthetized rat. Brain Res 491:1–14

Lang R, Zimmer R (1974) Neurogenic control of cerebral blood flow. Exp Neurol 43:143–161

Langfitt TW, Kassell NF (1968) Cerebral vasodilation produced by brain-stem stimulation: neurogenic control vs. autoregulation. Am J Physiol 215:90–97

Lauter JL, Herscovitch P, Formby C, Raichle ME (1985) Tonotopic organization in human auditory cortex revealed by positron emission tomography. Hearing Res 20:199–205

LeDoux JE, Thompson ME, Iadecola C, Tucker LW, Reis DJ (1983) Local cerebral blood flow increases during auditory and emotional processing in conscious rats. Science 221:576–578

McCasland JS, Woolsey TA (1988) High resolution 2-deoxyglucose mapping of functional cortical columns in mouse barrel cortex. J Comp Neurol 278:555–569

McKee JC, Denn MJ, Stone HL (1976) Neurogenic cerebral vasodilation from electrical stimulation of the cerebellum in the monkey. Stroke 7:179–186

MacKenzie ET, McCulloch J, O'Keane M, Pickard JD, Harper AM (1976) Cerebral circulation and norepinephrine: relevance of the blood–brain barrier. Am J Physiol 231:483–488

Maiese K, Pek L, Berger SB, Reis DJ (1992) Reduction in focal cerebral ischemia by agents acting at imidazole receptors. J Cerebr Blood Flow Metab 12:53–63

Meyer JS, Nomura F, Sakamoto K, Kondo A (1969) Effect of stimulation of the brain-stem reticular formation on cerebral blood flow and oxygen consumption. Electroencephalogr Clin Neurophysiol 26:125–132

Meyer JS, Teraura T, Sakamoto K, Kondo A (1971) Central neurogenic control of cerebral blood flow. Neurology 21:247–262

Milner TA, Aoki C, Pickel VM, Reis DJ (1989) Ultrastructural basis for cholinergic and catecholaminergic modulation of cerebral circulation. J Cerebr Blood Flow Metab 9 (Suppl 1):S481

Miura M, Reis DJ (1969) Cerebellum: a pressor response elicited from the fastigial nucleus and its efferent pathway in brain stem. Brain Res 13:595–599

Molnar L, Szanto J (1964) The effect of electrical stimulation of the bulbar vasomotor centre on the cerebral blood flow. Q J Exp Physiol 49:184–193

Mraovitch S, Seylaz J (1987) Metabolism-independent cerebral vasodilation elicited by electrical stimulation of the centromedian–parafascicular complex in rat. Neurosci Lett 83:269–274

Mraovitch S, Iadecola C, Ruggiero DA, Reis DJ (1985) Widespread reductions in cerebral blood flow and metabolism elicited by electrical stimulation of the parabrachial nucleus in rat. Brain Res 341:283–296

Mraovitch S, Lasbennes F, Calando Y, Seylaz J (1986a) Cerebrovascular changes elicited by electrical stimulation of the centromedian–parafascicular complex in rat. Brain Res 380:42–53

Mraovitch S, Pinard E, Seylaz J (1986b) Two neural mechanisms in rat fastigial nucleus regulating systemic and cerebral circulation. Am J Physiol 251:H153–H163

Nakai M (1985) An increase in cerebral blood flow elicited by electrical stimulation of the solitary nucleus in rats with cervical cordotomy and vagotomy. Jpn J Physiol 35:57–70

Nakai M, Iadecola C, Reis DJ (1982) Global cerebral vasodilation by stimulation of rat fastigial cerebellar nucleus. Am J Physiol 243:H226–H235

Nakai M, Iadecola C, Ruggiero DA, Tucker LW, Reis DJ (1983) Electrical stimulation of cerebellar fastigial nucleus increases cerebral cortical blood flow without change in local metabolism: evidence for an intrinsic system in brain for primary vasodilation. Brain Res 260:35–49

Nedergaard M, Gjedde A, Diemer NH (1986) Focal ischemia of the rat brain: autoradiographic determination of cerebral glucose utilization, glucose content and blood flow. J Cerebr Blood Flow Metab 6:414–424

Ngai AC, Ko KR, Mori S, Winn HR (1988) Effect of sciatic nerve stimulation on pial arterioles in rats. Am J Physiol 254:H133–H139

Pinard E, Lacombe P, Reynier-Rebuffel AM, Seylaz J (1985) Intrinsic and extrinsic mechanisms involved in the cerebrovascular reaction elicited by immobilization stress in rabbits. Brain Res. 340:305–314

Purves MJ (1972) The physiology of the cerebral circulation. University Press, Cambridge

Raichle ME (1987) Circulation and metabolic correlates of brain function in normal humans. In: Plum F (ed) Handbook of physiology: the nervous system, Vol. V. American Physiological Society, Bethesda, pp 643–674

Raichle ME (1991) The metabolic requirements of functional activity in the human brain: a positron emission tomography study. Adv Exp Med Biol 291:1–4

Raichle ME, Hartman BK, Eichling JO, Sharpe LG (1975) Central noradrenergic regulation of cerebral blood flow and vascular permeability. Proc Natl Acad Sci USA 72:3726–3730

Reinhard JF, Jr, Liebmann JR, Schlossberg AJ, Moskowitz MA (1979) Serotonin neurons project to small blood vessels in the brain. Science 206:85–87

Reis DJ (1979) Nervous control of cerebral blood flow in normal health and in relationship to cerebrovascular disease. In: Castaigne P, Lhermitte F, Gautier JC, Moriniere A (eds) Deuxieme Conference de la Salpétrière. Baillière, Paris, pp 197–224

Reis DJ, Iadecola C (1989) Central neurogenic regulation of cerebral blood flow. In: Seylaz J, Sercombe R (eds) Neurotransmission and cerebrovascular function. Elsevier, Amsterdam, pp 369–390

Reis DJ, Iadecola C (1991) Intrinsic central neural regulation of cerebral blood flow and metabolism in relation to volume transmission. In: Fuxe K, Agnati LF (eds) Volume transmission in the brain: novel mechanisms for neural transmission. Raven Press, New York, pp 523–538

Reis DJ, Sun M-K (1993) Sympathoexcitatory neurons of the rostral ventrolateral medulla may be central oxygen detectors. In: Sutton JR, Houston CS, Coates G (eds) Hypoxia and molecular medicine. Charles S Houston, Burlington VT, pp. 7–17

Reis DJ, Doba N, Nathan MA (1973) Predatory attack, grooming and consummatory behavior evoked by electrical stimulation of cerebellar nuclei in cat. Science 182:845–847

Reis DJ, Iadecola C, Nakai M (1985) Control of cerebral blood flow and metabolism by instrinsic neural systems in brain. In: Plum F, Pulsinelli W (eds) Cerebrovascular diseases, Raven Press, New York, pp 1–22

Reis DJ, Morrison S, Ruggiero DA (1988) The C1 area of the brain stem in tonic and reflex control of blood pressure. Hypertension 11 (Suppl):I8–I13

Reis DJ, Berger SB, Underwood MD, Khayata M (1991) Electrical stimulation of cerebellar fastigial nucleus reduces ischemic infarction elicited by middle cerebral artery occlusion in rat. J Cerebr Blood Flow Metab 11:810–818

Reis, DJ, Underwood MD, Berger SB, Khayata M, Zaiens NI (1989) Fastigial nucleus stimulation reduces the volume of cerebral infarction produced by occlusion of the middle cerebral artery in rat. In: Seylaz J, MacKenzie ET (eds) Neurotransmission and cerebrovascular function I. Elsevier Science Publishers, Amsterdam, pp 401–404

Reis DJ, Iadecola C, MacKenzie ET, Mori M, Nakai M, Tucker LW (1982) Primary and metabolically coupled cerebrovascular dilation elicited by stimulation of two intrinsic systems of brain. In: Heistad DD, Marcus ML (eds) Cerebral blood flow: effects of nerves and neurotransmitters. Elsevier, Amsterdam, pp 475–484

Ruggiero DA, Reis DJ (1988) Neurons containing phenylethanolamine-*N*-methyltransferase: a component of the baroreceptor reflex? In: Stolk JM, U'Prichard DC, Fuxe K (eds) Epinephrine in the central nervous system. Oxford University Press, New York, pp 291–307

Ruggiero DA, Cravo SL, Arango V (1989) Central control of the circulation by the rostral ventrolateral reticular nucleus: anatomical substrates. Prog Brain Res 81:49–79

Ruggiero DA, Ross CA, Anwar M, Park DH, Joh TH, Reis DJ (1985) Distribution of neurons containing phenylethanolamine *N*-methyltransferase in medulla and hypothalamus of rat. J Comp Neurol 239:127–154

Saeki Y, Sato A, Sato Y, Trzebski A (1989) Stimulation of rostral ventrolateral medullary neurons increases cortical cerebral blood flow via activation of the intracerebral neural pathway. Neurosci Lett 107:26–32

Sandor P, de Jong W, de Wied D (1986) Endorphinergic mechanisms in cerebral blood flow autoregulation. Brain Res 386:122–129

Sandor P, de Jong W, de Wied D (1988) Naloxone-like influence of TRH and ACTH-(4–7) on hypothalamic blood flow autoregulation in the rat. Peptides 9:215–219

Sandor P, de Jong W, Cox-van-Put J, de Wied D (1987) Influence of centrally administered alpha- and gamma-melanocyte-stimulating hormone on hypothalamic blood flow autoregulation in the rat. Brain Res 424:189–192

Saper CB (1984) Organization of cerebral cortical afferent systems in the rat. I. Magnocellular basal nucleus. J Comp Neurol 222:313–342

Saper CB (1985) Organization of cerebral cortical afferent systems in the rat. II. Hypothalamocortical projections. J Comp Neurol 237:21–46

Schmidt CF (1945) The present status of knowledge concerning the intrinsic control of the cerebral circulation and the effects of functional derangements in it. Fed Proc 3:131–136

Semba K, Reiner PB, McGeer EG, Fibiger HC (1988) Brain stem afferents to the magnocellular basal forebrain studied by axonal transport, immunohistochemistry, and electrophysiology in the rat. J Comp Neurol 267:433–453

Seylaz J, Pinard E, Mraovitch S (1986) Influence of specific intracerebral structures on the regulation of the cerebral circulation. In: Owman C, Hardebo JE (eds) Neural regulation of brain circulation. Elsevier, Amsterdam, pp 147–167

Seylaz J, Hara H, Pinard E, Mraovitch S, MacKenzie ET, Edvinsson L (1988) Effect of stimulation of the sphenopalatine ganglion on cortical blood flow in the rat. J Cerebr Blood Flow Metab 8:875–878

Stavraky GW (1936) Response of cerebral blood vessels to electrical stimulation of the thalamus and hypothalamic regions. Arch Neurol Psych 35:1002:1028

Sun M-K, Jeske IT, Reis DJ (1992) Cyanide excites medullary sympathoexcitatory neurons in rats. Am J Physiol 262:R182–R189

Tsubokawa T, Katayama Y, Ueno Y, Moriyasu N (1981) Evidence for involvement of the frontal cortex in pain-related cerebral events in cats: increase in local cerebral blood flow by noxious stimuli. Brain Res 217:179–185

Underwood MD (1988) Control of the cerebral circulation and metabolism by the rostral ventrolateral medulla: possible role in the cerebrovascular response to hypoxia. Doctoral thesis, Cornell University, New York

Underwood MD, Iadecola C, Reis DJ (1987) Neurons in C1 area of rostral ventrolateral medulla mediate global cerebrovascular responses to hypoxia but not hypercarbia. J Cerebr Blood Flow Metab 7 (Suppl 1):S226

Underwood MD, Berger SB, Khayata M, Reis DJ (1989) Fastigial nucleus stimulation reduces the volume of cerebral infarction produced by occlusion of the middle cerebral artery in rat. J Cerebr Blood Flow Metab 9 (Suppl 1):S32

Underwood MD, Bakalian MJ, Arango V, Smith RW, Mann JJ (1992a) Regulation of cortical blood flow by the dorsal raphe nucleus: topographic organization of cerebrovascular regulatory regions. J Cerebr Blood Flow Metab 12:664–673

Underwood MD, Iadecola C, Sved AF, Reis DJ (1992b) Stimulation of C1 area neurons globally increases regional cerebral blood flow but not metabolism. J Cerebr Blood Flow Metab 12:844–855

Underwood MD, Iadecola C, Reis DJ (1994) Lesions of the rostral ventrolateral medulla reduce the cerebrovascular response to hypoxia. Brain Res 635:217–223

Vertes RP (1988) Brain stem afferents to the basal forebrain in the rat. Neuroscience 24:907–935

Wahlestedt C, Golanov EV, Yamamoto S et al. (1993) Antisense oligodeoxynucleotides to the MNDAR1 receptor channel protect cortical neurons from excitotoxicity and reduce focal ischaemic infarctions. Nature 363:260–263

Yamamoto S, Golanov EV, Reis DJ (1993) Reductions in focal ischemic infarctions elicited from cerebellar fastigial nucleus do not result from elevations in cerebral blood flow. J Cerebr Blood Flow Metab 13:1020–1024

Concluding Comments

The authors in this section have presented detailed and scholarly reviews of factors relating to brain blood flow and its control. I will comment on only a few selected aspects.

Hyperperfusion

In his chapter on autoregulation, Heistad alludes to the potential ill effects of an oversupply of blood. Hyperperfusion is, of course, a relative term. In clinical situations, hyperperfusion refers to rather abruptly increased blood flow in situations in which flow had previously been reduced. Examples include carotid endarterectomy repair of a tightly stenotic large artery, with the creation of a normal-sized lumen which dramatically increases flow to that supply zone; repair of an arteriovenous malformation (AVM), where the enlarged feeding arteries accustomed to supplying a fast-flowing AVM with a rapid venous runoff now supply normal arteries unaccustomed to high flow rates; reperfusion of the territory of a blocked artery either spontaneously or by fibrinolysis; sudden augmentation of cerebral blood flow by repair of a congenital cardiac defect, thus greatly increasing cardiac output; acute increases in systemic blood pressure beyond the autoregulatory ability of the brain vasculature: this situation occurs in eclampsia, accelerated 'malignant' hypertension, acute nephritis and probably acutely in response to a variety of external circumstances, such as exposure to extreme cold, emotional stimuli, Vth nerve stimulation etc. (Caplan 1988b).

The major recognized and accepted danger of hyperperfusion is brain hemorrhage. Intracerebral hemorrhage is a known complication of all of the situations mentioned above. Hemorrhage is particularly apt to occur if previously underperfused brain tissue is exposed to augmented flow. During the period of ischemia, the capillaries and arterioles within the ischemic zone may suffer ischemic damage and are more permeable and vulnerable to breakage than normal, especially when exposed to an unaccustomed flow burden (Caplan 1988b). Another very important consequence of a rapid increase in perfusion is brain edema. When exposed to an overload, vessels become more permeable to water and interstitial edema develops. The edema can increase intracranial pressure and may lead to displacement and even herniation of the brain contents. Edema also complicates brain hemorrhage. Edema can occur after recanalization of an occluded artery. I have recently seen several patients who had severe hemispheral white matter edema develop after carotid endarterectomy (Breen et al. 1993). The frequency, extent and importance of edema formation in hyperperfused brains is not known and has not been well studied.

The other potential dangers of hyperperfusion are mainly theoretical fears that have been studied experimentally only in the situation of reperfusion of a previously ischemic zone. In some of the situations mentioned, such as after carotid endarterectomy, after AVM repair, after migrainous vasoconstriction (Cole and Aube 1990) the dramatic abruptly increased perfusion follows a period of real or potential brain ischemia. In this situation, 'reperfusion injury' (Siesjo 1981; Hallenbeck and Dutka 1990) can occur. Increased flow could theoretically bring into the ischemic zone substances that lead to further neuronal injury. These potential substances include calcium and sodium; ischemia alters cell membrane function and the function of a number of adenosine triphosphate (ATP)-dependent cation pump processes. The concentrations of sodium and calcium are usually higher extracellularly and K^+ is usually higher inside cells. When cell membranes are 'leaky', they cannot maintain normal extracellular to intracellular gradients. Increased entry of Na^+ and/or Ca^{2+} into cells can be the final blow that kills the cells (Siesjo 1988). Reperfusion and hyperperfusion can bring more Na^+ and Ca^{2+} to the scene, thus increasing the gradients of these substances; free oxygen radicals found in ischemic zones could circulate more and potentially injure cells (Kontos 1985; Southorn and Powis 1988); lactic acid forms as a result of bringing more metabolites into the ischemic zone, and the resulting lowered pH can injure tissues; excitatory neurotransmitters such as glutamate present in local tissues in increased amounts might potentially injure tissues; and leukocytes, especially neutrophils, are increased in local tissues (Hallenbeck and Dutka 1990) and can release substances which can damage cells (Weiss 1989). These metabolic changes are discussed in more detail in Section I. In regard to these chemical changes, reperfusion and hyperperfusion are two-edged swords. Although more potentially toxic substances and more metabolites are circulated into the previously ischemic zone, more chemicals are also brought out of the zone into the general circulation. In this respect, the augmented blood flow may have very positive effects on the local cell environment in the extracellular space.

Microvascular Flow and Microangiopathies

The contributions in this section are mainly concerned with blood flow in the larger basal and other macroscopic blood vessels, but most of the brain blood flow occurs through tiny blood vessels visible only through a microscope. Microvascular flow is more difficult to measure. Ames and co-workers (1968) introduced the concept of the 'no-reflow' phenomenon into the literature on microvascular flow. These authors postulated that local events occurring during ischemia acted to block capillaries and arterioles, thus preventing recirculation when reperfusion occurred. Capillaries could be blocked by the swelling of capillary cell membranes, local in-situ clotting stimulated by the release of thromboplastic agents, and increased blood viscosity. Impairment of microvascular reperfusion ('no-reflow') led to continued neuronal injury despite the restoration of flow through larger vascular channels.

Because microvessels are, by definition, considerably smaller than blood vessels visible grossly, the mechanics and physics of blood flow are quite different. Increased blood viscosity has much more effect on flow in the microvascular bed than in the macrocirculation. Erythrocytes and fibrinogen are the two most influential contributors to whole-blood viscosity (Grotta et al. 1982). Increased concentrations of either can considerably increase viscosity.

One of the most enigmatic and controversial cerebrovascular disorders is so-called Binswanger's disease (Blass et al. 1991; Caplan and Schoene 1978; Babikian and Ropper 1987; Caplan 1988a). In this condition there are diffuse changes in the white matter of the cerebrum, and sometimes the cerebellum and the basal gray matter. Lacunar infarcts are also invariably present, and the penetrating microvessels are thickened and their lumina are narrowed. The white-matter atrophy, demyelination and gliosis are almost surely due to an ischemic process, but the ischemia is different from classic infarcts. Occlusion of an artery usually leads to a discrete well-defined area of infarction within the territory supplied. Schoene and I (Caplan and Schoene 1978) had postulated a patchy but diffuse low-grade chronic ischemia as the cause of the white-matter changes. If multiple parallel penetrating microvessels were compromised, as they are in Binswanger's diseased brains, then any reduction in flow might reduce perfusion over a wide zone, ultimately leading to ischemic damage. Fisher (1989), in a review, emphasized that, in patients with Binswanger's disease studied at necropsy, the arteries are surprisingly seldom occluded or show evidence of prior thrombosis.

An alternative possibility to thrombosis is that, in these patients, an alteration in the rheology of flow caused by increased blood viscosity impedes flow in already compromised arteries. This hypothesis has never been tested in detail, but there is some preliminary suggestive supportive evidence. Schneider et al. (1987) studied hemorheological factors in 21 patients with presumed Binswanger's disease (diagnosed clinically and by imaging), 40 patients with lacunar infarcts and 275 normal controls who had no vascular risk factors. In the Binswanger's disease group, 19 of 21 (90%) had increased blood viscosity and many had increased erythrocyte aggregation and increased fibrinogen levels. The lacunar infarct patients had more hematological abnormalities than controls but significantly fewer abnormalities than the Binswanger's disease group (Schneider et al. 1987). More recently, my colleagues and I have begun to study our patients (>200) with neuroimaging and clinical features compatible with Binswanger's disease. Fibrinogen levels are almost invariably increased or in the upper range of normal. Clearly, viscosity and rheology need more study in this relatively common cause of vascular dementia. Lowering of viscosity is a potentially important treatment if the condition can be diagnosed early enough.

Diaschisis

Diaschisis is a term introduced originally by von Monakow to refer to decreased function occurring at a distance remote from an acute nervous system injury. The usual explanation is a form of shock or altered neuronal activity caused by decreased neuronal transsynaptic activation. Blood flow and metabolic studies of patients with focal strokes have frequently shown that areas remote from infarcts have decreased glucose metabolism and decreased blood flow. These remote metabolic and flow effects have also been called diaschisis (Slater et al. 1977; Feeney and Baron 1986). Decreased regional cerebral blood flow (rCBF), oxygen consumption ($CMRO_2$) and glucose consumption (CMRGlu) can occur in deep structures, especially the thalamus after an ipsilateral cortical infarct; in ipsilateral cortical structures after a deep basal ganglionic or thalamic infarct; in the contralateral cerebral hemisphere after a unilateral hemispheral infarct (Lagreze et al. 1987); in the ipsilateral pons and contralateral cerebellum after a unilateral cortical infarct (Martin and Raichle 1983; Baron 1985; Eckard et al. 1992; Fulham et al. 1992; Reivich 1992). Cerebellar and brain-stem infarcts are also probably associated with decreased hemispheral metabolism and blood flow, but have not been as well studied using positron emission tomography. These remote metabolic and blood flow changes are maximal during the acute phase of stroke, but can linger. In fact, transsynaptic thalamic atrophy is very common after unilateral hemispheral infarcts (Tamura et al. 1991).

Traditionally, remote blood flow effects are presumed to be secondary to neurogenically induced metabolic alterations produced by the decreased axonal input to the region. Such decreased metabolism has been well documented in patients with acute infarcts (Lenzi et al. 1982; Reivich 1992). Somewhat difficult to explain by this hypothesis is the observation that sometimes the remote effects increase progressively during the first weeks after infarction (Slater et al. 1977). The complexity of blood flow and its regulation has been captured in the chapters in this section. Might autoregulation be disrupted after a focal insult? Autoregulation has been shown to be abnormal in the contralateral hemisphere after a unilateral stroke (Paulson et al. 1972; Slater et al. 1977).

Infarcts are associated with the release of neurotransmitters and other vasoactive substances which can affect arteries directly or, as is more likely, act by influencing the trigeminovascular innervation, as described by Macfarlane and Moskowitz in their chapter in this section. Neural influences could then affect vascular tone and flow at a distance. Reis and Golanov, in their chapter, offer another potential explanation by documenting the influence of posterior circulation strokes on supratentorial flow. Stimulation of key nuclear structures in the cerebellum, brain stem and area postrema can affect blood flow at a distance. Insight into the phenomenology of diaschisis and its cause will undoubtedly teach much about the control of brain circulation.

Blood Volume Changes

Perfusion is related to blood pressure, blood flow, blood flow velocities and blood volume. These terms are sometimes confused and used interchangeably, but in reality they each describe a different measure. Blood flow velocities are now measured clinically using transcranial Doppler (TCD) (Caplan et al. 1990), whereas blood flow and blood volumes in various brain regions can be measured by positron tomography (Lenzi et al. 1982). Systemic blood pressure is clearly the simplest to measure using a bedside sphygmomanometer, whereas blood volume can only be measured by various radioisotopic techniques.

Alterations in these factors can produce different effects on blood vessels. It is well known and accepted that hypertension (elevated systemic blood pressure) has a deleterious effect on arteries and accelerates atherosclerosis in both large extracranial arteries and small intracranial perforating arteries. Hypertension can be caused by either increased vascular resistance, as mediated by the renin–angiotensin system, or increased volume. There is some evidence that there are racial differences in urinary sodium excretion and retention after sodium loads (Blaustein and Grim 1991). Blacks seem to have more high-volume hypertension than whites, and in blacks hypertension seems to respond better to diuretics than in whites (Freis et al. 1988). Japanese patients have a relatively high sodium intake and share with blacks a tendency towards hypertension, probably also of the high-volume type. Humans also have a ouabain-like compound which affects numerous sodium gradient-dependent processes and can increase blood volume (Hamlyn et al. 1991). Racial differences in the presence of this compound could affect the relative frequency of high-volume hypertension.

In the section on blood vessels, I discussed racial differences in the distribution of cerebrovascular disease. In general, blacks and individuals of Japanese, Chinese and Thai ancestry have more intracranial occlusive disease affecting medium-sized basal arteries, whereas white males have more extracranial occlusive arterial disease (Caplan et al. 1986a). Extracranial occlusive disease is often associated with coronary artery disease and peripheral limb arterial occlusive disease. My own present hypothesis of the racial differences in the distribution of occlusive disease is that increased blood volume creates stresses and effects on blood vessels at different sites and in different ways than increased vascular resistance. High-volume hypertension is also known to occur in women during pregnancy, during the use of oral contraceptives, in the puerperium and in diabetics and nephrotics. These situations are all associated with intracranial occlusive disease. Unfortunately, there are no data that presently establish or refute this hypothesis, which remains essentially untested.

Blood Pressure Treatment During Stroke

As shown by the discussion in this section, the relationships between blood pressure, blood flow and intracranial pressure are very complex. No clinician would argue with the assertion that hypertension is bad for the circulation and that very severe hypertension is very bad, yet there continues to be controversy about whether to lower blood pressure during acute strokes and, if so, to what extent and with what agents.

Moderate acute increases in blood pressure acutely increase blood flow. Just before carotid artery clamping, surgeons commonly inject a pharmacologic agent to increase blood pressure to attempt to increase blood flow during the relative ischemia that may follow clamping. Decreased blood pressure can diminish blood flow in regions that are critically perfused. As Heistad has reviewed in his section, autoregulation tends to keep blood flow relatively constant within a wide range of systemic blood pressures. However, when the blood pressure exceeds the autoregulatory range (60–160 mmHg) flow probably does change. Severe hypertension can have very deleterious effects on blood flow. Intracranial pressure also affects perfusion, since when the intracranial pressure is high the venous pressure in the cranium is also elevated, and so the systemic blood pressure must remain well above venous pressure to ensure an arteriovenous pressure gradient to maintain adequate perfusion. In practice, I advocate not treating hypertension acutely in patients with acute brain ischemia if the blood pressure elevation is moderate (170–150/85–95 range). If hypertension is severe (>180/100), then a modest reduction in blood pressure is probably indicated. If the intracranial pressure is high due to brain edema or hemorrhage, then even higher blood pressures are permitted to go untreated. I try to maintain adequate fluid and blood volume in all patients with acute ischemia. Neurologic function, monitored by frequent careful neurologic examinations, is a more important guide to treatment than the measured blood pressure.

Different agents used to treat systemic hypertension have different effects on cerebral blood flow and cerebral artery dilation (Kaplan 1992; Stumpf 1988). Nitroprusside given parenterally abruptly decreases blood pressure but may increase intracranial pressure. Diuretics can cause volume depletion and so lead to renal sodium retention when the blood pressure falls. Parenteral diazoxide, methyldopa and reserpine all have potent sedative effects. Labetalol, a competitive inhibitor of both α- and β-adrenergic receptors, is a very effective antihypertensive drug and has a vasodilator effect (Brott 1991). When labetalol is used, intravascular volume is not contracted, heart rate is not changed and cardiac output is usually not reduced (Brott 1991). Labetalol can be given orally or in an intravenous infusion. Nifedipine, a calcium entry blocker, captopril, an angiotensin-converting enzyme inhibitor, and clonidine, a centrally active sympatholytic drug, are also very effective, even when given orally. Clonidine also has a sedative effect and so is not as good a choice as the other drugs.

Newer diagnostic techniques, especially transcranial Doppler (Caplan et al. 1990) and SPECT (Caplan 1990) offer promise in their abilities to monitor blood flow velocities and blood flow in the ischemic zones before and after various drug treatments. The local, regional effects of the therapies, including antihypertensives, can be studied directly rather than hypothesized. The data derived have great promise in improving our understanding of drug effects on blood flow and metabolism.

Control of Blood Flow by Neurons, Especially in the Brain Stem and Cerebellum

The chapter by Reis and Golanov on neural structures and pathways that control the elevation or depression of brain blood flow provides an exciting look into the future. Clinicians have long noted the tendency of patients with posterior circulation ischemia to have impressive fluctuations in symptoms. This variability probably contributed to the designation 'vertebrobasilar insufficiency' (VBI) for ischemia in the posterior circulation (Millikan and Siekert 1955; Denny-Brown 1960). The general impression is that patients with posterior circulation ischemia fluctuate more than patients with anterior circulation ischemia, although I know of no data to support or refute this. Also, some clinicians have observed fluctuations in pulse, blood pressure and autonomic nervous system functions in patients with posterior circulation ischemia (Khurana 1982; Caplan et al. 1986b; Caplan et al. 1992). Some have even noted a tendency for sudden unexplained death in some patients with medullary ischemia (Caplan et al. 1986b; Bogousslavsky et al. 1990).

Reis and Golanov report detailed animal studies of various nuclei that have the ability to change brain blood flow, blood pressure and pulse rate. Strikingly, most of these structures lie within the brain stem and are localized to the medullary tegmentum and the midline cerebellar nuclei and thalamus. These are regions which are supplied by the vertebral and basilar arteries. The most active nuclei (rostral ventrolateral reticular nucleus, nucleus of the tractus solitarius, and nuclei in the dorsal medullary reticular formation) are all supplied by branches of the intracranial vertebral arteries. In patients with lateral medullary infarcts, fluctuations in pulse and blood pressure, as well as sudden death, are most often noted (Caplan et al. 1992; Caplan et al. 1986a). Might the findings detailed by Reis and Golanov explain the putative variability in clinical signs and autonomic function found in some patients with posterior circulation ischemia? Clearly, this issue needs more study and documentation in the clinic.

Blood flow and flow velocity studies using transcranial Doppler, PET and SPECT have begun to show impressive increases in blood flow and metabolism in various brain regions in relation to specific activities. For example, speech increases flow and metabolism in the left parasylvian and inferior frontal cortexes, and movement of the left arm is accompanied by increased metabolism and blood flow in the arm region of the right precentral region. Other stimuli, such as music, paintings, literature and touch, also elicit changes in the regions of the cerebral cortex that relate to these functions. This research has given many insights into brain physiology. The data of Reis and Golanov showing neurogenic mechanisms of this activation by sensory stimuli are intriguing and novel. Most theoreticians had explained this activation by localized biochemical changes, but Reis and Golanov provide another way to view this process.

Finally, Reis and Golanov have shown that it may be possible that the brain can protect itself in some ways from ischemic insults: the authors dub this 'central neurogenic protection'. Interestingly, the major actors may again be in the hindbrain. Teleologically, brain-stem ischemia is very threatening to the survival of vital functions and life, so that it would make sense that inborn protective mechanisms would reside in the brain stem. Elucidating the mechanisms of this self-protection may give clinicians new insights into the treatment of stroke patients with incomplete brain ischemia.

References

Ames AIII, Wright RL, Kowada M et al. (1968) Cerebral ischemia: II. The no-reflow phenomenon. Am J Pathol 52:437–453

Babikian V, Ropper AH (1987) Binswanger's disease: a review. Stroke 18:2–12

Baron JC (1985) Positron tomography in cerebral ischemia: a review. Neuroradiology 27:509–510

Blass JP, Hoyer S, Nitsch R (1991) A translation of Otto Binswanger's article (The delineation of generalized progressive paralyses). Arch Neurol 48:961–968

Blaustein MP, Grim CE (1991) The pathogenesis of hypertension: black–white differences. In: Saunders E (ed) Cardiovascular diseases in blacks. FA Davis, Philadelphia, pp 97–114

Bogousslavsky J, Khurana R, Deruaz JP et al. (1990) Respiratory failure and unilateral caudal brainstem infarction. Ann Neurol 28:668–673

Breen JC, Caplan LR, Belkin M et al. (1993) Severe white matter edema as part of the hyperperfusion syndrome after cardiac surgery. Presented at the International Congress of Advances in Brain Revascularization, Eilat, Israel, May 1993

Brott T (1991) Prevention and management of medical complications of the hospitalized elderly stroke patient. In: Biller J (ed) Clinics in geriatric medicine. Cerebrovascular disorders in the 1990s. WB Saunders, Philadelphia, pp 475–482

Caplan LR (1988a) Binswanger's disease. Curr Opin Neurol Neurosurg 1:57–62

Caplan LR (1988b) Intracerebral hemorrhage revisited. Neurology 38:624–627

Caplan LR, Schoene WC (1978) Clinical features of subcortical arteriosclerotic encephalopathy (Binswanger's disease) Neurology 28:1206–1215

Caplan LR (1990) Question-driven technology assessment: SPECT as an example. Neurology 41:187–191

Caplan LR, Gorelick PB, Hier DB (1986a) Race, sex, and occlusive cerebrovascular disease: a review. Stroke 17:648–655

Caplan LR, Pessin MS, Scott RM et al. (1986b) Poor outcome after lateral medullary infarcts. Neurology 36:1510–1513

Caplan LR, Brass LM, DeWitt LD et al. (1990) Transcranial Doppler ultrasound: present status. Neurology 40: 696–700

Caplan LR, Pessin MS, Mohr JP (1992) Vertebrobasilar occlusive disease. In: Barnett JHM, Mohr JP, Stein BM, Yatsu FM (eds) Stroke pathophysiology, diagnosis, and management, 2nd edn. Churchill Livingstone, New York, pp 443–515

Cole AJ, Aube M (1990) Migraine with vasospasm and delayed intracerebral hemorrhage. Arch Neurol 47:53–56

Denny-Brown D (1960) Recurrent cerebrovascular episodes. Arch Neurol 2:194–210

Eckard DA, Purdy PD, Bonte F (1992) Crossed cerebellar diaschisis and loss of consciousness during temporary balloon occlusion of the internal carotid artery. AJNR 13:55–57

Feeney D, Baron JC (1986) Diaschisis. Stroke 17:817–830

Fisher CM (1989) Binswanger's encephalopathy. A review. J Neurol 236:65–79

Freis ED, Reda DJ, Materson BJ (1988) Volume (weight) loss and blood pressure response following thiazide diuretics. Hypertension 12:244–250

Fulham M, Brook RA, Hallett M et al. (1992) Cerebrovascular diaschisis revisited. Neurology 42:2267–2273

Grotta J, Ackerman R, Correia J et al. (1982) Whole blood viscosity parameters and cerebral blood flow. Stroke 13:296–301

Hallenbeck JM, Dutka A (1990) Background review and current concepts of reperfusion injury. Arch Neurol 47:1245–1254

Hamlyn JM, Blaustein MP, Bova S et al. (1991) Identification and characterization of a ouabain-like compound from human plasma. Proc Natl Acad Sci USA 88:6259–6263

Kaplan NM (1992) Treatment of hypertensive emergencies and urgencies. Heart Dis Stroke 1:373–378

Khurana R (1982) Autonomic dysfunction in pontomedullary stroke. Ann Neurol 12:86

Kontos HA (1985) Oxygen radicals in cerebral vascular injury. Circ Res 57:508–516

Lagreze H, Levine RL, Pedula K et al. (1987) Contralateral flow reduction in unilateral stroke: evidence for transhemispheric diaschisis. Stroke 18:882–886

Lenzi GL, Frackowiak RSJ, Jones T (1982) Cerebral oxygen metabolism and blood flow in human cerebral ischemic infarction. J Cerebr Blood Flow Metab 2:321–335

Martin WR, Raichle M (1983) Cerebellar blood flow and metabolism in cerebral hemisphere infarction. Ann Neurol 14:168–176

Millikan C, Siekert R (1955) Studies in cerebrovascular disease: the syndrome of intermittent insufficiency of the basilar arterial system. Proc Staff Meet Mayo Clinic 30:61–68

Paulson O, Olesen J, Stig-Christensen M (1972) Restoration of autoregulation of cerebral blood flow by hypocapnia. Neurology 22:286–293

Reivich M (1992) Crossed cerebellar diaschisis. AJNR 13:62–64

Schneider R, Ringelstein EB, Zeumer H et al. (1987) The role of plasma hyperviscosity in subcortical arteriosclerotic encephalopathy (Binswanger's disease). J Neurol 234:67–73

Siesjo BJ (1981) Cell damage in the brain: a speculative hypothesis. J Cerebr Blood Flow Metab 1:155–185

Siesjo BK (1988) Historical overview: calcium, ischemia and death of brain cells. Ann NY Acad Sci 522:638–661

Slater R, Reivich M, Goldberg H et al. (1977) Diaschisis with cerebral infarction. Stroke 8:684–690

Southorn PA, Powis G (1988) Free radicals in medicine. I. Chemical nature and biologic reactions. Mayo Clin Proc 63:381–389

Stumpf JC (1988) Therapy review: drug therapy of hypertensive crises. Clin Pharmacol 7:582–591

Tamura A, Tahira Y, Nagashima H et al. (1991) Thalamic atrophy following cerebral infarction in the territory of the middle cerebral artery. Stroke 22:615–618

Weiss SJ (1989) Tissue destruction by neutrophils. New Engl J Med 320:365–376

SECTION VI

The Heart

Prefatory Comments

When authors of popular literature choose an organ to denote important human or inanimate features, the heart is easily the most popular. 'The heart of the matter' denotes the perception that the heart is the central, dominant organ, and 'affairs of the heart' and 'have a heart' attribute to that organ great emotional value. Yet most neurologists consider the heart to be a simple pump which exists mostly to perfuse the brain. They feel that the brain is obviously the most essential and important organ since it determines human character, intellect and personality – the things that make us 'human' – and controls all body movement and perception. Each specialist is guilty of some tunnel vision, and these claims are made in some jest to irritate rival specialists. Yet, who would argue that the brain and the heart are dominant organs? Diseases of these two organs account for over 50% of mortality and countless morbidity and suffering.

Recent studies in the laboratory and at the bedside are showing that the functions of these two organs are intricately interconnected and interdependent. Natelson (1985) even called for the creation of a new speciality, 'neurocardiology', in which the detailed relationships between the heart and the nervous system would be studied in more detail. In the clinic, these relationships can be categorized as:

1. Cardiac lesions or dysfunctions which cause secondary damage to the nervous system.
2. Nervous system pathology which causes secondary cardiovascular affects.
3. Diseases that affect both the brain and the heart concurrently or sequentially.
4. Iatrogenic situations in which what doctors do to treat disease of one organ affects the other.

Although the last two of these categories are very important, they will not be considered here. Drugs used to treat cardiac disorders often have central nervous system side-effects and cardiac surgery is followed by an extremely high rate of transient or persistent brain damage.

Atherosclerosis affects the coronary and cervicocranial arteries, usually concurrently and in parallel. All of us that care for stroke patients must be aware of the high incidence of myocardial infarction, which is the most important cause of mortality in stroke survivors. The coexistence of coronary and cerebrovascular disease has been covered in detail elsewhere (Furlan 1987; Rolak and Rokey 1990).

Cardiac diseases affect the brain in two different ways: by pump and perfusion failure, and by embolism. Cardiac arrest and arrhythmias often lead to severe brain damage, and many who survive cardiac resuscitation remain as brainless souls. Systemic hypoperfusion and the neurologic effects of cardiac pump failure will not be covered here. Instead, I wish to explore in more detail the issues surrounding cardiac-origin brain embolism. Newer technologies that yield insights into the heart and cardiac function, such as more advanced echocardiography, other cardiac imaging and radionuclide techniques and cardiac rhythm monitoring, have yielded new information about cardiac lesions that give rise to emboli to the brain. At the same time, increased knowledge of the physiology of coagulation and platelet function have allowed the study of clinical coagulation functions in patients with potential cardiac sources of embolism, many of whom are now treated with anticoagulants. The advent of better brain and cerebrovascular monitoring techniques, such as magnetic resonance imaging (MRI) and angiography (MRA) and transcranial Doppler ultrasound (TCD) has given neurologists and cardiologists better ways of detecting and studying brain embolism. Hanna and Furlan review in detail, in this section, much of the newer information about cardiac origin brain embolism, and I will also comment on this issue at the end of the section.

All of us at some time in life have experienced a telltale speeding of our heart rates when we are under stress or have strong emotional or romantic feelings. We know intuitively that brain events do affect the heart. During the last decade, many investigators and clinicians have begun to study this intricate brain–heart interface. Central nervous system lesions and events can cause cardiac lesions such as subendocardial hemorrhages and myocardial cell damage; various cardiac arrhythmias, some of which can be fatal; electrocardiographic changes, some characteristic of brain lesions; alterations in the pulse rate and blood pressure; and even sudden death. Ali and Levine, in this section, review and explore these various heart and brain relationships. Various anatomical brain regions, when lesioned or stimulated, produce different effects on the heart and cardiovascular system. I will also comment on some aspects of the brain–heart interaction at the end of the section.

References

Furlan AJ (ed) (1987) The heart and stroke. Springer-Verlag, London

Natelson B (1985) Neurocardiology. An interdisciplinary area for the 80s. Arch Neurol 42:178–184

Rolak LA, Rokey R (1990) Coronary and cerebral vascular disease. Mt. Futura Publishing Co., Kisco, NY

20. Cardiac Disease and Embolic Sources

J. P. Hanna and A. J. Furlan

Cardioembolic stroke accounts for approximately one in six ischemic strokes (Cerebral Embolism Task Force 1986, 1989). However, no clinical or laboratory finding assures the diagnosis. A large-vessel distribution cerebral infarct is considered to be of cardiac origin if the patient has a potential cardioembolic source without evidence of intrinsic cerebrovascular disease or other potential etiologies. Often, however, cardiac and cervicocranial occlusive disease coexist (Olsen et al. 1985; Bogousslavsky et al. 1986) and a small-vessel distribution brain infarct can be of cardioembolic origin (Jackson et al. 1984). Therefore, the diagnosis of cardioembolic stroke often remains speculative.

Many different cardiac sources can give rise to emboli (Table 20.1). Emboli may be composed of thrombus, calcific particles, tumor, air, nitrogen bubbles, fat or foreign bodies. Sources of cardiac embolism are divided into three categories, cardiac chamber, valve and paradoxic. This chapter will further explore these three categories of cardioembolic disease.

Table 20.1. Cardioembolic sources of stroke

Cardiac chamber
Ventricular
focal wall motion abnormality
myocardial infarction
left ventricular aneurysm
myxoma and other tumors
global wall motion abnormality
cardiomyopathies
Atrial
focal wall motion abnormality
myxoma and other tumors
atrial septal aneurysm
global wall motion abnormality
atrial fibrillation
sick sinus syndrome
dilated left atrium
Valvular
Aortic
calcific aortic stenosis
Mitral
mitral stenosis and incompetence
mitral valve prolapse
mitral annulus calcification
mitral valve strands
Prosthetic
All
endocarditis
Paradoxic
patent foramen ovale
ventricular septal defect
atrial septal defect

Cardiac Chambers

The cardiac chambers are the source of the majority of cardioembolic strokes. Non-valvular atrial fibrillation, myocardial infarction and ventricular aneurysms account for approximately

45%, 15% and 10% of cardioembolic stroke respectively (Cerebral Embolism Task Force 1986). The lack of warning, the absence of proven post-embolic therapy and the frequent substantial disability due to cardioembolic stroke emphasize the importance of preventive therapy for defined high-risk groups.

Pathogenesis of Intracavitary Thrombi and Subsequent Embolism

Rudolph Virchow, more than a century ago, defined a triad of antecedent conditions for intracardiac thrombosis: a zone of circulatory stasis, endothelial injury and a hypercoagulable state (Virchow 1856). After thrombus formation, the hydraulic forces of blood flow versus the adhesive forces of the thrombus determine the risk of subsequent embolization.

A zone of circulatory stasis is critical for the formation of intracavitary thrombus. The low shear rate that exists in areas of stasis promotes activation of the coagulation cascade, rather than platelets, leading to thrombus formation. Stagnant blood formed secondary to focal or global akinesis or dyskinesis, as in myocardial infarction (Visser et al. 1985; Meltzer et al. 1986) and atrial fibrillation (Castello et al. 1990), seems to be the essential component for thrombus formation.

Endothelial changes provide a nidus for triggering the coagulation cascade, platelet activation and platelet adhesion. After myocardial infarction the endothelial surface is denuded from the basal lamina within a few days (Johnson et al. 1979). Loss of the protective endothelial surface exposes circulating blood to the underlying type II collagen in the basement lamina, causing platelet activation and adhesion (Baumgartner and Haudenschild 1972). Along with the circulatory stasis induced by the regional mural dyskinesia in myocardial infarction, intracavitary clot formation is potentiated. Investigators using surface two-dimensional echocardiography have noted a one in six chance of mural thrombus formation during hospitalization for acute myocardial infarction (Weinreich et al. 1984).

The endocardial surfaces of hearts with dilated cardiomyopathy and ventricular aneurysms have also been studied. Autopsy studies of individuals with dilated cardiomyopathy reveal multiple patchy areas of fibrosis, replacing the endothelium in 57% of left ventricular sections (Roberts et al. 1987). Patients with left ventricular aneurysms have two distinct types of endocardial abnormality. The first type, designated type I, consists of extensive endocardial fibroelastosis. Little or no thrombus is associated with this abnormality. However, patients within this subgroup are prone to ventricular arrhythmias. Type II aneurysms have little or no fibroelastosis. Their endothelial surface consists of organized thrombus which is closely adherent to a thin aneurysmal wall. The thrombus serves to reconstitute the normal configuration of the ventricular cavity with time (Hochman et al. 1984).

The most controversial component of Virchow's triad is the presence of a hypercoagulable state related to cardioembolic disease. Investigators have noted various degrees of consumptive coagulopathy in patients following presumed cardiogenic cerebral embolism (Takano et al. 1991). Atrial fibrillation has also been associated with an increased clotting tendency (Gustafsson et al. 1990; Kumagai et al. 1990).

Local hypercoagulable states have also been implicated in thrombus formation. Residual thrombus and the surface of a fresh thrombus are highly thrombogenic (Francis et al. 1983). The retained thrombus creates intracavitary irregularities, leading to regions of increased turbulence and a subsequent increased shear rate, which promotes platelet activation and deposition. Platelet deposition on residual thrombus is two to four times greater than on deeply injured vessel walls (Lassila et al. 1990). The surface of the residual clot exposes fibrin-bound thrombin to the circulating blood, causing activation of both platelets and the coagulation cascade. Fibrin-bound thrombin, however, differs from freely circulating thrombin in several clinically important ways. First, bound thrombin is not easily accessible to the large heparin–antithrombin III complexes (Weitz et al. 1990). Secondly, fibrin II polymer, cleaved from fibrinogen by activated thrombin, is a natural inhibitor of heparin (Hogg and Jackson 1989). Lastly, activated platelets release platelet factor 4, another natural inhibitor of heparin. These factors may limit heparin's effectiveness in preventing further thrombus accumulation. Low molecular weight heparins and heparinoids may be suitable alternatives to heparin in situations with retained thrombus because of their diminutive size and lack of endogenous inhibitors.

The interplay of forces retaining or displacing an intracavitary thrombus has never been adequately investigated because of its complexity. However, clinical observation has yielded important information regarding the embolic potential of intracavitary thrombi. Left ventricular thrombi which, on echocardiography, protrude into the ventricular lumen and are mobile, are prone to systemic embolization following myocardial infarction (Visser et al. 1985). On the other hand, chronic left ventricular aneurysms often contain thrombi that seldom embolize (0.35% per year) (Lapeyre et al. 1985). Patients with dilated cardiomyopathy also develop left ventricular thrombi, but have a higher (1.6% per year) incidence of systemic embolism if not anticoagulated (Fuster et al. 1981).

These observations had led to speculation concerning why certain thrombi embolize. Early thrombi are more emboligenic because of their friability and mobility. Thrombi within chronic ventricular aneurysms are less prone to embolization for three major reasons. First, the thrombus has a broad attachment to the ventricular wall; secondly, the absence of contractile fibers within the aneurysm prevents propulsion into the ventricular lumen and disturbance of the attachment; and lastly, the clot, because of its position within the aneurysm, is isolated from hydraulic forces within the ventricular lumen. Thrombi in dilated cardiomyopathy are more prone to embolize for exactly the opposite reason (Cabin and Roberts 1980) (Fig. 20.1).

Cardiac Chambers: Sources of Thromboembolism

Atrial Fibrillation

Atrial fibrillation is associated with more than half of all presumed cardioembolic strokes (Cerebral Embolism Task Force 1986, 1989). Patients with chronic non-valvular atrial fibrillation have a 5% per year incidence of stroke (Wolf et al. 1978; Dunn et al. 1989). Paroxysmal atrial fibrillation may carry a lower risk than chronic atrial fibrillation (Petersen and Godtfredsen 1986). More controversial is the relationship between the duration of atrial fibrillation and stroke risk (Wolf et al. 1978; Dunn et al. 1989).

Not all strokes associated with atrial fibrillation are a direct consequence of the arrhythmia: in only 18% of patients with atrial fibrillation is the arrhythmia the sole potential cause (Bogousslavsky et al. 1990) and patients often have coexistent appropriately situated atherosclerosis or other cardiac sources (Weinberger et al. 1988; Bogousslavsky et al. 1990).

The true incidence of cerebral infarction associated with atrial fibrillation is often underestimated because of a significant incidence of asymptomatic brain ischemia. Twenty-six to 45% of patients with atrial fibrillation without a history of stroke harbor silent infarcts (Petersen et al. 1987; Feinberg et al. 1990). The toll that such 'silent infarcts' take on cognition over time may be significant.

The risk of stroke with atrial fibrillation is greatly influenced by associated conditions. Rheumatic mitral valve disease associated with atrial fibrillation carries an 18-fold increase in stroke incidence (Wolf et al. 1978; Halperin and Hart 1988). Atrial fibrillation associated with thyrotoxicosis carries a 14% stroke risk over a variable time period (Staffurth et al. 1977; Bar-Sela et al. 1981; Petersen and Hansen 1988). Five risk factors, three clinical and two echocardiographic, are independently associated with an increased stroke risk in non-valvular atrial fibrillation: recent congestive heart failure (last 3 months), hypertension, previous thromboembolism, left ventricular dysfunction and left atrial enlargement. The three clinical features carry a greater than 7% annual stroke risk. Individuals with two or more of these clinical features have a 17.6% annual stroke risk (SPAF Investigators 1992a). Symptomatic brain embolism recurs in 10% of patients with atrial fibrillation per year, and in approximately 13% of inadequately anticoagulated patients within 2 weeks of their initial infarct (Hart et al. 1983).

Lone atrial fibrillation, often excluded from randomized trials, carries a low rate of associated cardioembolism. When this diagnosis is limited to individuals less than 60 years old without clinical or echocardiographic history of associated cardiopulmonary disease, embolism occurs in only 1.3% of patients over 15 years (Kopecky et al. 1987). This surprisingly low stroke rate suggests that atrial fibrillation may be a marker of other less obvious causative conditions (Chesebro et al. 1990).

Myocardial Infarction

Systemic thromboembolism is a major cause of morbidity and mortality in the first few months following myocardial infarction. The true incidence of thromboembolism is hard to assess because of its silent nature. However, autopsy studies reveal the occurrence of thromboemboli in approximately 1 in 4 patients not receiving anticoagulants (Cooperative Clinical Trial 1973). Symptomatic brain infarction accounts for the

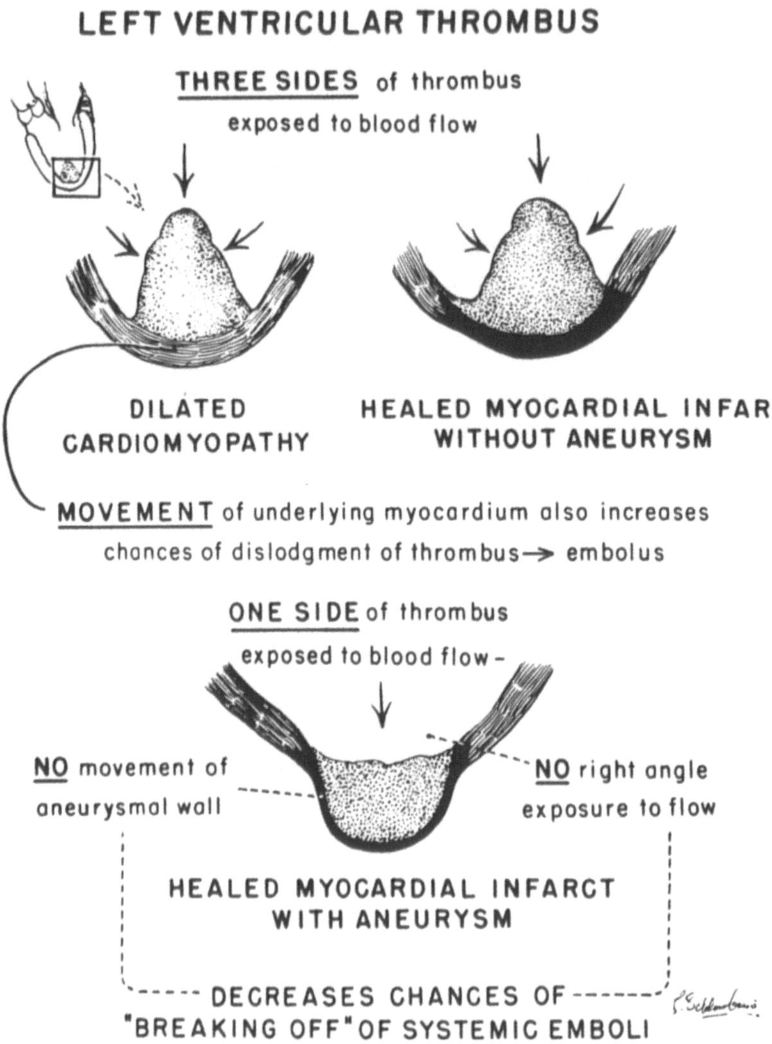

Fig. 20.1. Left ventricular thrombi in dilated cardiomyopathy and healed myocardial infarction without aneurysm are exposed on five sides to hemodynamic forces and are therefore more likely to embolize. Aneurysmal thrombi are isolated from flow except on one side and are attached to a non-contractile base, thereby decreasing the likelihood of embolism (Cabin and Roberts 1980).

majority of clinically identifiable emboli, with an incidence between 2.5% and 4.9% in patients not receiving anticoagulants (Wright et al. 1954; Working Party on Anticoagulant Therapy in Coronary Thrombosis 1969). Large anterior wall infarcts carry the highest risk of thrombus formation and subsequent cerebral embolism. The risk of embolism is highest in the first 3 months, and particularly during the first 10 days (Chesebro and Fuster 1986; Meltzer et al. 1986).

Transthoracic echocardiography has aided in identifying left ventricular mural thrombi ante mortem. Serial echocardiography identifies nearly all (96%) thrombi within 2 weeks of myocardial infarction (Kupper et al. 1989). One study identified mural thrombi formation in 34% (44 of 130) of anterior wall and only 1.5% (2 of 131) of inferior wall myocardial infarctions (Weinreich et al. 1984). The presence of severe apical dyskinesia/akinesia (Asinger et al. 1981) and a persistently abnormal left ventricular flow pattern (Delemarre et al. 1990) predicts future thrombus formation. The embolic potential of thrombi based upon their echocardiographic configuration has also been assessed. In one study, free mobility of the thrombus carried an 85% positive predictive value and an 89% negative predictive value for embolism. The absence of intraluminal protrusion was found to be 96% predictive of no future embolism (Visser et al. 1985).

Left Ventricular Aneurysm

The incidence of left ventricular aneurysm formation after myocardial infarction is 7%–10%, half of which contain mural thrombus (Schlicter et al. 1954; Dubnow et al. 1965). The incidence of embolism is about 5%, 80% of which occur within 6 weeks of myocardial infarction (Darling et al. 1967). Retrospective analysis of patients with chronic left ventricular aneurysms not receiving anticoagulant therapy indicates an embolic event rate of only 0.35 per 100 patient-years (Lapeyre et al. 1985). The low incidence of embolism reflects that thrombi in chronic left ventricular aneurysms are isolated from hydraulic forces within the lumen, and have a broad attachment to the aneurysm wall. The lack of contractile fibers within the aneurysmal sac also prevents the propulsion of contents into the outflow tract (Cabin and Roberts 1980).

Cardiomyopathies

Cardiomyopathies are generally divided into three subtypes: dilated, hypertrophic and restrictive. Hypertrophic cardiomyopathy leads to cardioembolism, usually secondary to associated atrial fibrillation (Furlan et al. 1984). Dilated and restrictive cardiomyopathies lead to thromboembolism secondary to poor contractility, congestive heart failure and associated rhythm disturbances. Dilated cardiomyopathy is by far the most common type of cardiomyopathy. In one study, more than 60% of individuals with idiopathic dilated cardiomyopathy had an echocardiographically defined mural thrombus, and more than 60% of these had either clinical or necropsy evidence of systemic emboli (Roberts et al. 1987). The reason for the high rate of thromboembolism is unknown. However, a narrow contractile thrombus attachment and exposure to intraluminal hydraulic forces probably play a role.

Cardiac Myxomas

Primary cardiac neoplasms are rare causes of stroke. However, because of advances in echocardiography more are now being detected. Cardiac myxomas comprise between 40% and 65% of primary cardiac neoplasms (Straus and Merliss 1945; Dein et al. 1987). Most cardiac myxomas reside in the left atrium. However, right atrial, biatrial and left ventricular myxomas have been described (Sutton et al. 1980; Marvasti et al. 1984; Dein et al. 1987).

Myxomas usually occur in isolation, and following surgical excision the rate of recurrrence is low (1%–3%). However, various subgroups with a worse prognosis have been identified. One subgroup of myxomas (5 of 85 in one series) has a 'complex' of unusual findings, including multiple pigmented skin lesions (lentiginosis), skin myxomas, myxoid fibroadenomas of the breast and primary pigmented nodular adrenocortical disease. Patients with this myxoma 'complex' present at a younger age, more frequently have multicentric tumors (53%) and recurrences (22%), and 35% have a primary relative with at least one component of the disorder. Another subgroup of patients with myxomas present with multifocal lesions. These individuals also present at a young age, and have a high rate of recurrence (33%) (McCarthy et al. 1986).

The risk of systemic embolism in patients with cardiac myxomas is between 35% and 45% (Silverman et al. 1962; Sandok et al. 1980). Twenty-seven per cent of individuals with left atrial myxomas had symptoms of brain ischemia (Sandok et al. 1980). However, autopsy and angiographic studies often show multiple asymptomatic emboli (Price et al. 1970). Both tumor and thrombotic emboli occur. The tumor's friability and intraluminal presence probably contribute to the high rate of systemic embolization.

The most common neurologic complication of myxoma is brain ischemia secondary to embolism. Myxomatous emboli can occlude the vessel lumen without wall invasion, or invade the vessel wall with both luminal stenosis and aneurysm formation. Aneurysm rupture, although rare, has been reported (Price et al. 1970). The least common neurologic manifestations of cardiac myxoma are leptomeningeal and intracerebral metastasis (New et al. 1970; Rankin and Desousa 1978).

Sick Sinus Syndrome

Sick sinus syndrome, a term originally used to define defects in the elaboration or conduction of sinus impulses, is the second most common cardiac arrhythmia associated with cardioembolic stroke. This rhythm abnormality is manifest on ECG as chaotic atrial activity, changing P-wave contour and bradycardia interspersed with multiple and recurrent ectopic beats and runs of artrial and nodal tachycardia (Lown 1967).

Systemic emboli occur in 14%–48% of individuals with sick sinus syndrome (Radford and Julian 1974; Rosenqvist et al. 1985). Patients with tachyarrhythmias are more prone to embolize

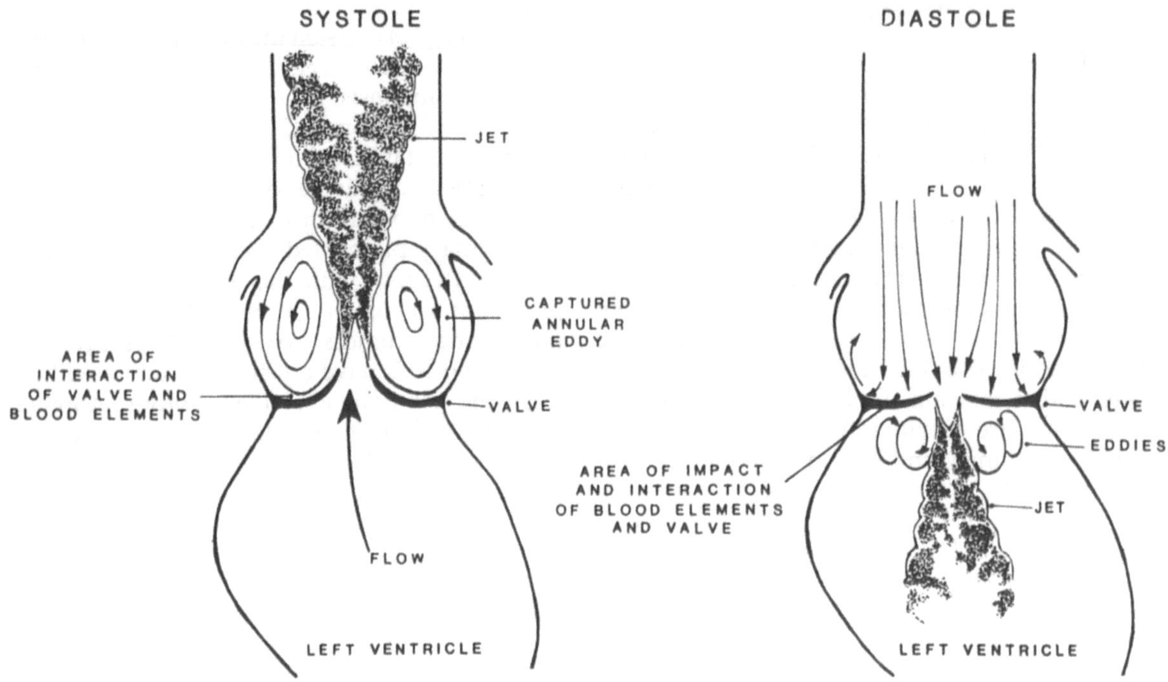

Fig. 20.2. Flow dynamics surrounding the aortic valve in systole (left) and diastole (right). Platelets activated in the high shear forces of the jet are exposed for prolonged periods to the valve surface after capture in annular eddies, thereby predisposing to thrombus formation (Stein et al. 1977).

than those with bradyarrhythmias (Bathen et al. 1978). Increased age has also been identified as a possible predictor of embolism (Rosenqvist et al. 1985).

Left Atrial Enlargement and Spontaneous Contrast Echoes

Left atrial enlargement remains a controversial risk factor for cardioembolic stroke, whether in isolation or in association with atrial fibrillation. Some of the controversy is related to difficulties in assessing the size of the left atrium at the time of embolism. This three-dimensional structure, the size of which is dependent upon circulatory volume status, is usually assessed with two-dimensional echocardiography at a time remote from the clinical event. These methodologic difficulties have led to conflicting results concerning the risk of cardioembolism (Wiener et al. 1987; Burchfiel et al. 1990).

Technological advances have introduced transesophageal echocardiography (TEE) into the diagnostic test arena (Tegeler and Downes 1991). TEE is particularly sensitive in detecting spontaneous contrast echoes (SCE) (Zenker et al. 1988), which are thought to represent the microaggregates of platelets and erythrocytes forming in areas of intracavitary circulatory stasis (Mahoney et al. 1989). The presence of left atrial SCEs has been correlated with a 1.6-fold increased incidence of embolism (Black et al. 1991). Patients with non-valvular atrial fibrillation may be at increased stroke risk when SCEs are present (Chimowitz et al. 1993). The presence of SCEs has also been correlated with increasing left atrial size (Castello et al. 1990; Lee et al. 1991). Although no prospective investigation of individuals with left atrial enlargement with and without SCEs has been done, individuals with the left atrial enlargement and SCEs are probably at an increased risk for cardioembolic stroke.

Valvular Embolic Sources

Valvular heart disease is associated with a 10%–20% risk of cardioembolic stroke (Cerebral Embolism Task Force 1986) and also with other sources of emboli, making assessment of associated risk difficult. Valvular emboli are usually composed of either white platelet-rich thrombus or dystrophic calcification. The size of valvular thromboemboli is usually smaller than those of mural origin, making the clinical spectrum of presentation different from that of intramural thrombi (Hoffman et al. 1990).

Pathogenesis of Valvular Thromboemboli

Cardiac valves function by allowing the passage and retention of blood within specific chambers while maintaining non-thrombotic surfaces. The passage of blood from a cardiac chamber through its outlet is dependent upon the cross-sectional area of the orifice, the pliability of the valve leaflets and the regularity of the valvular surface. Valvular stenosis is associated with progressive commissural adhesion and dystrophic calcification of the valve leaflets. Progressive valvular outlet obstruction is associated with increasing turbulence. Although low-intensity turbulence occurs in normal valves, especially with increasing cardiac output, the intensity of turbulence is markedly increased in the jet distal to a stenotic valve (Stein and Sabbah 1976). Platelets are activated in regions of increased turbulence and shear stress, and turbulence at the valve orifice is directly related to the amount of thrombus that forms (Stein and Sabbah 1974).

The pattern of flow in the region of a stenotic valve is composed of a central jetstream surrounded by annular eddies created between the outflow tract walls and the mainstream (Macagno and Hung 1967) (Fig. 20.2). These eddies allow blood elements to remain in contact with valve surfaces longer than in regions of laminar flow. Therefore, platelets activated in the turbulence of the jetstream are allowed prolonged contact with dystrophic valvular surfaces, leading to adhesion, further platelet activation and thrombus formation (Stein et al. 1977).

Valvular incompetence permits the prolonged presence of blood in the valvular region. If the valvular elements are diseased, then platelets are exposed for an extended period to thrombogenic tissue. Valvular incompetence is often associated with stenosis, further escalating the risk for platelet activation and thrombus formation.

Valvular surfaces are of major importance in promoting or preventing thrombus formation. Endogenous and prosthetic valves are limited in their effectiveness by surface thrombogenicity. Despite normal flow dynamics, platelets are activated by contact with artificial surfaces.

Platelet and fibrinogen kinetics have been used to predict thromboembolic risk in patients with valve lesions. Individuals with mitral valve prolapse and thromboembolism have decreased platelet survival compared to individuals with mitral valve prolapse without evidence of thromboemboli (Steele et al. 1979b). The shortened half-life of platelets probably reflects an increase in mitral valve surface abnormalities. Following prosthetic cardiac valve surgery, shortened platelet half-life and fibrinogen consumption with the production of fibrinopeptide A have been correlated with an increased incidence of thromboembolism (Weily et al. 1974; Pengo et al. 1989). Comparison of fibrinopeptide A levels, thought to reflect in-vivo thrombin formation, showed a twofold increase in individuals with mechanical valves compared to controls (Pengo et al. 1989). Fibrinopeptide A levels in those with bioprosthetic valves were not significantly different. Fibrinopeptide A levels were also correlated with the level of oral anticoagulation. These findings again reflect the added thrombogenicity of artificial valvular surfaces.

Valvular Disease: Sources of Embolism

Mitral Stenosis and Insufficiency

Rheumatic heart disease carries the greatest risk of embolism of any cardiac condition. Individuals with rheumatic mitral valve disease carry a one in five chance of clinically detectable embolism during the course of their illness (Coulshed et al. 1970; Fleming and Bailey 1971). Such individuals often have numerous cardiac abnormalities, which can provide an embolic source.

Mitral stenosis and incompetence often coexist, but one of the conditions predominates clinically. Of 500 individuals with mitral valve disease, 66% were predominantly stenotic, 21% were predominantly incompetent and 13% were mixed. The rate of clinically detectable embolism was 25%, of which two-thirds were cerebral. Thirty per cent of individuals had multiple emboli. Among individuals with clinically detectable emboli, 93% had predominantly mitral stenosis and 7% predominantly mitral incompetence (Fleming and Bailey 1971). Therefore, 35% of individuals with predominant mitral stenosis and 8% of those with predominant incompetence had clinically identifiable emboli. Associated atrial fibrillation predisposes to embolism (Coulshed et al. 1970).

Aortic Stenosis

Progressive calcific aortic stenosis may occur as a consequence of a congenital bicuspid aortic valve or rheumatic heart disease, or be idiopathic in nature. Regardless of etiology, the progressive nature of the disease seems to be related to the positive feedback loop created between flow turbulence and valve stenosis. Pathologic studies show microthrombi in 53% (10 of 19) of individuals with aortic stenosis. Thrombi are composed of hyalinized fibrin with platelet clumps and occasional erythrocytes and leukocytes. The valve surface shows some degree of fibrosis, with marked calcification (Stein et al. 1977). Both fibrin–platelet and calcific emboli are thought to arise from stenotic aortic valves. Most emboli from calcified aortic valves are small and lead to little morbidity or mortality (Brockmeier et al. 1981). However, an occasional fatal brain embolism does occur (Mills et al. 1978).

Mitral Valve Prolapse

Mitral valve prolapse is present in 5%–10% of the general population (Brown et al. 1975; Procacci et al. 1975). This common cardiac abnormality is associated with stroke, especially in patients less than 45 years of age (Barnett et al. 1980), and is the only detectable cause for stroke in 5%–30% of young adults (Barnett et al. 1980); Smith and McNight 1981).

Pathologically the valve appears voluminous, opaque and thickened. The posterior cusp is usually larger. Microscopically, the affected cusps show replacement of the fibrosa with myxomatous tissue and fibroelastic thickening of the endocardium. Fibrin–platelet thrombi are commonly found on the atrial surface of the posterior leaflet (Pomerance 1969).

Subpopulations of individuals with mitral valve prolapse have distinctively different risks for cardioembolic stroke. Individuals with mitral valve prolapse without auscultatory findings of a midsystolic click or murmur are at low risk for embolic complications (Nishimura et al. 1985). Thickened, redundant mitral valve leaflets and myxomatous degeneration of multiple valves carries a higher incidence of embolic complications (Chandraratna et al. 1983; Barletta et al. 1985; Nishimura et al. 1985).

Platelet kinetics have also been used to predict cardioembolic potential in individuals with mitral valve prolapse. Platelet survival time is significantly decreased in individuals who have had clinical embolism, compared to those with mitral valve prolapse without embolism (Steele et al. 1979a).

Like atrial fibrillation, mitral valve prolapse is often associated with other potential cardioembolic sources. Supraventricular tachyarrhythmias occur in a substantial number of patients with mitral valve prolapse (Winkle et al. 1975; Schwartz et al. 1977), the risk of infective endocarditis is about five times greater (Clemens et al. 1982; MacMahon et al. 1987) and Patent foramen ovale and atrial septal aneurysm occur with greater frequency (Schneider et al. 1990).

Mitral Annulus Calcification

The mitral annulus is part of the fibrous skeleton of the heart. It is divided into two parts, an anteromedial one-third giving rise to the anterior cusp and a posterolateral two-thirds forming the posterior cusp of the mitral valve.

Mitral annulus calcification occurs in approximately 10% of patients in large autopsy series (Simon and Liu 1954), but the prevalence is higher in women and the elderly. The pathophysiologic mechanism of mitral annulus calcification is probably multifactorial. Increased forces on the mitral valve related to increased left ventricular afterload or outflow obstruction and left ventricular hypertrophy predispose to the formation of mitral annulus calcification. An association with Marfan's syndrome and other connective tissue diseases implicates primary changes in the cardiac skeleton (Grossman et al. 1968). Abnormal calcium-phosphate metabolism has also been associated with an increased prevalence of mitral annulus calcification (Nestico et al. 1983).

Calcification occurs in the subvalvular recess, predominantly associated with the posterior cusp. These calcifications coalesce into a solid bar, sometimes attaining 2 cm in maximal diameter. Numerous calcific thrombi are seen adhering to the inner surface of the bar. Occasionally large excrescences form, protruding into the atrial and ventricular cavities (Kirk and Russell 1969).

Mitral annulus calcification has been associated with a variety of cardiac disorders. Conduction abnormalities include sinus node dysfunction, atrioventricular conduction blocks, atrial fibrillation and bundle branch blocks. Two mechanisms account for these arrhythmias. First, the mitral annulus is closely situated anatomically to the bundle of His and the AV node. Secondly, the cardiac conduction system is part of the cardiac skeleton and may be involved in a degenerative process similar to the mitral annulus (Nestico et al. 1984). Mitral annulus calcification has also been associated with mitral stenosis, mitral insufficiency and infectious endocarditis, all with a potential for the development of cardioembolism.

Mitral annulus calcification can cause cardioembolic stroke secondary to calcific emboli (Ridolfi and Hutchins 1976; Fulkerson et al. 1979). Because of its frequent association with other causes for brain infarction, the risk for cardioembolic stroke due to mitral annular calcification alone is unknown. However, it may be more often a marker for atherosclerosis than a source for cardioemboli (Furlan et al. 1984).

Mitral Valve Strands

Transesophageal echocardiography (TEE) has greatly helped clinicians to identify potential cardioembolic sources. A recent TEE study of 50 individuals with stroke and no defined embolic source on transthoracic echocardiography showed that 22% (11 of 50) had mitral valve strands, which were strongly correlated with mitral valve thickening or calcification (Lee et al. 1991).

The histology of mitral valve strands is unknown. Possibilities include fibrin strands, thrombi and torn chordae. One report of a mitral valve replacement with previously diagnosed mitral valve strands revealed Lambl's excrescences (Lee et al. 1991). Pathologically, these are filiform processes found on degenerated heart valves (Heath et al. 1961), and have been associated with cardioemboli (Harris and Adelson 1965; Abu Nassar and Porter 1971). The clinical importance of mitral valve strands still needs to be established.

Prosthetic Cardiac Valves

Thromboembolism accounts for more than half of prosthetic valve complications (Edmunds 1987). The central nervous system is the site of 80%–90% of all clinically detectable thromboembolic events (Edmunds 1982), and between 20% and 30% of individuals will have recurrent emboli if they are not adequately anticoagulated (Wilson et al. 1978; Kohler 1982). The fatality rate of these events approaches 15%–25% (Edmunds 1987).

As with endogenous valves, prosthetic valves are limited functionally by their hemodynamic characteristics and the thrombogenicity of the surfaces. The additional factor of valvular half-life must also be considered. Bioprosthetic valves include glutaraldehyde-preserved porcine valves and homograft valves. The advantages of bioprosthetic valves are their relatively natural hemodynamics and low material thrombogenicity. The limitation of this class of valves is prosthetic half-life. Degeneration of the valve slows with increasing age (Antunes and Santos 1984; Magilligan et al. 1985). Deterioration of bioprosthetic valves causes loss of endothelium, with the exposure of large regions of subendothelial collagen increasing the valves' thrombogenic surfaces (Riddle et al. 1981).

The thromboembolic rate from bioprosthetic valves, although low, is still greater than that from normal valves. Bioprosthetic aortic valves have a 1% per year chance of thromboembolic complications. When bioprosthetic valves are present in the mitral or mitral and aortic positions, the risk of thromboembolism rises to 2% per year without anticoagulation. The rates decrease to 0.5% and 1% per year respectively with anticoagulation (Chesebro and Fuster 1992).

Mechanical valves are currently available in the caged ball, tilting disc and bileaflet types. All three types have distinctly different hemodynamics, leading to different degrees of turbulence, flow obstruction and insufficiency. These shortcomings, along with an increased material–blood incompatibility, increase the thromboembolic rate. Although more thrombogenic, mechanical valve failure secondary to degeneration is rare (Whittlesey and Geha 1991).

The incidence of clinically detectable thromboembolism in patients with mechanical valves also depends on valve location. Aortic mechanical valves have a 4% annual incidence of systemic embolism, which is reduced to 2% per year with adequate anticoagulation. Mitral mechanical valves have a 5% per year incidence of systemic emboli, again approximately halved with adequate anticoagulation (Chesebro and Fuster 1992).

The pathophysiologic events that predispose to thromboembolism begin during surgery. Prosthetic materials and damaged perivalvular tissue cause platelet activation as soon as circulation resumes. A Dacron sewing ring, which is common to all prosthetic valves, forms a nidus for platelet activation and adhesion (Dewanjee et al. 1984). Prosthetic materials also activate Factor XII, Hagemann factor, initiating the intrinsic pathway of the coagulation cascade (Edmunds 1987). Valvular design and manufacturing determine the amount of turbulent and stagnant flow leading to platelet and coagulation cascade activation with subsequent thrombus formation.

Platelet activation is increased in virtually all

prosthetic valves, with the possible exception of aortic homografts (Harker and Schlichter 1970; Steele et al. 1975). Both bioprosthetic and mechanical valves decrease platelet survival (Harker and Schlichter 1970). Mechanical ball valves also decrease platelet adhesiveness secondary to platelet trauma induced by the valves (Dale 1977).

Plasma proteins have also been evaluated in prosthetic valve replacement to aid in predicting thrombus formation. Levels of β-thromboglobulin, a platelet protein released after platelet activation, are elevated in patients with prosthetic valves. Levels are higher in multiple mechanical valve replacement than in single mechanical valve replacement, and with porcine valve implantation, but not to the levels found with single mechanical valve replacement (Pumphrey and Dawes 1981). Fibrinopeptide A, a degradation product of fibrinogen activation by thrombin, is elevated in individuals with mechanical prosthetic heart valves on coumadin. Fibrinopeptide A levels are inversely correlated with level of anticoagulation and may represent a more physiologic means for assessing the dose of anticoagulants (Pengo et al. 1989).

Other factors contribute to the incidence of thromboembolism in individuals with prosthetic cardiac valves. The most important identifiable risk factor in most studies is inadequate anticoagulation. Other commonly cited predisposing conditions include recent surgery, operation before the mid-1970s, left ventricular failure, atrial fibrillation, left atrial thrombus and previous embolus (Farah et al. 1984; Roux et al. 1984).

Infectious Endocarditis

Neurologic complications occur in approximately 30%–40% of individuals with bacterial endocarditis and the rates for prosthetic and endogenous valves are similar (Jones et al. 1969; Pruitt et al. 1978; Salgado et al. 1989). A neurologic symptom is the presenting complaint in up to 29% of patients (Garvey and Neu 1978).

Bacterial endocarditis has been classically divided into acute or fulminant and subacute, based upon the aggressiveness of the infectious organism. Perhaps it is more appropriate to refer to the etiologic agent, for example, *Staphylococcus aureus* endocarditis, which better describes the predicted clinical course and treatment.

The causative organisms and predisposing cardiac lesions are changing. Rheumatic heart disease was once the most prevalent cardiac disease predisposing to endocarditis, but now mitral valve prolapse, congenital heart disease and intravenous drug abuse are among the most commonly associated conditions in the young (Lachman et al. 1975; Johnson et al. 1975; Openshaw 1976), whereas degenerative calcific valvular lesions, including aortic stenosis and mitral annular calcification and prosthetic valves, are predisposing conditions in the elderly (Calderwood et al. 1985; McKinsey et al. 1987). Along with the changes in predisposing conditions, the organisms responsible for endocarditis have also changed. Streptococcal and staphylococcal species remain the most common pathogens (Terpenning et al. 1957). However, fungal and enterobacteriaceae are becoming more common, perhaps secondary to immunosupression, changing valvular substrate and antimicrobial use.

The characteristic pathologic feature of bacterial endocarditis is the presence of vegetations on the valve cusps or leaflets. The mitral valve, followed by the aortic valve, are most commonly involved. Simultaneous vegetations on both the aortic and mitral valves are common.

Histologically, vegetations consist of a mass of amorphous fibrin, platelets and erythrocytes, along with masses of organisms and inflammatory cells. The organisms are often buried deep within the vegetation, which might help to explain the difficulty in controlling these infections. Vegetations range in size from several millimeters to several centimeters and their size and friability have been implicated in their potential for embolism (Buja 1987).

The neurologic complications of infectious endocarditis are related to the number, size, location and infectivity of the emboli. Patients with infectious endocarditis may present with a stroke, encephalopathy, mycotic aneurysm, intracerebral abscess, meningitis, seizures or headache. In one study, approximately 70% (45 of 64) of neurologic complications occurred before the initiation of antimicrobial therapy. The remaining 30% (19 of 64) of individuals developed their neurologic complications within a median time of 4 days (range 1–24) after the implementation of antimicrobial therapy (Salgado et al. 1989).

Stroke is the most common neurologic manifestation of infectious endocarditis, occurring in 17%–26% of individuals in two large studies (Pruitt et al. 1978; Salgado et al. 1989). The middle cerebral arteries and their branches are the most likely lodging sites for emboli, occurring in

38 of 44 (Jones et al. 1969) and 34 of 38 (Pruitt et al. 1978) instances. The remainder of the patients had vertebrobasilar circulation symptoms.

Intraparenchymal hemorrhage occurs in approximately 5% of individuals with infectious endocarditis (Pruitt et al. 1978; Salgado et al. 1989), and occurs secondary to three distinct pathophysiologic mechanisms. Cerebral septic embolism with subsequent erosive arteritis is the cause of both early intraparenchymal hemorrhage and mycotic aneurysm formation. Hemorrhagic conversion of a bland cerebral infarction completes the triad (Hart et al. 1987).

Experimental studies of cerebral septic emboli document the development of septic arteritis within 1 day following embolism. Vessel destruction is thought to occur due to the escape of organisms from the vessel lumen to Virchow–Robin spaces via the vasa vasorum of the occluded vessel segment. Organisms then invade the adventitia of the vessel, eroding more medial layers until the internal elastic lamina is reached. By 24 hours after embolization aneurysm formation is evident in animals with staphylococcal emboli without antibiotic therapy (Molinari 1972; Molinari et al. 1973). Without antimicrobial therapy, all animals with either *Staphylococcus aureus* or *Escherichia coli* arteritis died from intracranial hemorrhage, whereas only 1 of 12 animals with streptococcal arteritis developed a mycotic aneurysm (Molinari et al. 1973).

Perhaps the most frequent cause of intraparenchymal hemorrhage associated with endocarditis is isolated erosive arteritis. Septic erosion of the arterial wall without subsequent aneurysm formation was found in 4 of 5 autopsied patients with endocarditis. As in the animal model, most hemorrhages occurred in endocarditis secondary to virulent organisms within 48 hours of admission (Hart et al. 1987).

Mycotic aneurysms occur in 5% (Pankey 1962) to 12% (Pruitt et al. 1978) of individuals with endocarditis and proven mycotic aneurysms are the cause of intracranial hemorrhage in 1.7% of patients with endocarditis (Hart et al. 1987). More than half of the aneurysms that rupture do so without warning (Salgado et al. 1987), but aneurysm rupture is less likely following a course of antimicrobial therapy (Salgado et al. 1987).

Hemorrhagic infarction accounts for about 10% (2 of 20) of all brain infarcts in endocarditis (Pruitt et al. 1978; Salgado et al. 1989). The pathophysiologic mechanism is probably embolus lysis, with subsequent reperfusion and hemorrhage into necrotic tissue.

Acute encephalopathy is the second most frequent neurologic complication of endocarditis (Jones et al. 1969; Pruitt et al. 1978; Salgado et al. 1989). Multifocal septic microemboli with subsequent microinfarction and abscess formation is the pathophysiologic mechanism. An autopsy study of 45 patients with neurologic complications of endocarditis showed multiple microinfarcts in 23% and microabscesses in 26% (Pruitt et al. 1978).

Non-bacterial Thrombotic Endocarditis

Non-bacterial thrombotic endocarditis is a rare source of cerebral embolism in the general population. In two large autopsy series it was found in 0.71% (99 of 13 913) and 1.6% (65 of 4096) of cases (Deppisch and Fayemi 1976; Biller et al. 1982). Non-bacterial thrombotic endocarditis is associated with both debilitating illnesses and cancer: a 7% incidence of marantic endocarditis in individuals with cancer has been reported (Croft and Wilkinson 1963). Mucin-secreting lung and gastrointestinal adenocarcinomas are the most commonly associated cancers.

Pathologically, non-bacterial thrombotic endocarditis is characterized by the accumulation of small vegetations of fibrin and blood elements along valve leaflets. The mitral and aortic valves are most commonly involved (Deppisch and Fayemi 1976; Kooiker et al. 1976; Biller et al. 1982). The masses are sterile and small (1–5 mm) and histologically there is no inflammatory response, unlike with infectious endocarditis.

Embolism has been reported in 63% (62 of 99) of non-bacterial thrombotic endocarditis patients. Of these, 53% (33 of 62) involved the central nervous system (Biller et al. 1982), and most of these patients had multiple brain emboli. Approximately one-half of infarcts are hemorrhagic. As with other forms of emboli, the middle cerebral artery territory is the major lodging site (Biller et al. 1982).

Libman–Sacks Endocarditis

Libman–Sacks endocarditis is found in individuals with systemic lupus erythematosus. Pathologically, 1–4 mm vegetations of amorphous material are found on the inferior surfaces of the mitral and tricuspid valves. These vegetations may embolize and cause brain infarcts (Futrell and Millikan 1989; Kitagawa et al. 1990).

Paradoxic Cardioembolism

Pathophysiologic Mechanisms of Paradoxic Embolism

The necessary components for paradoxic embolism consist of a systemic venous circulation embolus and a right-to-left cardiac shunt. The most commonly implicated source of paradoxic emboli is venous thrombosis, which is often difficult to establish: even in patients with documented pulmonary emboli, a venous source is identified in only 50% (Rosenow et al. 1981). Right atrial myxomas have also been associated with presumed paradoxic embolism (Butler et al. 1986). Air embolism occurring during surgery or the removal of venous instrumentation in the seated position may also cause paradoxic emboli. Nitrogen bubble embolism occurring during decompression is yet another source (Moon et al. 1989; Wilmshurst et al. 1989).

Classically, right-to-left shunting was thought to occur only with persistently elevated right heart pressures. However, contrast echocardio graphy has demonstrated the presence of right-to-left shunting in other circumstances. The most common situation in which shunting occurs is during each cardiac cycle. The contraction of the right atrium precedes that of the left. This lead time allows for the transient development of a positive pressure gradient between the right and left heart. Provided a communication exists between the two chambers, the shunting of blood occurs transiently from the pulmonary to the systemic circulation.

Another common circumstance in which right-to-left shunting of blood occurs is following release from Valsalva-like maneuvers (e.g. coughing, straining, micturition). During Valsalva, positive airway pressure causes a decrease in the venous return secondary to a diminished pressure gradient between the venous circulation and the thoracic cavity. After release of the positive airway pressure, the increase in venous return increases right atrial pressure, causing a transient reversal of the pressure gradient between the right and left atria.

A right atrial mass may also cause shunting by obstructing the flow of blood into the right atrium from the inferior vena cava. If a passageway to the left atrium exists inferior to the mass, then preferential streaming of venous return may occur, with the development of local pressure gradients and subsequent shunting (Langholz et al. 1991).

Right-to-Left Communications

Patent Foramen Ovale and Atrial Septal Aneurysm

Patent foramen ovale is present in 30% of individuals from 20 to 50 years of age who come to autopsy (Hagen et al. 1984). The size of the foramen ranges from 1 to 10 mm in diameter (mean 4.5 mm). Contrast echocardiography using bubbles injected intravenously detects a patent foramen ovale in 10%–15% (Lechat et al. 1988; Webster et al. 1988) without provocative maneuvers in young individuals. Young patients with stroke, when compared to matched controls, have a 3.3- to fourfold increased incidence of patent foramen ovale detectable by contrast echocardiography without provocative procedures (Lechat et al. 1988; Webster et al. 1988).

Atrial septal aneurysm is a localized outpouching of the fossa ovalis region of the atrial septum that has been identified in 1% of autopsies (Silver and Dorsey 1978). Individuals with atrial septal aneurysms have an increased incidence of stroke (Pearson et al. 1991b). Two mechanisms have been postulated. First, thrombus may form in the aneurysm and subsequently embolize. Echocardiography studies (Schneider et al. 1990), however, detected an associated thrombus in only 10% (2 of 23) of atrial septal aneurysms. Secondly, atrial septal aneurysms are associated with patent foramen ovale and right-to-left shunting in 70%–83% of patients (Schneider et al. 1990; Pearson et al. 1991b). This suggests that atrial septal aneurysm is sometimes a marker for patent foramen ovale.

Mitral valve prolapse is also associated with patent foramen ovale. When patients with stroke and mitral valve prolapse with patent foramen ovale were compared to those without a patent foramen ovale, a twofold increase in stroke incidence was found (Lechat et al. 1988).

Ventricular Septal Defects

Ventricular septal defects are the most common congenital cardiac anomaly, comprising 20%–30% of congenital heart defects and occurring in approximately 1 in 1000 live births (Keith et al. 1971). They are frequently associated with other structural defects, especially cyanotic heart disease. The membranous portion of the ventricular septum is the location of 90% of ventricular septal defects. The size of the defect, although extremely variable, helps determine the functional significance. Defects greater than 1 cm in diameter may allow significant left-to-right flow, with the subsequent development of pulmonary hypertension. With time, pulmonary pressures increase causing increased right ventricular pressures and the reversal of flow across the shunt (Hoffman and Rudolph 1965). Cardiac systole sequencing may also predispose to right-to-left shunts. If the right ventricle contracts before the left, a transient flow reversal across the shunt occurs.

The incidence of stroke in individuals with ventricular septal defects attributable to paradoxic embolism is unknown. However, two recent cases of presumed paradoxic embolic stroke have been reported (Shuiab 1989; Silka and Rice 1991). Besides paradoxic emboli, another cardioembolic cause for stroke in individuals with ventricular septal defects is infectious endocarditis (Keith et al. 1971).

Atrial Septal Defects

Atrial septal defects (ASD) are among the most common of congenital heart anomalies, with an incidence of 5%–15%. Three types of ASD occur: ostium primum, ostium secundum and sinus venosus. The location of the ASD allows subtype determination. Sinus venosus ASDs occur near the entry of the superior vena cava; ostium secundum ASDs are midseptal, resulting from incomplete closure of the septum primum; and ostium primum ASDs are located near the atrioventricular junction. Ostium secundum ASDs comprise about 90% of all ASDs. As with patent foramen ovale, atrial septal defects may predispose to paradoxic emboli. Confirmed cases of embolus within an atrial septal defect have been reported (Nellessen et al. 1985), but associated atrial arrhythmias and bacterial endocarditis sometimes make the identification of the source of an embolus difficult.

References

Abu Nassar SG, Porter JC (1971) Incidental papillary endocardial tumor: its potential significance. Arch Pathol 92:370–377

Antunes MJ, Santos LP (1984) Performance of glutaraldehyde-preserved porcine bioprosthesis as a mitral valve substitute in a young population group. Ann Thorac Surg 37:387–392

Asinger RW, Mikell FL, Elsperger J, Hodges M (1981) Incidence of left-ventricular thrombosis after acute transmural myocardial infarction. New Engl J Med 305:297–302

Barletta GA, Gagliardi R, Benvenuti L, Fantini F (1985) Cerebral ischemic attacks as a complication of aortic and mitral valve prolapse. Stroke 16:219–223

Barnett HJM, Boughner DR, Taylor DW, Cooper PE, Kostuk WJ, Nichol PM (1980) Further evidence relating mitral-valve prolapse to cerebral ischemic events. New Engl J Med 302:139–144

Bar-Sela S, Ehrenfeld M, Eliakim M (1981) Arterial embolism in thyrotoxicosis with atrial fibrillation. Arch Intern Med 141:1191–1192

Bathen J, Sparr S, Rokseth R (1978) Embolism in sinoatrial disease. Acta Med Scand 203:7–11

Baumgartner HR, Haudenschild (1972) Adhesion of platelets to subendothelium. Ann NY Acad Sci 201:22–36

Biller J, Challa VR, Toole JF, Howard VJ (1982) Nonbacterial thrombotic endocarditis. Arch Neurol 39:95–98

Black IW, Hopkins AP, Lee LC, Walsh WF (1991) Left atrial spontaneous echo contrast: a clinical and echocardiographic analysis. J Am Coll Cardiol 18:398–404

Bogousslavsky J, Hachinski VC, Boughner DR, Fox AJ, Vinuela F, Barnett HJM (1986) Cardiac and arterial lesions in carotid transient ischemic attacks. Arch Neurol 43:223–228

Bogousslavsky J, Van Melle G, Pegli F, Kappenberger (1990) Pathogenesis of anterior circulation stroke in patients with nonvalvular atrial fibrillation: the Lausanne stroke registry. Neurology 40:1046–1050

Brockmeier LB, Adolph RJ, Gustin BW, Holmes JC, Sacks JG (1981) Calcium emboli to the retinal artery in calcific aortic stenosis. Am Heart J 101:32–37

Brown OR, Kloster FE, DeMots H (1975) Incidence of mitral valve prolapse in the asymptomatic normal. Circulation 51&52 (Suppl 2):77 Abstract

Buja LM (1987) The heart. In: Robbins S, Kumar V (eds) Basic pathology, 4th edn. WB Saunders, Philadelphia, pp 312–350

Burchfiel CM, Hamermeister KE, Krause-Steinrauf H et al. (1990) Left atrial dimension and risk of systemic embolism in patients with a prosthetic heart valve. J Am Coll Cardiol 15:32–41

Butler MJ, Adams HP, Hiratzka LF (1986) Recurrent cerebral embolism from a right atrial myxoma. Ann Neurol 19:608

Cabin HS, Roberts WC (1980) Left ventricular aneurysm, intra-aneurysmal thrombus and systemic embolus in coronary heart disease. Chest 77:586–590

Calderwood SB, Swinski LA, Waternaux CM, Karchmer AW, Buckley MJ (1985) Risk factors for the development of prosthetic valve endocarditis. Circulation 72:31–37

Castello R, Pearson AC, Labovitz AJ, Wallace P (1990) Atrial spontaneous contrast in patients undergoing transesophageal echocardiography: prevalence and clinical applications. Am J Cardiol 65:603–606

Cerebral Embolism Task Force (1986) Cardiogenic brain embolism. Arch Neurol 43:71–84

Cerebral Embolism Task Force (1989) Cardiogenic brain embolism: the second report of the Cerebral Embolism Task Force. Arch Neurol 46:727–743

Chandraratna PA, Nimalasurya A, Duncan P et al. (1983) Echocardiographic identification of a subset with increased cerebrovascular abnormalities in mitral valve prolapse. J Am Coll Cardiol 1:607

Chesebro JH, Fuster V (1986) Antithrombotic therapy for acute myocardial infarction: mechanism and prevention of deep venous, left ventricular, and coronary artery thromboembolism. Circulation 74(Suppl 3):1–10

Chesebro JH, Fuster V (1992) Valvular heart disease and prosthetic heart valves. In: Fuster V, Verstraete M (eds)

Thrombosis in cardiovascular disorders. WB Saunders, Philadelphia, pp 191–214
Chesebro JH, Fuster V, Halperin JL (1990) Atrial fibrillation risk marker for stroke. New Engl J Med 323:1556–1558
Chimowitz M, DeGeorge M, Poole M et al. (1993) Left atrial spontaneous contrast is a risk factor for stroke. Stroke 24:188
Clemens JD, Horwitz RI, Jaffe CC, Feinstein AR, Stanton BF (1982) A controlled evaluation of the risk of bacterial endocarditis in persons with mitral valve prolapse. New Engl J Med 307:776–781
Cooperative Clinical Trial (1973) Anticoagulants in acute myocardial infarction – results of a cooperative clinical trial. JAMA 225:724–729
Coulshed N, Epstein EJ, McKendrick CS, Galloway RW, Walker E (1970) Systemic embolism in mitral valve disease. Br Heart J 32:26–34
Croft PB, Wilkinson M (1963) Carcinomatous neuromyopathy: its incidence in patients with carcinoma of the lung and carcinoma of the breast. Lancet 1:184–188
Dale J (1977) Reduced platelet adhesiveness in patients with prosthetic ball valves: relation to adenosine diphosphate and mechanical trauma. Am Heart J 94:562–567
Darling RC, Austen WG, Linton RR (1967) Arterial embolism. Surg Gynecol Obstet 124:106–114
Dein JR, Frist WH, Stinson EB et al. (1987) Primary cardiac neoplasms. J Thorac Cardiovasc Surg 93:502–511
Delemarre BJ, Visser CE, Bot H, Dunning AJ (1990) Prediction of apical thrombus formation in acute myocardial infarction based on left ventricular spatial flow pattern. J Am Coll Cardiol 15:355–360
Deppisch LM, Fayemi AO (1976) Non-bacterial thrombotic endocarditis – clinicopathologic correlations. Am Heart J 92:723–729
Dewanjee MK, Trastek VF, Tago M, Kaye MP (1984) Radioisotopic techniques for noninvasive detection of platelet deposition in bovine-tissue mitral-valve prostheses and in vitro quantification of visceral microembolism in dogs. Invest Radiol 19:535–542
Dubnow MH, Burchell HB, Tituss JL (1965) Postinfarction ventricular aneurysm: a clinicomorphologic and electrocardiographic study of 80 cases. Am Heart J 70:753–760
Dunn M, Alexander J, de Silva R et al. (1989) Antithrombotic therapy in atrial fibrillation. Chest 95(Suppl):118s–127s
Edmunds LH (1982) Thromboembolic complications of current cardiac valvular prostheses. Ann Thorac Surg 34:96–106
Edmunds LH (1987) Thrombotic and bleeding complications of prosthetic heart valves. Ann Thorac Surg 44:430–445
Farah E, Enriques-Sarano M, Vahanian A (1984) Thrombotic and embolic complications in mechanical and biologic aortic prosthesis. Eur Heart J 5(Suppl D):43–47
Feinberg WM, Seeger JF, Carmody RF, Anderson DC, Hart RG, Pearce LA (1990) Epidemiologic features of asymptomatic cerebral infarction in patients with nonvalvular atrial fibrillation. Arch Intern Med 150:2340–2344
Fleming HA, Bailey SM (1971) Mitral valve disease, systemic embolism and anticoagulants. Postgrad Med J 47:599–604
Francis CW, Markham RE, Barlow GH, Florack TM, Dobrzynski DM, Marder VJ (1983) Thrombin activity of fibrin thrombi and soluble plasmic derivatives. J Lab Clin Med 102:220–230
Fulkerson PK, Beaver BM, Auseon JC, Graber HL (1979) Calcification of the mitral annulus – etiology, clinical association, complications and therapy. Am J Med 66:967–977
Furlan AJ, Craciun AR, Raju NR, Hart N (1984) Cerebrovascular complications associated with idiopathic hypertrophic subaortic stenosis. Stroke 15:282–284
Fuster V, Gersh BJ, Giuliani ER, Tajik AJ, Brandenburg RO, Frye RL (1981) The natural history of idiopathic dilated cardiomyopathy. Am J Cardiol 47:525–531
Futrell N, Millikan C (1989) Frequency, etiology, and prevention of stroke in patients with systemic lupus erythematosus. Stroke 20:583–591
Garvey GJ, Neu HC (1978) Infectious endocarditis – an evolving disease. Medicine 57:105–127
Grossman M, Knott AP, Jacoby WJ (1968) Calcified annulus fibrosus with mitral annulus fibrosus in the Marfan's syndrome. Arch Intern Med 121:561–563
Gustafsson C, Blomback M, Britton M, Hamsten A, Svensson J (1990) Coagulation factors and the increased risk of stroke in nonvalvular atrial fibrillation. Stroke 21:47–51
Hagen PT, Scholz DG, Edwards WD (1984) Incidence and size of patent foramen ovale during the first 10 decades of life: an autopsy study of 965 normal hearts. Mayo Clin Proc 59:17–20
Halperin JL, Hart RG (1988) Atrial fibrillation and stroke: new ideas, persisting dilemmas. Stroke 19:937–941
Harker LA, Schlichter SL (1970) Studies of platelet and fibrinogen kinetics in patients with prosthetic heart valves. New Engl J Med 283:1302–1305
Harris LS, Adelson L (1965) Fatal coronary embolism from a myxomatous polyp of the aortic valve. Am J Clin Pathol 43:61–65
Hart RG, Coull BM, Hart D (1983) Early recurrent embolism associated with nonvalvular atrial fibrillation: a restrospective study. Stroke 14:688–693
Hart RG, Kagen-Hallet K, Joerns SE (1987) Mechanisms of intracranial hemorrhage in infectious endocarditis. Stroke 18:1048–1056
Heath D, Best PV, Davis BT (1961) Papilliferous tumors of the heart valves. Br Heart J 23:20–24
Hochman JS, Platia EB, Healy-Bulkley B (1984) Endocardial abnormalities in left ventricular aneurysms – a clinicopathologic study. Ann Intern Med 100:29–35
Hoffman JIE, Rudolph AM (1965) The natural history of ventricular septal defects in infancy. Am J Cardiol 16:634–653
Hoffman T, Kasper W, Meinertz T, Geibel A, Just H (1990) Echocardiographic evaluation of patients with clinically suspected arterial emboli. Lancet 336:1421–1424
Hogg, PJ, Jackson CM (1989) Fibrin monomer protects thrombin from inactivation by heparin–antithrombin III: implications for heparin efficacy. Proc Natl Acad Sci USA 86:3619–3623
Jackson AC, Boughner DR, Barnett HJM (1984) Mitral valve prolapse and cerebral ischemic events in young patients. Neurology 34:784–787
Johnson DH, Rosenthal A, Nadas AS (1975) A forty-year review of bacterial endocarditis in infancy and childhood. Circulation 51:581–588
Johnson RC, Crissman RS, DiDio LJA (1979) Endocardial alterations in myocardial infarction. Lab Invest 40:183–193
Jones HR, Siekert RG, Geraci JE (1969) Neurologic manifestations of bacterial endocarditis. Ann Intern Med 71:21–28
Keith JD, Rose V, Collins G, Kidd BSL (1971) Ventricular septal defect – incidence, morbidity, and mortality in various age groups. Br Heart J 33S:81–87
Kirk RS, Russell JGB (1969) Subvalvular calcification of the mitral valve. Br Heart J 31:684–692
Kitagawa Y, Gotoh F, Koto A, Okayasu H (1990) Stroke in systemic lupus erythematosus. Stroke 21:1533–1539
Kohler RL (1982) Recurrent embolic cerebral infarction in anticoagulation. Neurology 32:283–285
Kooiker JC, Maclean JM, Sumi SM (1976) Cerebral embolism,

marantic endocarditis, and cancer. Arch Neurol 33:260–264
Kopecky SL, Gersh BJ, McGoon MD et al. (1987) The natural history of lone atrial fibrillation. New Engl J Med 317:669–674
Kumagai K, Fukunami M, Ohmori M, Kitabatake A, Kamada T, Hoki N (1990) Increased intracardiovascular clotting in patients with chronic atrial fibrillation. J Am Coll Cardiol 16:377–380
Kupper AJF, Verheugt FWA, Peels CA, Galema TW, Roos JP (1989) Left ventricular thrombus incidence and behavior studied by serial two-dimensional echocardiography in acute anterior myocardial infarction: left ventricular wall motion, systemic embolism and oral anticoagulation. J Am Coll Cardiol 13:1514–1520
Lachman AS, Bramwell-Jones DM, Lakier JB, Pocock WA, Barlow JB (1975) Infectious endocarditis in the billowing mitral leaflet syndrome. Br Heart J 37:326–330
Langholz D, Louie EK, Konstadt SN, Rao TLK, Scanlon PJ (1991) Transesophageal echocardiographic demonstration of distinct mechanisms for right to left shunting across a patent foramen ovale in the absence of pulmonary hypertension. J Am Coll Cardiol 18:1112–1117
Lapeyre AC, Steele PM, Kazmier FJ, Chesebro JH, Vlietstra RE, Fuster V (1985) Systemic embolism in chronic left ventricular aneurysm: incidence and the role of anticoagulation. J Am Coll Cardiol 6:534–538
Lassila R, Badimon JJ, Vallabhajasula S, Badimon L (1990) Dynamic monitoring of platelet deposition on severely damaged vessel wall in flowing blood. Effects of different stenosis on thrombus growth. Arteriosclerosis 10:306–315
Lechat P, Mas JL, Lascault G et al. (1988) Prevalence of patent foramen ovale in patients with stroke. New Engl J Med 318:1148–1152
Lee RJ, Bartzokis T, Yeoh T, Grogin HR, Choi D, Schnittger I (1991) Enhanced detection of intracardiac sources of cerebral emboli by transesophageal echocardiography. Stroke 22:734–739
Lown B (1967) Electrical reversion of cardiac arrhythmias. Br Heart J 29:469–489
Macagno EO, Hung TK (1967) Computational and experimental study of a captive annular eddy. J Fluid Mech 28:43–64
McCarthy PM, Piehler JM, Schaff HV et al. (1986) The significance of multiple, recurrent, and 'complex' cardiac myxomas. J Thorac Cardiovasc Surg 91:389–396
McKinsey DS, Ratts TE, Bisno AL (1987) Underlying cardiac lesions in adults with infective endocarditis: a changing spectrum. Am J Med 82:681–688
MacMahon SW, Roberts K, Kramer-Fox R, Zucker DM, Roberts RB, Devereux RB (1987) Mitral valve prolapse and infectious endocarditis. Am Heart J 113:1291–1298
Magilligan DJ, Lewis JW Jr, Tilley B et al. (1985) The porcine bioprosthetic valve: twelve years later. J Thorac Cardiovasc Surg 89:499–507
Mahony C, Sublett KL, Harrison MR (1989) Resolution of spontaneous contrast with platelet disaggregatory therapy. Am J Cardiol 63:1009–1010
Marvasti MA, Obeid AI, Potts JL, Parker FB (1984) Approach in the management of atrial myxoma with long-term follow-up. Ann Thorac Surg 38:53–58
Meltzer RS, Visser CA, Fuster V (1986) Intracardiac thrombi and systemic embolization. Ann Intern Med 104:689–698
Mills P, Leech G, Davies M, Leatham A (1978) The natural history of non-stenotic bicuspid aortic valve. Br Heart J 40:951–957
Molinari GF (1972) Septic cerebral embolism. Stroke 3:117–122
Molinari GF, Smith L, Goldstein MN, Satran R (1973) Pathogenesis of cerebral mycotic aneurysms. Neurology 23:325–332
Moon RE, Camporesi EM, Kisslo JA (1989) Patent foramen ovale and decompression sickness in divers. Lancet 1:513–514
Nellessen U, Daniel WG, Matheis G, Oelert H, Depping K, Lichtlen PR (1985) Impending paradoxical embolism from atrial thrombus: correct diagnosis by transesophageal echocardiography and prevention by surgery. J Am Coll Cardiol 5:1002–1004
Nestico PF, DePace NL, Kotler MN et al. (1983) Calcium-phosphorus metabolism in dialysis patients with and without mitral annular calcium. Am J Cardiol 51:497–500
Nestico PF, Depace NL, Morganroth J, Kotler MN, Ross J (1984) Mitral annular calcification: clinical, pathophysiologic and echocardiographic review. Am Heart J 107:989–996
New PFJ, Price DL, Carter B (1970) Cerebral angiography in cardiac myxoma. Radiology 96:262–266
Nishimura RA, McGoon MD, Shub C, Miller FA, Ilstrup DM, Tajik J (1985) Echocardiographically documented mitral-valve prolapse: long-term follow-up of 237 patients. New Engl J Med 313:1305–1309
Olsen TS, Skriver EB, Herning M (1985) Cause of cerebral infarction in the carotid territory: its relation to the size and the location of the infarct and to the underlying vascular disease. Stroke 16:459–466
Openshaw H (1976) Neurological complications of endocarditis in persons taking drugs intravenously. West J Med 124:276–281
Pankey GA (1962) Acute bacterial endocarditis at the University of Minnesota Hospitals, 1939–1959. Am Heart J 64:583–591
Pearson AC, Labovitz AJ, Tatineni S, Gomez CR (1991a) Superiority of transesophageal echocardiography in detecting cardiac source of embolism in patients with cerebral ischemia of uncertain etiology. J Am Coll Cardiol 17:66–72
Pearson AC, Nagelhout D, Castello R, Gomez GR, Labovitz AJ (1991b) Atrial septal aneurysm and stroke: a transesophageal echocardiographic study. J Am Coll Cardiol 18:1223–1229
Pengo V, Peruzzi P, Baca M et al. (1989) The optimal therapeutic range for oral anticoagulant treatment as suggested by fibrinopeptide A (FpA) levels in patients with heart valve prostheses. Eur J Clin Invest 19:181–184
Petersen P, Godtfredsen J (1986) Embolic complications in paroxysmal atrial fibrillation. Stroke 17:622–626
Petersen P, Hansen JM (1988) Stroke in thyrotoxicosis with atrial fibrillation. Stroke 19:15–18
Petersen P, Madsen EB, Brun B, Pedersen F, Gyldensted C, Boysen G (1987) Silent cerebral infarction in chronic atrial fibrillation. Stroke 18:1098–1100
Pomerance A (1969) Ballooning deformity (mucoid degeneration) of atrioventricular valves. Br Heart J 31:343–351
Price DL, Harris JL, New PF, Cantu RC (1970) Cardiac myxoma – a clinicopathologic and angiographic study. Arch Neurol 23:558–567
Procacci PM, Savran SV, Schreiter SL, Bryson AL (1975) Clinical frequency and implications of mitral valve prolapse in the female population. Circulation 51 & 52 (Suppl 2):78 Abstract
Pruitt AA, Rubin RH, Karchmer AW, Duncan GW (1978) Neurologic complications of bacterial endocarditis. Medicine 57:329–343
Pumphrey CW, Dawes J (1981) Elevation of plasma β-thrombo globulin in patients with prosthetic cardiac valves. Thromb

Res 22:147–155
Radford DJ, Julian DG (1974) Sick sinus syndrome. Experience of a cardiac pacemaker clinic. Br Med J 3:504–507
Rankin LI, Desousa AL (1978) Metastatic atrial myxoma presenting as intracranial mass. Chest 74:451–452
Riddle JM, Magilligan DJ, Stein PD (1981) Surface morphology of degenerated bioprosthetic valves four to seven years following implantation. J Thorac Cardiovasc Surg 81:279–287
Ridolfi RL, Hutchins GM (1976) Spontaneous calcific emboli from calcific mitral annulus fibrosus. Arch Pathol Lab Med 100:117–120
Roberts WC, Siegel RJ, McManus BM (1987) Idiopathic dilated cardiomyopathy: analysis of 152 necropsy patients. Am J Cardiol 60:1340–1355
Rosenow EC III, Osmundson PJ, Brown ML (1981) Pulmonary embolism. Mayo Clin Proc 56:161–178
Rosenqvist M, Vallin H, Edhag O (1985) Clinical and electrophysiologic course of sinus node disease: five-year follow-up study. Am Heart J 109:513–522
Roux M, Ponzio J, Brunet M, Garaix J (1984) Systemic thromboembolic accidents in the early post-operative period in patients with prosthetic valve. Eur Heart J 5(Suppl D):27–31
Salgado AV, Furlan AJ, Keys TF (1987) Mycotic aneurysm, subarachnoid hemorrhage, and indications for cerebral angiography in infectious endocarditis. Stroke 18:1057–1060
Salgado AV, Furlan AJ, Keys TF, Nichols TR, Beck GJ (1989) Neurologic complications of endocarditis: a 12-year experience. Neurology 39:173–178
Sandok BA, von Estorff I, Giuliani ER (1980) CNS embolism due to atrial myxoma – clinical features and diagnosis. Arch Neurol 37:485–488
Schlichter J, Hellerstein HK, Katz LN (1954) Aneurysm of the heart: a correlative study of one hundred and two proved cases. Medicine 33:43–86
Schneider B, Hanrath P, Vogel P, Meinertz T (1990) Improved morphologic characterization by transesophageal echocardiography: relation to cerebrovascular events. J Am Coll Cardiol 16:1000–1009
Schwartz MH, Teichholz LE, Donoso E (1977) Mitral valve prolapse. A review of associated arrhythmias. Am J Med 62:377–389
Shuiab A (1989) Cerebral infarction and ventricular septal defect. Stroke 20:957–958
Silka MJ, Rice MJ (1991) Paradoxic embolism due to altered hemodynamic sequencing following transvenous pacing. Pace 14:499–503
Silver MD, Dorsey JS (1978) Aneurysms of the septum primum in adults. Arch Pathol Lab Med 102:62–65
Silverman J, Olwin JS, Graettinger JS (1962) Cardiac myxoma with systemic embolization. Circulation 26:99–103
Simon MA, Liu SF (1954) Calcification of the mitral valve. Am Heart J 48:497–505
Smith DL, McKnight TE (1981) TIAs, completed strokes and mitral valve prolapse. South Med J 74:1454–1456
SPAF Investigators (1992a) Predictors of thromboembolism in atrial fibrillation: I. Clinical features of patients at risk. Stroke 23:620–621
SPAF Investigators (1962b) Predictors of thromboembolism in atrial fibrillation: II. Echocardiographic features of patients at risk. Stroke 23:621
Staffurth JS, Gibberd JC, Tang-Fui SNG (1977) Arterial embolism in thyrotoxicosis with atrial fibrillation. Br Med J 2:688–690
Steele P, Weily H, Davies H et al. (1975) Platelet survival time following aortic valve replacement. Circulation 51:358–362
Steele P, Rainwater J, Vogel R (1979a) Platelet suppressant therapy in patients with prosthetic cardiac valves. Circulation 60:910–913
Steele P, Weily H, Rainwater J, Vogel R (1979b) Platelet survival time and thromboembolism in patients with mitral valve prolapse. Circulation 60:43–45
Stein PD, Sabbah HN (1974) Measured turbulence and its effects on thrombus formation. Circ Res 35:608–614
Stein PD, Sabbah HN (1976) Turbulent flow in the ascending aorta of humans with normal and diseased aortic leaflets. Circ Res 39:58–65
Stein PD, Sabbah HN, Pitha JV (1977) Continuing disease process of calcific aortic stenosis. Am J Cardiol 39:159–163
Straus R, Merliss R (1945) Primary tumor of the heart. Arch Pathol Lab Med 39:74–78
Sutton MG, Mercier LA, Giuliani ER, Lie JT (1980) Atrial myxomas – a review of clinical experience in 40 patients. Mayo Clin Proc 55:371–376
Takano K, Yamaguchi T, Kato H, Omae T (1991) Activation of coagulation in acute cardioembolic stroke. Stroke 22:12–16
Tegeler CH, Downes TR (1991) Cardiac imaging in stroke. Stroke 26:13–18
Terpenning MS, Buggy BP, Kauffman GA (1957) Infective endocarditis: clinical features in young and elderly patients. Am J Med 83:626–634
Virchow R (1856) Gesammelte Abhandlungen zur Wissenschaftlichenmedezin. Meidinger Sohn, Frankfurt, pp 219–732
Visser CA, Kan G, Meltzer RS, Dunning AJ, Roelandt J (1985) Embolic potential of left ventricular thrombus after myocardial infarction: a two-dimensional echocardiographic study of 119 patients. J Am Coll Cardiol 5: 1276–1280
Webster MWI, Smith HJ, Sharpe DN et al. (1988) Patent foramen ovale in young stroke patients. Lancet 2:11–12
Weily HS, Steele PP, Davies H, Pappas G, Geenton G (1974) Platelet survival in patients with substitute heart valves. New Engl J Med 290:534–537
Weinberger J, Rothlauf E, Materese E, Halperin J (1988) Noninvasive evaluation of the extracranial carotid arteries in patients with cerebrovascular events and atrial fibrillations. Arch Intern Med 148:1785–1788
Weinreich DJ, Burke JF, Pauletto FJ (1984) Left ventricular mural thrombi complicating acute myocardial infarction: long-term follow-up with serial echocardiography. Ann Intern Med 100:789–794
Weitz JI, Hudoba M, Massel D, Maraganore J, Hirsh J (1990) Clot-bound thrombin is protected from inhibition by heparin–antithrombin III but is susceptible to inactivation by antithrombin III–independent inhibitors. J Clin Invest 86:385–391
Whittlesey D, Geha AS (1991) Selection and complications of cardiac valvular prostheses. In: Baue AE, Geha AS, Laka H, Hammond GL, Naunheim KS (eds) Glenn's thoracic and cardiovascular surgery, 5th edn. Appleton and Lange, Norwalk, Conn. Vol 2, pp 1719–1728
Wiener I, Hafner R, Nicolai M, Lyons H (1987) Clinical and echocardiographic correlates of systemic embolization in nonrheumatic atrial fibrillation. Am J Cardiol 59:177
Wilmhurst PT, Byrne JC, Webb-Peploe MM (1989) Relation between interatrial shunts and decompression sickness in divers. Lancet 2:1302–1305
Wilson WR, Geraci JE, Danielson GK et al. (1978) Anticoagulant therapy in central nervous system complications in patients with prosthetic valve endocarditis. Circulation 57:1004–1007

Winkle RA, Lopes MG, Fitzgerald JW, Goodman DJ, Schroeder JS, Harrison DC (1975) Arrhythmias in patients with mitral valve prolapse. Circulation 52:73–81

Wolf PA, Dawber TR, Thomas HE, Kannel WB (1978) Epidemiologic assessment of chronic atrial fibrillation and risk of stroke: the Framingham study. Neurology 28:973–977

Working Party on Anticoagulant Therapy in Coronary Thrombosis to the Medical Research Council (1969) Assessment of short-term anticoagulant administration after cardiac infarction. Br Med J 1:335–342

Wright IS, Marple CD, Beck DF (1954) Myocardial infarction, its clinical manifestations and treatment with anti-coagulants: a study of 1031 cases. Grune and Stratton, New York, pp 1–672

Zenker G, Ergel R, Kramer G et al. (1988) Transesophageal two dimensional echocardiography in young patients with cerebral ischemic events. Stroke 19:348–354

21. Heart and Brain Relationships

A. S. Ali and S. R. Levine

History and Perspective

In his book *De Anatome Cerebri* Thomas Willis (1664) recognized that the brain and the cardiovascular system (CVS) are interdependent. The possible connection between neurogenic cardiac arrhythmias and disturbances in blood pressure was first noted by Cushing in patients with transtentorial brain herniation. In his paper on the occurrence of ventricular premature contractions (VPCs) caused by chloroform anesthesia, Levy, in 1913, pointed out the possibility of nervous influences on cardiovascular regulation. In animal studies, Cannon (1931) showed that the most highly encephalized areas of the brain regulate both sensory and autonomic input. Leriche et al. (1931) found that high thoracic sympathectomy had a beneficial effect in patients with angina pectoris. Erickson (1939) reported the first systematic study of the problem of rhythm changes during seizures. Peaked inverted T waves on ECG were first described by Byer et al. (1947) in a patient with subarachnoid hemorrhage (SAH).

In autopsy studies, Koskelo (1964) described the occurrence of subendocardial petechial hemorrhages in patients with SAH. Van Buren (1958) noted that rhythm disturbances were seen in patients with temporal lobe seizures and were not always related to movement. Walsh et al. (1972) reported the case of a 4-year-old boy who had epileptic discharges associated with a shorted R-R interval, and on occasion these led to supraventricular tachycardia.

It has become clear that the brain closely oversees and regulates cardiovascular control and houses the supranuclear mechanisms to influence autonomic nervous system function. The major clinical importance of these observations has a bearing on the unexpected increase in cardiac events following stroke (Oppenheimer et al. 1990).

Clinical Basis of the Brain–Heart Interaction

The common clinical problems of stroke and SAH vividly depict the interrelationship between the brain and the heart (Benedict and Loach 1978). Approximately 50%–100% of patients with SAH have ECG changes (Bass Rea and Dunbar 1992), up to 20% of which may be life-threatening. They commonly occur in the first 24 hours after the ictus. In 13 patients with SAH, Pollick et al. (1988) demonstrated the occurrence of reversible wall motion abnormalities, serum cardiac enzyme changes and ECG changes. Patients with wall motion abnormalities tended to have higher creatinine phosphokinase (CPK) levels, higher heart rate, a more common occurrence of inverted T waves and a more severe neurological insult. The mortality of these patients was higher than in those patients without ECG changes.

Rhythm Disturbances in Patients with SAH

ECG changes in patients with SAH include peaked P waves, shortened PR interval, long QTc interval, T-wave inversion and, occasionally, new Q waves. The most common ECG changes are ST-T changes and prominent U waves. ECG rhythm changes in SAH in descending order of frequency are sinus tachycardia, multifocal ventricular extrasystoles, couplets, frequent supraventricular extrasystoles, non-sustained ventricular tachycardia, asystole and sinus arrhythmia (Stober et al. 1988). The incidence of these arrhythmias was higher in patients with prior cardiac disease, although this was not statistically significant. The most common arrhythmias on the first day of SAH were sinus tachycardia, sinus bradycardia and frequent supraventricular extrasystoles. The incidence of asystoles peaks on the 3rd and the 9th days. In this study a positive correlation was demonstrated between the severity of the neurological deficits and the severity of the cardiac arrhythmia.

Torsade de pointes, a particularly malignant form of arrhythmia, is seen in up to 3.8% of SAH patients on Holter monitoring (n=132) (Dipascuale et al. 1988). Patients with this arrhythmia had it during the first 24 hours after the bleeding. The QT interval was prolonged in all patients with torsade de pointes and all of them had hypokalemia. Ten to 50% of patients with SAH have a transient prolongation of their QT interval (Dipascuale et al. 1988). The underlying pathophysiology is probably an imbalanced sympathetic stimulation. A prolonged QT interval is the substrate for many ventricular arrhythmias. A large proportion (42%) of patients had a prolonged QT interval but only a small number had torsade de pointes (Dipascuale et al. 1988). This led these authors to hypothesize that a prolonged QT interval may play a critical role in the genesis of malignant ventricular arrhythmias only in the presence of other factors, such as sympathetic stimulation, catecholamine excess and hypokalemia.

A detailed analysis of the time course of hypokalemia and arrhythmias in patients with SAH (Stober et al. 1988) revealed that, although all types of arrhythmias were more common in patients with hypokalemia, the differences were only significant for multifocal ventricular extrasystoles. Their study demonstrated a highly significant relationship between hypokalemia and midbrain symptoms and fatal outcome.

Traggot et al. (1988) evaluated the incidence of various arrhythmias in patients with intracerebral hemorrhage (ICH), with particular emphasis on the presence or absence of brain-stem compression. They found a positive correlation between patients with evidence of brain-stem compression and ECG evidence of sinus arrhythmia, multifocal VPCs, couplets and ventricular tachycardia. They found ECG changes in almost all patients. They hypothesized that transtentorial herniation destroyed the hypothalamic centers bilaterally. Progressive ischemia of the brain stem manifests as respiratory arrest, bradycardia and hypertension when the ischemia reaches the lower pons. When brain ischemia advances to the lower medulla there is an abrupt change from bradycardia to tachycardia. Sinus arrhythmia, seen in patients with intracranial tension more than 50%–70% greater than systolic blood pressure, is abolished by bilateral vagotomy and i.v. atropine. These data suggest that some arrhythmias seem to be vagally mediated. The apparent paradox of a higher incidence of arrhythmias in patients with cortical damage compared to those with brain-stem damage can be explained by the fact that most brain-stem infarcts are unilateral, leaving part of the cardiovascular regulatory mechanism intact.

A different pathogenic mechanism was proposed by Stober et al. (1988), who tried to relate the localization of the cerebral aneurysm to clinical outcome. The posterior hypothalamus is a primary area involved in regulation of the autonomic nervous system. Most patients with a cerebral aneurysm rupture in the region of the anterior or posterior communicating arteries develop a hypothalamic lesion. Peripheral middle cerebral artery aneurysm rupture leads to a hypothalamic lesion in only one-third of all patients. The hypothalamic lesions are predominantly caused by vasospasm: microhemorrhages are found in paraventricular and supraoptic areas. A higher incidence of multifocal extrasystoles and couplets is found in anterior communicating artery aneurysmal ruptures than in other locations. The lack of myocardial lesions in controls with elevated intracranial pressures due to varying causes further strengthens this hypothesis. The degree of CNS involvement, as evidenced by a clouded consciousness, correlates with plasma renin levels (Stober et al. 1988). The role of hypothalamic lesions in the elevated levels of plasma renin remains to be clarified, since a high sympathetic tone could also account for this finding.

Pollick et al. (1988) postulate that the presence of small hematomas or severe vasospasm of the vessels supplying the hypothalamus could contribute to the autonomic nervous system abnormalities seen in these patients.

Depolarization Disturbances

These include ST segment elevation, ST segment depression and T-wave abnormalities. Most studies investigating these changes in patients with SAH eliminated those with known ischemic heart disease from the analysis, and therefore removed a group in whom such changes are more likely to occur. ECG changes seen in patients with SAH are probably the result of cardiac damage. Among the patients who died in this study, none had documented ventricular arrhythmias that contributed to their demise.

Prognostic Value of ECG Alterations in Patients with Aneurysmal SAH

In a prospective study of patients with SAH, Brouwers et al. (1989) found that the outcome of patients could not be reliably predicted by ECG changes alone. The neurological grade on admission was more predictive of outcome. On baseline CT scan ECG changes did not correlate with the amount of blood in the cisterns, cerebral ischemia or infarction. They did find an association between ECG changes in fast rhythm disturbances or ischemia or both and poor outcome. It was felt by the authors that the predictive value of the ECG is probably an indirect reflection of the CNS damage. Cardiac disease did not contribute directly to mortality.

Sinus tachycardia may be related to death 7 days after SAH. Changing ECG abnormalities are a bad prognostic sign, as opposed to patients with a normal ECG or those with a persistently abnormal but unchanging ECG.

Autopsy Data

A higher incidence of changes in the cardiac muscle fibers from myofibrillar to granular staining pattern for succinate dehydrogenase was found in patients with intracranial lesions (Kolin and Norris 1984). A myofibrillar staining pattern is characterized by the presence of very fine granules of violet formazan of nitroblue tetrazolium (NBT) deposited between normal myocardial fibers. These granules in damaged fibers are coarse and irregularly distributed, giving a 'granular' staining pattern. The presence of granular muscle fibers in the central and external thirds of the left ventricular wall was considered to be evidence of transmural myocardial damage. There is a higher incidence of such damage in patients with intracranial lesions. Patients who died of cerebral hemorrhage within a brief period of time did not show these changes, implying that they take time to evolve. Further, the lesions were not seen in patients who died later than 2 weeks after the insult.

Connor (1968, 1970) found an 8% incidence of focal myocytolysis in the hearts of patients dying from a variety of intracranial lesions. It has been postulated that transmural myocardial damage may constitute a morphological basis for the functional myocardial damage observed in stroke patients, whereas myocytolysis implies the destruction of myocardial fibers. The changes due to transmural myocardial damage appear to resolve with time.

In an autopsy series of 54 patients, Doshi and Dwyer (1980) found that patients dying after SAH had both hypothalamic and myocardial lesions. The hypothalamic lesions included perivascular hemorrhage, microinfarcts and macroinfarcts. They were believed to be responsible for the derangement of the autonomic nervous system, resulting in secondary myocardial damage. The subendocardial region of the left ventricle appears on careful histological examination to be the area most frequently involved with myocytolysis, myofibrillar degeneration and fuchsinophilic degeneration.

The 3 patients who were autopsied in Pollick et al.'s (1988) study revealed changes due to myocardial damage in the same areas which had wall motion abnormalities seen by two-dimensional echocardiography at premortem examination.

In the 7 patients autopsied in the study by Brouwers et al. (1989) there was a concordance between ECG changes and myocardial changes.

These autopsy data suggest that, although some of the ECG changes seen in patients with SAH may be 'neurogenic', they also have a structural basis, i.e. they are associated with damaged myocardium.

Cardiac Involvement in Stroke

Dimant and Grob (1977) described ECGs from 100 consecutive stroke patients taken within the first 3 days of hospital admission. The control group was patients admitted for other medical problems (e.g. carcinoma of the colon). They found a higher incidence of ST depression and prolongation of the QT interval (seven times more common) and a fourfold higher incidence of T-wave inversion and VPCs in the stroke group. In their population, however, the incidence of coronary artery disease was much higher in the subgroup of patients with stroke; the most common underlying pathology, common to both conditions, was atherosclerosis.

In stroke patients, the occurrence of arrhythmias may be independent of the presence of underlying heart disease or the level of sympathetic activity as reflected by plasma norepinephrine (NE) levels (Myers et al. 1982). Serious arrhythmias were more common in patients with hemispheric infarction than in those with brainstem infarction. The incidence of arrhythmias in stroke patients was higher than in the control group. Serum CPK elevation was equally common in those with and without arrhythmias. The incidence of arrhythmias was higher in older stroke patients and was also higher in older controls without stroke than in younger controls.

As both cardiovascular disease and stroke are often manifestations of the same basic pathogenetic process, it is still possible that the higher incidence of arrhythmias in those with stroke is related to the presence of covert heart disease. Alternatively, this could be related to a different pathogenetic mechanism, such as raised intracranial pressure, in those with stroke (Meyers et al. 1982). In a series of patients with acute brain infarcts, among those who died and had premortem ECG changes, none had any evidence of coronary disease at autopsy (Fentz and Gormsen 1962). These data do not support the theory that ECG changes in patients with stroke are a result of covert heart disease, but the possibility is not completely excluded. A systematic study designed to delineate the contribution of covert heart disease to the ECG changes in patients with stroke is clearly needed.

Most of the studies cited are lacking in that they do not take into account an old ECG as comparison. Lavy et al. (1974) compared the ECGs of patient who had had a prior ECG and found that 18 of 25 patients without pre-existing heart disease had had an ischemic cerebral event, and of these 8 (44%) had either a recent onset arrhythmia by ECG or changes of myocardial ischemia.

Cardiac Rhythm Disturbances in Stroke Patients

The most common rhythm abnormality in patients with thromboembolic stroke is atrial fibrillation, which occurs with a frequency of up to 9% (Norris et al. 1978). It is thought that the high incidence of atrial fibrillation in these patients could be a result of the arrhythmia's role in the pathogenesis of embolic stroke. The incidence of new arrhythmias in patients with stroke varies from 25% (Goldstein 1979) to 39% (Lavy et al. 1974). Other arrhythmias seen include sinus tachycardia, VPCs and premature atrial contractions. There is a very low incidence of ventricular arrhythmias in these patients.

The incidence of cardiac arrhythmias in stroke patients with Holter monitoring may be as high as 60% (Rem et al. 1985). In a similar study (Norris et al. 1978) this incidence was 24%.

Prognostic Value of ECG Changes in Stroke Patients

The mortality of patients with ECG changes in the study by Norris et al. (1978) was 69% compared to no mortality in those without ECG changes. The incidence of sudden death in patients with stroke is approximately 6%, and it is reasonable to postulate that arrhythmias are a possible cause. However, in their paper on early mortality following stroke, Silver et al. (1984) found that cardiac deaths tend to occur towards the end of the first month following stroke. These deaths appear also to involve those patients with slight functional impairments. This study is limited in that it studied mortality in patients who survived until hospital admission; a significant number of stroke patients could have died suddenly as a result of a cardiac arrhythmia, and thus never reached the hospital alive.

Plasma Creatinine Phosphokinase (CPK)

At least 29% of patients with acute stroke have elevated CPK levels (Dimant and Grob 1977) T-wave inversion, ST depression and conduction defects correlate with a high CPK level and the high incidence of ECG abnormalities in patients with a high CPK level could reflect underlying vascular disease in the heart.

In their work on CPK levels and stroke, Norris et al. (1978) did an initial study which did not involve fractionation of the various cardiac isozymes, after which they discovered that a significant number of patients with elevated CPK had other reasons for this elevation, e.g. intramuscular injection. In a second phase, Norris et al. (1979) studied the incidence of elevation of creatine phosphokinase MB isozyme (CPK-MB) and other fractions in patients with stroke; 68% of stroke patients had elevated CPK levels and ECG changes consistent with myocardial damage. These patients had transmural infarction (n=3), evolving repolarizing ECG changes (n=5) and ECG evidence of left ventricular hypertrophy (LVH) with strain; 90% of the patients with a positive CPK-MB had cardiac arrhythmias. Raised CPK levels with evidence of ischemia on ECG portended an increased chance of death. There was no difference in the incidence of raised cardiac enzymes in patients with cerebral ischemia and hemorrhage.

Unlike with myocardial infarction, the temporal trend in CPK values is different in patients with stroke. The enzymes rise slowly and reach a peak 4–5 days after the event. The temporal timing of the stroke and the rise in enzymes and the level of elevation (suggesting a small area of myocardial necrosis) suggest that the elevation is a result of myocardial damage after the stroke, as opposed to strokes occurring in patients who have embolism due to a large myocardial infarction and wall motion abnormalities. The elevation of the creatine phosphokinase BB isozyme (CPK-BB) fraction is very transient and has an early peak.

Cardiovascular Regulation in Lesions of the CNS

Intracranial pathology and its association with cardiovascular regulation has been partly discussed in the preceding sections. In addition to stroke and various types of intracranial hemorrhage, tumors of the basal forebrain and paraseller structures are also associated with a relatively high incidence of ECG changes (Talman 1985). This suggests that critically placed lesions, regardless of their underlying cause, can contribute to altered cardiovascular control of the heart.

Lesions of the stellate ganglion and stimulation of this structure yield interesting insights into the mechanisms of cardiovascular regulation. Direct stimulation of the stellate ganglion sometimes shows changes compatible with myocardial ischemia. Stimulation of the right stellate ganglion produces ST elevation and T-wave inversion. On the other hand, peaked T waves, ST segment depression and QT prolongation are seen with stimulation of the left stellate ganglion. It is not clearly established whether stimulation of any one side of the hypothalamus in turn produces activation of the stellate ganglion or the cardiac nerve of any one side.

Grossman (1976) reported a woman with SAH and arrhythmias. Blockade of her left stellate ganglion with 1% lidocaine completely abolished her arrhythmias. Direct stimulation of the left cardiac nerve produces a higher incidence of arrhythmias than right-sided stimulation (Armour et al. 1972). Relative overactivity of the left stellate ganglion, as seen with cooling/ removal of the right stellate ganglion, leads to a higher incidence of arrhythmias.

Sympathetic Activity in Stroke Patients

There are significant elevations in plasma catecholamines following cerebral infarction (Meyet al. 1981). The location of the stroke did not affect plasma catecholamine levels, and the extent of elevation of norepinephrine (NE) was greater than epinephrine (E). An extra-adrenal source for the NE is likely.

Behavioral Aspects of Brain–Heart Relationships

It is well established that certain behaviors such as anger, anxiety, fear and sleep can trigger myocardial ischemia (Verrier and Dickerson 1991). Coronary vasoconstriction, probably catecholamine-mediated, may follow classically conditioned aversive stimuli. Twofold alterations in coronary blood flow can be measured during sleep. Surges in coronary blood flow and heart rate accompany rapid eye movement (REM) sleep periods (Kirby and Verrier 1989). The CNS sites for producing REM-induced changes in cardiac perfusion are not established.

Autonomic and Myocardial Changes in Experimental Stroke

Rats with experimental lesions of the middle cerebral artery (MCA) were compared with rats with common carotid lesions (CCA) and sham animals in a study by Cechetto et al. (1989). There was a steady decline in blood pressure in animals with CCA occlusion, CCA and MCA occlusion and in sham lesions. Animals with MCA lesions alone did not show this decline, and 180 minutes following stroke, rats with MCA lesions had higher NE and E levels than animals in other groups. At autopsy, four of the six rats with MCA lesions had abnormal hearts, with evidence of subendocardial hemorrhage or ischemic damage. The incidence of myocardial damage was higher in this group than in all others in the experiment.

In this experimental model catecholamine levels are higher only if the insular cortex is involved in the infarct. It is possible that high catecholamine levels are secondary to unloading of the baroreceptor reflex arc as a result of the cerebral lesions.

High Norepinephrine Levels and Their Role in Myocardial Damage

A high NE level reflects stimulation of the sympathetic nervous system. A high level of sympathetic nervous activity could result in myocardial damage (Tables 21.1 and 21.2) as a result of many possible pathogenetic mechanisms (Table 3).

Table 21.1 Myocardial damage due to sympathetic hyperactivity

Excessively high myocardial oxygen demand
Coronary vasoconstriction
Relative tissue ischemia as a result of above
Direct cardiotoxicity from catecholamines. Indirect evidence to support this hypothesis includes: ● Intravenous infusion of NE induces myocardial toxicity (Schenk and Moss 1966; Szakacs and Melhman 1960) ● Subjects with central lesions and high plasma epinephrine and norepinephrine levels on admission have a poorer outcome (Benedict and Loach 1978) ● In experimental models of SAH in mice it has been shown that reserpine pretreatment (which depletes monoamine stores) reduces the myocardial damage associated with the central lesion ● Maximal damage is seen in cells next to the intracardiac nerve terminals (Greenhoot and Reichenbach 1969) ● The role of the vagus nerve in the pathogenesis of myocardial damage is unsettled. Electrical stimulation of the vagus causes myocardial damage preventable by pretreatment with atropine (Manning et al. 1937)

Table 21.2. Pathogenesis of ECG changes

Increased levels of circulating catecholamines
Involvement of the hypothalamus
Prevention of ECG changes with β-blocking drugs

Table 21.3. Possible catecholamine-dependent mechanisms for cardiac damage (Oppenheimer et al. 1990)

The catecholamine-induced ECG changes may reflect myocytolysis/cardiac injury induced
Activation of calcium channels
Production of free radicals
Generation of adrenochrome

Table 21.4. Relationship between lesion location and arrhythmia type

Location of lesion	Type of arrhythmia	Reference
Hemispheric	VPC and AF	Norris et al. 1978)
Left frontal hematomas	QT and T changes	Yamour et al. (1980)
Temporoparietal hematomas	VPC and sinus bradycardia	Yamour et al. (1980)
Thalamic/basal ganglia ICH	Sinus tachycardia	Yamour et al. (1980)
Brain-stem ICH	AF and APCs	Yamour et al. (1980)

AF=atrial fibrillation; APC=atrial premature contractions (PACs)

Role of Local Catecholamines

In the setting of SAH the role of locally released catecholamines appears to be more important than that of circulating catecholamines (Pollick et al. 1988). Baboons with a completely denervated heart prior to an increase in their intracranial pressure had a completely normal myocardium. Baboons with incomplete or no cardiac denervation developed evidence of myocardial damage after an increase in intracranial pressure.

In their model of stimulation of the rat insular cortex, Oppenheimer et al. (1991) have data supporting the concept of the importance of local catecholamine excess. The pattern of ECG changes suggested a sequential involvement of the conducting system. The predominant sympathetic outflow in the rat is to the conducting system, suggesting that catecholamine-induced damage to the conducting system may be responsible for these changes. Autopsy data in rats showed septal hemorrhage in the area of the left bundle of His, which is the area supplied by intracardiac branches of the sympathetic nerves, further supporting this hypothesis.

Stimulation of certain central centers results in the release of large amounts of catecholamines (especially NE) from the adrenal medulla. In humans these centers lie in the most lateral and posterior parts of the hypothalamus. Factors causing stimulation of these regions include increases in intracranial pressure, ischemia, the irritant effect of blood or a combination of these factors. Experimental data suggest that high NE levels predispose to non-uniform depolarization of cardiomyocytes, leading to a predisposition for arrhythmias. Experimental data also suggest that, in the face of high NE levels, there is ischemia of primarily the subendocardial area of the left vertricle (LV). Pathological changes supporting this hypothesis include evidence of subendocardial hemorrhage, mononuclear cell infiltrate, focal necrosis and cellular degeneration (Szakacs and Melhman 1960; Schenk and Moss 1966). In certain cases such changes may be reversible and may not be seen at postmortem (Syverd 1991).

Location of the Cerebral Lesion and Incidence of ECG Changes

Clinical Data

There is a higher incidence of VPCs and atrial fibrillation in patients with hemispheric stroke than in those with brain-stem involvement (Norris et al. 1978, 1979). Left frontal hematomas have an association with prolongation of the QT interval and T-wave abnormalities. Opercular lesions may also predispose to ECG changes due to involvement of the supranuclear pathways of the sympathetic descending tracts. Table 4 summarizes the type of arrhythmias seen, based on location of the stroke.

Experimental Data

Experimental data suggest that there is asymmetry for supranuclear heart rate control. Inactivation of the left cerebral hemisphere with intracarotid amobarbital increases heart rate, whereas injection into the right hemisphere causes bradycardia. In experiments on rats, Hachinski et al. (1992) found that right MCA lesions did not show the decline in blood pressure after experimental stroke seen in control animals and to a lesser extent in animals with left MCA lesions. There was no change in plasma NE levels in sham and LMCA occlusion groups; however, there was a significantly higher NE level in animals with RMCA lesions. There was a significant lengthening of QT intervals in animals with RMCA lesions. There was a significant lengthening of QT intervals in animals with right MCA lesions. Sympathetic nervous system activity was significantly higher in animals with right MCA lesions. These authors suggest that right-sided lesions have a higher incidence of autonomic instability, and since a greater degree of autonomic instability is associated with QT prolongation, greater chance for sudden death. Patients with right-sided cerebral lesions involving the insula may need to be more closely monitored.

Cardiac Arrhythmias and Sudden Cardiac Death: A Possible Role for the Insular Cortex

Oppenheimer et al. (1990, 1991) used a unique technique of phasic microstimulation of the rat insular cortex synchronized with the ECG to study the role of the insular cortex in arrhythmias. Specifically, the stimulation was linked to the T wave. Using this technique, profound ECG alterations associated with cardiac pathology and death from asystole were seen. The ECG changes after insular cortex stimulation include P-wave changes, PR interval prolongation, QRS widening (indicating a change in the pacemaker site), prolonged atrioventricular conduction and lengthening of the intraventricular conduction time. Following this, VPCs, complete heart block and marked bradycardia followed by asystolic arrest were seen. These changes were stereotyped and reproducible. Similar ECG changes were seen in humans on infusion of NE. NE levels were higher only in those rats undergoing insular stimulation. The absence of an elevation in E levels in the presence of high NE levels suggests a neural origin for the elevation in catecholamines. This is similar to the catecholamine changes seen in humans after injury to the brain.

Myocytolysis was found in only those rats undergoing insular cortex stimulation and was not seen in any of the control animals. Recently Cechetto et al. (1989) demonstrated raised catecholamine levels and myocytolysis after experimental stroke involving the sensorimotor cortex and parts of the insula. It has been suggested that, under these circumstances, inhibitory influences to the cardioactive area of the insula release catecholamines, resulting in increased sympathetic drive to the heart. It is possible that specific involvement of the insula by stroke or seizure activity can predispose to ECG changes and arrhythmogenesis. A possible clinical outcome of these observations could be that patients with insular cortex involvement might be more prone to sudden death, and may therefore require more intensive cardiac monitoring. Speculation regarding the mechanisms of sudden death have also been put forward by these authors, in that 25% of all sudden deaths are secondary to asystole. Therefore, it is possible that insular stimulation could play a role in the pathogenesis of sudden death.

Mechanism of Heart Rate Changes and Catecholamine Levels Following Insular Cortex Stimulation in the Rat

Insular cortex stimulation produces NE changes from all three parts of the cortex, granular, dysgranular and degranular (Oppenheimer et al. 1990, 1991). Microstimulation of the posterior insular cortex in the tachycardia zones probably selectively activates the cardiac nerves. Stimulation of tachycardia sites produced increases in NE levels associated with changes in heart rate. There were no changes in NE levels after stimulation of bradycardia sites or control sites. Tachycardia resulting from insular cortex stimulation is a result of sympathetic stimulation alone in the ketamine- and chloralose-anesthetized rats.

Bradycardia is mediated by reduction of sympathetic tone in chloralose-anesthetized rats and by an increase in vagal activity in those anesthetized by ketamine. In conscious rabbits heart rate changes are due to a combination of sympathetic and parasympathetic changes, and the mechanism appears to be intermediate between the two different mechanisms seen in rats. One hour of tachycardia obtained by insular cortex stimulation decreases the intracardiac NE levels proportional to the degree of heart rate increase.

Organization of the Rat Insular Cortex

Tachycardia sites are located in the rostral posterior insular cortex (Oppenheimer et al. 1990);

bradycardia sites are located in the caudal posterior insula. Pressor responses are elicited from the rostral anterior insula, whereas the depressor zone is probably more caudally placed. Both bradycardia and tachycardia were abolished by β-blockade. A possible explanation for this invokes a central role for atenolol, and yet another explanation suggests that both tachycardia and bradycardia are sympathetically mediated in this model.

Effects of Stimulation of the Human Epileptic Insular Cortex

Intraoperative stimulation of the brains of epileptics produces cardiovascular effects (Oppenheimer et al. 1992b). Left insular cortex stimulation resulted in significantly more frequent bradycardia and depressor responses and right insular cortex stimulation more commonly caused tachycardia and pressor responses. There appeared to be a right-sided predominance for sympathetic effects. An anterior–posterior distribution of responses was established. The amplitude of the bradycardia was significantly greater after stimulation of the left posterior than the left anterior insular cortex. ECG tracing did not change during stimulation. It should be noted that these patients had been taking anticonvulsant medication on a chronic basis.

Injection of amobarbital into the left carotid artery is associated with tachycardia, with the converse noted following right carotid artery injection (Zamrini et al. 1990). These data suggest that the right brain underlies tachycardia and the left hemisphere controls bradycardia responses.

Hypothalamus and Cardiovascular Regulation

The effects of stimulation of the supraoptic nucleus (SON) and the paraventricular nucleus (PAH) were studied by Ciriello and Calaresu (1980). They described the effects of such stimulation in chloralose-anesthetized normal, spinal and vagotomized cats. The hypothalamus was stimulated from 11 to 14 mm rostral to the interaural line and from 0.5 to 3.5 mm lateral to the midline.

Stimulation in vagal animals at the PAH and SON produced a cardioacceleratory response with a short latency. Similar stimulation in the region of the SON and medial forebrain bundle elicited a similar response, except that this was followed by an immediate poststimulus bradycardia. Administration of a β-blocker did not alter this response but atropine administration and bilateral vagotomy abolished it.

In sympathetically stimulated animals the cardiovascular responses elicited were increases in both heart rate and arterial pressure. The administration of propranolol abolished the heart rate response but did not change the magnitude of the pressor response.

The effect of stimulation of these areas on the response elicited by stimulation of the carotid sinus nerves revealed that the SON and PAH play a role in influencing responses elicited by stimulation of the carotid sinus nerve alone. Vagal influence on the cardiovascular system appears to be mediated via a different mechanism.

Histological localization of the areas stimulated was achieved by injection of iron at the tip of the stimulating electrode, the presence of which was subsequently studied by histochemical methods. Stimulation of the PAH in the dorsolateral area most readily caused cardioacceleration and arterial hypertension. This site corresponds to the location of the paraventriculospinal tract. Sites in the ventrolateral region of the PAH mediated the vagally influenced aspects of the cardioacceleratory response (i.e. vagal inhibition). These sites might be the location of the neurons that project to the medullary centers.

Cardiovascular responsive sites were also localized at a site dorsomedial to the supraoptic nucleus. One of the major descending tracts from the PAH courses through this region, raising the possibility that these responses are a result of stimulation of these fibers.

As a result of their observations, Ciriello and Calaresu (1980) concluded that centers more rostral to the SON and PAH, although possible, are unlikely. Their study further suggests that the SON and PAH maintain a tonic inhibitory influence on the cardiac component of the carotid sinus reflex.

Role of the Adrenergic and Cholinergic Pathways of the Hypothalamus in Cardiovascular Regulation

Central cholinergic stimulation evokes hypertension, and a major role in this response is played by the hypothalamus (Brezenoff and Guiliano 1982; Philippu 1988). In studies involving the injection of carbachol into the ventromedial nucleus (VMH), Brezenoff (1972) found that an increase in heart rate and arterial pressure was induced in anesthetized rats. The same injection into the dorsomedial nucleus produced a fall in arterial pressure and heart rate. Similar responses have been noted following injection of catecholamines into these areas.

The interaction of the catecholamine and cholinergic systems in the CNS is not yet completely defined. It is thought that release of catecholamines in the CNS is modulated by acetylcholine receptors. Part of the central effects of cholinergic stimulation is mediated by presynaptic cholinergic receptors. The release of NE is increased by stimulation of nicotinic receptors and blockade of muscarinic receptors.

In their studies on rats, Valadao et al. (1990) found an interaction between cholinergic and adrenergic pathways at the level of the VMH. The α_1 and β receptors of the VMH are involved in the cardiovascular responses induced by carbachol. In the case of nicotine, the pressor and tachycardic responses were blocked by pretreatment with a β-blocker, whereas prazosin blocked only the pressor response.

Hypothalamic Input to Medulla

Single-cell stimulation studies reveal that there is a connection between cells in the hypothalamus and the brain-stem medulla. The specific area of the hypothalamus studied was the hypothalamic defense area. Stimulation of this elicited inhibitory responses from two-thirds of the cells in the lateral tegmental field of the medulla (LTFM). Most cells responding to this stimulation also receive inputs from the baroreceptor.

LTFM neurons cannot be activated antidromically by stimulation of the T2 region of the intermediolateral column. The current theory of the defense reaction states that the response is mediated by neurons in the rostral ventrolateral medulla (Hilton 1965). Antidromic studies reveal that the LTFM is connected to the rostral ventrolateral medulla (RVLM). It is possible that LTFM plays a role in the basal discharge from neurons of the RVLM, and may therefore play a role in basal sympathetic tone. It is also possible that the LTFM neurons may play a role in patterns of cardiovascular responses.

Brain Stem

The postsynaptic actions of the carotid sinus nerve (SN), aortic nerve (AN), superior laryngeal nerve (SLN) and vagal nerve (VN) on the neurons of the nucleus tractus solitarius (NTS) has been studied in vivo. Three distinct patterns of postsynaptic responses are evoked by SN stimulation: an excitatory postsynaptic potential (EPSP), an EPSP-IPSP sequence and an inhibitory postsynaptic potential (IPSP), observed separately in individual neurons. This diversity of response was represented in cells proven to receive baroreceptor input by inflation of a balloon-tipped catheter within the ipsilateral carotid sinus. Virtually none of the neurons identified as baroreceptive exhibited a pulse-related discharge. A variety of influences to SN, SLN and VN stimulation were observed in neurons receiving baroreceptor afferent information. This wide convergence of input implies that this region of the brain stem is important in the integration of cardiovascular reflexes.

Therapeutic Considerations in Brain-Induced Myocardial Damage

It may be possible to reduce the incidence of life-threatening arrhythmias by influencing factors that contribute to cardiac electrical instability (Levine et al. 1987). There is no clearly established treatment following stroke to reduce fatal neurogenic arrhythmias. Research has focused on β-blockade, morizine, disopyramide, inhibitory neuropeptides and chemical or mechanical sympathectomy (Levine et al. 1987).

Conclusions and Future Directions

There are specific regions of the brain that control and influence cardiac function. These include the orbitofrontal cortex, insular cortex, hypothalamus and autonomic nuclei in the brain stem. Acute stroke, either ischemic or hemorrhagic (Cropp and Manning 1960) can lead to cardiac death from arrhythmias or myocardial infarction. Neurologists and cardiologists should collaborate to design prospective studies of carefully monitored acute stroke patients to further clarify the basis for and treatment of neurogenic arrhythmias, an important, and still somewhat mysterious cause of death following stroke.

References

Armour JA, Hageman GR, Randall WC (1972) Arrhythmias induced by local cardiac nerve stimulation. Am J Physiol 223: 1068–1075

Bass-Rea J, Dunbar S (1992) Neurogenic electrocardiographic abnormalities in subarachnoid hemorrhage. Focus Crit Care 19:50–54

Benedict CR, Loach AB (1978) Sympathetic nervous system activity in patients with subarachnoid hemorrhage. Stroke 9:237–244

Brezenoff HE (1972) Cardiovascular responses to intrahypothalamic injection of carbachol and certain cholinesterase inhibitors. Neuropharmacology 11:637–644

Brezenoff HE, Guiliano R (1982) Cardiovascular control by cholinergic mechanism in the central nervous system. Ann Rev Pharmacol Toxicol 22:351–381

Brouwers P, Wijdicks D, Hasan M et al. (1989) Serial electrocardiographic recording in aneurysmal subarachnoid hemorrhage. Stroke 20:1162–1167

Byer E, Ashman R, Toth LA (1947) Electrocardiograms with large, upright T-waves and long Q-T intervals. Am Heart J 33:796–906

Cannon WB (1931) Again the James–Lange and the thalamic theories of emotion. Psychol Rev 38:281–295

Cechetto DF, Wilson JX, Smith KE, Silver MD, Hachinski VC (1989) Autonomic and myocardial changes in middle cerebral artery stroke models in the cat. Brain Res 502:296–305

Ciriello J, Calaresu FR (1980) Role of the paraventricular and supraoptic nuclei in central cardiovascular regulation in the cat. Am J Physiol 239:137–142

Connor RCR (1968) Heart damage associated with intracranial lesions. Br Med J 3:29–31

Connor RCR (1970) Fuchsinophilic degeneration of myocardium in patients with intracranial lesions. Br Heart J 32:81–84

Cropp GJ, Manning GW (1960) Electrocardiographic changes simulating myocardial ischemia and infarction in association with spontaneous intracranial hemorrhage. Circulation 2:25–38

Dimant J, Grob D (1977) Electrocardiographic changes and myocardial damage in patients with acute cerebrovascular accidents. Stroke 8:448–455

Dipascuale G, Pinelli G, Andreoli A, Mannini GL, Grazi P, Tognetti F (1988) Torsade de pointes and ventricular flutter–fibrillation following spontaneous cerebral subarachnoid hemorrhage. Int J Cardiol 18:163–172

Doshi RN, Dwyer GA (1980) Clinicopathologic study of patients following a subarachnoid hemorrhage. J Neurosurg 52:295–301

Erickson T (1939) Cardiac activity during epileptic seizures. Arch Neurol Psych 41:511–518

Fentz V, Gormsen J (1962) Electrocardiographic patterns in patients with cerebrovascular accidents. Circulation 25:253–259

Goldstein DS (1979) The electrocardiogram in stroke: relationship to pathophysiological type and comparison with prior tracings. Stroke 10:253–259

Greenhoot JH, Reichenbach DD (1969) Cardiac injury and subarachnoid hemorrhage: a clinical, pathological and physiological correlation. J Neurosurg 30:521–531

Grossman MA (1976) Cardiac arrhythmias in acute central nervous system disease. Arch Intern Med 136:203–207

Hachinski VC, Oppenheimer SM, Wilson JX, Guiraudon C, Cechetto DF (1992) Asymmetry of sympathetic consequences of experimental stroke. Arch Neurol 49:697–702

Hilton SM (1965) Hypothalamic control of the cardiovascular responses in fear and rage. In: Scientific Basis of Medicine Annual Reviews, Athlone Press, London pp 217–238

Kirby DA, Verrier RL (1989) Differential effects of sleep stage on coronary hemodynamic function. Am J Physiol 256:H1378–H1383

Kolin A, Norris JW (1984) Myocardial damage from acute cerebral lesions. Stroke 15:990–993

Koskelo P, Punsar S, Sipila W (1964) Subendocardial hemorrhage and EKG changes in intracranial bleeding. Br Med J 1:1419–1480

Lavy S, Yaar I, Melamed E, Stern S (1974) The effect of stroke on cardiac functions as observed in an intensive care stroke unit. Stroke 5:775–780

Leriche R, Herman L, Fontaine R (1931) Ligature de la coronaire gauche et fonction de colur apes enervation sympathetique. Cr Soc Biol (Paris) 107:547–548

Levine SR, Patel VM, Welch KMA, Skinner JE (1987) Are heart attacks really brain attacks? In: Furlan AJ (ed) The heart and stroke (clinical medicine and the nervous system). Exploring mutual cerebrovascular and cardiovascular issues. Springer-Verlag, London, pp 185–216

Levy AG (1913) The exciting cases of ventricular fibrillation in animals under chloroform anaesthesia. Heart 14:319

Manning GW, Hall GE, Banting FG (1937) Vagal stimulation and the production of myocardial damage. Can Med Assoc J 37:314–318

Meyers MG, Norris JW, Hachinski VC, Weingart ME, Sole MJ (1982) Cardiac sequelae of acute stroke. Stroke 13:838–842

Myers M, Norris JW, Hachinski VC (1981) Plasma norepinephrine in stroke. Stroke 12:200–204

Norris JW, Froggatt GM, Hachinski VC (1978) Cardiac arrhythmias in acute stroke. Stroke 4:392–396

Norris JW, Hachinski VC, Myers MG, Callow J, Wong T, Moore RW (1979) Serum cardiac enzymes in stroke. Stroke 10:548–553

Oppenheimer SM, Cechetto DF, Hachinski VC (1990) Cerebrogenic cardiac arrhythmias: cerebral ECG influences and their role in sudden death. Arch Neurol 47:513–519

Oppenheimer SM, Wilson JX, Guiraudon C, Cechetto DF (1991) Insular cortex stimulation produces lethal cardiac arrhythmias: a mechanism of sudden death. Brain Res

550:115–121

Oppenheimer SM, Saleh TM, Wilson JX et al. (1992a) Plasma and organ catecholamine levels following stimulation of the rat insular cortex. Brain Res 569:221–228

Oppenheimer SM, Gelb A, Girvin JP, Hachinski VC (1992b) Cardiovascular effects of human insular cortex stimulation. Neurology 42:1727–1732

Philippu A (1988) Regulation of blood pressure by central neurotransmitters and neuropeptides. Rev Physiol Biochem Pharmacol 111:1–111

Pollick C, Cujec B, Parker S, Tator C (1988) Left ventricular wall motion abnormalities in subarachnoid hemorrhage. JACC 12:600–605

Rem JA, Hachinski VC, Boughner DR, Barnett HJM (1985) Value of cardiac monitoring and echocardiography in TIA and stroke patients. Stroke 16:950–956

Schenk EA, Moss AJ (1966) Cardiovascular effects of sustained norepinephrine infusions II. Morphology. Circ Res 18:605–615

Silver FL, Norris JW, Lewis AJ, Hachinski VC (1984) Early mortality following stroke: a prospective review. Stroke 15:492–496

Stober D, Anstat P, Sen S, Schimrik J, Jager H (1988) Cardiac arrhythmias in subarachnoid hemorrhage. Neurochirguick 93:37–44

Syverd G (1991) Electrocardiographic changes and intracranial pathology. J 59(3):229–232

Szakacs JE, Melhman B (1960) Pathologic changes induced by norepinephrine. Am J Cardiol 5:619–627

Talman WT (1985) Cardiovascular regulation and lesions of the central nervous system. Ann Eurol 18:1–12

Traggot S, Semi S, Thomas A, Ludwig B (1988) Correlation of cardiac arrhythmias with brain stem compression in patients with intracerebral hemorrhage. Stroke 19:688–692

Valadao AS, Saad WA, De Arruda Camargo LA et al (1990) Interaction between cholinergic and adrenergic pathways of the hypothalamic ventromedial nucleus on cardiovascular regulation. J Autonom Nerv Syst 30:239–246

Van Buren J (1958) Some autonomic concomitants of ictal automatisms. Brain 81:505–528

Verrier RL, Dickerson LW (1991) Autonomic nervous system and coronary bloodflow changes related to emotional activation and sleep. Circulation 83 (Suppl II):II-81–II-89

Walsh G, Masland W, Glodenshohn E (1972) Relationship between paroxysmal atrial tachycardia and paroxysmal cerebral discharges. Bull Neurol Soc 37:28–35

Willis T (1664) The anatomy of the brain. In: The remaining medical works of Dr. Thomas Willis, vol 2. S.P. Esq, London

Yamour BJ, Sridharan MR, Rice JF, Flowers NC (1980) Electrocardiographic changes in cerebrovascular hemorrhage. Am Heart J 99:294–300

Zamrini EY, Meador KJ, Loring DW (1990) Unilateral cerebral inactivation produces differential left/right heart rate responses. Neurology 40:1408–1411

Concluding Comments

Herein I will amplify on some of the issues covered in these two chapters, and will briefly touch on some related subjects.

Cardiac-Source Embolism

Heterogeneity of Embolic Particles

Conceptions of the causes and frequency of cardiac-origin embolism have changed dramatically during the past few decades. During the middle of the 20th century, two large studies attributed only 3% (Aring and Merritt 1935) and 8% (Whisnant et al. 1971) of strokes to cardiac-origin embolism. In contrast, the Harvard Stroke Registry, a study completed before echocardiography was introduced into clinical practice, attributed 31% of strokes to cardiac-origin embolism (Mohr et al. 1978). Originally three criteria were needed before cardiac-origin embolism could be diagnosed reliably. These included a known cardiac lesion with high embologenic potential (in practice, only acute myocardial infarction and rheumatic mitral stenosis with atrial fibrillation were accepted); sudden, maximal-at-onset neurologic signs; and systemic embolism. We now know that many other cardiac lesions can serve as the source of emboli. Hanna and Furlan categorize and review the various cardiac disorders known or suspected to cause embolism. About 20% of emboli do not cause maximal-at-onset deficits but instead have fluctuations and worsenings during the first 48 hours (Caplan 1993a). Systemic embolism is commonly found at necropsy but is recognized clinically in only 2% of patients with cardiogenic brain embolism (Mohr et al. 1978; Caplan 1993a).

The focus has now changed. Echocardiography and coagulation studies, as described by Hanna and Furlan, are beginning to make it clear that different lesions have different risks of embolization and that different types of material enter the circulation to form emboli. Analysis of some of the anatomical and physical properties of the lesions and blood flow and heart function in the neighborhood of these lesions can help predict the probability of the release of embolic particles. The body's reaction to the abnormality, e.g. deposition of fibrin–platelet material or fibrin–red blood cell clots is also very important in predicting the frequency and nature of embolism.

Physicians have customarily chosen treatments for patients with brain embolism solely on the basis of the origin of the embolic material, e.g. coumadin is usually given to patients with cardiac-origin embolism, and aspirin or other agents that modify platelet function are given for intra-arterial embolism. However, embolic materials from each site are varied. Emboli arising from the aorta or large arteries could consist of calcific particles, cholesterol crystals, pieces of atheromatous plaque, white fibrin–platelet thrombi or red fibrin-dependent erythrocyte-rich clots. Cardiac lesions give rise to an even more varied heterogenous lot of particles, including calcium from valves and the mitral annulus, bacterial vegetations, marantic non-bacterial valvular vegetations, myxoma cells, white fibrin–platelet thrombi and red thrombi. For some of these particles, e.g. myxoma, calcific particles and fibrotic non-bacterial vegetations, there is no known effective treatment.

Red thrombi are rich in erythrocytes mixed with fibrin. They tend to form in areas of stagnation and low flow and adhere to damaged surfaces, which activates the coagulation cascade. Heparin and coumadin are thought to be effective in preventing red thrombus formation, propagation and embolization. In contrast, white thrombi are rich in fibrin and platelets and form in fast-moving streams in which platelets are activated. Aspirin and other agents such as ticlopidine, which alter platelet functions, are thought to be effective in preventing white clot formation and adherence to abnormal surfaces. Hanna and Furlan point out that red thrombi predominate in patients with atrial fibrillation, myocardial infarction, ventricular aneurysms and cardiomyopathies. White thrombi form in patients with valvular disease, especially calcific valvulitis and bacterial and marantic endocarditis. Prosthetic valves are associated with both platelet and red thrombi, perhaps accounting for the effectiveness of combined coumadin and dipyridamole in preventing emboli (Sullivan et al. 1971).

Use of Ultrasound and Other Imaging for Embolus Detection

In any example of embolism the embolic material must arise from a source, at some point travel free in the circulation, and stop, at least temporarily, in a recipient artery. Hanna and Furlan have discussed study of the cardiac sources and how clarification of the morphology and dynamics of the heart lesions can help predict the nature of embolic matter. Now, newer ultrasound and imaging capabilities can give information about emboli in the circulation and at recipient sites. Transcranial Doppler (TCD) monitoring of intracranial arteries, e.g. the MCAs, has yielded both quantitative and qualitative data about embolism. The dose of emboli was found to be much higher than had been expected. Most neurologists and cardiologists had expected that release of embolic material was a very infrequent event with a high hit rate. When an embolus hits a recipient site it was expected that damage to the organ supplied would occur. Recent studies using TCD in a variety of situations shows that the inoculum is very high, often many particles a minute (Padayachee et al. 1986; Spencer et al. 1990; Tegeler et al. 1991; Caplan 1993b). Very rarely do most emboli cause damage.

Qualitative information is also possible. The amplitude of sound reflected and detected by ultrasound depends on the nature and size of embolic matter and the length of time it rests at any site. In experimental models, various sizes and types of emboli can be reliably detected (Russell et al. 1991; Russell 1992; Markus and Brown 1993). Duplex scanning of the extracranial arteries can also detect and potentially separate different embolic particles in the circulation. Color Doppler imaging has also been used to detect the nature of lesions within the ophthalmic artery circulation to the eye (Lieb et al. 1991a, b). I recently cared for a patient with multiple possible causes of her central retinal artery (CRA) occlusion. Color Doppler showed a calcific embolus in the CRA, indicating that her mitral annulus calcification was the probable embolic source. Ultrasound shows great promise in aiding the detection and identification of embolic materials.

Other imaging devices are also useful. CT (computed tomography) often images acute red thrombi that have embolized to major intracranial arteries such as the MCA by showing increased density within the vessel on non-enhanced scans (Yock 1981; Gacs et al. 1983; Tomsick et al. 1990; Wolpert et al. 1993). This phenomenon has usually been called the hyperdense MCA sign. I have also seen calcium on CT in intracranial arteries, indicating the likelihood of calcific emboli. Magnetic resonance imaging (MRI), especially the newer techniques, also shows promise in differentiating the nature of emboli, since the magnetic resonances of particles should differ. In the future, physicians should be able to choose rational therapy for patients with embolism depending on the nature of the material, not just on the organs of origin (Caplan 1991; Caplan 1993b).

Aorta as Source of Emboli

The embolic source that has in the past been neglected is the aorta. Atherosclerotic disease is usually most severe within the aorta and often occurs in the aortic arch. Ulcerated plaques are often found in the aortic arch at necropsy, especially in patients with unexplained strokes (Amarenco et al. 1992). Embolization from these aortic plaques is especially likely after arteriography and after cross-clamping of the aorta during cardiac surgery. The recent advent of transesophageal echocardiography (TEE) has made it possible to detect some aortic lesions. Tunick and colleagues (1991a,b), in two different studies, were able to identify protruding mobile aortic atheromas as sources of embolism. Atheromas are also detectable by direct application of ultrasound to the aorta during cardiac surgery before cross-clamping of the aorta (Marshall et al. 1989). Using this technique, clamping can avoid regions of severe disease.

Why do Echocardiograms Fail to Detect Thrombi and Other Sources?

Ultrasound (TEE, duplex scans of extracranial arteries and TCD) can show embolic sources and recipient arteries. However, there is still a high false negative rate for echocardiography. In my opinion, three explanations account for the majority of missed detections: (1) we now know that transesophageal studies have a much higher yield than transthoracic echocardiograms. When contrast is not used, atrial septal defects and patent foramen ovale can be missed. Failure of detection of sources in the past was probably most often explained by suboptimal studies; (2) embolism is a dynamic process: thrombi form and then leave their site of origin when they embolize and looking at the heart shortly after an embolus departed will not show the thrombus (DeWitt et al. 1988; Caplan 1991, 1993a). Sequential studies have shown changes in intracardiac thrombi over time (Yasaka et al. 1990); (3) emboli are often very small. In one reported patient, a 2–3 mm shotgun embolus occurred in the ICA siphon and the MCA and caused a devastating stroke (Kase et al. 1981) Small lesions often escape detection. At times, clinicians must treat on the suspicion of an embolic source without being able to document the actual lesion.

Cardiac and Cardiovascular Effects of Brain Lesions

Myocardial Damage

As noted by Ali and Levine in their contribution, the two most common lesions found in the heart in patients dying with acute central nervous system lesions are patchy regions of myocardial necrosis and subendocardial hemorrhages. Myofibrillar necrosis can involve individual cardiac muscle cells or may occur in focal areas of necrosis. The cell changes can range from eosinophilic staining of cells with preserved striations to the transformation of myocardiocytes into dense eosinophilic bands. These changes are usually referred to as myocytolysis (Schlesinger and Reiner 1955; Greenshoot and Reichenbach 1969; Weidler 1974; Samuels 1984). Often, there is a neutrophilic or monocytic response in regions of necrosis.

Microscopy reveals that these cardiac muscle cells die in a hypercontracted state. Early calcium entry with calcifications is found in regions of myocytolysis, as discussed by Ali and Levine. They have reviewed evidence of a role for catecholamines in causing the lesion. The cardiac lesion is probably identical to the situation in the brain during ischemia. Excitotoxins drive brain cells, causing injury. Similar chemical responses probably drive cardiac muscle cells, causing hypercontractility and changes in calcium channels which allow early calcium entry into cells. Myocytolysis was probably the first reported example of 'excitotoxic injury' (as described by Choi in Section I) although it was not originally recognized as such.

Cardiovascular and Blood Pressure Reactions

Ali and Levine have emphasized changes in the heart caused by brain lesions and brain stimulation. The pulse and blood pressure often change independently of cardiac effects. I have recently seen a number of patients with intracerebral hemorrhage presumed to be due to acute elevations in blood pressure that followed severe stress, anger, and frustration. These individuals were known to be previously normotensive, hypertensive when examined directly after their ICH, and normotensive at follow-up. The hemorrhages were located in the putamen, thalamus and cerebral lobes in typical sites for hypertensive ICH, and no other cause was found. Wilson (1955) commented on factors that could precipitate ICH: 'emotional experience, joy, anger, fear, or apprehension, may disturb the action of the heart, trivial though the incident may be – an address at a public meeting, trouble with a cook and so on'.

Intracerebral hemorrhage, presumed to be caused by abrupt increases in blood pressure, has also followed situations that involved manipulation or stimulation of the Vth cranial nerve. These include intracranial posterior fossa surgery to decompress the Vth nerve in patients with trigeminal neuralgia (Haines et al. 1978);

percutaneous stereotaxic treatment for trigeminal neuralgia (Caplan 1988); and dental treatment for toothache (Barbas et al. 1987; Caplan 1988). Moskowitz (1984) has reviewed, elsewhere and in this volume, the trigeminal innervation of the intracranial blood vessels. Changes in blood pressure and pulse can be evoked in experimental animals by trigeminal stimulation.

The anatomy and physiology of blood pressure control by the central nervous system have not been as well studied as the brain regions that affect the heart. We simply do not know the vicissitudes of blood pressure during normal everyday activities, during and after severe emotional stress and stimuli, and when having toothache or visiting the dentist. The advent of continuous ambulatory blood pressure monitoring devices may help fill this informational void.

Seizures

Ali and Levine mention that cardiac rhythm and ECG changes have been occasionally noted after epileptic attacks. I have seen several patients during the past two decades in whom cardiac arrhythmias were undoubtedly caused by central nervous system seizure discharges. One such patient was a woman in her 40s who was admitted to hospital because her recurrent attacks of tachycardia were disabling and were refractory to the cardiac antiarrhythmic drugs that had been used. She never had a recognized seizure. She described to me attacks that she had had for many years. She would see a short train always off to one side of her vision. The train would move and then she would have a very brief instant when she lost touch with her environment. Sometimes, she had rapid heartbeat just after seeing the train and sometimes the tachycardia preceded or came concurrently with the train. EEG showed focal spikes and the attacks disappeared after diphenylhydantoin treatment. Cardiologists, in my experience, seldom inquire about seizures or any paroxysmal experiences in their patients with arrhythmias.

Pulmonary Edema

Acute pulmonary edema, often sudden in onset and sometimes fatal, has been described in patients who have subarachnoid hemorrhage (SAH) and posterior circulation ischemia and hemorrhages (Norris and Hachinski 1987). The occurrence of pulmonary edema correlates well with the presence of sudden-onset and severe increases in intracranial pressure. Weir (1978) studied the occurrence of pulmonary edema in patients with SAH who died. Among the 70% of his fatal cases of SAH who developed pulmonary edema, all had the sudden onset of coma. Respiratory symptoms were reported within a short interval after the onset of headache and neurologic symptoms. Weir (1978) attributed the occurrence of pulmonary edema to a sudden severe increase in intracranial pressure, which in turn caused massive autonomic stimulation. Experimental data from studies in cats confirm this hypothesis (Hoff and Nishimura 1978).

Centrally mediated sympathetic nervous system discharges produce intense systemic vasoconstriction (Norris and Hachinski 1987). Sympathetic discharges can be triggered by sudden increases in intracranial pressure which cause secondary effects in the hypothalamus and brain stem, and by direct involvement of descending sympathetic fibers traveling in the lateral brain stem tegmentum. The intense vasoconstriction leads to very high resistance in the systemic circulation, which in turn provokes sudden shifts of blood volume into the lower-resistance pulmonary circulation. The sudden increase in volume in the pulmonary vascular bed leads to increased pulmonary capillary pressure, pulmonary hypertension, development of pulmonary edema and rupture of pulmonary capillaries, with lung hemorrhage. The pulmonary edema fluid has a high protein content and can develop despite normal cardiac function and hemodynamics (Theodore and Robin 1975).

Sudden Death

Sudden catastrophic events are well known to precipitate death. So-called 'voodoo death' is an example of literally being frightened to death (Engel 1976, 1978). One mechanism of sudden death, ventricular fibrillation, can be elicited by stimulation of the cardiac sympathetic nerves in normal and ischemic hearts (Talman 1985). Stress causes intense CNS stimulation which in turn triggers autonomic activation. Left-sided sympathetic activation may have more powerful effects than right-sided stimulation (Schwartz et al. 1976; Haws and Burgess 1978; Norris and Hachinski 1987). Cardiac vulnerability plays a role since the threshold for ventricular fibrillation is reduced in ischemic myocardium (Lown 1979).

Another mechanism of sudden death is cardiac standstill: asystole. Sudden vagotonic stimulation can cause bradycardia and cardiac standstill. This response mediated by the vagal nerve has often been called the 'duck-diving' reflex, since ducks often have transient sinus arrests after suddenly submerging in water. In humans, the effects of vagal stimulation on the development of arrhythmias is not clear (Talman 1985). Patients with lateral medullary and lateral pontine infarcts involving the lateral tegmental structures have a high frequency of autonomic dysregulation, such as labile blood pressure, syncope, flushing and tachycardia. These patients may also die suddenly without an obvious explanation (Caplan et al. 1986).

References

Amarenco P, Duyckaerts C, Tzourio C, Henin D, Bousser M-G, Hauw J-J (1992) The prevalence of ulcerated plaques in the aortic arch in patients with stroke. N Engl J Med 326:221–225

Aring C, Meritt H (1935) Differential diagnosis between cerebral hemorrhage and cerebral thrombosis. Arch Intern Med 56:435–456

Barbas N, Caplan LR, Baquis GD et al. (1987) Dental chair intracerebral hemorrhage. Neurology 37:511–512

Caplan LR (1991) Of birds and nests and brain emboli. Rev Neurol 147:265–273

Caplan LR (1993a) Stroke: a clinical approach, 2nd edn. Butterworths, Boston

Caplan LR (1993b) Brain embolism revisited. Neurology 43:1281–1287

Caplan LR (1988) Intracerebral hemorrhage revisited. Neurology 38:624–627

Caplan LR, Pessin MS, Scott M, Yarnell P (1986) Poor outcome after lateral medullary infarcts. Neurology 36:1510–1513

DeWitt LD, Pessin MS, Pandian NG, Pauker SG, Sonnenberg FA, Caplan LR (1988) Benign disappearance of ventricular thrombus after embolic stroke. Stroke 19:393–396

Engel GL (1976) Psychologic factors in instantaneous cardiac death. New Engl J Med 294:664–665

Engel GL (1978) Psychologic stress, vasodepressor (vasovagal) syncope and sudden death. Ann Intern Med 89:403–412

Gacs G, Fox AJ, Barnett JH, Vinuela F (1983) CT visualization of intracranial arterial thromboembolism. Stroke 14:756–763

Greenshoot JH, Reichenbach DD (1969) Cardiac injury and subarachnoid hemorrhage. J Neurosurg 30:521–531

Haines S, Maroon J, Janetta P (1978) Supratentorial intracerebral hemorrhage following posterior fossa surgery. J Neurosurg 49:881–886

Haws CW, Burgess MJ (1978) Effects of bilateral and unilateral stellate stimulation on cardiac refractory periods at sites of overlapping innervation. Circ Res 42:195–198

Hoff JT, Nishimura M (1978) Experimental neurogenic pulmonary edema in cats. J Neurosurg 18:383–389

Kase CS, White R, Vinson T et al. (1981) Shotgun embolus to the middle cerebral artery. Neurology 31:458–461

Lieb WE, Cohen SM, Merton PA et al. (1991a) Color Doppler imaging of the eye and orbit. Technique and normal vascular anatomy. Arch Ophthalmol 109:527–531

Lieb WE, Flaherty PM, Sergott RC et al. (1991b) Color Doppler imaging provides accurate assessment of orbital blood flow in occlusive carotid artery disease. Ophthalmology 98:548–552

Lown B (1979) Sudden cardiac death: the major challenge confronting contemporary cardiology. Am J Cardiol 43:313–328

Markus HS, Brown M (1993) Differentiation between different pathological cerebral embolic materials using transcranial Doppler in an in vitro model. Stroke 24:1–5

Marshall WG, Barzilai B, Kouchoukos N, Saffitz J (1989) Intraoperative ultrasonic imaging of the ascending aorta. Ann Thorac Surg 48:339–344

Mohr JP, Caplan LR, Melski J (1978) The Harvard Cooperative Stroke Registry: a prospective registry. 2:11–22

Moskowitz MA (1984) The neurobiology of vascular head pain. Ann Neurol 16:157–168

Norris JW, Hachinski V (1987) Cardiac dysfunction following stroke. In: Furlan AJ (ed) The heart and stroke. Springer Verlag, London, pp 171–183

Padayachee TS, Gosling RG, Bishop CC et al. (1986) Monitoring middle cerebral artery blood velocity during carotid endarterectomy. Br J Surg 73:98–100

Russell D (1992) The detection of cerebral emboli using doppler ultrasound. In: Newell DW, Aaslid R (eds) Transcranial Doppler. Raven Press, New York, pp 207–213

Russell D, Madden KP, Clark WM et al. (1991) Detection of arterial emboli using Doppler ultrasound in rabbits. Stroke 22:253–258

Samuels M (1984) Electrocardiographic manifestations of neurologic disease. Semin Neurol 4:453–459

Schlesinger MJ, Reiner L (1955) Focal myocytolysis of heart. Am J Pathol 31:443–459

Schwartz PJ, Stone HL, Brown AM (1976) Effects of unilateral stellate ganglion blockage on the arrythmias associated with coronary occlusion. Am Heart J 92:589–599

Spencer MP, Thomas GI, Nicholls SC, Sauvage LR (1990) Detection of middle cerebral artery emboli during carotid endarterectomy using transcranial Doppler ultrasonography. Stroke 21:415–423

Sullivan J, Harkin D, Gorlin R (1971) Pharmacologic control of thromboembolic complication of aortic valve replacement. New Engl J Med 284:1391–1394

Talman WT (1985) Cardiovascular regulation and lesions of the central nervous system. Ann Neurol 18:1–12

Tegeler CH, Hitchings LP, Leighton VJ, Burke GL, Downes TR, Stump DA et al. (1991) Carotid ultrasound emboli monitoring in stroke: initial clinical experience. J Neuroimag 1:61

Theodore J, Robin ED (1975) Pathogenesis of neurogenic pulmonary edema. Lancet 2:749–751

Tomsick T, Brott T, Barsan W et al. (1990) Thrombus localization with emergency cerebral computed tomography. Stroke 21:180

Tunick PA, Culliford AT, Lamparello PJ, Kronzon I (1991a) Atheromatosis of the aortic arch as an occult source of multiple systemic emboli. Ann Intern Med 114:391–392

Tunick PA, Perez JL, Kronzon I (1991b) Protruding atheromas in the thoracic aorta and systemic embolization. Ann Intern Med 115:423–427

Weidler DJ (1974) Myocardial damage and cardiac arrythmias after intracranial hemorrhage: a critical review. Stroke 5:759–764

Weir BK (1978) Pulmonary edema following fatal aneurysmal rupture. J Neurosurg 49:502–507

Whisnant J, Fitzgibbons J, Kurland L et al. (1971) Natural history of stroke in Rochester, Minnesota, 1945–1954. Stroke 2:11–22

Wilson SAK (1955) Neurology, 2nd edn. Butterworths, London, pp 1367–1383

Wolpert SM, Bruckmann H, Greenlee R et al. (1993) Neuroradiologic evaluations of patients with acute stroke treated with recombinant tissue plasminogen activator. AJNR 14:3–13

Yasaka M, Yamaguchi T, Miyashita T, Tsuchiya T (1990) Regression of intracardiac thrombus after embolic stroke. Stroke 21:1540–1544

Yock D (1981) CT appearance of cerebral emboli. J Comput Assist Tomogr 5:190–196

Section VII

Clinical Research: Epidemiology Statistics, Databases and Trials

Prefatory Comments

The final section of this book considers information gained from clinical stroke research. Until the past few decades, the vast bulk of clinical research in stroke was performed by individual clinician-investigators and small groups. Most studies involved descriptions of findings and phenomenology in various vascular disorders, clinical correlations with pathologic, radiologic or laboratory results, and observations of the course, outcome and treatment of individuals or groups of patients. Recently these so-called 'anecdotal' studies and reports have appropriately advanced to research that is larger in scope, often involving multiple centers and large numbers of patients or populations in the community, and is quantitative. Stroke clinical research has become a 'numbers game'. During the last two decades the three major central themes have been epidemiological studies, accumulations of large databases and registries, and clinical therapeutic trials.

Epidemiological Studies

Some epidemiological studies have sought to identify various data that are statistically related to the occurrence of stroke, various cerebrovascular lesions and various specific subtypes of stroke. These data items vary. Some are demographic, such as age, race and sex; some are past and present medical conditions, such as hypertension, diabetes and coronary artery disease; others are genetic, consisting of illnesses occurring in family members; some items include habits such as cigarette smoking, alcohol use, amount of coffee consumed, exercise, etc. If risk factors were identifiable before a stroke and predicted the likely occurrence of stroke, perhaps modification of these factors or close surveillance for cerebrovascular lesions in individuals at risk might help prevent stroke. Even if identified after a stroke, modification and surveillance might help prevent recurrences. These epidemiological studies have prevention as their main goals. Louis Pasteur (1884) has been quoted as saying, 'When meditating over a disease, I never think of finding a remedy for it, but instead a means of preventing it'. Clearly, prevention of an event is preferred to picking up the pieces afterwards.

Epidemiologists obtain risk factor data by studying large groups of patients with the condition under investigation (vascular lesion, stroke, or stroke subtype, such as subarachnoid hemorrhage). The incidence of various factors in these cohorts is compared with the incidence of the same factors in comparable control groups without the condition to be studied. More difficult, but in some ways preferable, is to study large population-based cohorts prospectively to determine what factors predict the subsequent occurrence of stroke or any other condition being studied. The Framingham study is an excellent example of such a population study which has yielded important information about risk factors (Wolf et al. 1991a,b). Now even the children and grandchildren of the original index cases from the Framingham study are being followed.

A number of criteria are used to determine whether data items qualify as valid or important risk factors. These include temporal sequence: the risk factor precedes the event; consistency: multiple studies of representative and different groups show similar results; strength of association: the incidence of the event in persons with the putative risk factor is considerably higher than the incidence without the risk factor; dose response: increasing levels of exposure to a risk factor, e.g. cigarettes, corresponds to a gradient increase in the event or disease; and biologic plausibility: the risk factor relationship makes sense and is consistent with other scientific data and theory (Gorelick and Kelly 1992).

Other types of epidemiologic studies quantify the incidence and prevalence of various conditions in various populations and geographic regions. The results provide an estimate of the number of individuals who have a stroke each year (morbidity incidence), how many die of stroke each year (mortality incidence), and how many stroke patients are present during that year (disease prevalence). Different countries and geographical regions can be compared. The incidence and prevalence of different stroke subtypes can also be measured if the patients had had sufficient evaluation to allow subtype diagnosis. The measurements can be repeated periodically, often yearly, so that trends and changes can be monitored. During the past decades mortality from stroke has gradually and consistently declined until very recently. Analysis of the prevalence of risk factors and concomitant changes in health practices and medical treatment, e.g. the introduction of effective antihypertensive treatment, can be pursued to attempt to explain the changes in disease incidence, mortality and prevalence. In his contribution in this section, Gorelick provides data on risk factors, stroke incidence and prevalence.

Data Banks and Registries

The introduction of computers into medicine during the early 1970s made possible the storage, retrieval and analysis of large volumes of clinical, epidemiologic and laboratory data. At first, very large expensive 'temperamental' computers were used which required sophisticated technical personnel and maintenance and, despite this care, the computers were often 'down' and data were sometimes unintentionally erased. The recent widespread availability of personal microcomputers has greatly facilitated computerized databases. Complex data of all sorts can now be quickly entered, stored, analyzed, transmitted and disseminated, even to distant locations.

During the early 1970s, J.P. Mohr, Howard Bleich, Robert Goldstein and I began the first large prospective computer-based stroke or neurological registry, the Harvard Stroke Registry (HSR) (Mohr et al. 1978). The project was stimulated by a failed attempt to create a computerized diagnostic program that could help teach the diagnosis, evaluation and management of patients with acute stroke (Goldstein et al. 1973). Previously, Bleich, a nephrologist with computer expertise, had successfully created and reported a computer-based diagnosis and management teaching program for use in patients with acid–base and electrolyte problems (Bleich 1971). The program consisted of computer interaction with the user (usually a medical student, resident or junior staff physician); the output taught the user the data needed, calculated the types and amounts of fluid solutions to be given, diagnosed the biochemical abnormality and referred the user to appropriate references to explain how and why the computer had arrived at the diagnosis and management advice. Bleich persuaded me into creating a similar program for stroke diagnosis. With the help of Goldstein, then a Harvard medical student with an MIT computer background and now a nephrologist, and Bleich, a program was created that was designed to estimate the probabilities that a given stroke patient had each of the various stroke subtypes (embolism, intracerebral hemorrhage, subarachnoid hemorrhage etc.). To calculate the probabilities we used Bayes theorem methodology which required the incidence of each of the stroke subtypes in the population, and the frequency of the various data items (demographics, past medical history, blood pressure, symptoms, signs etc.) in each subtype group. We very soon learned that these types of data were simply not available. How often did patients with brain embolism, brain hemorrhage or lacunar infarcts have headache before, during or after stroke onset? What was the frequency of vomiting, seizures, loss of consciousness etc. in each stroke subtype? At that time, CT was not yet available and existing data were entirely clinicopathological, and so limited to only large fatal strokes. Few quantitative data were available about newly described subtypes such as lacunes.

Since the quantitative data were not available anywhere in the literature, I asked five well known and experienced stroke experts to estimate the various frequencies. Estimates varied very widely, and I became painfully aware of our lack of information and knowledge and was determined to acquire the data. At that time, I was at the Beth Israel Hospital in Boston. I enlisted the help of J.P. Mohr, who was active on the stroke service at the Massachusetts General Hospital. Mohr, Bleich and I met often to decide on what data items to collect, to agree on definitions to be used and to construct data entry forms. All stroke patients at the two Boston hospitals were examined personally by Mohr or myself, and we filled out the forms ourselves. Bleich and his colleague, Warner Slack, were responsible for entering the data into the large computer, data formatting and analysis. In all, 694 cases were entered and the data analyzed (Mohr et al. 1978).

Nearly half the patients in the HSR did not have CT scans. Both Mohr and I left Boston. I created the Michael Reese Stroke Registry (MRSR) in Chicago in the hope of adding more laboratory data, especially CT (Caplan et al. 1983). The population studied in the MRSR was quite different from that in the HSR. In Chicago, a majority of the stroke patients were black, whereas in Boston nearly all of the patients were white. Dan Hier and Phil Gorelick, the authors of the chapters in this section, were colleagues of mine in the MRSR. These first two registries involved two hospitals and one large hospital in one city, respectively. In 1978, the National Institute of Neurological and Communicative Disorders and Stroke (NINCDS) initiated a data bank project, the Pilot Stroke Data Bank, which was soon followed by the Stroke Data Bank (SDB). Each of these registries included multiple centers in five different cities (Kunitz et al. 1984; Foulkes et al. 1988). The major objectives of the SDB were to obtain data on clinical course and outcome; to standardize clinical evaluations; to identify useful prognostic factors; and to provide baseline data useful for planning therapeutic trials (Foulkes et al. 1988).

Concurrently, Al Heyman and his colleagues at Duke University had created and published a very different type of registry (Heyman et al. 1979). Data from patients with TIAs were entered into a computer program along with the results of the evaluation and outcome. Physicians caring for new patients with TIA could be furnished online with a 'prognostigram' which included the evaluation and outcome results of patients in the Duke TIA registry who shared designated features with that patient. For example, if the patient had episodes of transient monocular blindness and had a cervical bruit, all patients with these two findings were retrieved and the frequency of various degrees of carotid artery stenosis and results of treatment of these patients were printed out. This pattern-matching strategy allowed physicians to quickly use the accumulated experience from similar past cases.

Table 1. Physician aims in the HSR, MRSR and SDB

Define and quantify stroke risk factors and associations
Study the utility of various historical, examination and laboratory features in the diagnosis and prognosis of the various stroke subtypes
Facilitate recognition and analysis of the features of uncommon syndromes (made possible because of the size of the data banks)
Study the course and prognosis of stroke syndromes and subtypes
Design useful diagnostic algorithms and paradigms
Analyze clinical–radiologic and laboratory correlations
Describe the frequency of various common features in patients with various stroke subtypes and of strokes at various brain sites

The HSR, MRSR and SDB all had similar aims (Caplan 1985; Mohr 1986; Table 1). Some other stroke registries, such as the Lausanne Stroke Registry in Switzerland (Bogousslavsky et al. 1988) and the Austin Hospital Registry in Melbourne, Australia (Chambers et al. 1983), had very similar compositions and aims. The results from these studies allowed international sharing and comparison of stroke data. Also, during the late 1970s and 1980s, data banks and registries of quite a different sort began to spring up in widely different locations. These were population-based and were intended more for storing and analyzing epidemiologically valid data. The largest and most geographically diverse of these cohorts is the Community Hospital-based Stroke Programs, which includes an aggregate of over 4000 patients studied in the states of New York, North Carolina and Oregon (Yatsu et al. 1986; Coull et al. 1990). Other community-based projects developed in south Alabama (Gross et al. 1984), the Lehigh Valley in Pennsylvania (Alter et al. 1985; Friday et al. 1989), in Oxfordshire in the United Kingdom (Bamford et al. 1988) and elsewhere. An important outcome of these registries has been the need to standardize definitions, disability and deficit criteria and scores, and data to be studied. Statistical methods of analyzing the data were also developed. The use of multiple observers and multiple centers stimulated study of the consistency and interobserver reliability of the data (Koudstaal et al. 1986, 1989; Shinar et al. 1985, 1987).

Therapeutic Trials

During the latter part of the 20th century, one of the most important developments has been the proliferation of therapeutic trials to scientifically study treatment. In the field of stroke, these trials can be simplistically divided into two groups. The bulk of trials, many during the earlier decades, studied various treatments in patients characterized by the tempo of brain ischemia. These trials were mostly prospective and randomized, and observers were usually blinded to the treatment given. The trials were placebo-controlled or compared various treatments or various dosages of the same treatment. Prominent examples of such trials are the Canadian Cooperative Study of Aspirin and Sulfinpyrazone in Threatened Stroke (1978); the American–Canadian Cooperative Study Group Trial of Dipyridamole and Aspirin in Cerebral Ischemia (1983); the 'AICLA' study of aspirin and dipyridamole in secondary prevention of brain ischemia (Bousser et al. 1983); the Dutch TIA trial (1991) comparing two doses of aspirin; and the ticlopidine–aspirin trial (Hass et al. 1989). Unfortunately, in most such trials in which patients were chosen by tempo (e.g. TIA) or deficit, e.g. minor ischemic stroke, the etiology of the ischemia was not well studied or reported.

The other group of trials is mostly newer and considers the treatment of specific etiologies and conditions. These include the North American Symptomatic Carotid Endarterectomy Trial (1991) and the European Carotid Surgery Trial (1991), each of which studied endarterectomy in patients with ischemic symptoms and various severities of carotid artery stenosis; the EC/IC Bypass study of surgical vascular shunts in patients with severe carotid territory occlusive disease (1985); and the Boston Area Anticoagulation Trial for Atrial Fibrillation (1990) and the Stroke Prevention in Atrial Fibrillation Trial (1991), each of which studied aspirin and warfarin prophylaxis in patients with atrial fibrillation. These trials are, in my opinion, more relevant to physicians in practice, since they analyze the effectiveness of treatment in patients with defined conditions such as atrial fibrillation and carotid artery disease. Modern technology now readily allows non-invasive detection of these various conditions.

In order to perform these trials, the methodology of patient selection and exclusion, data and category definitions, interobserver reliability, calculation of the power and needed numbers of various treatment trials, and of statistical analysis and even meta-analysis of pooled trial data has been developed and gradually improved. Gorelick, in his contribution, reviews the subject of therapeutic trials.

References

Alter M, Sokel E, McCoy RL et al. (1985) Stroke in the Lehigh Valley: incidence based on a community-wide hospital register. Neuroepidemiology 4:1–15

American–Canadian Cooperative Study Group (1983) Persantin aspirin trial in cerebral ischemia. Stroke 14:99–103

Bamford J, Sandercook P, Dennis M et al. (1988) A prospective study of acute cerebrovascular disease in the community: the Oxfordshire community stroke project: 1981–1986. I. Methodology, demography, and incident cases of first-ever stroke. J Neurol Neurosurg Psych 51:1373–1380

Bleich HL (1971) The computer as a consultant. New Engl J Med 284:141–146

Bogousslavsky J, van Melle G, Regli F (1988) The Lausanne Stroke Registry: analysis of 1000 consecutive patients with first stroke. Stroke 19:1083–1092

Boston Area Anticoagulation Trial for Atrial Fibrillation Investigators (1990) The effect of low-dose warfarin on the risk of stroke in patients with non-rheumatic atrial fibrillation. New Engl J Med 323:1505–1511

Bousser MG, Eschwege E, Hagenau M et al. (1983) Controlled cooperative trial of secondary prevention of cerebral ischemic accidents caused by atherosclerosis, using aspirin and dipyridamole. Presse Med 12:3049–3057

Canadian Co-operative Study Group (1978) A randomized trial of aspirin and sulfinpyrazone in threatened stroke. New Engl J Med 299:53–59

Caplan LR (1985) Stroke data banks, then and now. In: Courbier R (ed) Basis for a classification of cerebrovascular disease. Excerpta Medica, Amsterdam, pp 152–162

Caplan LR, Hier DB, D'Cruz I (1983) Cerebral embolism in the Michael Reese Stroke Registry. Stroke 14:530–536

Chambers BR, Donnan GA, Bladin PF (1983) Patterns of stroke. An analysis of the first 700 consecutive admissions to the Austin Hospital stroke unit. Aust NZ J Med 13:57–64

Coull BM, Brockschmidt JK, Howard G et al. (1990) Community hospital-based stroke programs in North Carolina, Oregon, and New York IV. Stroke diagnosis and its relation to demographics, risk factors, and clinical status after stroke. Stroke 21:867–873

Dutch TIA Trial Study Group (1991) A comparison of two doses of aspirin (30 mg vs. 283 mg a day) in patients after a transient ischemic attack or minor ischaemic stroke. New Engl J Med 325:1261–1266

EC-IC Bypass Study Group (1985) Failure of extracranial-intracranial arterial bypass to reduce the risk of ischemic stroke. New Engl J Med 313:1191–1200

European Carotid Surgery Trialists Collaborative Group (1991) MRC European Carotid Surgery Trial: interim results for symptomatic patients with severe (70–99%) or with mild (1–29%) carotid stenosis. Lancet 337:1235–1243

Foulkes MA, Wolf PA, Price TR et al. (1988) The Stroke Data Bank: design, methods, and baseline characteristics. Stroke 19:547–554

Friday G, Lai SM, Alter M et al. (1989) Stroke in the Lehigh Valley: racial/ethnic differences. Neurology 39:1165–1168

Goldstein RJ, Caplan LR, Bleich HL (1973) Computer assisted stroke diagnosis. Clin Res:537

Gorelick PB, Kelly MA (1992) Alcohol as a risk factor for stroke. Heart Dis Stroke 1:255–258

Gross CR, Kase CS, Mohr JP, Cunningham SC (1984) Stroke in South Alabama: incidence and diagnostic features. Stroke 15:249–255

Hass WK, Easton JD, Adams HP et al. (1989) Ticlopidine Aspirin Stroke Study Group: a randomized trial comparing ticlopidine hydrochloride with aspirin for the prevention of stroke in high-risk patients. New Engl J Med 321:501–507

Heyman A, Burch JG, Rosati R, Haynes C, Utley C (1979) Use of a computerized information system in the management of patients with transient cerebral ischemia. Neurology 29:214–221

Koudstaal PJ, van Gijn J, Staal A et al. (1986) Diagnosis of transient ischemic attacks: improvement of interobserver agreement by a check-list in ordinary language. Stroke 17:723–728

Koudstaal P, Gerritsma JGM, van Gijn J et al. (1989) Clinical disagreement on the diagnosis of transient ischemic attacks: is the patient or the doctor to blame? Stroke 20:300–311

Kunitz SC, Gross CR, Heyman A et al. (1984) The Pilot Stroke Data Bank: definition, design, and data. Stroke 15:740–746

Mohr JP (1986) Editorial – Stroke data banks. Stroke 17:171–172

Mohr JP, Caplan LR, Melski JW et al. (1978) The Harvard Cooperative Stroke Registry: a prospective registry of cases hospitalized with stroke. Neurology 28:754–762

North American Symptomatic Carotid Endarterectomy Trial (NASCET) collaborators (1991) Beneficial effect of carotid endarterectomy in symptomatic patients with high-grade carotid stenosis. New Engl J Med 325:445–453

Pasteur L (1884) Address to the fraternal association of former students of the Ecole Centrale des Artes et Manufactures, Paris

Shinar D, Gross CR, Mohr JP et al. (1985) Interobserver variability in assessment of the neurologic and history and examination in the Stroke Data Bank. Arch Neurol 42:557–565

Shinar D, Gross CR, Hier DB et al. (1987) Interobserver reliability in the interpretation of computed tomographic scans of stroke patients. Arch Neurol 44:149–155

Stroke Prevention in Atrial Fibrillation Investigators (1991) The Stroke Prevention in Atrial Fibrillation Study. Final results. Circulation 84:527–539

Wolf PA, D'Agostino RB, Belanger AJ, Kannel WB (1991a) Probability of stroke: a risk profile from the Framingham study. Stroke 22:312–318

Wolf PA, Abbott RD, Kannel WB (1991b) Atrial fibrillation as an independent risk factor for stroke: The Framingham Study. Stroke 22:983–988

Yatsu FM, Becker C, McLeroy KR et al. (1986) Community Hospital-based stroke programs: North Carolina, Oregon and New York. I: Goals, objectives and data collection procedures. Stroke 17:276–284

22. Epidemiology and Trials

P. B. Gorelick

Stroke is a major cause of mortality and disability. It is estimated that there are over 150 000 deaths attributable to stroke, over 2 million stroke survivors and approximately 400 000 stroke patients discharged annually from acute care hospitals in the USA (American Heart Association 1991; Wolf et al. 1992). About half of all individuals who survive a stroke have significant persisting neurological impairment and physical disability (Brandstater 1990). Furthermore, the economic costs of stroke total billions of dollars.

Although stroke looms as a major cause of morbidity and mortality, epidemiologic studies have elucidated treatable risk factors and medical and surgical therapies for the primary and secondary prevention of stroke. In this chapter we will review the distribution and determinants of stroke, as well as medical and surgical prevention strategies that have emanated from controlled clinical trials.

Stroke Distribution

Mortality

Overall, stroke mortality increases with age, is high among certain racial or ethnic groups such as blacks and Asians, and maybe higher among men, except in the oldest age group. Stroke death rates have shown a steady decline in the USA since the early 1900s (Wolf et al. 1986). A rate of decline of about 1% per year was detected until the late 1960s and early 1970s, when death rates from stroke began to fall by 5%–7% per year. Overall, total deaths and death rates from stroke declined more rapidly than for any other component of cardiovascular mortality. From 1968 to 1975 the age-adjusted US stroke death rate decline was most prominent for non-white men and women (Soltero et al. 1978)

International studies of stroke mortality demonstrated similar trends (Bonita et al. 1990). For the time period 1970–1985, stroke mortality rates declined in most countries for both men and women, and were usually greater for women. Furthermore, the rate of decline was higher than that for coronary heart disease in those countries in which there was a decline in both categories. Since 1979, however, the annual rate of decline of stroke mortality has begun to slow considerably in the USA (Cooper et al. 1990). The slowing in the absolute rate of decline has been estimated to be 57% for white men, 58% for white women, 44% for black men and 62% for black women. Table 22.1 (part 1) summarizes stroke mortality rates in the USA during key time periods by race and sex.

The reason for the accelerated decline in US stroke mortality in the early 1970s is uncertain, but may have resulted from improved antihypertensive therapy, treatment of other cardiovascular risk factors, decrease in stroke incidence or

Table 22.1. Stroke distribution

(1) Age-adjusted stroke mortality rates in USA by race–sex group[1]

	White		Black	
Year	Men	Women	Men	Women
1960	80.3	68.7	141.0	139.3
1970	68.8	56.2	122.5	107.9
1980	41.9	35.2	77.5	61.7
1986	31.1	27.2	58.9	47.6

[1]Rate per 100 000. Reproduced with permission from the American Heart Association, Stroke (1990) 21:1262–1267

(2) Overal incidence rates by stroke subtype, race and sex for south Alabama, 1980[2]

	Infarction Type				Hemorrhage Type	
Race–sex	Atherothrombotic	Unspecified	Embolic	Lacunar	Parenchymatous	Subarachnoid
White						
Men	0	75.3	100.4	41.8	16.7	0
Women	0	73.9	44.3	14.8	7.4	22.2
Black						
Men	36.6	95.2	109.8	29.3	36.6	0
Women	22.8	182.8	51.4	51.4	28.6	34.3

[2]Rate per 100 000. Reproduced with permission from the American Heart Association, Stroke (1984) 15:249–255

(3) Average annual incidence rates of cerebral infarction (by sex), intracerebral hemorrhage and subarachnoid hemorrhage for Rochester, Minnesota during various periods 1945–1984[3]

Category of stroke	1945–49 Rate	1950–54 Rate	1955–59 Rate	1960–64 Rate	1965–69 Rate	1970–74 Rate	1975–79 Rate	1980–84 Rate
Cerebral infarction								
Men	209	181	213	177	180	139	124	143
Women	155	187	161	114	82	82	68	88

	1945–54 Rate	1955–64 Rate	1965–74 Rate	1975–84 Rate
Intracerebral hemorrhage	20	14	9	14
Subarachnoid hemorrhage	10	12	12	10

[3]Rate per 100 000. Reproduced with permission from the American Heart Association, Stroke (1989) 20:577–582

(4) Age-specific prevalence rates for stroke from the National Survey of Stroke (NSS) and Health Interview Survey (HIS)[4]

Age	NSS (1976)	HIS (1977)
All Ages	743	1238
<45	66	390*
45–64	998	2070
65+	5063	7000

[4]Rate per 100 000. Reproduced with permission from the American Heart Association, Stroke (1981 Suppl 1) 12:I-59–I-68
*data refer to ages 20–44

case fatality, or some other factor (Klag et al. 1989). Many believe that the use of antihypertensives explains only a small proportion of the accelerated rate of decline in stroke mortality. Riggs (1991) has proposed a competitive and deterministic mortality model to explain the observed decline. Although the precise explanation for the deceleration in stroke mortality in the late 1970s remains unknown, national trends show that continued improvement in blood pressure control has not occurred and that the leveling off in stroke and all-cause mortality has coincided with decreasing social and economic conditions for many (Cooper et al. 1990).

When considering secular trends for mortality by stroke subtype, it appears that death rates for brain infarction, intraparenchymal hemorrhage and subarachnoid hemorrhage generally show a decrease over time (Bonita et al. 1983; Garraway et al. 1983; Ueda et al. 1981). Possible explanations for these trends are reviewed by Garraway et al. (1983), Ingall et al. (1989) and Bozzola et al. (1992).

Incidence

Stroke incidence increases with age and is generally higher among men than women (Kurtzke 1985). The average annual incidence of all strokes and atherothrombotic brain infarctions generally doubles in each successive decade. In the Framingham study the incidence of all strokes and atherothrombotic brain infarctions specifically was about 30% higher in men than in women (Wolf et al. 1992). However, the relative frequency of stroke by subtype did not differ substantially by sex. Population data from South Alabama estimate that age-adjusted stroke incidence is substantially higher among US blacks (208/100 000) than whites (109/100 000). Black women had the highest overall age-adjusted stroke incidence rates and the highest incidence rate for lacunar stroke. Blacks also had a higher incidence rate for hemorrhages (Table 22.1 (part 2)) (Gross et al. 1984). These data and recent declines in stroke mortality in the southeastern USA may herald a breakup of the concentration of excess stroke in this region (Lanska 1991).

Population studies from Rochester, Minnesota, provide data on secular trends for the incidence of stroke in North America (Broderick et al. 1989). Between 1950 and 1954 and 1975 and 1979 the average annual incidence rate of stroke declined by 46% from 213 to 115/100 000 population. This was true for all age and sex groups, but it occurred earlier in women than in men. However, between 1975 and 1979 and 1980 and 1984 the incidence rate of stroke was 17% higher. When classified by stroke subtype, the incidence rate for brain infarction generally declined until the 1980s, when there was a substantial increase, fluctuated for intracerebral hemorrhage and changed little for subarachnoid hemorrhage (Table 22.1 (part 3)). Stroke incidence in the 1980s has also increased in Sweden (Terent 1988).

Further study of the Rochester and other populations is needed to provide precise explanations for these trends in stroke incidence (Kuller 1989). The impact of such factors as the availability of CT scans; levels of blood pressure of treated and non-treated hypertensive residents over time; the available pool of individuals at high risk for stroke due to the decline in coronary heart disease incidence and mortality; the successful treatment of diabetes mellitus; and the possible increase in stroke secondary to the greater use of newer surgical and diagnostic procedures needs to be clarified.

Recently, Whisnant (1992) has reported the results of the estimation of incidence of ischemic stroke by individual risk factor from 1960 to 1984 in Rochester, Minnesota. For those persons with hypertension, ischemic heart disease (IHD), diabetes mellitus and normal blood pressure there were increasing ischemic stroke incidence rates with age, and men had higher rates than women. There was a decline in ischemic stroke to 1979, an increase in 1980–1984 for persons with either hypertension or IHD, and no variation over time in persons with diabetes or cardiac-source embolism. There was no evidence that the introduction of CT scanning contributed significantly to these trends. It was concluded that the increase in incidence rate of ischemic stroke among persons with IHD was related to their increased survival. The explanation for the rise in those with hypertension remains to be clarified.

Prevalence

Stroke prevalence generally parallels stroke incidence and mortality in that there is exponential growth with advancing age (Baum and Robins 1981) (Table 22.1 (part 4)). Prevalence rates are indicators of the magnitude of a disease and are used to estimate the allocation of public health resources. There are over 2 million stroke survivors in the USA.

Stroke Determinants

In this section we will review stroke risk factors. These will be classified by stroke subtype when there is sufficient epidemiologic information to make such a designation. Much of the risk factor information to be presented emanates from the Framingham study. The reader should keep in mind that, although there is a general consensus about many stroke risk factors, there are variations such that some factors are not generalizable to the entire population, or even to specific racial or ethnic groups.

Consensus Statements

Although the presence of stroke risk factors does not guarantee that stroke will occur and the absence of any known risk factor does not ensure that stroke will not occur, stroke risk factors most certainly influence the *probability* of stroke occurrence (WHO/MNH Task Force on Stroke and Other Cerebrovascular Disorders 1989). Consensus statements about the status of stroke risk factors have been developed by the American Heart Association Stroke Council (Dyken et al. 1984) and WHO (WHO/MNH Task Force on Stroke and Other Cerebrovascular Disorders 1989). As can be seen from Table 2, risk factors for stroke may be subclassified into the following categories: single risk factors, multiple risk factors, well-documented and less well-documented risk factors, and treatable or non-treatable risk factors. An in-depth review of each risk factor is beyond the scope of this chapter, and emphasis will be placed primarily on major stroke determinants.

Table 22.2. Risk factors for stroke

Single risk factors
Well-documented risk factors
Treatment not feasible or value not established
Age and gender
Familial factors
Race
Diabetes mellitus
Prior stroke
Aysmptomatic carotid bruits
Treatable
Hypertension
Cardiac disease
Transient ischemic attacks
Elevated hematocrit
Sickle-cell disease
Less well-documented risk factors
Treatment not feasible or value not established
Geographic location
Season and climate
Socioeconomic factors
Treatable but value not established
Elevated blood cholesterol and lipids
Cigarette smoking
Alcohol consumption
Oral contraceptive use
Physical inactivity
Obesity
Multiple risk factors
Framingham profile
Systolic blood pressure
Serum cholesterol
Glucose tolerance
Cigarette smoking
Electrocardiogram
Left ventricular hypertrophy
Paffenbarger and Williams criteria
Cigarette smoking
Systolic blood pressure
Low ponderal index
Body height
A parent dead
Not a varsity athlete

Atherothrombotic Brain Infarction (ABI)

This discussion will focus on atherogenic host and environmental risk factors for ABI. In the Framingham study ABI was defined to include infarction secondary to large-vessel atherothrombosis as well as lacunar infarction. ABI comprised the most common stroke subtype, accounting for 44% of strokes (Wolf et al. 1992). As already mentioned, there is an exponential growth of stroke with age, and the incidence of all strokes generally and ABI specifically was about 30% higher in men than women.

Hypertension

International studies show that hypertension, whether it is systolic, diastolic or combined systolic and diastolic, is an important risk factor for ischemic stroke (WHO/MNH Task Force on Stroke and Other Cerebrovascular Disorders 1989). In the Framingham study history of hypertension was a powerful predictor of brain infarction in men and women, but the regression coefficient of effect was smaller in the oldest age group (65–94 years) (Wolf et al. 1992). Furthermore, both systolic and diastolic blood pressure levels were powerful independent predictors of ABI, and there was no evidence that women tolerated hypertension better than men or that diastolic elevations were more important than systolic elevations. The incidence of stroke and, to a lesser extent, ABI is increased with isolated systolic hypertension (Kannel et al. 1981; SHEP Cooperative Research Group 1991).

Heart Disease

Included in this category are coronary heart disease, congestive heart failure, left ventricular hypertrophy and arrhythmias such as atrial fibrillation. Heart disease is an important factor that predisposes to ischemic stroke independent of the effect of hypertension. For example, congestive heart failure is associated with a ninefold increase in risk of acute brain infarct, and nonvalvular atrial fibrillation with a fivefold increased incidence of stroke (Kannel et al. 1983; Wolf et al. 1987; Wolf et al. 1991). Atrial fibrillation increases in its relative importance as a predisposing condition for stroke as one ages.

Diabetes Mellitus

Diabetes mellitus is a risk factor for ischemic stroke of the large-artery type but is of questionable significance in other stroke subtypes (WHO/MNH Task Force on Stroke and Other Cerebrovascular Disorders 1989). For ABI the impact of diabetes mellitus is greater for women than for men, and is an independent contributor to incidence only in older women (Wolf et al. 1992).

Race

Overall, Asians of Japanese and Chinese descent and US blacks have higher stroke rates than white Americans. In addition, there may be a preponderance of intracranial to extracranial occlusive cerebrovascular disease among Asians and African Americans as compared to white Americans (Gorelick 1993, in press). The reasons for the racial/ethnic disparity of stroke rates and location of occlusive cerebrovascular disease are not known, but may be environmentally rather than genetically determined (Caplan et al. 1986; Reed 1990).

Family History

Genetic or heredofamilial factors have not yet been proven conclusively to be independent predictors of stroke. There are some studies that link family history of stroke to stroke in first-degree relatives (Welin et al. 1987).

Obesity

Obesity, a risk factor for heart disease, has been shown to be an independent contributor to ABI incidence in men 35–64 years of age and in women 65–94 years of age in the Framingham cohort (Wolf et al. 1992). However, the aggregate of international studies yields inconclusive results about obesity as an independent stroke risk factor (WHO/MNH Task Force on Stroke and Other Cerebrovascular Disorders 1989).

Blood Lipids

The relationship of blood lipids to brain infarction will require further clarification. In certain segments of western societies there is evidence linking hypercholesterolemia and increased low-density lipoprotein to ischemic stroke. However, this is not true of all study populations (e.g. Asians) (WHO/MNH Task Force on Stroke and Other Cerebrovascular Disorders 1989). As additional study data are accrued and analyzed in aggregate, we will soon have a more definitive understanding of the relationship between blood lipids and ABI (Salonen et al. 1988; Iso et al. 1989).

Hematocrit and Fibrinogen

Data from the Framingham study suggest a relationship between high normal hematocrit level and increased incidence of brain infarction (Wolf et al. 1992). The relationship holds only for men 35–64 years of age, and remains significant after controlling for the confounding effects of hypertension and cigarette smoking. Several other studies have linked serum fibrinogen level to cardiovascular disease incidence, including stroke (Wilhelmsen et al. 1984; Welin et al. 1987). Additional studies are needed to further substantiate this relationship.

Hyperuricemia

Although the number of studies are few, there are international data linking hyperuricemia to stroke (WHO/MNH Task Force on Stroke and Other Cerebrovascular Disorders 1989). The evidence is suggestive but not conclusive.

Alcohol Consumption

In a comprehensive and provocative review of the relationship between alcohol consumption and stroke, Camargo (1989) proposed a J-shaped curve to describe the association between moderate customary alcohol consumption and the relative risk of ischemic stroke among predominantly white populations. Little, if any, such association was found among Japanese, and possibly black, populations. Thus, the epidemiologic pattern of ischemic stroke and alcohol consumption might be similar to that of coronary heart disease and alcohol consumption in that there may be excess risk at high levels of alcohol consumption and protective effects at moderate levels of consumption. The pathophysiologic mechanisms by which alcohol use might contribute to ischemic stroke or exert a protective effect have been reviewed elsewhere (Gorelick 1989).

Cigarette Smoking

Cigarette smoking has emerged more recently as a risk factor for ischemic stroke. Data from the Honolulu Heart Program (Abbott et al. 1986), Framingham (Wolf et al. 1988) and others (Gorelick et al. 1989) support this contention. There may be a substantial reduction in stroke risk with cessation of cigarette smoking. The mechanisms by which smoking might increase the risk of ischemic stroke are reviewed by Wolf (1986).

Oral Contraceptives

Longstreth and Swanson (1984) reviewed case-control and cohort studies of oral contraceptive use and stroke. They concluded that oral contraceptives increased the risk of ischemic stroke and that most strokes occurred in women older than 35 years with other risk factors such as smoking and hypertension. 'Low-dose' oral contraceptive pills are thought to be associated with a lower risk of stroke than higher-dose oral contraceptives.

Exercise

Although the physiologic benefits of exercise may be substantial, there is no clear relationship between physical activity and reduction of stroke risk. However, lack of rigorous physical activity early in life may be a precursor of fatal stroke later in life (Paffenbarger and Williams 1967).

Asymptomatic Carotid Bruit

Asymptomatic carotid bruit is a marker of advanced systemic atherosclerosis, myocardial infarction and increased stroke risk (Heyman et al. 1980; Wolf et al. 1981). However, if stroke occurs it is often in a vascular territory outside that which was ipsilateral to the asymptomatic carotid bruit, and by a stroke mechanism other than atherothrombosis. Recent hospital-based studies have emphasized the importance of the degree of asymptomatic carotid stenosis as a determinant of stroke risk. With carotid stenosis ≤75% the stroke rate is relatively low (1.3% annually), whereas with carotid stenosis >75% the stroke rate may be two-to threefold greater (3.3% annually), with many of the ischemic events occurring ipsilateral to the stenosed artery (Norris et al. 1991). The rate of cardiac events and vascular death also follows this type of dose-response pattern of increasing rates with increasing severity of carotid stenosis.

Lacunar Infarction

Risk factors for lacunar infarction have not been studied with the same epidemiologic rigor as some other stroke subtypes. Available data suggest that risk factors for lacunar infarction may be typical atherogenic host factors (Tuszynski et al. 1989; Arboix et al. 1990; Chamorro et al. 1991; Sacco et al. 1991). Hypertension is present in about 70%–80% of lacunar infarct patients, diabetes mellitus in about 25%–35%, and heart disease in about 25%.

Intraparenchymal Hemorrhage (IPH)

The discussion of risk factors for IPH is restricted primarily to those factors associated with primary IPH, or those intraparenchymal bleeds not related to structural lesions such as intracranial neoplasm or vascular malformation. Although there is a consensus about some determinants of IPH, others vary by study population or are controversial.

Hypertension is the leading risk factor for IPH (Brott et al. 1986; Kagan et al. 1985; Ueda et al. 1988). Studies that have challenged this view have generally included secondary causes of IPH, such as bleeding diathesis and vascular malformations.

Alcohol consumption is also considered to be a risk factor for both IPH and subarachnoid hemorrhage (Stemmermann et al. 1984; Donahue et al. 1986; Camargo 1989). Camargo (1989) has proposed that there is a positive direct linear association between moderate regular alcohol consumption and intracranial hemorrhage. With reduction in alcohol consumption there may be reduction in the risk of subsequent hemorrhage. The pathophysiologic mechanisms by which alcohol may promote IPH are reviewed by Gorelick (1989).

A number of drugs have been proposed as determinants of IPH. A variety of street drugs, including cocaine, amphetamines and diet pills, 'T's and blues' and phencyclidine have been associated with IPH (Gorelick 1990a) and may cause vasculopathy or vasculitis, or induce hypertension. Another class of drugs, anticoagulants, may also be associated with IPH, especially when there is long-term use (Landefeld and Goldman 1989).

In the US Physicians Health Study (1989), concern was raised about a trend for more intracranial hemorrhages in the aspirin treatment group than in the placebo group. In practical terms the expected risk of IPH with aspirin use for the primary prevention of cardiovascular disease is small, whereas the potential for reduction in myocardial infarction is substantial. Furthermore, the reported rate of IPH in some aspirin trials for the secondary prevention of ischemic stroke is almost the same as the corresponding placebo treatment arms of these studies.

Other risk factors for IPH are as follows: age appears to be a risk factor in some reports (Bozzola et al. 1992); studies of Japanese show that low serum cholesterol, low intake of animal fat and protein, high levels of uric acid and proteinuria may be predictors of IPH (Stemmermann et al. 1984; Kagan et al. 1985; Ueda et al. 1988), race (Shi et al. 1989), seasons with colder ambient temperature (Shinkawa et al. 1990) and cigarette smoking in some geographical regions (Kagan et al. 1985) may also be associated with IPH. The role of diabetes mellitus and obesity is yet to be clarified.

Subarachnoid Hemorrhage (SAH)

Although the incidence of SAH rises steadily with age, there is a disproportionate number of younger adults with SAH, and therefore the young carry a heavier burden with SAH than with other stroke subtypes (Longstreth et al. 1985). Furthermore, age-specific rates for SAH are higher for women (Harmsen et al. 1979; Bonita and Thomson 1985).

Although the evidence is somewhat limited, hypertension is the most widely accepted risk factor for SAH (Longstreth et al. 1985; Harmsen et al. 1990). Other factors that may predict the occurrence of SAH are current and former oral contraceptive use (Longstreth and Swanson 1984), acute current and heavy cigarette smoking (Longstreth et al. 1991), alcohol consumption (Camargo 1989; Gorelick 1989), street drugs (Kaku and Lowenstein 1990), and smokers who exercise (Longstreth et al. 1991). Factors that may be associated with cerebral aneurysm include coarctation of the aorta, polycystic kidney disease, Marfan's syndrome, pseudoxanthoma elasticum, Ehlers–Danlos syndrome, familial factors, atherosclerosis, fibromuscular dysplasia, arteriovenous malformation and Moya Moya disease (Longstreth et al. 1985).

Prevention

Prevention of stroke by the treatment of modifiable atherogenic host and environmental risk factors is likely to be the most effective means of reducing morbidity and mortality (Wolf et al. 1992). The decline in stroke death rates in the USA since 1968 supports this belief, and can be explained at least in part by a reduction in the incidence and severity of stroke which many attribute to the improved detection and treatment of hypertension (Whisnant 1984). Public health strategies to avoid stroke by modification of risk factors amenable to drugs, diet or other interventions include the 'high-risk' approach, whereby individuals in the population with high levels of the risk factor are identified and usually treated with strategies and drugs to achieve substantial reductions in the risk factor; or 'mass' approach, whereby lifestyle modification is used to achieve modest reductions in the level of the risk factor in every individual in the population (Dunbabin and Sandercock 1990). Ideally, for risk factor modification to be most effective the risk factor should be common in the population, and the association between the risk factor and the target disease should be strong and dose-related.

The 'high-risk' approach can be achieved by a massive and expensive screening process which may still fail to prevent at least 50% of strokes (Dunbabin and Sandercock 1990). The 'mass' approach is advanced through health education and economic measures to discourage exposure to various risk factors. The two approaches are complementary and, if combined, could result in substantial reductions in stroke. Furthermore, they might be combined with other prophylactic measures such as antiplatelet therapy.

Modification of Host and Environmental Risk Factors

As mentioned previously blood pressure elevation, both systolic and diastolic, is strongly associated with stroke. In most treatment trials the diastolic component of blood pressure has been emphasized. It has been estimated that the risk of stroke increases throughout the diastolic blood pressure range 70–110 mmHg and for a 7.5 mmHg usual difference in diastolic blood pressure there is at least 46% less stroke (Collins et al. 1990; MacMahon et al. 1990). A prolonged difference in usual diastolic blood pressure of 5–6 mmHg could lead to a stroke risk reduction of 35%–40%. Treatment of isolated systolic hypertension is also important, and may be associated with a 36% reduction in incidence of total stroke (SHEP Cooperative Research Group 1991). Thus, there are substantial preventative gains for the treatment of systolic and diastolic blood pressure. Furthermore, the benefits of treatment may occur promptly and treatment response may be observed not only with medications but with dietary alterations such as decreasing sodium and alcohol intake and increasing potassium intake (Dunbabin and Sandercock 1990). Even modest reductions in blood pressure (2–3 mmHg), which can be achieved by dietary means, may result in substantial avoidance of vascular-related deaths (Rose 1981).

Cigarette smoking is a common lifestyle habit that increases the risk of stroke and does so in a dose-responsive manner (Shinton and Beevers 1989). With cessation of smoking there is a reduction in stroke risk to that of non-smokers within about 2–5 years (Wolf et al. 1988). Among white men it is estimated that up to 16 000 strokes per year are attributed solely to cigarette smoking (Klag and Whelton 1987). Thus, cessation of cigarette smoking is expected to significantly reduce the risk of stroke.

The relationship between serum cholesterol and cerebral infarction requires further study and clarification. However, epidemiologic evidence is accumulating for a dose-dependent relationship between serum cholesterol level and non-hemorrhagic stroke (Iso et al. 1989). The population attributable risk among whites with predominantly occlusive strokes and elevated serum cholesterol may be substantial (Dennis and Warlow 1987). Thus, with serum cholesterol reduction the savings in cardiac and ischemic cerebrovascular morbidity and mortality could be significant.

Heavy alcohol consumption is another lifestyle factor that is prevalent in many populations and may elevate the risk of stroke (Gorelick 1989). Observational epidemiologic studies estimate a dose-dependent effect and a reduction in stroke risk with cessation or reduction of alcohol consumption. Heavy alcohol consumption is a discretionary factor that is modifiable through both the 'mass' and the 'high-risk' approaches, and its treatment could result in significant stroke reduction in some populations and high-risk groups.

Because heart disease is such an important precursor of stroke, prevention and treatment of coronary heart disease, hypertensive heart disease and atrial fibrillation are anticipated to reduce stroke incidence (Wolf et al. 1992). Prophylactic treatment of patients with non-valvular atrial fibrillation with antithrombotic agents has proved to be an important factor in stroke risk reduction (Petersen et al. 1989; Anderson 1990). In addition, our understanding of predictors of thromboembolism in non-valvular atrial fibrillation has been refined by the identification of clinical (e.g. recent congestive heart failure, history of hypertension, previous thromboembolism) and echocardiographic (e.g. left ventricular dysfunction, left atrial size) risk factors (Stroke Prevention in Atrial Fibrillation Investigators 1992a,b).

Modification of Stroke Risk by Aspirin and Related Agents: Secondary Prevention

A substantial body of clinical trial data supports the effectiveness of aspirin in the reduction of cerebral and other ischemic events and vascular death in patients with TIA or mild ischemic stroke (Canadian Cooperative Study Group 1978; Bousser et al. 1983; American–Canadian Cooperative Study Group 1985; European Stroke Prevention Study Group 1987; UK TIA Study Group 1988; SALT Collaborative Group 1991; Dutch TIA Trial Study Group 1991). Although the study members include patients with different types of ischemic stroke mechanisms, in aggregate the studies show approximately a 22%–33% reduction in the risk of major outcomes such as stroke, death or myocardial infarction in varying combinations. The most efficacious dose of aspirin therapy has so far not been determined (Barnett 1990). Furthermore, the charge that aspirin is not effective in women may be an artefact of insufficient numbers of women followed for inadequate periods of time in the clinical trials (Dyken 1983). The efficacy of aspirin as a preventative in those with moderate or severe stroke remains in question (Gent 1987).

A new antiplatelet agent, ticlopidine, may be more effective than aspirin and shows substantial benefits for women (Gent et al. 1989; Hass et al. 1989). However, most other antiplatelet agents have either not proved effective or confer no additional benefits compared to aspirin alone (Dyken 1983).

Modification of Stroke Risk by Aspirin: Primary Prevention

Taken in aggregate, two major randomized primary prevention trials, one of US physicians and the other of British doctors (Peto et al. 1988; Steering Committee of the Physicians Health Study Research Group 1989) showed a statistically significant reduction in non-fatal myocardial infarction of about 33%, no reduction in the risk of stroke and some increase in disabling strokes in those subjects treated with aspirin (Hennekens et al. 1988). However, the increase in disabling strokes was based on small numbers and the association is not supported by data from secondary-prevention trials. Although there is as yet no final conclusion on the association between aspirin and hemorrhagic stroke, it is unlikely that aspirin poses a significant risk (Barnett 1990).

Modification of Stroke Risk by Surgery

Carotid endarterectomy (CE) trials have clarified the role of CE in symptomatic carotid stenosis (Winslow et al. 1988). Two large-scale controlled trials have shown that CE is the treatment of choice when there is high-grade (70%–99%) ipsilateral symptomatic carotid stenosis in patients with TIA or mild stroke (North American Symptomatic Carotid Endarterectomy Trial Collaborators 1991; European Carotid Surgery Trialists's Collaborative Group 1991). The absolute risk reduction of any ipsilateral stroke after 2 years may be 17% with CE, compared to medical treatment alone. For low grades of symptomatic carotid stenosis (≤29%) the surgical risks probably outweigh the benefits, whereas for moderate grades of stenosis (30%–69%) the answer is not yet available and is the subject of ongoing clinical trials. Similarly, asymptomatic carotid stenosis trials, including the Asymptomatic Carotid Atherosclerosis Study (ACAS), Asymptomatic Carotid Stenosis Veterans Administration Study and Carotid Artery Stenosis with Asymptomatic Narrowing: Operation Versus Aspirin (CASANOVA), are under way and preliminary results are expected as early as 1993 (Gorelick 1990b). A surgical trial of extracranial-to-intracranial arterial bypass for patients with symptomatic atherosclerotic narrowing or occlusion of the internal carotid artery or middle cerebral artery failed to show that surgery was superior to aspirin and other medical therapy (EC/IC Bypass Study Group 1985).

References

Abbott RD, Yin Y, Reed DM, Yano R (1986) Risk of stroke in male cigarette smokers. New Engl J Med 315:717–720

American–Canadian Cooperative Study Group (1985) Persantine aspirin trial in cerebral ischemia. II. Endpoint results. Stroke 16:406–415

American Heart Association (1991) Heart and stroke facts. American Heart Association, Dallas, Texas

Anderson DC (1990) Progress report of the Stroke Prevention in Atrial Fibrillation Study. Stroke (Suppl III):III-12–III-17

Arboix A, Marti-Vilalta JL, Garcia JH (1990) Clinical study of 227 patients with lacunar infarcts. Stroke 21:842–847

Barnett HJM (1990) Aspirin in stroke prevention. An overview. Stroke (Suppl IV) 21:IV-40–IV-43

Baum HM, Robins M (1981) Survival and prevalence. Stroke (Suppl 1) 12:I-59–I-68

Bonita R, Thomson S (1985) Subarachnoid hemorrhage: epidemiology, diagnosis, management and outcome. Stroke 16:591

Bonita R, Beaglehole R, North JDK (1983) Subarachnoid hemorrhage in New Zealand: an epidemiological study. Stroke 14:342

Bonita R, Stewart A, Beaglehole R (1990) International trends in stroke mortality. Stroke 21:989–992

Bousser MG, Eschwege E, Haguenan M et al. (1983) 'AICLA' controlled trial of aspirin and dipyridamole in the secondary prevention of athero-thrombotic cerebral ischemia. Stroke 14:5–14

Bozzola FG, Gorelick PB, Jensen JM (1992) Epidemiology of intracranial hemorrhage. Neuroimag Clin North Am 2:1–10

Brandstater ME (1990) An overview of stroke rehabilitation. Stroke (Suppl) 21:II-40–II-42

Broderick JP, Phillips SJ, Whisnant JP, O'Fallen WM, Bergstrahl EJ (1989) Incidence rates of stroke in the eighties: the end of the decline in stroke? Stroke 20:577–582

Brott T, Thalinger K, Hertzberg V (1986) Hypertension as a risk factor for spontaneous intracerebral hemorrhage. Stroke 17:1078

Camargo CA (1989) Moderate alcohol consumption and stroke. The epidemiological evidence. Stroke 20:1611

Canadian Cooperative Study Group (1978) A randomized trial of aspirin and sulfinpyrazone in threatened stroke. New Engl J Med 229:53–59

Caplan LR, Gorelick PB, Hier DB (1986) Race, sex and occlusive cerebrovascular disease: a review. Stroke 17:648–655

Chamorro A, Sacco RL, Mohr JP et al. (1991) Clinical–computed tomographic correlations of lacunar infarction in the

Stroke Data Bank. Stroke 22:175–181
Collins R, Peto R, MacMahon S et al. (1990) Blood pressure, stroke and coronary heart disease: Part II. Effects of short-term reductions in blood pressure and overview of the unconfounded randomised drug trials in an epidemiologic context. Lancet 335:827–838
Cooper R, Sempos C, Hsieh S-C, Kovar MG (1990) Slowdown in the decline of stroke mortality in the United States, 1978–1986. Stroke 21:1274–1279
Dennis MS, Warlow CP (1987) Stroke: incidence, risk factors and outcome. BR J Hosp Med 37:194–198
Donahue RP, Abbott RD, Reed DM et al. (1986) Alcohol and hemorrhagic stroke. The Honolulu Heart Program. JAMA 255:2311
Dunbabin DW, Sandercock PAG (1990) Preventing stroke by modification of risk factors. Stroke 21 (Suppl IV):IV-36–IV-39
Dutch TIA Trial Study Group (1991) A comparison of two doses of aspirin (30 mg vs 283 mg a day) in patients after a transient ischemic attack or minor ischemic stroke. New Engl J Med 325:1261–1266
Dyken ML (1983) Anticoagulant and platelet-antiaggregating therapy in stroke and threatened stroke. Neurol Clin 1:223–242
Dyken ML, Wolf PA, Barnett HJM et al. (1984) Risk factors in stroke. A statement for physicians by the subcommittee on Risk Factors and Stroke of the Stroke Council. Stroke 15:1105–1111
EC/IC Bypass Study Group (1985) Failure of extracranial–intracranial arterial bypass to reduce the risk of ischemic stroke: results of an international randomized trial. New Engl J Med 313:1191–1200
European Carotid Surgery Trialists' Collaborative Group (1991) MRC European carotid surgery trial: interim results for symptomatic patients with severe (70%–99%) or with mild (0%–29%) carotid stenosis. Lancet 337:1235–1243
European Stroke Prevention Study Group (1987) The European Stroke Prevention Study (ESPS): principal endpoints. Lancet 2:1351–1354
Garraway WM, Whisnant JP, Drury I (1983) The changing pattern of survival following stroke. Stroke 14:699–703
Gent M (1987) Single studies and overview analyses: is aspirin of value in cerebral ischemia? Stroke 18:541–544
Gent M, Blakely JA, Easton JD et al. (1989) The Canadian–American Ticlopidine Study (CATS) in thromboembolic stroke: results of a North American randomized trial. Lancet 1:1215–1220
Gorelick PB (1989) The status of alcohol as a risk factor for stroke. Stroke 20:1607–1610
Gorelick PB (1990a) Stroke from alcohol and drug abuse. A current social peril. Postgrad Med 88:171–178
Gorelick PB (1990b) Carotid endarterectomy: a neurologist's perspective. J Neurosurg Anesthesiol 2:203–205
Gorelick PB (1993) Distribution of occlusive cerebrovascular lesions: effects of age, race and sex. Stroke 24(Suppl I):I-16–I-19
Gorelick PB, Robin MB, Langenberg P, Hier DB, Costigan J (1989) Weekly alcohol consumption, cigarette smoking and the risk of ischemic stroke: results of a case-control study at three urban medical centers in Chicago, Illinois. Neurology 39:339–343
Gross CR, Kase C, Mohr JP, Cunningham SC, Baker WE (1984) Stroke in South Alabama: incidence and diagnostic features– a population based study. Stroke 15:249–255
Harmsen P, Berglund G, Larsson O et al. (1979) Stroke registration in Goteborg, Sweden, 1970–75. Incidence and fatality rates. Acta Med Scand 206:337
Harmsen P, Rosengren A, Tsipogianni A et al. (1990) Risk factors for stroke in middle aged men in Goteborg, Sweden. Stroke 21:223
Hass WK, Easton JD, Adams HP et al. (1989) A randomized trial comparing ticlopidine hydrochloride with aspirin for prevention of stroke in high-risk patients. New Engl J Med 321:501–507
Hennekens CH, Peto R, Hutchison GB, Doll R (1988) An overview of the British and American aspirin studies. New Engl J Med 318:923–923
Heyman A, Wilkinson WE, Heyden S et al. (1980) Risk of stroke in asymptomatic persons with cervical bruits: a population study in Evans County, Georgia. New Engl J Med 302:838
Ingall TJ, Whisnant JP, Wiebers DO, O'Fallon WM (1989) Has there been a decline in subarachnoid hemorrhage mortality? Stroke 20:718–724
Iso H, Jacobs DR, Wentworth D et al. (1989) Serum cholesterol levels and six-year mortality from stroke in 350 977 men screened for the Multiple Risk Factor Intervention Trial. New Engl J Med 320:904–910
Kagan A, Popper JS, Rhoads GG et al. (1985) Dietary and other risk factors for stroke in Hawaiian Japanese men. Stroke 16:390
Kaku DA, Lowenstein DH (1990) Emergence of recreational drug abuse as a major risk factor for stroke in young adults. Ann Intern Med 113:821
Kannel WB, Wolf PA, McGee DL et al. (1981) Systolic blood pressure, arterial rigidity and risk of stroke: the Framingham study. JAMA 245:1225–1229
Kannel WB, Wolf PA, Verter J (1983) Manifestations of coronary disease predisposing to stroke: the Framingham study. JAMA 250:2942–2946
Klag MJ, Whelton PK (1987) Risk of stroke in male cigarette smokers. (Letter) New Engl J Med 316:628
Klag MJ, Whelton PK, Seidler AJ (1989) Decline in US stroke mortality. Demographic trends and antihypertensive treatment. Stroke 20:14–21
Kuller LH (1989) Incidence rates of stroke in the eighties: the end of the decline in stroke? Stroke 20:841–843
Kurtzke JF (1985) Epidemiology of cerebrovascular disease. In: McDowell F, Caplan LR (eds) Cerebrovascular survey report. The National Institute of Neurological and Communicative Disorders and Stroke, pp 1–34
Landefeld CS, Goldman L (1989) Major bleeding in outpatients treated with warfarin: incidence and prediction of factors known at the start of outpatient therapy. Am J Med 87:144
Lanska DJ (1991) The geographic distribution of stroke mortality in the United States: 1940–1980. Neurology 41 (Suppl 1):285
Longstreth WT, Swanson PD (1984) Oral contraceptives and stroke. Stroke 15:747–750
Longstreth WT, Koespel TD, Yerby MS et al. (1985) Risk factors for subarachnoid hemorrhage. Stroke 16:377
Longstreth WT, Nelson L, Koespel TD et al. (1991) Risk factors for subarachnoid hemorrhage. Stroke 22:25
MacMahon S, Peto R, Cutler J et al. (1990) Blood pressure, stroke and coronary heart disease: Part I. Effects of prolonged differences in blood pressure – evidence from nine prospective observational studies corrected for the regression dilution bias. Lancet 335:765–774
Norris JW, Zhu CZ, Bornstein NM, Chambers BR (1991) Vascular risks of asymptomatic carotid stenosis. Stroke 22:1485–1490
North American Symptomatic Carotid Endarterectomy Trial Collaborators (1991) Beneficial effect of carotid endarterectomy in symptomatic patients with high-grade stenosis. New Engl J Med 325:445–453

Paffenbarger RS Jr, Williams JL (1967) Chronic disease in former college students. V. Early precursors of fatal stroke. Am J Public Health 57:1290

Petersen P, Boysen G, Godtfredson J, Andersen ED, Andersen B (1989) Placebo controlled, randomized trial of warfarin and aspirin for prevention of thromboembolic complications in chronic atrial fibrillation: the Copenhagen AFASAK Study. Lancet 1:175–179

Peto R, Gray R, Collins R et al. (1988) Randomized trial of prophylactic daily aspirin in British male doctors. Br Med J 296:313–316

Reed DM (1990) The paradox of high risk of stroke in populations with low risk of coronary heart disease. Am J Epidemiol 108:161–175

Riggs JE (1991) The decline of mortality due to stroke: a competitive and deterministic perspective. Neurology 41:1335–1338

Rose G (1981) Strategy of prevention: lessons from cardiovascular disease. Br Med J 282:1847–1850

Sacco SE, Whisnant JP, Broderick JP, Phillips SJ, O'Fallon WM (1991) Epidemiological characteristics of lacunar infarcts in a population. Stroke 22:1236–1241

Salonen R, Seppanen K, Rauramaa R, Salonen JT (1988) Prevalence of carotid atherosclerosis and serum lipid levels in eastern Finland. Atherosclerosis 8:788–792

SALT Collaborative Group (1991) Swedish aspirin low-dose trial (SALT) of 75 mg aspirin as secondary prophylaxis after cerebrovascular ischemic events. Lancet 338:1345–1349

SHEP Cooperative Research Group (1991) Prevention of stroke by antihypertensive drug treatment in older persons with isolated systolic hypertension: final results of the Systolic Hypertension in the Elderly Program (SHEP). JAMA 265:3255–3264

Shi F, Hart RG, Sherman DG et al. (1989) Stroke in the People's Republic of China. Stroke 20:1581

Shinkawa A, Ueda K, Hasuo Y et al. (1990) Seasonal variation in stroke in Hisayama, Japan. Stroke 21:1262

Shinton R, Beevers G (1989) Meta-analysis of relation between cigarette smoking, and stroke. Br Med J 298:787–794

Soltero I, Liu K, cooper R, Stamler J, Garside D (1978) Trends in mortality from cerebrovascular diseases in the United States, 1960–1975. Stroke 9:549–555

Steering Committee of the Physicians Health Study Research Group (1989) Final report on the aspirin component of the ongoing Physicians Health Study. New Engl J Med 321:129–135

Stemmermann GN, Hayashi T, Resch JA et al. (1984) Risk factors related to ischemic and hemorrhagic cerebrovascular disease at autopsy: the Honolulu Heart Study. Stroke 15:23

Stroke Prevention in atrial Fibrillation Investigators (1992a) Predictors of thromboembolism in atrial fibrillation: I. Clinical features of patients at risk. Ann Intern Med 116:1–5

Stroke Prevention in Atrial Fibrillation Investigators (1992b) Predictors of thromboembolism in atrial fibrillation: II. Echocardiographic features of patients at risk. Ann Intern Med 116:6–12

Terent A (1988) Increasing incidence of stroke among Swedish women. Stroke 19:598–603

Tuszynski MH, Petito CK, Levy DE (1989) Risk factors and clinical manifestations of pathologically verified lacunar infarctions. Stroke 20:990–999

Ueda K, Omae T, Hirota Y et al. (1981) Decreasing trend in incidence and mortality from stroke in Hisayama residents, Japan. Stroke 12:154

Ueda K, Hasuo Y, Kiyohara Y et al. (1988) Intracerebral hemorrhage in a Japanese community, Hisayama: incidence, changing pattern during long-term follow-up, and related factors. Stroke 19:48

UK-TIA Study Group (1988) United Kingdom transient ischemic attack (UK-TIA) aspirin trial: interim results. BMJ (Clin Res) 296:316–320

Welin I, Svardsudd K, Wilhelmsen L et al. (1987) Analysis of risk factors for stroke in a cohort of men born in 1913. New Engl J Med 317:521–526

Whisnant JP (1984) The decline of stroke. Stroke 15:160–168

Whisnant JP (1992) Changing incidence and mortality rates for stroke. J Stroke Cerebrovasc Dis 2:42–44

WHO/MNH Task Force on Stroke and Other Cerebrovascular Disorders (1989) Stroke 1989. Recommendations on stroke prevention, diagnosis, and therapy. Stroke 20:1407–1431

Wilhelmsen L, Svardsudd K, Korban-Bengsten K et al. (1984) Fibrinogen as a risk factor for stroke and myocardial infarction. New Engl J Med 311:501–505

Winslow CM, Solomon DH, Chassin MR et al. (1988) The appropriateness of carotid endarterectomy. New Engl J Med 318:721–727

Wolf PA (1986) Cigarettes, alcohol and stroke. New Engl J Med 315:1087–1089

Wolf PA, Kannel WB, Sorlie P, McNamara P (1981) Asymptomatic carotid bruit and risk of stroke. The Framingham Study. JAMA 245:1442

Wolf PA, Kannel WB, McGee DL (1986) Epidemiology of strokes in North America. In: Barnett HJM, Stein BM, Mohr JP, Yatsu F (eds) Stroke: pathophysiology, diagnosis and management. Churchill Livingstone, New York, pp 19–29

Wolf PA, Abbott RD, Kannel WB (1987) Atrial fibrillation: a major contributor to stroke in the elderly: the Framingham study. Arch Intern Med 147:1561–1564

Wolf PA, Abbott RD, Kannel WB (1991) Atrial fibrillation, an independent risk factor for stroke: the Framingham study. Stroke 22:983–988

Wolf PA, Belanger AJ, D'Agostino RB (1992) Management of risk factors. Neurol Clin 10:177–191

Wolf PA, D'Agostino RB, Kannel WB, Bonita R, Belanger AJ (1988) Cigarette smoking as a risk factor for stroke: the Framingham study. JAMA 259:1025–1029

23. Stroke Data Banks and Stroke Registries

D. B. Hier

Stroke Registries

During the past 30 years, stroke registries and data banks have merged as powerful instruments of clinical stroke research (Mohr 1986). Typically, stroke registries are one of three types: population-based, case-based and cohort-based.

In population-based registries, all strokes in a defined population are recorded. Population-based registries permit the collection of epidemiologically valid data about stroke incidence, and can be used to calculate attack rates by age, sex or race. If properly designed, they can identify secular trends in stroke incidence. They also allow comparisons in stroke incidence by geographical region. Most data collection instruments for population-based registries are brief and lack detail. Validation of stroke type by neuroimaging is often lacking. In 1971, the World Health Organization (WHO) collaborated with 15 centers in 10 countries of Africa, Asia and Europe to establish pilot population-based stroke registries. By 1974, 6395 cases of stroke had been registered worldwide (Hatano 1976). This study provided important data on stroke incidence (Table 23.1). These early studies encouraged further population-based studies (Table 23.2).

Case-based registries do not provide epidemiologically valid data on stroke incidence. They are usually hospital based and seek to identify all strokes admitted to a given hospital (Table 23.3). The Harvard Stroke Registry (Mohr et al. 1978), the Pilot Stroke Data Bank (Kunitz et al. 1984) and the Stroke Data Bank (Foulkes et al. 1988) are examples of case-based registries. In most, extensive data are collected on each case, and cases are often subjected to detailed clinical and radiological evaluations. In some case-based registries, patients are followed longitudinally for extended periods of time. Case-based registries can provide valuable information about stroke presentation, course and outcome.

Table 23.1 World Health Organization pilot stroke registries (Hatano 1976)

Center	Country	Stroke incidence/ 100 000	
		Men	Women
Ulan Bator	Mongolia	30	30
Zarifin	Israel	100	120
Dublin	Ireland	120	130
Espoo	Finland	120	130
Zagreb	Yugoslavia	120	80
Fukuoka	Japan	120	50
Osaka	Japan	120	110
N. Karelia	Finland	180	190
Copenhagen	Denmark	200	230
Saku	Japan	250	180
Akita	Japan	290	230

Table 23.2. Population-based studies of stroke incidence (sorted in ascending order)

City	Country	Total pop	No. of events	Study duration	Incidence M	F	All	Race	Reference
S. Alabama	USA	103 358	139	1980	139	88	109	White	Gross et al. 1984
Helsinki	Finland	NS	17629	1970–80	NS	NS	139		Tilvis et al. 1987
Rochester	USA	51 267	993	1955–69	165	146	154		Matsumoto et al. 1973
Oxfordshire	England	105 476	675	1981–86	150	171	160		Bamford et al. 1988
Lehigh Valley	USA	580 000	1850	1982–86	NS	NS	167		Alter et al. 1985
Dijon	France	140 000	418	1985–87	170	126	145		Giroud et al. 1989
S. Alabama	USA	103 358	139	1980	172	236	208	Black	Gross et al. 1984
Espoo-Kauniai	Finland	113 000	286	1972–73	193	206	199		Waltimo et al. 1980
Fargo	USA	94 037	408	1965–66	215	219	217		Alter et al. 1970
Area Latina	Italy	400 000	2245	1983–85	220	152	185		Giampaoli et al. 1989
Copenhagen	Denmark	19 698	848	1976–88	267	161	214		Lindenstrom et al. 1992
Akita	Japan	6427	109	1975–81	280	200	NS		Kojima et al. 1990
Malmo	Sweden	232 000	524	1989	280	244	225		Jerntorp and Berglund 1992
Moscow	Russia	208 921	1538	1972–74	NS	NS	245		Scmidt et al. 1988
Evans County	USA	3102	94	1960–62	530	170	NS	White	Heyman et al. 1971
Evans County	USA	3102	94	1960–62	600	500	NS	Black	Heyman et al. 1971

NS, not specified

Cohort-based registries involve the prospective observation of a defined cohort of patients. These studies typically involve the periodic re-examination of patients to detect stroke occurrence. Cohort-based studies usually enrol smaller numbers of subjects than population-based studies, and are generally not useful in identifying or delineating rarer forms of stroke. Unlike population-based studies, patients in cohort-based studies are scheduled for re-examination. Cohort-based studies provide epidemiologically valid data on stroke incidence. They can also be used to quantify accurately risk factors for stroke occurrence, morbidity and recurrence. Examples of cohort-based studies are the Mayo Clinic cohort (Davis et al 1987), the Framingham study (Wolf et al. 1991a, b) and the Copenhagen Heart Study (Lindenstrom et al. 1992).

Working Definitions of Stroke Subtypes

The establishment of stroke registries has been instrumental in the establishment of generally accepted criteria for stroke subtype definitions. The Harvard Stroke Registry (Mohr et al. 1978) developed widely used definitions for large-artery thrombosis, lacune, embolism, intracerebral hemorrhage and subarachnoid hemorrhage. The Pilot Stroke Data Bank (Kunitz et al. 1984) developed diagnostic criteria for atherosclerosis, embolism, infarction of unproven etiology, lacune, intracerebral hemorrhage and subarachnoid hemorrhage. These definitions were further refined in the full-phase Stroke Data bank (Foulkes et al. 1988) into nine diagnostic categories that included atherosclerosis, cardiac embolism, infarction with normal angiogram, infarction of unknown cause, lacunar infarction, infarction due to tandem arterial pathology (artery-to-artery embolism), intracerebral hemorrhage, subarachnoid hemorrhage and other.

Development of Data Collection Instruments

Stroke registries have led to the development of improved instruments for data collection. In the most ambitious stroke registry to date, the Stroke Data Bank developed 18 different data collection forms with 52 pages (Table 23.4.) For each patient, certain forms were unique (e.g. neurological examination or CT scan). Each form has a unique patient identification assigned to it, so that all the forms of the 1805 patients in the Stroke Data Bank could be linked into a large relational database (Foulkes et al. 1988). This permitted analysis of the relational database by commercially available statistical analysis software, such as SAS.

Table 23.3. Case-based stroke registries: stroke by etiology (figures are percentages)

Registry	Country	Reference	*n*	All infarcts	All hemorrhages	ATH	LAC	EMB	INFNOS	ICH	SAH	OTH	CFR
S. Alabama	USA	Gross et al. 1984	160	85.00	14.00	6.00	13.00	26.00	40.00	8.00	6.00	1.00	22.00
Riyadh	Saudi Arabia	Yaqub et al. 1992	200	87.00	11.00	NS	21.00	NS	66.00	6.50	4.50	2.00	8.00
Framingham	USA	Wolf et al. 1991b	366	85.90	11.90	57.30	NS	28.60	NS	5.70	6.20	1.90	NS
Dijon	France	Giroud et al. 1989	418	80.00	17.00	NS	12.00	NS	68.00	12.00	5.00	0.00	21.50
Hamburg	Germany	Spitzer et al. 1989b	441	75.90	20.50	17.80	0.27	18.63	39.20	14.50	6.00	0.00	8.79
Oxfordshire	England	Bamford et al. 1988	642	84.90	10.30	NS	20.80	NS	64.10	10.30	NS	4.80	17.80
Harvard	USA	Mohr et al. 1978	694	84.00	16.00	34.00	19.00	31.00	NS	10.00	6.00	NS	NS
Shatin	Hong Kong	Kay et al. 1992	777	68.30	27.10	NS	18.50	NS	49.80	27.10	NS	4.50	25.40
Lausanne	Switzerland	Bogousslavsky et al. 1988	1000	86.10	10.90	42.70	14.70	20.40	8.30	10.90	NS	6.00	5.90
Pilot SDB	USA	Kunitz et al. 1984	1158	61.00	19.00	15.00	9.00	17.00	20.00	9.00	10.00	0.00	12.70
SDB	USA	Foulkes et al. 1988	1805	71.00	26.00	6.00	19.00	14.00	32.00	13.00	13.00	3.00	11.60
CHSP	USA	Becker et al. 1986	2390	60.00	10.00	NS	NS	NS	60.00	10.00	NS	30.00	19.70
Lehigh Valley	USA	Alter et al. 1985	2621	89.80	10.30	NS	9.40	NS	80.40	9.00	1.30	1.30	14.00

LAC, lacune; ATH, atherosclerosis; EMB, embolism; ICH, intracerebral hemorrhage; SAH, subarachnoid hemorrhage; INFNOS, infarction, not otherwise specified; OTH, other stroke; CFR, 30-day case fatality rate; NS, not specified or not recorded.

Table 23.4 Data collection instruments for Stroke Data Bank

Form	Single/ Multiple	Completed By	Number of Pages
Patient identification	S	PN	1
Daily flow sheet	S	PI	1
Background information	S	PN	1
Social history	S	PN	2
Medical history	S	PI	1
Neurologic history	S	PI	1
Neurological examination	M	PI	3
Functional assessment	M	PN	2
CT scan	M	PI	5
Angiography	M	PI	2
Evolving stroke	S	PI	1
Pure motor worksheet	S	PI	1
Complications	S	PI	7
Death information	S	PN	1
Autopsy	S	PI	3
Summary of hospitalization	S	PN	8
Diagnosis of stroke	S	PN	2
Follow-up	M	PI	8
Recurrent stroke	M	PI	2
Total			52

S, Only single form completed; M, Multiple forms may be completed on different days; PI, Principal investigator; PN, Project nurse.

Interobserver Agreement

Stroke registries have fostered studies of interobserver agreement for history, physical examination, CT scan interpretation and diagnosis in stroke patients (Table 23.5). As part of the Stroke Data Bank project interobserver agreement was examined for history, neurological examination, CT scan interpretation and stroke diagnosis in 17 patients (Shinar et al. 1985, 1987; Gross et al. 1986), and was found to be modest for most items of the neurological history and examination (Shinar et al. 1985). For interpretation of CT scans (Shinar et al. 1987), interobserver agreement was found to be excellent for number of lesions, whether the scan was normal, side of the lesion and lesion density. Agreement was less good for lesion size, edema and mass effect. Good levels of agreement were found for location of lesions within the brain. In a study of interobserver agreement for stroke type, agreement was good for major category of stroke (subarachnoid hemorrhage, intracerebral hemorrhage, infarction), especially after laboratory tests were made available to the neurologists.

Pooling of Data From Several Hospitals

Multicenter data banks have been established, which allows the collection of large amounts of data about stroke, including acute outcome, long-term outcome and risk of recurrence. The Community Hospital-Based Stroke Programs pooled data from community hospitals from the coastal plains of North Carolina, Monroe County of New York and the state of Oregon. During a 2-year period data from 4132 stroke patients was collected (Becker et al. 1986; Yatsu et al. 1986).

Table 23.5. Inter-observer agreement for selected variables in stroke data bank

Variable	κ statistic
Headache at onset	0.36
Vomiting at onset	0.35
Coma at onset	0.32
Prior stroke	0.31
Prior TIAs	0.11
Deficit present on awakening	0.11
Dysphagia	0.71
Weakness (location)	0.67
Language disorder	0.54
Dysarthria	0.53
Ataxia	0.45
Visual fields	0.40
Alertness	0.38
Sensory deficits	0.32
Dementia suspected	0.34
Depression suspected	0.15
Cervical bruit	−0.05
Number of CT scan lesions	0.56
Side of CT scan lesion	0.65
CT lesion size	0.37
Edema on CT	0.34
Mass effect on CT	0.52
Stroke etiology (9 categories)	0.64*
Stroke etiology (4 categories)	0.76**

*Infarction cause unknown, subarachnoid hemorrhage, intracerebral hemorrhage, other stroke, cardiac embolism, infarction, normal angiogram, tandem arterial pathology, atherosclerosis
**Infarction, subarachnoid hemorrhage, intracerebral hemorrhage and other stroke.

Note: The κ statistic is a measure of agreement corrected for change agreements. A value of 1.00 reflects a perfect level of agreement whereas a level of 0.00 reflects agreement at the level of chance.

Table 23.6. Risks factors for stroke from cohort-based registries

	Odds ratio for stroke for subjects with risk factor		
	Rochester (Davis et al. 1987)	Framingham (men) (Wolf et al. 1991b)	Framingham (women) (Wolf et al. 1991b)
Age per 10 yr	1.60	1.66	1.91
Male	2.00	–	–
History of hypertension	4.00	–	–
History of TIA	3.90	–	–
Hypertensive heart disease	2.20	2.32	2.34
CAD	2.20	1.68	1.54
CHF	1.70	–	–
DM	1.70	1.40	1.72
AF	–	1.83	3.16
Systolic BP per 10 mmHg	–	1.91	1.68
Hypertensive treatment	–	1.39	–
Cigarettes	–	1.67	1.70

CAD=Coronary artery disease; CHF=Congestive heart failure; DM=Diabetes Mellitus; AF=Atrial fibrillation;

This project demonstrated the possibility of merging data from a large number of hospitals using a common protocol. In the Stroke Data Bank, data on 1805 patients from four academic teaching hospitals were pooled into a single data bank (Foulkes et al. 1988).

Identification of Risk Factors for Stroke

Stroke registration in well defined cohorts of healthy patients followed prospectively for stroke is a powerful tool for the identification of risk factors for stroke. A cohort of 1804 residents of Rochester, Minnesota, was followed for 13 years (Davis et al. 1987). There were 110 first ischemic strokes in the cohort during this time. Risk factors included age, male sex, heart disease, diabetes, hypertension and TIAs (Table 23.6). In the Framingham study, a cohort of 5070 participants was followed for 34 years (Wolf et al. 1991a,b). Factors associated with an increased risk of stroke included coronary artery disease (twofold increase), hypertension (threefold increase), congestive heart failure (fourfold increase) and atrial fibrillation (fivefold increase).

Racial and Gender Differences in Stroke Type

Stroke registration offers a method to analyze racial and sex differences in stroke incidence and type. Stroke is more frequent in men than women, and more frequent in blacks than whites (see Table 23.2). These differences cannot be explained by gender or racial differences in the incidence of risk factors for stroke: there appear to be true gender and race differences in stroke incidence. Racial differences also exist for the distribution of stroke type. Extracranial atherosclerosis is more common in whites and intracranial atherosclerosis is more common in blacks (Gorelick et al. 1984). In the Lehigh Valley Stroke Registry (Friday et al. 1989), lacunar stroke was more common in blacks whereas embolic stroke was more common in whites. The Shatin Stroke Registry (Kay et al. 1992) of 777 strokes in Hong Kong suggests that intracerebral hemorrhage may be two to three times more frequent in the Hong Kong Chinese than in westerners.

Stroke Recurrence

Stroke registration has permitted an identification of risk factors associated with recurrence. The Lehigh Valley Stroke Registry registered 1850 ischemic strokes (Alter et al. 1987; Sobel et al. 1989). In a subset of 243 cases, 59 with recurrent stroke were identified. The odds ratio of recurrent stroke was increased by TIA (5.4), myocardial infarction (2.7), heart disease (3.1), hypertension (1.4) or diabetes (1.5). The Stroke Data Bank (Sacco et al. 1989; Hier and Edelstein 1991) has delineated some of the risk factors associated with a high risk of either early or late ischemic stroke recurrence (Table 23.7).

Table 23.7. Two-year cumulative risk for stroke recurrence after initial ischemic infarct

Predictor	Risk (%)
Male	13.4
Female	14.7
White	13.9
Black	13.7
Hypertension by history	15.9
Diabetes	19.8
Myocardial infarction	19.6
Valvular heart disease	13.7
artrial fibrillation	15.7
prior stroke	20.7
atherosclerotic stroke	21.9
embolic stroke	14.7
lacunar stroke	14.6
infarct of cause unknown	10.5
supratentorial infarct	15.2
infratentorial infarct	8.0
All cases (*n*=1273)	14.1

Survival After Stroke

Stroke registries have yielded valuable information about the prognosis for survival after stroke (Table 23.8). In the study by Chambers et al. (1987), 1013 patients with acute stroke were registered and followed for 2–8 years. For brain infarction, adverse predictors of long-term survival were reduced level of activity at hospital discharge, advanced age, male sex, heart disease and hypertension. In the Auckland Registry (Bonita et al. 1988) 635 stroke patients were followed longitudinally for 3 years. Adverse predictors of survival included increasing age, institutionalization at time of stroke onset, previous stroke, loss of consciousness due to stroke and severe motor deficit. Based on the 4219 patients in the Community Hospital-Based Stroke Program, Howard et al. (1986) identified level of consciousness as the best predictor for survival at 1 year for intracerebral hemorrhage and level of consciousness, and age, cardiac disease and prior stroke as predictors of survival at 1 year after brain infarction.

Stroke registries have identified causes for early mortality after stroke. Silver et al. (1984) studied 212 deaths within the first 30 days of stroke: herniation (usually occurring within the first week after stroke) accounted for 86 deaths (42 infarcts and 44 hemorrhages). Cardiac deaths were associated with brain infarction and accounted for 26 deaths. Pulmonary embolism (4), pneumonia (31) and sepsis (5) were other important causes of death in the first 30 days after stroke. Population-based studies such as those of Rochester, Minnesota (Garraway et al. 1983), permit analysis of secular trends in stroke survival. A slight improvement in 3-year survival after brain infarction from 42% in 1945–49 to 45% in 1970–74 was noted in the Rochester study.

Accumulation of Data on Specific or Less Common Stroke Syndromes

Stroke registration allows the accumulation of detailed data on specific stroke subtypes, such as brain embolism (Bogousslavsky et al. 1991; Kittner et al. 1992; Timsit et al 1992; Caplan et al. 1983), lacunar infarction (Gandolfo et al. 1986; Giroud et al. 1991a) and intracerebral hemorrhage (Giroud et al. 1991b). It has also provided a method for acquiring data on some less common stroke syndromes (Table 23.9). For example, caudate hemorrhage accounts for less than 10% of intracerebral hemorrhage (Stein et al. 1984). Through stroke registration, enough patients can be accumulated to accurately describe the clinical syndrome, its presentation and its prognosis. Data from three different registries were used to delineate the syndrome of anticoagulant-related intracerebral hemorrhage. Data from two different hospital registries have been used to better describe the occurrence of headache after stroke (Gorelick et al. 1986). Stroke registries have also been useful in defining stroke mechanisms in younger patients (Bogousslavsky and Pierre 1992).

Table 23.8. 30-day case fatality rate by etiology

		Reference	*n*	ATH	LAC	EMB	INFCU	ICH	SAH	OTH
SDB	USA	Foulkes 1986	1805	18.0	1.0	16.0	12.0	34.0	31.0	15.0
Lausanne	Switzerland	Bogousslavsky et al. 1988	889	5.0	1.0	5.0	NS	20.0	NS	NS
Pilot SDB	USA	Kunitz et al. 1984	1158	11.9	1.1	7.3	10.3	34.0	28.9	NS

ATH, atherosclerosis; LAC, lacune; EMB, embolism; INFCU, infarction cause unknown; ICH, intracerebral hemorrhage; SAH, subarachnoid hemorrhage; OTH, other stroke; NS, not significant

Table 23.9. Less usual causes of infarction, based on 891 cases of cerebral infarction in Lausanne Stroke Registry (Bogousslavsky et al. 1988)

	Number of cases
Fibromuscular dysplasia	4
Intacranial aneurysm	1
Intracranial angiopathy	3
Cerebral venous thrombosis	2
Polycythemia	2
Thrombocythemia	4
Myeloma with hyperviscosity	1
Migraine	20
Giant-cell arteritis	1
Churg–Strauss disease	1
Sarcoid	1
Syphilis	2
Herpes zoster	1
Systemic lupus erythematosus	3
Isolated angiitis of nervous system	4
Mitochondrial cytopathy	1

Diurnal and Seasonal Variations in Stroke Incidence

Stroke registration has proved to be a useful technique for detecting seasonal and diurnal variations in stroke incidence. A morning increase between the hours of 10 a.m and 12 noon was found in the Stroke Data Bank (Marler et al. 1989). In the Oxfordshire Stroke Registry (Wroe et al. 1992) a morning peak in incidence for all stroke types occurred between 8 a.m and 10 a.m. The cause of this diurnal variation is uncertain.

Seasonal variations in stroke incidence have also been noted. The highest monthly incidence for intracerebral hemorrhage is in November–December, and the lowest in July–August (Capon et al. 1992).

Secular Trends in Stroke Incidence

Stroke registration over long periods of time has allowed the identification of secular trends in stroke incidence. Over the 20-year period 1960–1979 the rate of hemispheric infarction in Rochester, Minnesota, dropped from 101 to 55/100 00 and the rate of brain-stem/cerebellar infarction dropped from 19 to 11/100 000 (Turney et al. 1984). In Rochester, the incidence rate for all strokes dropped steadily from 209 in 1945–1949 to 115/100 000 in 1975–1979. However, in the period 1980–84 the stroke incidence rose to 135/100 000. This rise has been attributed to the introduction of computed tomography and the detection of less severe strokes (Broderick et al. 1989). Terent (1989) used the Stroke Registry in Soderhamn, Sweden, to study secular trends in stroke survival and recurrence in two different time periods (1975–1979 compared to 1983–1987). It was noted that 3-year survival increased from 49% to 66% but the recurrence rate remained stable at 10% per year in both study periods. In Finland, Tilvis et al. (1987) found that the stroke incidence rate declined in the period 1970–1980 from 221/100 000 to 139. However, all of this decline was due to a decrease in hemorrhagic stroke. In contrast to Tilvis, Lindenstrom et al. (1992) found no evidence of a decrease in stroke incidence in Copenhagen between 1976 and 1988.

Interhospital Comparisons

When stroke registration is done by consistent methods at various hospitals, comparisons between stroke type outcome, diagnostic methods and treatments can be made between hospitals. The Harvard Stroke Registry allowed a comparison between stroke diagnoses at the Massachusetts General Hospital and the Beth Israel Hospital in Boston (Table 23.10), and showed significant interhospital differences in stroke type frequencies (Melski et al. 1989). Analysis showed that the differences in the frequency of subarachnoid hemorrhage and lacunar infarction were due to differences in patient population, whereas differences in the frequency of atherosclerosis and embolism were due to differences in the selection of diagnostic tests, especially the use of cerebral angiography.

Table 23.10. Stroke diagnosis frequencies at two teaching hospitals

Diagnosis	Massachusetts General Hospital n=348 (%)	Beth Israel Hospital n=303 (%)
Atherothrombosis	27.0	40.0
Embolism	37.0	24.0
Lacune	16.0	25.0
Parenchymatous hemorrhage	13.0	8.0
Subarachnoid hemorrhage	7.0	3.0

Development of Clinical Prediction Rules

Clinical prediction rules are a relatively new area of research that seeks to use statistical methods to predict disease outcome. Stroke registries are particularly good sources of data for the development of such rules. Using data from 94 patients in the Pilot Stroke Data Bank, Tuhrim et al. (1988) developed rules to predict outcome after intracerebral hemorrhage. Eight univariate factors were found to predict hemorrhage outcome: Glascow Coma Scale score, brain-stem-cerebellar deficits, pulse pressure, hemorrhage size, gaze palsies, weakness, systolic pressure and stroke course. A final decision utilizing only pulse pressure, Glascow Coma Scale score and hemorrhage size correctly predicted the outcome for 92% of patients at 30 days.

Data from stroke registries can also be used in the development of expert systems to assist in the diagnosis and treatment of stroke patients (Hier et al. 1985). These are computer programs that seek to emulate the behavior of human experts. Data stored in large patient registries can be used to create decision support tools for such systems (Hier and Edelstein 1991). Probabilistic data in the Hamburg Stroke Data Bank has been used to develop an expert system to classify strokes by subtype using Bayes' theorem (Spitzer et al 1989a).

The Future of Stroke Registries

Further enhancements in stroke registries are likely to occur in the near future. As hospital information systems become more sophisticated, stroke registration is likely to become more automated. Diagnostic and clinical information will be directly downloaded from the hospital patient information system into a stroke registry without requiring data re-entry. Stroke registries of the future will acquire both artificial intelligence and multimedia features. Multimedia will allow registration not only of textual data but also graphical and audio data for each patient, including patient video, neuroimages and patient audio. Artificial intelligence methods will strengthen the potential to derive new information from large patient databases. Artificial intelligence methods will allow the stroke registries of the future to perform not only as a repository for patient data but also as an assistant to the physicians performing stroke consultations.

References

Alter M, Sobel E, McCoy RL et al. (1985) Stroke in the Lehigh Valley: incidence based on a community-wide hospital register. Neuroepidemiology 4:1–15

Alter M, Sobel E, McCoy RL et al. (1987) Stroke in the Lehigh Valley: risk factors for current stroke. Neurology 37:503–507

Alter M, Christoferson, Resch J et al. (1970) Cerebrovascular disease: frequency and population selectivity in an upper midwestern community. Stroke 1:454–465

Bamford J, Sandercock P, Dennis M et al. (1988) A prospective study of acute cerebrovascular disease in the community: the Oxfordshire Community Stroke Project 1981–86. 1. Methodology, demography and incident cases of first-ever stroke. J Neurol Neurosurg Psych 51:1373–1380

Becker C, Howard G, McLeroy KR et al. (1986) Community Hospital-Based Stroke Programs: North Carolina, Oregon, and New York. II. Description of study population. Stroke 17:285–293

Bogousslavsky J, Pierre P (1992) Ischemic stroke in patients under age 45. Neurol Clin 10:113–124

Bogousslavsky J, Van Melle G, Regli F (1988) The Lausanne Stroke Registry: analysis of 1000 consecutive patients with first stroke. Stroke 19:1083–1092

Bogousslavsky J, Cachin C, Regli F et al. (1991) Cardiac sources of embolism and cerebral infarction – clinical consequences and vascular concomitants: the Lausanne Stroke Registry. Neurology 41:855–859

Bonita R, Ford MA, Stewart AW (1988) Predicting survival after stroke: a three-year follow-up. Stroke 19:669–673

Broderick JP, Phillips SJ, Whisnant JP et al. (1989) Incidence rates of stroke in the eighties: the end of the decline in stroke? Stroke 20:577–582

Caplan LR, Hier DB, D'Cruz I (1983) Cerebral embolism in the Michael Reese Stroke Registry. Stroke 14:530–536

Capon A, Demeurisse G, Zheng L (1992) Seasonal variation of cerebral hemorrhage in 236 consecutive cases in Brussels. Stroke 23:24–27

Chambers BR, Norris JW, Shurvell BL, Hachinski VC (1987) Prognosis of acute stroke. Neurology 37:221–225

Davis PH, Dambrosia JM, Schoenberg BS et al. (1987) Risk factors for ischemic stroke: a prospective study in Rochester, Minnesota. Ann Neurol 22:319–327

Foulkes MA, Wolf PA, Price TR et al. (1988) The Stroke Data Bank: design, methods, and baseline characteristics. Stroke 19:547–554

Friday G, Lai SM, Alter M et al. (1989) Stroke in the Lehigh Valley: racial/ethnic differences. Neurology 39:1165–1168

Gandolfo C, Moretti C, Dall'Agata D et al. (1986) Long-term prognosis of patients with lacunar syndromes. Acta Neurol Scand 74:224–229

Garraway WM, Whisnant JP, Drury I (1983) The changing pattern of survival following stroke. Stroke 14:699–703

Giampaoli, S, Menottie A, Righetti G, Verdecchia A (1989) Register for cerebrovascular events in the Area Latina, Italy. Clinical and epidemiological data during three years of surveillance. Ital J Neurol Sci 10:499–503

Giroud M, Beuriat P, Vion P et al. (1989) Stroke in a French prospective population study. Neuroepidemiology 8:97–104

Giroud M, Gras P, Milan C et al. (1991a) Natural history of lacunar syndromes. Contributions of the Dijon registry. Rev Neurol 147:556–572

Giroud M, Milan C, Beuriat P et al. (1991b) Incidence and survival rates during a 2-year period of intracerebral and subarachnoid hemorrhages, cortical infarcts, lacunes, and

transient ischemic attacks – The Stroke Registry of Dijon. Int J Epidemiol 20:892–899
Gorelick PB, Caplan LR, Hier DB et al. (1984) Racial differences in the distribution of anterior circulation occlusive disease. Neurology 34:54–59
Gorelick PB, Hier DB, Caplan LR, Langenberg P (1986) Headache in cerebrovascular disease. Neurology 36:1445–1450
Gross CR, Kase CS, Mohr JP et al. (1984) Stroke in south Alabama: incidence and diagnostic features – a population based study. Stroke 15:249–255
Gross CR, Shinar D, Mohr JP et al. (1986) Interobserver agreement in the diagnosis of stroke type. Arch Neurol 43:893–898
Hatano S (1976) Experience from a multicentre stroke register: a preliminary report. Bull WHO 54:541–552
Heyman A, Karp HR, Heyden S et al. (1971) Cerebrovascular disease in the biracial population of Evans County, Georgia. Arch Intern Med 128:949–955
Hier DB, Edelstein G (1991) Deriving clinical prediction rules from stroke outcome research. Stroke 22:1431–1436
Hier DB, Atkinson GD, Perline R et al. (1985) Can a patient data base help build a stroke diagnostic expert system? Med Informatics 10:75–81
Howard G, Walker MD, Becker C et al. (1986) Community Hospital-Based Stroke Programs: North Carolina, Oregon, and New York. III. Factors influencing survival after stroke: proportional hazards analysis of 4219 patients. Stroke 17:294–299
Jerntorp P, Berglund G (1992) Stroke registry in Malmo, Sweden. Stroke 23:357–361
Kay R, Woo J, Kreel L et al. (1992) Stroke subtypes among Chinese living in Hong Kong: the Shattin Stroke Registry. Neurology 42:985–987
Kittner SJ, Sharkness CM, Sloan MA et al. (1992) Infarcts with a cardiac source of embolism in the NINDS Stroke Data Bank: neurologic examination. Neurology 42:299–302
Kojima S, Omura T, Wakamatsu W et al. (1990) Prognosis and disability of stroke patients after 5 years in Akita, Japan. Stroke 21:72–77
Kunitz SC, Gross CR, Heyman A et al. (1984) The Pilot Stroke Data Bank: definition, design, and data. Stroke 15:740–746
Lindenstrom E, Boysen G, Nyboe J, Appleyard M (1992) Stroke incidence in Copenhagen, 1976–1988. Stroke 23:28–32
Marler JR, Price TR, Clark GL et al. (1989) Morning increase in onset of ischemic stroke. Stroke 20:473–476
Matsumoto N, Whisnant JP, Kurland LT, Okazaki H (1973) Natural history of stroke in Rochester, Minnesota, 1955 through 1969: an extension of a previous study, 1945 through 1954. Stroke 4:20–29
Melski JW, Caplan LR, Mohr JP et al. (1989) Modeling the diagnosis of stroke at two hospitals. MD Comput 6:157–163
Mohr JP (1986) Stroke Data Banks. Stroke 17:171–172
Mohr JP, Caplan LR, Melski JW et al. (1978) The Harvard Cooperative Stroke Registry: a prospective registry. Neurology 28:754–762
Sacco RL, Foulkes MA, Mohr JP et al. (1989) Determinants of early recurrence of cerebral infarction: the Stroke Data Bank. Stroke 20:983–989
Scmidt EV, Smirnov VE, Ryaboa VS (1988) Results of the seven-year prospective study of stroke patients. Stroke 19:942–949
Shinar D, Gross CR, Mohr JP et al. (1985) Interobserver variability in assessment of the neurological history and examination in the Stroke Data Bank. Arch Neurol 42:557–565
Shinar D, Gross CR, Hier DB et al. (1987) Interobserver reliability in the interpretation of computed tomographic scans of stroke patients. Arch Neurol 44:149–155
Silver FL, Norris JW, Lewis JL, Hachinski VC (1984) Early mortality following stroke: a prospective review. Stroke 15:492–496
Sobel E, Alter M, Davanipour Z et al. (1989) Stroke in the Lehigh Valley: combined risk factors for recurrent ischemic stroke. Neurology 39:669–672
Spitzer K, Thie A, Caplan LR, Kunze K (1989a) The MICRO-STROKE expert system for stroke type diagnosis. Stroke 20:1352–1356
Spitzer K, Becker V, Thie A, Kunze K (1989b) The Hamburg Stroke Data Bank: goals, design, and preliminary results. J Neurol 236:139–144
Stein RW, Kase CS, Hier DB et al. (1984) Caudate hemorrhage. Neurology 34:1549–1554
Terent A (1989) Survival after stroke and transient ischemic attacks during the 1970s and the 1980s. Stroke 20:1320–1326
Tilvis R, Autio L, Mahonen Y et al. (1987) The incidence and prognosis of cerebrovascular disease in hospital patients in Helsinki, Finland, in the decade 1970–1980. Acta Med Scand 221:267–273
Timsit SG, Sacco RL, Mohr JP et al. (1992) Early clinical differentiation of cerebral infarction from severe atherosclerotic stenosis with cardioembolism. Stroke 23:486–491
Tuhrim S, Dambrosia JM, Price TM et al. (1988) Prediction of intracerebral hemorrhage survival. Ann Neurol 24:258–263
Turney TM, Garraway WM, Whisnant JP (1984) The natural history of hemispheric and brain-stem infarction in Rochester Minnesota. Stroke 15:790–794
Waltimo O, Kaste M, Aho K, Kotila M (1980) Outcome of stroke in the Espoo-Kauniainen area, Finland. Ann Clin Res 12:326–330
Wolf PA, Abbott RD, Kannel WB (1991a) Artrial fibrillation as an independent risk factor for stroke: the Framingham Study. Stroke 22:983–988
Wolf PA, D'Agostino RB, Belanger AJ, Kannel WB (1991b) Probability of stroke: a risk profile from the Framingham Study. Stroke 22:312–318
Wroe SJ, Sandercock P, Bamford J et al. (1992) Diurnal variation in incidence of stroke: Oxfordshire community stroke project. Br Med J 304:155–157
Yaqub VA, Shamena AR, Kolawole TM, Patel PJ (1992) Cerebrovascular disease in Saudi Arabia. Stroke 22:1173–1176
Yatsu F, Becker C, McLeroy KR et al. (1986) Community Hospital-Based Stroke Programs: North Carolina, Oregon, and New York. I. Goals, objectives, and data collection procedures. Stroke 17:276–284

Concluding Comments

Epidemiological Studies

Epidemiological studies are able to show relationships and associations in the cohort of patients studied, but the associations cannot be generalized unless the patients or populations studied are representative of all those with the condition in question. In a series of young patients with stroke, drugs and oral contraceptives are often correlated with stroke occurrence but age is not. In contrast, stroke occurrence increases with age in studies that include geriatric patients. Properly selected control groups are important if relationships between a factor and a condition are being studied. For example, in a study in France of stroke in the young, 34% of the female stroke patients took oral contraceptives, an impressive proportion (Gautier et al. 1989). However, at that time in France 32% of women of the same age took oral contraceptives, so that the use of these agents was not shown to be a risk factor for stroke in the group of patients studied. An association between a factor and a condition does not necessarily prove a cause: being female strongly correlates with being pregnant, but clearly does not cause the condition. At times, the associated factor may simply be a marker for another factor not studied. Coffee consumption could theoretically be a marker for lack of sleep, cigarette use, job stress etc. Some factors might correlate inversely with others, e.g. individuals who jog and exercise regularly seldom smoke cigarettes. Studies could be designed to collect data about these posited relationships to see if the association represents a true risk factor or merely a marker for something else.

A striking finding in some epidemiological studies and registries is the difference in strokes in various racial groups (Caplan et al. 1986; Caplan and Cooper 1987; Viriavejakul 1990; Feldmann et al. 1990; Cooper et al. 1991). The racial differences are both quantitative and qualitative. Blacks have more strokes and more disability from stroke than whites. Individuals of Chinese, Japanese and Thai origin also have a relatively high rate of strokes compared to white populations. In Chinese, Japanese, Thai and black individuals, occlusive lesions involve mostly the intracranial large arteries (e.g. the middle, anterior and posterior cerebral arteries, the vertebral and distal basilar arteries and the posterior inferior cerebellar artery), whereas in whites, especially men, occlusive arterial lesions predominate in the extracranial arteries (the proximal internal carotid and vertebral arteries).

There are also important sex differences in the distribution of occlusive disease (Caplan et al. 1986). In nearly all studies of patients with severe internal carotid occlusive disease or those having carotid artery surgery, two-thirds of the patients are men. Women have more intracranial occlusive disease and less extracranial disease than men. Although hypertension is more prevalent in black, Chinese, Japanese and Thai cohorts than in whites, the presence of hypertension does not account for the racial differences. The race and sex differences uncovered in registry and epidemiological studies represent a very important finding that should stimulate investigations to explain the finding. These racial differences may represent an important clue to the nature and causes of vascular occlusive disease.

Registries

Stroke registries have many potential functions, some of which have not been fully explored or realized. Anatomical information gained from neuroimaging results (CT, MRI) was entered on to paper grids in the MRSR and the SDB. The anatomical data can be digitized and correlated with the presence of various signs and symptoms. The relationship between the brain lesion and the clinical findings may be complex. In a study of patients with right cerebral strokes, the occurrence of certain findings depended not only on the site of the lesion but also on lesion type (hemorrhage vs. infarct), lesion size and additional involvement or sparing of other regions and other cerebral lobes (Hier et al. 1983). Computers provide a unique potential for aiding anatomicoclinical correlation. Spitzer et al. (1989a) used a rule-based system and pattern matching to create a computer program to localize lesions anatomically but did not use anatomical data from imaging tests.

Computers also allow investigators to study complex multifactorial relationships. In the HSR, vomiting proved to be an important indicator of intracerebral hemorrhage if the location of the stroke was also considered. Patients with posterior circulation infarcts and hemorrhages each had a high rate of vomiting, but patients with anterior circulation infarcts, irrespective of whether caused by thrombotic or embolic mechanisms, had a very low frequency of vomiting (2%), whereas anterior circulation hemorrhage was often accompanied by vomiting. Similarly, registries can identify patterns or syndromes by searching for four or five components. The syndrome of hemorrhage into the far posterior portion of the putamen was identified in the HSR by the combination of fluent Wernicke-type aphasia associated with hemisensory symptoms and minor hemiparesis. All such patients had hematomas that began in the retrolenticular portion of the posterior limb of the internal capsule and the putamen and spread into the temporal lobe isthmus white matter. The large numbers of patients in registries allows for the detection of uncommon patterns and syndromes and for the clarification of the findings in unusual lesions or syndromes.

Computer Diagnostic and Treatment Programs

Computers have not customarily been used online to help with diagnosis, prognosis or treatment decisions. Heyman and colleagues (1979) were probably the first to use a registry to provide physicians with information that might help with management. They stored in their computer database information about patients with TIAs who presented to physicians at Duke University Hospital during the 4-year period 1973–1977. Physicians could request the computer to print out information from patients sharing specific features with their patient, e.g. bruit, number, length of TIAs and presence of risk factors. The results of evaluations in these patients, such as angiography results, outcome and effect of treatment, were also printed out in the 'prognostigram'. This information helped physicians to take advantage of experience at their own center, since studies have shown that results and complication rates vary widely among centers, hospitals and surgeons. Similar hospital-based registries could help surgeons and physicians access their past cases quickly. There are many potential uses of such data.

Hier, in his contribution, mentions the development of expert systems: the HSR was born because of early experience with such a computer diagnostic system. Diagnostic programs have been used in internal medicine (Miller et al. 1982; Barnett et al. 1987; Bankowitz et al. 1989), and Spitzer et al. (1989b) devised a computer diagnostic program for stroke mechanism diagnosis using data from the HSR, MRSR and SDB. However, computer-assisted diagnosis and management have not often been used in neurology. Such programs can make junior physicians and students aware of the methodology of diagnosis; teach what data are needed to arrive at the diagnosis; and quickly provide references appropriate to the clinical problem and diagnosis. In many areas of the world there is rarely quick access to stroke specialists. Computer programs could help non-specialists arrive at triage decisions and might aid in the evaluation process.

Even in stroke centers, diagnostic programs based on data from registries can serve as checks for the experts to be certain they have not overlooked important possibilities or weighted the data erroneously. We tested the original computer diagnostic program, using HSR data, against the diagnosis of HSR physicians (mostly Caplan and Mohr). The computer diagnosis matched the clinicians' final diagnosis 84% of the time (Goldstein et al. 1976). When compared with the final diagnosis after all testing, the doctors were correct in their diagnosis about 80% of the time, and the computer also had an 80% average. However, the correctly diagnosed cases were not the same. The computer correctly diagnosed cases that the physicians missed. The physicians correctly recognized diagnoses that the computer missed. I think the doctor and the computer together do better than either alone. Diagnostic programs can use Bayesian analysis, branching logic algorithms and pattern-matching strategies to arrive at the probabilities of various diagnoses. The computer can also be taught rules that will govern management advice. These programs can not only serve as excellent teaching devices but may also improve patient management. Decision analysis programs are now popular in departments of internal medicine to aid physicians in making difficult therapeutic decisions (Pauker and Kassirer 1975, 1980). Data from registries and data banks can provide real numbers to aid in decision analysis.

Treatment Trials

Prospective randomized double-blind therapeutic trials are undoubtedly the fashion of the last 20 years, and have become almost sacrosanct. Who could say a negative word about trials? Few would argue about the importance of the EC/IC Bypass Trial, the carotid endarterectomy trials and the trials of prophylactic stroke prevention for patients with atrial fibrillation. The results have clearly changed the way physicians practice – I think for the better, but trials cannot provide answers to all important therapeutic dilemmas. There are limitations and problems with some of the trials that are conducted, the major limitation being their absolute need for numbers; quantities of patients are required to derive statistically valid results. Diseases or conditions that are not common are thus excluded from most trials since insufficient numbers can be generated. The need for numbers usually means multiple centers and high expenses, and it is usually necessary to study general conditions and to classify problems into broad categories. Physicians would have difficulty collecting sufficient patients with stenosis of the middle cerebral artery (MCA) who had TIAs, but would have no difficulty studying all patients with TIAs or brain ischemia. Numbers are achieved at the cost of specificity. Is the finding that aspirin helps reduce the incidence of strokes by 30% in patients who have had TIAs helpful for the practicing neurologist? The neurologist has one patient with MCA stenosis and another with fibromuscular dysplasia, and another with atrial fibrillation, all of whom have had TIAs. Unless the data were collected prospectively and analyzed for these conditions, the general answer will not help the physician treat a specific patient with any of the conditions described. I have always been a 'splitter' not a 'lumper': I believe treatment has to be aimed at the individual patient; few neurologists or other doctors are in a position to need to choose a single treatment for thousands of patients. To be useful, trials must provide data about treatment decisions faced by physicians in practice. Data must be as specific as possible.

Although not as scientifically and statistically valid, there is some worth in retrospective analysis of various treatment results. Using large registries, searches can be made about treatment and outcome in patients with well defined vascular and brain lesions. These results are similar to the 'prognostigram' described by Heyman et al. (1979). Anecdotal treatment results are not comparable to well-designed trials but at present, in regard to most conditions, that is all that is available. Using registries with fully evaluated patients and good follow-up details will clearly increase the numbers, allow wider access to the data and formalize the anecdotal treatment results. Such a practice is better than individual memories of our past 10 cases, or alternatively, our best or worst case experience.

References

Bankowitz R, McNeil M, Challinar SM et al. (1989) A computer-assisted medical diagnostic consultation service. Ann Intern Med 110:824–832

Barnett GO, Cimino JJ, Hupp JA, Hoffer EP (1987) DXplain: an evolving diagnostic decision-support system. JAMA

258:67–74
Caplan LR, Cooper ES (1987) Cerebrovascular disease in blacks. JAMA 79:32–36
Caplan LR, Gorelick PB, Hier DB (1986) Race, sex and occlusive cerebrovascular disease: a review. Stroke 17:648–655
Cooper ES, Kuller LH, Saunders E et al. (1991) Cardiovascular disease and stroke in African-Americans and other racial minorities in the US. Stroke 22:551–569
Feldmann E, Daneault N, Kwan E et al. (1990) Chinese–white differences in the distribution of occlusive cerebrovascular disease. Neurology 40:1541–1545
Gautier JC, Pradat-Diehl P, Loran PL et al. (1989) Accidents vasculaires cérébraux des sujets jeunes. Une étude de 133 patients agés de 9 à 45 ans. Rev Neurol 145:437–442
Goldstein RJ, Caplan LR, Bleich HL et al. (1976) Computer stroke registry and diagnosis program. Neurology 25:356
Heyman A, Birch JG, Rosati R et al. (1979) Use of a computerized information system in the management of patients with transient cerebral ischemia. Neurology 29:214–221
Hier D, Mondlock J, Caplan LR (1983) Behavioral abnormalities after right hemisphere stroke. Neurology 33:337–344
Miller RA, Pople HE Jr, Myers JD (1982) Internist-1, an experimental computer-based diagnostic consultant for general internal medicine. New Engl J Med 307:468–476
Pauker SG, Kassirer JP (1975) Therapeutic decision making: a cost-benefit analysis. New Engl J Med 293:229–234
Pauker SG, Kassirer JP (1980) The threshold approach to clinical decision making. New Engl J Med 320:1109–1117
Spitzer K, Thie A, Caplan LR, Kunze K (1989a) The TOPOSCOUT expert system for stroke localization. Stroke 20:1195–1201
Spitzer K, Thie A, Caplan LR, Kunze K (1989b) The MICROSTROKE Expert system for stroke type diagnosis. Stroke 20:1353–1356
Viriavejakul A (1990) Stroke in Asia: an epidemiological consideration. Clin Neuropharm 13 (Suppl 3):526–533

Subject Index

MIX
Papier aus verantwortungsvollen Quellen
Paper from responsible sources
FSC® C105338

If you have any concerns about our products,
you can contact us on
ProductSafety@springernature.com

In case Publisher is established outside the EU,
the EU authorized representative is:
Springer Nature Customer Service Center GmbH
Europaplatz 3, 69115 Heidelberg, Germany

Printed by Libri Plureos GmbH
in Hamburg, Germany